Student Tested, Faculty Approved

98% of 4LTR Press adopters said they will recommend 4LTR Press to a colleague

73% of 4LTR Press adopters have seen improved student engagement or better grades

4LTR
P·R·E·S·S

87% of students believe 4LTR Press solutions are designed more closely to the way they study

92% of students find 4LTR Press solutions a better value than other textbooks

4ltrpress.cengage.com

4LTR Press solutions available in:

AMERICAN GOVERNMENT • ASTRONOMY • BUSINESS COMMUNICATION • CHEMISTRY • CHILDHOOD AND ADOLESCENT DEVELOPMENT • COMMUNICATION • CONSUMER BEHAVIOR • CORPORATE FINANCE CRIMINAL JUSTICE • EARTH SCIENCES • ENGLISH COMPOSITION • FINANCIAL ACCOUNTING • FUNDAMENTALS OF STATISTICS • GEOLOGY • GLOBAL BUSINESS • HUMAN DEVELOPMENT • INTERMEDIATE MACROECONOMICS • INTRODUCTION TO BUSINESS • INTRODUCTION TO LAW ENFORCEMENT • MACROECONOMICS MANAGEMENT INFORMATION SYSTEMS • MANAGERIAL ACCOUNTING • MARKETING MANAGEMENT MARRIAGE AND FAMILY • MATH • MICROECONOMICS • MONEY & BANKING • NON-MAJORS BIOLOGY OPERATIONS MANAGEMENT • ORGANIZATIONAL BEHAVIOR • PERSONAL FINANCE • PHYSICS • PRE-ALGEBRA PRINCIPLES OF MANAGEMENT • PRINCIPLES OF MARKETING • PROFESSIONAL SELLING • PSYCHOLOGY SOCIOLOGY • U.S. HISTORY • WORLD HISTORY

LIFE 2010–2011 Edition
John H. Postlethwait
Janet L. Hopson

Senior Acquisitions Editor: Peggy Williams

Editor-in-Chief: Michelle Julet

Publisher: Yolanda Cossio

Developmental Editor: Dana Freeman, B-books, Ltd.

Assistant Editor: Elizabeth Momb

Product Development Manager, 4LTR Press: Steven E. Joos

Senior Project Manager, 4LTR Press: Michelle Lockard

Editorial Assistant: Alexis Glubka

Senior Marketing Manager: Tom Ziolkowski

Brand Development Manager, 4LTR Press: Robin Lucas

Marketing Coordinator: Elizabeth Wong

Marketing Communications Manager: Linda Yip

Production Director: Amy McGuire, B-books, Ltd.

Senior Content Project Manager: Hal Humphrey

Senior Media Editor: Amy Cohen

Senior Print Buyer: Karen Hunt

Production Service: B-books, Ltd.

Art Director: John Walker

Internal Designer: Beckmeyer Design

Cover Image: © Frans Lantig/Corbis

Photography Manager: Deanna Ettinger

Photo Researchers: Dana Freeman, B-books, Ltd., and Charlotte Goldman

Library of Congress Control Number: 2009943808

SE Package ISBN-13: 978-0-538-74134-7
SE Package ISBN-10: 0-538-74134-1

Brooks/Cole
20 Davis Drive
Belmont, CA 94002
USA

Cengage Learning products are represented in Canada by Nelson Education, Ltd.

For your course and learning solutions, visit **academic.cengage.com**
Purchase any of our products at your local college store or at our preferred online store **www.CengageBrain.com**

Printed in the United States of America
1 2 3 4 5 6 7 13 12 11 10

LIFE Brief Contents

LIFE Contents

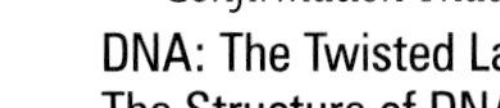

PART 2 Evolution and Biodiversity

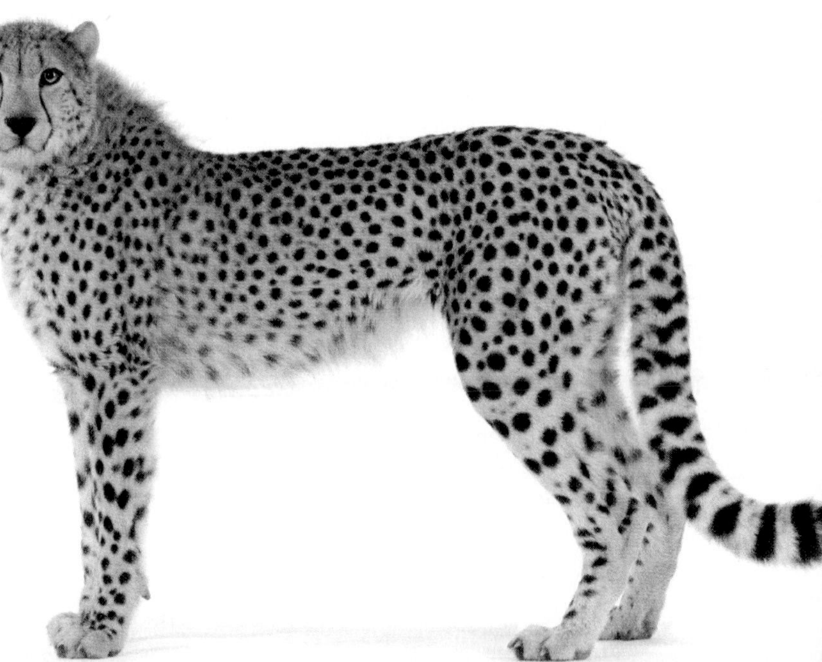

PART 3 The Living World

CARBON
DIOXIDE

CHAPTER 1

Learning Outcomes

- LO[1] Recognize the characteristics of life
- LO[2] Identify the characteristics of life relating to energy use
- LO[3] Identify the characteristics of life relating to reproduction
- LO[4] Identify the characteristics of life relating to evolution
- LO[5] Identify the characteristics of life relating to environment
- LO[6] Explain how biologists study life
- LO[7] Understand how biologists can help solve problems

What Is Life?

"After centuries of study and thought, biologists have come to a consensus about when an entity is alive, formerly alive, or nonliving."

Full Circle from Earth to Mars

A huge headline in *Time* magazine screamed unambiguously: "LIFE ON MARS." In August 1996, NASA scientists created an international media frenzy by claiming that a 4-pound, potato-shaped rock contained the remains of Martian life.

What do you know?

A car uses energy, is highly organized, and is motile . . . but it's not "alive." Why not?

The blackened lump in question is a meteorite from Mars that entered Earth's atmosphere as a brilliant shooting star about 13,000 years ago during our planet's Ice Age. Prehistoric hunters and farmers probably watched that arrival blaze across the night sky. But the stone fell anonymously to the frozen surface of Antarctica and lay undisturbed for 13 millennia until scientists discovered it in 1984. Twelve years later, a team from Johnson Space Center in Houston published evidence that the charred chunk contains microscopic wormlike structures formed nearly 4 billion years ago (see Fig. 1.1). This evidence, they claimed, was compatible with life on Mars. Within months, many scientists had rebutted the claim: Nothing but mineral structures, they said.

Debate still simmers, even years later. But an answer may be on its way. NASA began a series of Mars missions in 1998 that will retrieve Martian soil and rocks and return them to Earth for study. And recent photographic data suggest that Mars may have liquid water below its surface capable of sustaining life. Once NASA has retrieved the sample of Martian soil, biologists will immediately begin searching it for signs that organisms lived—or still live—on our neighboring red planet. Proof of life on Mars could be one of the most exciting discoveries in the history of life science.

If there was ever a time in our history to understand the central enigma beneath the search for extraterrestrial organisms, it is now and it is this: *What is life?* This chapter begins our multimedia exploration of biology's central puzzle. As you read along, you'll learn why it's so important for each of us to understand life, and how biologists study every aspect of living organisms, their environments, and their interrelationships. And you'll find the answer to these questions:

- What characteristics do all living things share?
- Which characteristics relate to gathering and using energy?
- Which characteristics relate to reproduction?
- Which characteristics relate to evolving and adapting?
- Which characteristics relate to the physical environment?
- How do biologists study life and how will we explore it in this course?
- How can the study of life help us solve societal and environmental problems?

LO[1] Characteristics of Living Organisms

The search for life on Mars and other planets requires that we recognize life when we see it to identify the general properties of life on our planet, and then speculate about the similarities and differences scientists might encounter while searching on Mars and elsewhere in the solar system and universe. After centuries of study and thought, biologists have come to a consensus about when an entity is alive, formerly alive, or nonliving.

The problem of recognizing life is clearly fundamental to the search for alien organisms. But it has other important applications, too. Today's biologists and physicians have unprecedented abilities. These include sustaining the human body and individual organs on life-support machines, freezing human and animal embryos for later use, and changing and merging hereditary traits of microbes, plants, and animals. Perhaps one day this list will extend to generating life in a test tube and to creating hybrids between computers and living things. To manipulate life's most fundamental properties, biological engineers need to know exactly what the boundaries are, how far they can be stretched, and what changes would be desirable, practical, and worth pursuing.

At the same time, the public needs to be fully aware of the benefits and risks of manipulating life so they can be informed watchdogs and consumers of these biotechnologies. In the broadest sense, every course in biological science and every experiment, no matter how simple or complex, probes the question "What is life?" It takes this entire course of study to provide an answer. The citizen who learns about biology—including the college student majoring in some other field like business, physical education, psychology, or English—also discovers a realm of intricacy and beauty that helps them understand their environment, their health, their day-to-day functioning, their children's growth and development, and the issues they see in the news, including cancer treatments, impotence drugs, habitat destruction, species loss, and global climate change.

D. MCKAY, K. THOMAS-KEPRTA, R. ZARE, NASA

Figure 1.1 **Fossilized Martian Bacterium?**

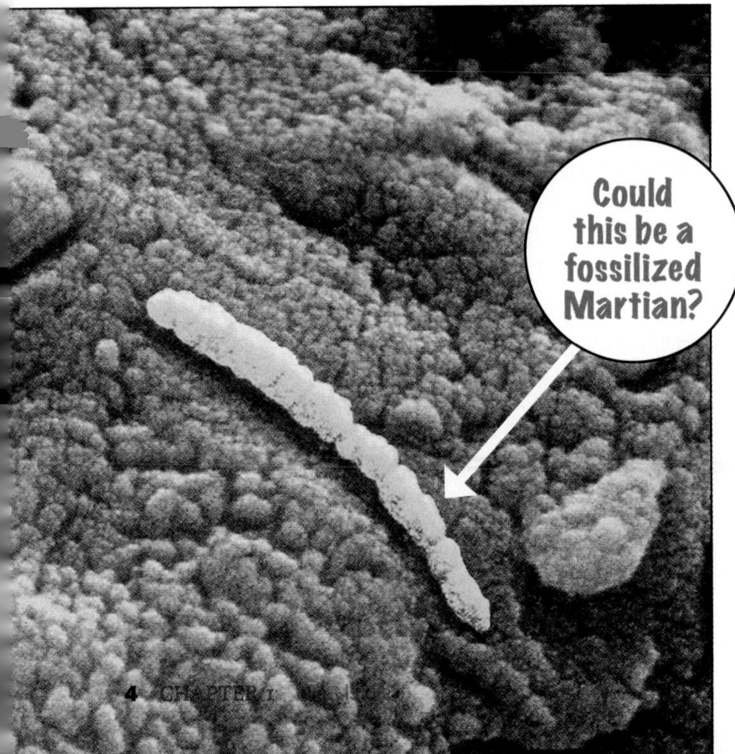

What, then, is life? You may be surprised to learn that no one, not basic biologist, bioengineer, or planetary protector, has a thumbnail definition that lays out the essence of the living state. Instead, they focus on the characteristics of life as a collective, descriptive definition.

Life's Characteristics

Think for a minute about puppies, roses, dinosaur bones, and motorcycles. Why those things? Puppies are clearly alive, frolicking, rolling around, and begging for dog biscuits, as directed by their highly organized brains. A rose is obviously alive, too, although it hardly moves and never devours dog treats. Still, a rose plant soaks up sunshine and soil nutrients, and it makes beautiful flowers with seeds that can produce a new generation of roses. A dinosaur bone is certainly not alive, but its close appearance to the bones of living animals confirms that it was once alive. And the motorcycle? It's highly organized and requires energy from the environment. It moves under its own power and responds by going faster when the throttle is turned up. It must be alive, too, right? Of course not. The last time we checked, motorcycles couldn't reproduce, at least not without the help of an assembly line and some good mechanics.

Comparisons like these underlie the list of characteristics that living things share (Table 1.1). Living systems have internal order, or a high degree of organization. To maintain that order, living things carry out metabolism: they use energy to transform and organize materials. Living things also use energy to move under their own power, a trait biologists call motility. They also use energy to react to outside stimuli, a trait called responsiveness.

While living things do all of the above, they also do more. Living things have the ability for self-replication or reproduction. They show growth and development or the expansion of young organisms in size and complexity. Living things are related by heredity; that is, organisms give rise to like organisms (not dinosaurs from roses or roses from puppies). Finally, living things evolve or change over many generations, and they adapt or change to better fit shifting environments.

We can look around the nonliving world and see many of life's characteristics in action: Waves move, flames use energy, crystals grow. Only living organisms, however, display *all* the characteristics we just discussed at some point during their individual life cycle or species history. For example, rose petals move as the rosebud unfolds; the rose plant captures and uses energy from the sun; the plant originally emerged from a seed, then grew, and developed flowers. Rose bushes can evolve and adapt to changing climates. All in all, they're alive.

Themes That Recur as We Explore Life

Some of the life characteristics we just listed occur again and again as themes that guide our exploration of biology. Living things use energy at several levels of biological organization: in the smallest cells, in individual organisms, and in large groups of organisms called biological communities. The members of a species reproduce and adapt to the environment over generations by means of evolution, and these themes help explain why organisms act as they do. A fun part of biology is learning how people have explored and discovered exactly what organisms do. These discoveries almost always use a special system of investigation called the scientific method. We'll see the five themes of energy, reproduction, environment, evolution, and the scientific method interwoven throughout all of our discussions, and these themes will help to organize the sprawling subject at hand, biology.

Table 1.1

Characteristics of Life

LIFE CHARACTERISTIC	PROPERTY
1. Order	Each structure or activity lies in a specific relationship to all other structures and activities.
2. Metabolism	Organized chemical steps break down and build up molecules, making energy available or building needed parts.
3. Motility	Using their own power, organisms move themselves or their body parts.
4. Responsiveness	Organisms perceive the environment and react to it.
5. Reproduction	Organisms give rise to others of the same type.
6. Development	Ordered sequences of progressive changes result in an individual acquiring increased complexity.
7. Heredity	Organisms have units of inheritance called genes that are passed from parent to offspring and control physical, chemical, and behavioral traits.
8. Evolution	Populations of organisms change over time, acquiring new ways to survive, to obtain and use energy, and to reproduce.
9. Adaptations	Specific structures, behaviors, and abilities suit life-forms to their environment.

LO2 Life Characteristics Relating to Energy

order
a precise arrangement of structural units and activities; also, in taxonomy, a taxonomic group comprising members of similar families

Once NASA scientists suspected that the potato-shaped meteorite found in Antarctica was actually from Mars, they started studying slices of it, looking for hints of life on Mars and perhaps new answers to the question "What is life?" They first sought to confirm that the meteorite indeed originated on Mars. They did this by showing that tiny bubbles trapped inside the rock contain air with the same chemical composition as the atmosphere of Mars—a mix that was measured directly by the Viking mission in 1976 and is distinctly different from Earth's. The mixture of chemicals in the rock itself suggested that the meteorite probably formed 4.5 billion years ago, shortly after Mars solidified as a planet. So how did a rock that old from Mars get here, only to be discovered on the Antarctic ice cap in 1984? NASA geologists and astronomers surmise that a huge asteroid slammed into the Martian surface about 16 million years ago, blasting dirt and rocks high enough into the atmosphere that some escaped. This material orbited the sun independently for millions of years, and some of it eventually got tugged firmly enough by Earth's gravity to streak into our atmosphere as fiery meteorites that fell to the surface. The NASA researchers were excited by the prospect of looking for signs of life in the Martian rock. But what approach would they take to detecting the unmistakable signatures of living things in this traveling chunk of Mars?

Order

First, they could look for order or structural and behavioral complexity and regularity, because living things possess a degree of order far greater than that of the finest Swiss clock, the fastest racing car, or anything else in the nonliving world. The eye of a fly, for example, and the spiral-packed seeds of a sunflower head both consist of highly organized units repeated and arranged in precise geometric arrays. We know that order is important because disorder quickly leads to death in a living thing: most weapons of murder, in fact—clubs, knives, guns, and poisons—will disorganize you beyond repair.

Knowing that order is a hallmark of life, NASA investigators searched

organism
an individual that can independently carry out all life functions

organ system
a group of organs that carries out a particular function in an organism

organ
a body structure composed of two or more tissues that together perform a specific function

Figure 1.2
Order Reigns at Every Level in the Living World

Organism: An individual, independent living entity

Organ System: A group of body parts that carries out a particular function in an organism

Organ: A structure consisting of two or more tissues that performs specialized functions within an organism

Tissue: A group of similar cells that carries out a particular function in an organism

Cell: The simplest entity that has all the properties of life

Organelle: A structure within a cell that performs a specific function

Molecule: A cluster of atoms held together by chemical bonds

their slices of the Mars rock for organized structures and found tiny, regular tubes (see Fig. 1.1). The researchers became convinced that these forms are "microfossils," the small, preserved bodies of ancient organisms, and published a scientific paper claiming so. Many other biologists, however, disagree that the so-called microfossils are evidence of past or present Martian life because they are smaller than the smallest known organisms on Earth. Scientists have concluded that life as we know it could not survive in a package any smaller than a sphere 200 nm (200 billionths of a meter) in diameter. The tubules in the Mars meteorite, however, were half that long and one-tenth that wide. For this reason, many think they must be simply mineral formations. The tubules indeed looked highly ordered, but that alone doesn't confirm that they were once alive.

A Hierarchy of Order

Martian organisms, if they ever existed, are a near-total mystery, including the degree of order they might possess. But organisms on Earth have an order that is readily apparent at several levels. Biologists define an **organism** as an independent individual possessing the characteristics of life. The elephant in Figure 1.2 is an individual organism. Each organism, in turn, is made up of **organ systems**, groups of body parts arranged so that together they carry out a particular function within the organism. The skeletal system, for example, supports an elephant's body.

Organ systems are made up of **organs**, sets of two or more tissues that together perform specialized functions for the organ system. An example is a single bone that supports part of an elephant's leg. Each organ is made up of **tissues**, groups of similar cells that carry out the function of the organ. For example, bone tissue—made up of several kinds of cells functioning collectively—provides physical support to the elephant's leg. Tissues are made up of **cells**, the simplest entities that have all the properties of life. Cells contain within them small structures known as **organelles**, which perform the functions necessary for the life of the cell. Finally, organelles consist of **biological molecules**, the building blocks of all biological structure and activity. The tubules in the Martian meteorite yielded no visual evidence of organelles. Researchers, however, did find some subunits of biological molecules that on Earth can be associated with organelles. No one is sure, of course, whether the biological molecules were contaminants from Earth organisms or arrived in the meteorite itself.

Metabolism

Scientists looking at Mars rocks could—and did—look for evidence of energy use. Living things maintain order in their organelles, cells, and organs through **metabolism**: they take energy from the environment and use it, along with materials, in a series of consecutive chemical steps, for repair, maintenance, and growth. By taking energy and materials from the environment and using them for repair, growth, and other survival processes, metabolism helps to combat the disorganization that occurs with time. If you scrape your knee in a fall, for example, metabolism in your cells helps to repair the damage and to generate new, healthy skin, nerves, and blood vessels.

In 1976, NASA sent a lander called Viking to Mars to photograph the planet's surface and test a scoop of Martian soil. The highlights of that mission were several experiments conducted remotely and designed to detect something in the soil—something alive, perhaps—that could use energy to metabolize (to transform and organize materials) as living things do on Earth. The Viking mission created quite a stir because the "right" byproducts (including carbon dioxide, CO_2) were released during the experiments—as if life were present and actively metabolizing. However, the *ways* the products were generated cast doubt on a biological origin. Most biologists are now convinced that the results from the 1976 Mars experiments were due solely to nonliving soil chemistry and not to metabolism by living cells.

The chemical reactions of metabolism require water, for reasons we'll see in Chapter 3. The Viking mission found no surface water on Mars. But later missions did document geological features (such as deep channels, flood plains, wave patterns in sand, and erosion in sedimentary rocks) consistent with standing and flowing water in the past. Recent findings even suggest that water may have flowed within the past few thousand years. These discoveries have encouraged biologists to believe that life may once have flourished on Mars and that remnant populations may still survive today in areas that harbor small amounts of liquid water.

Motility

There was a NASA joke that while the Viking lander was focusing in on chemical evidence for microbes in the Martian soil, it might miss bigger evidence like footprints or little green aliens walking by. Self-propelled movement, or **motility**, would certainly have been as good an indicator of life on Mars as it is here. Even organisms as simple as bacteria can move on their own. Plants, which cannot move from place to place, do show various subtle movements based on growth. For example, the little organelles that capture sunlight in plant cells are in constant motion. The flowers of some plants open in the morning, trace the sun's arc through the sky, then close at night. Animals, of course, have elevated movement to an art form in their pursuit of food, displays of dominance, and escape from enemies.

Responsiveness

If you poke a sea slug, it withdraws. If you turn a houseplant around, its leaves move imperceptibly until, in a day or two, they're once again oriented toward a window. Organisms are **responsive**: they respond to changes in their environment involving temperature, food, water, enemies, mates, or other elements. The reaction to the change can be

tissue
a group of cells of the same type performing the same function within the body

cell
the basic unit of life; cells are bounded by a lipid-containing membrane and are generally capable of independent reproduction

organelle
a complex cytoplasmic structure with a characteristic shape that performs one or more specialized functions

biological molecules
molecules derived from living systems; the four major types are carbohydrates, lipids, proteins, and nucleic acids

metabolism
(Gr. *metabole,* to change) the series of chemical reactions by which cells acquire and use energy and that contribute to repair, growth, and other survival processes

motility
the self-propelled movement of an individual or its parts

responsiveness
the tendency of a living thing to sense and react to its surroundings

reproduction the method by which individuals give rise to other individuals of the same type

asexual reproduction a type of reproduction in which new individuals arise directly from only one parent

sexual reproduction a type of reproduction in which new individuals arise from the mating of two parents

instantaneous: A moth hears the high-pitched whine of a swooping bat and zigzags away on a midnight breeze, and a Venus flytrap snaps shut on a tiny, unsuspecting frog. The response can be gradual, as well. A trumpeter swan detects the shortening days of autumn and responds by feeding more heavily and then migrating south. Or, a daffodil reacts to the lengthening days of spring by forming flowers.

By metabolizing, moving, and responding, organisms obtain energy and materials from the environment and use them to maintain order in their bodies. Nevertheless, aging inevitably sets in—whether in hours or days for a microbe, a century for a tortoise, or a millennium for a bristlecone pine. An aged organism can no longer stem the resulting disorganization and death. But life continues to exist because organisms reproduce.

LO3 Life Characteristics Relating to Reproduction

Like begets like—that's a central feature of life. If Martian organisms do exist, and if NASA brings them back to Earth, could the aliens reproduce here and multiply out of control? If so, how might we recognize and stop the process? Let's look at the life characteristic of reproduction, and two related ones, development and heredity.

Reproduction

Organisms give rise to others of the same kind—roses to roses, robins to robins—by means of a defining life process, **reproduction**, or the means by which individuals give rise to other individuals of the same type. In **asexual reproduction**, a single parent produces offspring identical to it and each other. One-celled microbes, for example, reproduce asexually by splitting into two identical daughter cells (Fig. 1.3a). Most complex organisms reproduce by **sexual reproduction**, with genetic information coming from two parents and combining in offspring that are very similar but not identical to the parent or each other. Organisms sometimes go to great lengths for sexual reproduction to occur. A visiting bee must pry open a snapdragon flower's petals to feed from the nectar within (Fig. 1.3b). In doing so it will pick up pollen (sex cells) from the flower; some is visible as a fine dust on its abdomen. If the bee visits another flower of the same type, it can then deliver the pollen to the other flower's sex organ, facilitating the combining of genes and the flower's reproduction.

Development

Young organisms usually start out smaller and simpler in form than their parents. The offspring then grow in size and increase in complexity, a process

Figure 1.3
Organisms Reproduce

(a) Asexual reproduction

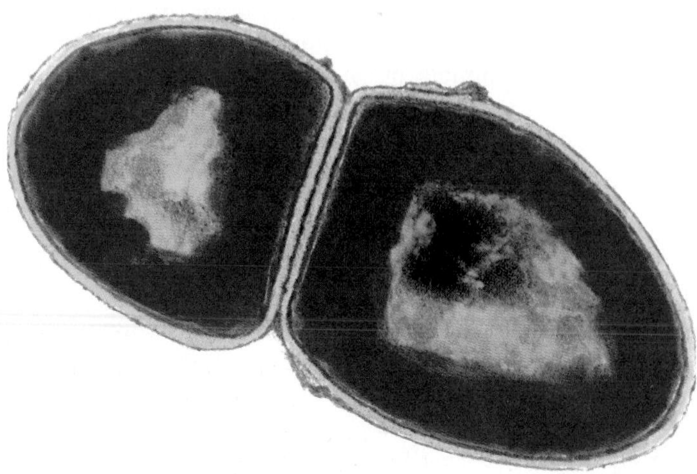

(b) Sexual reproduction

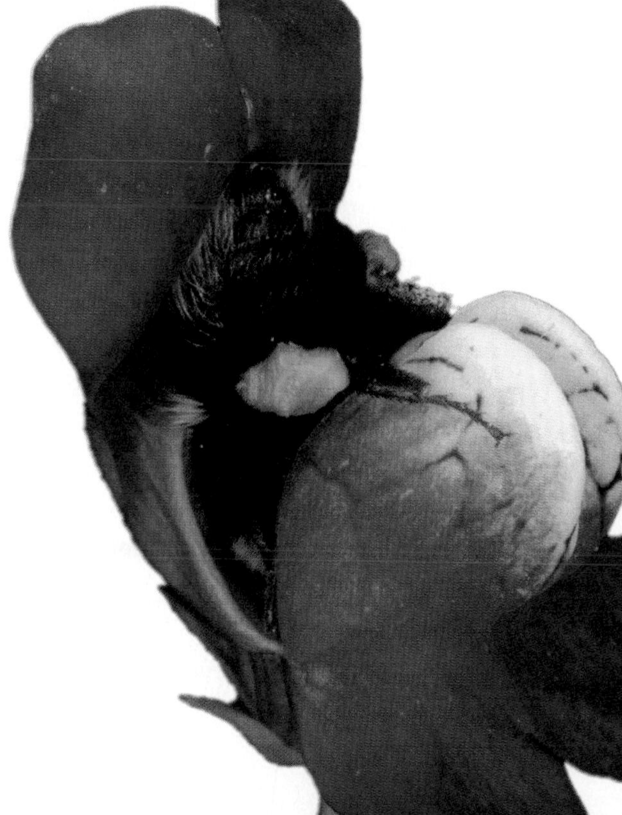

known as development. Eventually, the organism may reach sexual maturity and become a parent itself.

Heredity

One of the most intriguing questions in all of biology is how a fertilized egg develops into the millions of cells of various types that function as a viable organism. The answer lies in the remarkable process of heredity, the transmission of genetic characters from parents to offspring.

Do you know any sets of identical twins? Twins are proof that some type of hereditary information directs each individual's development with such amazing precision—so much that two separate organisms can go through all the steps of growth and increasing complexity over years of time and still wind up looking virtually alike. Contrast identical twins with their different-looking brothers and sisters, however, and you can see that hereditary information must also contain variations so that the offspring in one family can have similar noses but different heights, or similar eyes but different hair color. Biologists have identified the units of inheritance that control an organism's traits and call them genes.

Genes, made of a remarkable molecule called DNA (deoxyribonucleic acid), determine such things as whether a person's hair is red, black, brown, blond, or gray. As we will see, genes also direct the day-to-day metabolic activities within cells. If Martian organisms exist, would they have DNA? DNA is so crucial to Earthly life that some of the upcoming NASA tests of the Mars sample will look for evidence of DNA and similar molecules.

Frogs and elephants are more similar to each other than to mushrooms or grass.

LO4 Life Characteristics Relating to Evolution

Four and a half billion years ago, Earth and Mars were newly formed planets and these neighbors—the third and fourth "rocks" from the sun—were probably very similar. The radius of Mars is only about half that of Earth, but the two planets have similar compositions and both probably had stable water near the planet surfaces for much of their history. Biologists think that if life did arise on Mars, it was probably similar to early life here. Fossil evidence tells us that life on our planet has changed or evolved over the millennia. So would the same be true for Martian life? NASA will be looking for any minute evidence of fossils or changes in cell-like structures in the Martian soil and rock samples.

Life Changes over Time

Over time, life forms change. Biologists call this descent with modification evolution, and it is based on changes in the frequencies of genes within populations over time. In part, we can tell that life evolves from our analysis of the fossilized imprints of early organisms. The older a fossil, the less similar it is likely to be to present-day forms. This dissimilarity is good evidence not only of change but also of *continued* change in living species. Using fossils, DNA analysis, and other evidence of changes in gene frequencies, biologists can trace an organism's family tree. For example, modern day house cats and tigers (young branches on the feline tree) are closely related. Reaching farther back in time, cats and dogs would share a common ancestor, and going back to life's early history, cats would share a common ancestor with snakes, fish, beetles, mushrooms, trees, and bacteria. Tracing the evolutionary tree back to life's first beginnings on Earth, all organisms would eventually share a common ancestor in the distant past.

development
the process by which an offspring increases in size and complexity from a zygote to an adult

heredity
the transmission of genetic characters from parents to offspring

gene
(Gr. *genos,* birth or race) the biological unit of inheritance that transmits hereditary information from parent to offspring and controls the appearance of a physical, behavioral, or biochemical trait; a gene is a specific discrete portion of the DNA molecule in a chromosome that encodes an rRNA molecule

evolution
changes in gene frequencies in a population over time

Classification of Living Things

The process of evolution and the tracing of family lineages back in time can help explain the immense diversity of life, which some biologists estimate at upwards of 50 million species. To help make sense of this vast diversity, biologists have created a system for categorizing organisms into groups according to their similarities. Brightly colored tropical frogs, for example, are more similar to bullfrogs than to elephants, but frogs and elephants are more similar to each other than to mushrooms or

species
a taxonomic group of organisms whose members have very similar structural traits and who can interbreed with each other in nature

genus
(pl. *genera*) a taxonomic group of very similar species of common descent

family
a taxonomic group comprising members of similar genera

order
a precise arrangement of structural units and activities; also, in taxonomy, a taxonomic group comprising members of similar families

class
a taxonomic group comprising members of similar orders

phylum
(pl. *phyla*) a major taxonomic group just below the kingdom level, comprising members of similar classes, all with the same general body plan; equivalent to the division in plants

division
a taxonomic group of similar classes belonging to the same phylum, which is often called a division in the kingdoms of plants or fungi

kingdom
a taxonomic group composed of members of similar phyla, i.e., Animalia, Plantae, Fungi, and Protista

domain
a taxonomic group composed of members of similar kingdoms

Figure 1.4
A Hierarchy of Cat Species

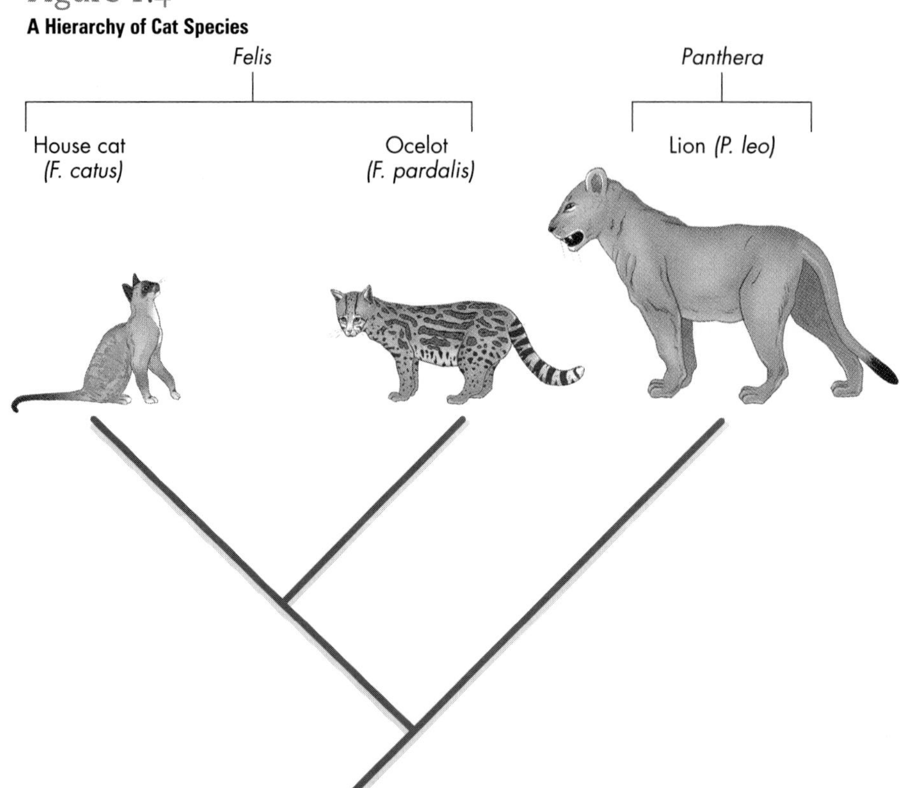

grass. If they discover Martian or other extraterrestrial organisms, biologists can begin to catalog them as well, perhaps based on our own existing system for Earth organisms.

Species

Species are groups of individuals with similar structures that descended from the same initial group and that have the potential to breed successfully with one another in nature. House cats are one species, and ocelots, small jungle cats whose habitats range from steamy Amazon rain forests to the dry chaparral of Texas, are a different but related species (see Fig. 1.4).

Genus

A **genus** (plural, *genera*) contains several related and similar species. Biologists refer to each species by a two-part name beginning with a term denoting the genus, followed by a separate term denoting the species. This two-word system for naming genus and species is called "binomial nomenclature." For example, the house cat *Felis catus* is related to but clearly distinct from the ocelot in the same genus *Felis* but the different species *pardalis*. (Together, then, the official name is *Felis pardalis*.) A lion, which is obviously still a cat but quite different in size, coloration, and habits from house cats and ocelots, is in a different genus, *Panthera*, and its species name is *Panthera leo*. After once mentioning the complete two-part name, biologists often abbreviate the genus, referring to *F. catus* or *P. leo*. You've probably seen the term *E. coli* in newspaper articles about outbreaks of food poisoning. In this case, the "E." stands for the unwieldy bacterial genus name *Escherischia*.

Just as biologists group related species into genera, they also group similar genera into **families**, similar families into **orders**, similar orders into **classes**, similar classes into **phyla** (or, in plants, **divisions**), similar phyla into **kingdoms**, and similar kingdoms into **domains**. Biologists recognize just three domains, each containing millions of life-forms (Fig. 1.5). Two of the domains, Bacteria and Archaea, consist of microscopic, mostly single-celled organisms that differ in fundamental ways (see Chapter 11 for details). The domain Bacteria includes the species that cause strep throat, for example, and that recycle decaying matter in soil and at the bottom of ponds. Members of the domain Archaea often live in harsh environments that are very hot, cold, acidic, or salty, such as thermal springs, salty lakes, or Antarctic ice. After Earth formed, environments like these would have been quite common, and the Archaea alive today probably share similarities with some of Earth's earliest organisms. (The name "Archaea" reflects the supposed "archaic" nature of these cells.)

Mars and Earth had similar beginnings, but Mars is very different now, with its frigid surface temperatures of 15°F during the day and 125°F at night. (The Martian day, by the way, is 24 hours and 37 minutes long.) Some biologists think that if Martian organisms once existed, or exist now, they would probably resemble members of the domain Archaea. Those terrestrial organisms therefore are a model in many ways for the search for extraterrestrial life. In addition, recent research on soil samples from many meters below Earth's surface turned up living Archaea that had been dormant for over a million years and then started dividing and growing again when provided with appropriate nutrients. This, too, has exciting implications for the Martian soil samples NASA is planning to retrieve.

adaptation
(L. *adaptere,* to fit) a particular form of behavior, structure, or physiological process that makes an organism better able to survive and reproduce in a particular environment

Earth's third domain of living organisms, Eukarya (*eu*=true *karya*=nucleus), consists of larger, more complex cells containing a nucleus, a special compartment that contains the cell's DNA. The domain Eukarya contains four kingdoms: Plantae, Animalia, Fungi, and Protista (Fig. 1.5). You're familiar with the plant and animal kingdoms, of course. Fungi include mushrooms, molds, and yeast. Protists are less familiar because they are often microscopic, and include amebas and other organisms with a single, but complex, cell.

Figure 1.5
The Domains of Life

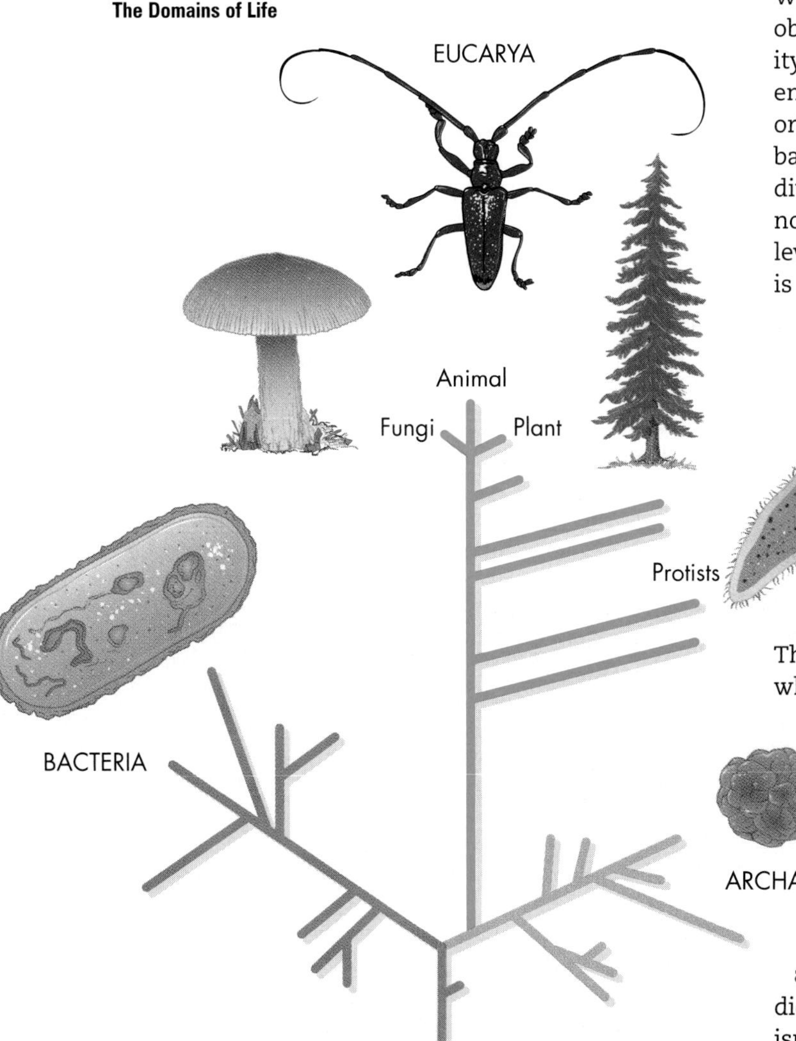

The Unity and Diversity of Life

With 50 million living species or more, life on Earth is obviously diverse, but the fact that all this multiplicity arose from a single group of ancestral cells present at the dawn of life gives it *unity* as well—unity of origin, of cell structure, of genetic material, and of basic day-to-day functioning. How can life be both diverse and unified? What mechanisms can foster not only vast diversity of form but also unity at the level of genes, cells, and basic function? The answer is evolution, the unifying theme for all life science.

Adaptation

Different species have different ways of extracting energy and materials from their surroundings. Think, for a minute, about organisms living in the bitter cold environment of Antarctica in the icy continent's McMurdo Dry Valleys. This area receives very little moisture, and temperatures are below freezing nearly all year round. The rocks that line the valleys, however, are somewhat porous, and they trap water. As sunlight hits the rocks, they can warm up above freezing, and the water can become liquid. Amazingly, some bacterial species have become adapted to live *within* the rocks, extracting energy from the sunshine, minerals from the rocks, and molecules necessary for life from the air. Specializations that help an organism adapt to its own special way of life are called **adaptations**. Some biologists hypothesize that if life did evolve on Mars in the distant past, Martian organisms may have become adapted to ways of living very similar to Earth's Antarctic life-forms.

natural selection the increased survival and reproduction of individuals better adapted to the environment

Many adaptations relate to taking in energy and materials, but there are others, too, that improve an organism's ability to grow, to reproduce, to move, to live in a group, or to attract a mate more successfully.

Natural Selection

Adaptations usually arise by **natural selection**, a "weeding out" process that depends on individuals' hereditary differences in their abilities to reproduce and to obtain energy. To see how this works, let's go back to the example of cells living in the McMurdo Dry Valleys. Some of these bacteria may have hereditary factors that allow them to manufacture sugars at a slightly lower temperature than other nearby cells. In such a bitterly cold climate, this ability could help these particular bacterial cells survive and reproduce under conditions in which other cells become dormant or die. As time passes, this advantage may mean that most or all of the cells in this better-surviving population and their descendants have the favorable trait while the nearby cells, lacking the trait, are out-competed. In 1859, Charles Darwin published the landmark book *On the Origin of Species,* and coined the term *natural selection,* for the mechanism underlying change in living species over time. He called it "natural selection" because nature was "selecting" the individuals with the most suitable variations to survive and become the parents of the next generation. A contemporary of Darwin's, Alfred Russell Wallace, also published papers naming natural selection as the prime mechanism of evolution.

During the century and a half since Darwin's and Wallace's work, biologists have delved deeply into both the principles of, and the evidence for, evolution by natural selection. Some of the most important advances pinpointed the sources of variation from which nature selects individuals: Variations usually arise through mutations or alterations in gene structure. Largely because of modern molecular genetics, today's biologists can explain both the diversity of life and the unity of its origins and shared characteristics. They can account for life's remarkable diversity because different environments require unique adaptations. Looking at foxes, for example, some of the characteristics that help an arctic fox survive in the frozen north are very different from those that help a desert fox survive in the desert. The arctic fox has small ears and a short muzzle that help conserve body heat and a white coat that serves as camouflage against snow and ice (Fig. 1.6a). In contrast, the desert fox has large ears and a long muzzle that radiate extra body heat and a tan coat that disappears against the surrounding sand and rocks (Fig. 1.6b). An arctic fox in the desert would overheat, stand out against the earth-toned background, and likely exhaust itself before catching rabbits or other prey. A desert fox in the arctic would have frost-bitten ears, would stand out against the snowy background, and would probably die of exposure or hunger before catching enough prey. The diversifying action of the environment, working through natural selection over millions of years, could have produced the remarkable variety of life-forms that have existed on our planet.

Likewise, evolution—descent with changes over time—can account for life's unity, including the common characteristics we're surveying in this chapter. Biologists can trace the ancestry of both arctic and desert foxes back to a single fox species that lived in North America millions of years ago. They can trace that fox back to a common ancestor with all dogs. They can trace this group, the canids, back to a common ancestor with all mammals, and so on back to the origin of life itself. As we will see in more detail in later chapters, the metabolic machinery within all living cells is very similar despite wide differences in cell

Figure 1.6
Fox Adaptations to Arctic and Desert
(a) (b)

shape, size, and function. Biologists think all organisms alive today inherited this underlying machinery from cells that appeared at the dawn of life, and this helps explains the characteristics they share today.

Evolution by natural selection is so grand an organizing principle for all biology that it will resurface repeatedly in this book, in the many magazine and newspaper articles you may read on life science, and in any future biology courses you may take.

LO5 Life Characteristics Relating to Environment

Was life widespread on Mars early in its history? People used to believe the planet was covered in canals built by little green aliens until modern space probes showed that at least some of the "canals" were the remains of giant river beds. We know from the geological evidence of flowing water and the current absence of liquid water on the planet's surface that Mars was once wet and must have changed drastically to its current extreme aridness. If life was present and if it survived at all, then it would have had to change too. Ecology is the branch of biology that studies the relationships between living organisms and their environment, and it is an interesting and pervasive part of exploring life.

The Hierarchy of Life

Organisms interact with their living and nonliving environments at several different levels. These levels extend the continuum of order we discussed earlier: molecules to cells to organisms. Take, for example, life in the African savanna environment—a splendid collection of plants, animals, fungi, microbes, and habitats that includes elephants, acacia trees, tussock grasses, and arid plains. The hierarchy of life on the savanna proceeds from small to large in the following sequence: Organisms, as we saw earlier, are individual, independent living things; an elephant is an organism and so is an acacia tree. Groups of a particular type of organism that live in the same area and actively interbreed with one another are called populations, for example, an elephant herd or a field of grass. All the populations that live in a particular area, including the plants, animals, and other organisms that share the savanna, for example, make up a community. The living community together with its nonliving physical surroundings is called an ecosystem. The savanna ecosystem includes elephants, the egrets that pick insects off their skin, and the coarse grass they chew and trample, as well as the water in clouds, the sandy soil underfoot, and the hot African sunshine. All the ecosystems of the Earth make up the biosphere, that portion of the Earth on which life exists, including every body of water; the atmosphere to a height of about 10 km (6 mi); the Earth's crust to a depth of many meters; and all living things within this collective zone. The biosphere encompasses unimaginably remote places that nevertheless still teem with life, such as the deepest parts of the ocean floor, deep-sea vents spewing superheated water, the frozen ice of Antarctic lakes, and porous hot rocks a mile and a half (2.7 km) down toward the center of the Earth. A huge variety of organisms exists in these and the more familiar and hospitable forests, meadows, lakes, marshes, and grasslands we know in our own surroundings. The surprising abundance of life-forms in Earth's extreme environments gives scientists good reason to think that Mars could also harbor life. The search for life throughout the solar system and beyond is fueled by our understanding of life's diversity and tenacity.

ecology
the scientific study of how organisms interact with their environment and with each other and of the mechanisms that explain the distribution and abundance of organisms

population
a group of individuals of the same species living in a particular area

community
two or more populations of different interacting species occupying the same area

ecosystem
a community of organisms interacting with a particular environment

biosphere
(Gr. *bios,* life + *sphaira,* sphere) that part of the planet that supports life; includes the atmosphere, water, and the outer few meters of the Earth's crust

If you inflated a balloon to represent Earth, then the thickness of its taut rubber skin would be proportional to the biosphere.

We've seen that organisms take their energy from the environment and that over time evolution fits organisms to their environments. Because this is just as true for Earth's extreme environments as for its temperate ones, there is good reason to think that at least some environments on ancient Mars were inhabited by living organisms.

LO6 How Biologists Study Life

So far, we've examined the basic life characteristics on our planet. But we haven't talked much about how biologists coax, prod, and pry the secrets of nature from living organisms. It's important to keep in mind that behind every fact and concept in this course, there were people in laboratories or field stations engaged in the often joyful

scientific method a series of steps for understanding the natural world based on experimental testing of a hypothesis, a possible mechanism for how the world functions

hypothesis a possible answer to a question about how the world works that can be tested by means of scientific experimentation

and exciting, but sometimes tedious and frustrating, pursuit of knowledge about living things.

Natural Causes and Uniformity of Nature

A lightning bolt flashes in a cloud-darkened sky. A man stands on a street corner shouting that alien invaders have placed probes in his head. Modern scientists assume that events like these are due to natural causes. The ancient Greeks, on the other hand, believed that thunderbolts arose when the god Zeus hurled them at the Earth and that mental illness was due to evil spirits. Today's scientists do not yet fully understand what causes Alzheimer's disease, for instance, or the El Niño climate fluctuation. But they firmly believe these are based upon natural causes they will someday discover by applying the scientific process.

Scientists consider the fundamental laws of nature to be uniform and to operate the same way at all places and at all times. For example, biologists assume that the fixed speed of light, the laws of gravity, and the properties of chemical elements work the same way in Ohio today as they did in East Africa 1 million years ago or on Mars 3 billion years ago. The events that led to life's origin and diversity on Earth, and perhaps on Mars and other planets, occurred long before humans lived to observe them. Yet biologists are confident that today's natural laws functioned the same way at the dawn of time, as life began, and all during its evolution.

Scientific Method:

1. identify a problem
2. propose a hypothesis
3. make a prediction
4. test the prediction
5. draw a conclusion

The Power of Scientific Reasoning

The search for life on Mars reveals the two kinds of scientific reasoning biologists use. In one type, the biologist collects specific cases and then generalizes from them to arrive at broad principles. For example, after observing that cells can live inside rocks in Antarctica's frigid deserts, biologists proposed a generalization: that Mars rocks could perhaps harbor life, as well. The instant when the scientist's mind leaps from previously isolated facts to a broad, unifying generalization is a creative, intuitive, exciting moment every bit as original as writing a sonnet or sculpting a form from clay.

The second type of reasoning starts with general principles and then goes in the opposite direction to analysis of specific cases. For example, biologists knew that living organisms had turned up in places previously thought to be lifeless. Based on that knowledge, they reasoned that living cells might be found within the permanent ice that covers Antarctic lakes. They took ice samples, tested them, and found that, indeed, bacteria do survive around dark specks of dust incorporated in the ice. These particles absorb sunlight, warm up, and melt a small halo of water around themselves. This moisture and the pale Antarctic summer sunlight are enough to provide a brief growing period each year for the bacteria.

These two types of reasoning—from the specific to the general, and from the general to the specific—help shape how scientists think, but they aren't unique to science. What is unique to the scientific process is a particular approach to testing generalizations. The steps may sound regimented, but they're really just organized common sense.

Testing Generalizations: The Scientific Method

You may not know it, but you already use scientific reasoning. Say you come home late one night and flip on the light switch in the hall, but the overhead fixture stays dark. You think to yourself, "I guess the bulb's burned out." On the basis of that hypothesis (guess), you predict that a bulb you know to be working (because it lights up a nearby floor lamp) will fix the hall light. You test your hypothesis by screwing the working bulb into the hall socket and flip the switch again. Still no light. You have just disproven your initial hypothesis ("burned-out bulb") and need a new hypothesis ("broken switch" or "broken socket"). Biologists use this approach of hypothesis and testing—the **scientific method**—in much the same way.

First, they *ask a question* or identify a problem to be solved based on observations of the natural world. Your observation was a dark hall, and your question was, "Why won't the light turn on?"

Second, they *propose* a **hypothesis**, a possible answer to the question or a potential solution to the problem. A hypothesis is a guess. Yours was "burned-out bulb."

Then they *make* a **prediction**, a statement of what they will observe in a specific situation if the hypothesis is correct. You predicted that a working bulb would fix the problem.

They *test the prediction* by performing an experiment or making further observations. You tested a bulb in a floor lamp to make sure it glowed, then screwed that working bulb into the hall light socket and flipped the switch again. The floor lamp provided a **control**, a standard for comparison based on keeping all factors the same except for the one being tested. The hall light socket provided the **experimental** situation—the carefully planned and measured test of the hypothesis.

Finally, they *draw a conclusion*. If the hypothesis predicts incorrectly, then they must discard it as wrong. In your case, you said, "Nope. Not the bulb." If the hypothesis predicts correctly—let's say the light did go on—then they devise more tests to see whether the hypothesis might still be incorrect in some way. If they can never design a situation that shows the hypothesis to be wrong, then they begin to accept it. (Here's where we differ from scientists in our daily lives. If the bulb goes on, we think, "Solved!" and go about our business. We don't dream up more tests for *why* it worked!)

The Scientific Method at Work

How do biologists approach scientific puzzles? Let's review how they applied the scientific method to the question, "Does life exist on Mars?"

Searching the Mars Rock for Signs of Life

NASA researchers investigating the Mars rock found in Antarctica began by posing a question: Are there signs of past life inside this Martian meteorite? Next they stated an assumption they were making: that Martian life is or was similar to Earthly life. Without knowing anything about possible life on Mars, they had to make *some* assumptions about what it would be like and what they'd be looking for, and so logically they chose the characteristics of life on Earth. (Explicitly stating underlying assumptions is an important, but sometimes overlooked, part of the scientific method.) Then they created a hypothesis that life did exist at one time in the Martian rock and left fossilized remains. Based on that hypothesis and their assumption, they carried out several tests: They looked for shapes similar to fossilized Earthly bacteria; they searched for traces of chemicals similar to those formed by life on Earth; and they studied grains of a magnetic substance in the Mars rock similar to grains found inside certain bacteria that can orient the cells to Earth's magnetic field. Their hypothesis predicts each of these factors, and indeed, the researchers found these items in the rock.

prediction
in the scientific method, an experimental result expected if a particular hypothesis is correct

control
a check of a scientific experiment based on keeping all factors the same except for the one in question

experimental
during the application of the scientific method, the phase involving the carefully planned and measured test of the hypothesis

We saw that to interpret any experiment, scientists need a control or known standard for comparison. In tests of the Martian rock, NASA researchers found little tubular structures that look like fossilized bacteria (see Fig. 1.1). For the control, they chose fossilized Earthly bacteria of a size and shape similar to the tubular structures and in rocks of a similar age and composition to the Martian meteorite. Despite their best choices thus far, however, all of the fossilized bacteria from Earth have been much larger than the Martian tubular "fossils." Because of such comparisons with Earthly controls, most biologists doubt that the tubules in the Martian rock are fossilized cells, since there would have been so little space inside them for genes and the machinery of life as we know it.

In their second test, the NASA scientists did find molecules in the Martian meteorite similar to those produced by Earthly life. Their hypothesis predicted this finding. Remember, though, that *while an incorrect prediction disproves a hypothesis, a correct prediction does not automatically prove it to be correct.* For example, the chemicals they found in the Mars rock can be produced by living organisms, but they can also be produced by purely chemical means, as well. In our previous light bulb example, screwing in a working bulb didn't fix the overhead fixture. But if it had, it would have been *consistent* with the hypothesis (burned-out bulb) yet still not *proven* it. Other hypotheses could have been true, as well—for example, maybe the original bulb still worked, but was not screwed in tightly enough. How would you test *that* hypothesis?

In their third test, the NASA researchers found tiny crystals of an iron compound called magnetite that some bacteria on Earth use as an internal compass. Crystals of this exact shape and size are not known to be formed outside of living cells. Other shapes are, however, and many researchers think that the magnetite crystals in the Mars rock may have a nonliving origin. Again, the NASA team's hypothesis predicted correctly, but they were not able to rule out other competing hypotheses.

theory
a general hypothesis that is repeatedly tested but never disproved

The search for life in the Mars rock illustrates several points about the scientific method, but differs in a couple of key ways from how biologists often learn about the natural world:

1. The scientists could not produce controls or do direct experiments, since they were observing a unique specimen.
2. Scientists usually carry out the same experiment or observation many times before drawing a firm conclusion. To do this would require searching for life on many different Earthlike planets and/or getting many samples from Mars. Martian soil and rocks may help provide controls and duplicate the experiments.

A Word About Theories

Eventually, a theory can emerge from a broad general hypothesis that is tested repeatedly but never disproved. But what is a theory? A theory is a general principle about the natural world, like the theory of gravity, the cell theory, or the theory of evolution. People often say "It's just a theory," meaning something that's an untested idea. But scientists don't use "theory" in that way; to them, a theory is a highly tested and never disproven principle that explains a large number of observations and experimental data.

The scientific method is a powerful tool for understanding the natural world, but it does not apply to matters of religion, politics, culture, ethics, or art. These valuable systems for approaching the world rely on different lines of inquiry and experience. There will always be a place for scientific reasoning, though, because so many of the world's complex problems have underlying biological bases, and we can't solve them without biological facts and principles.

Who knew?

While an automobile might exhibit some of the characteristics of life, it does not exhibit all of the characteristics found in a living organism. It is not able to acquire energy and nutrients from its environment (it has no metabolism), and it cannot grow or reproduce.

LO7 Biology Can Help Solve World Problems

The search for life elsewhere in the universe is an obvious application of biology and of the question, "What is life?" But biology can do more than just prepare us to look for life on other worlds. It can contribute solutions to a long and growing list of problems here on Earth.

In the chapters that follow, you'll explore how biology is helping to solve world problems. We are, in fact, in the midst of a revolution in the biological sciences, with exciting new information surfacing weekly in the fights against cancer, heart disease, AIDS, infertility, and obesity. Researchers are making rapid advances in gene manipulation to create new drugs, crops, and farm animals; in exercise physiology to improve human performance; in the diagnosis of genetic diseases; and in the transplantation of organs, including brain tissue. The discoveries are so frequent and fast-moving, in fact, that many of them will appear only on this book's companion Web site and not in this text. Across all frontiers of biological science, at all levels of life's organization—from molecules to the biosphere—scientists are learning the most profound secrets of how living things survive day to day and reproduce new generations. You're about to embark on an adventure of exploration and discovery that will not only excite your imagination and enrich your appreciation of the natural world, but will also allow you to contribute intelligently to the difficult choices all human societies must make in the future.

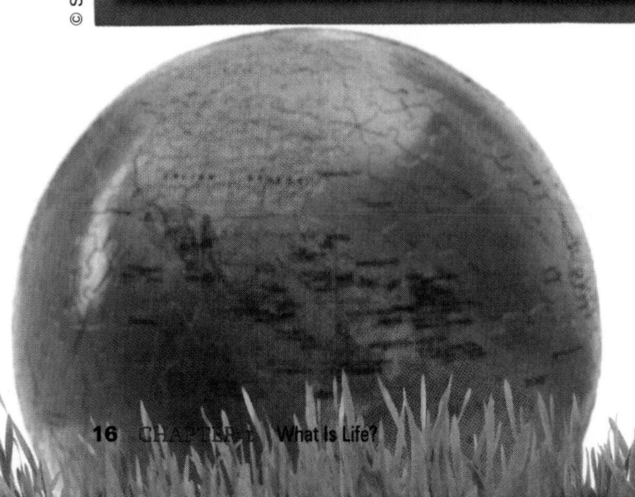

87% of students surveyed feel that LIFE is a better value than other textbooks.

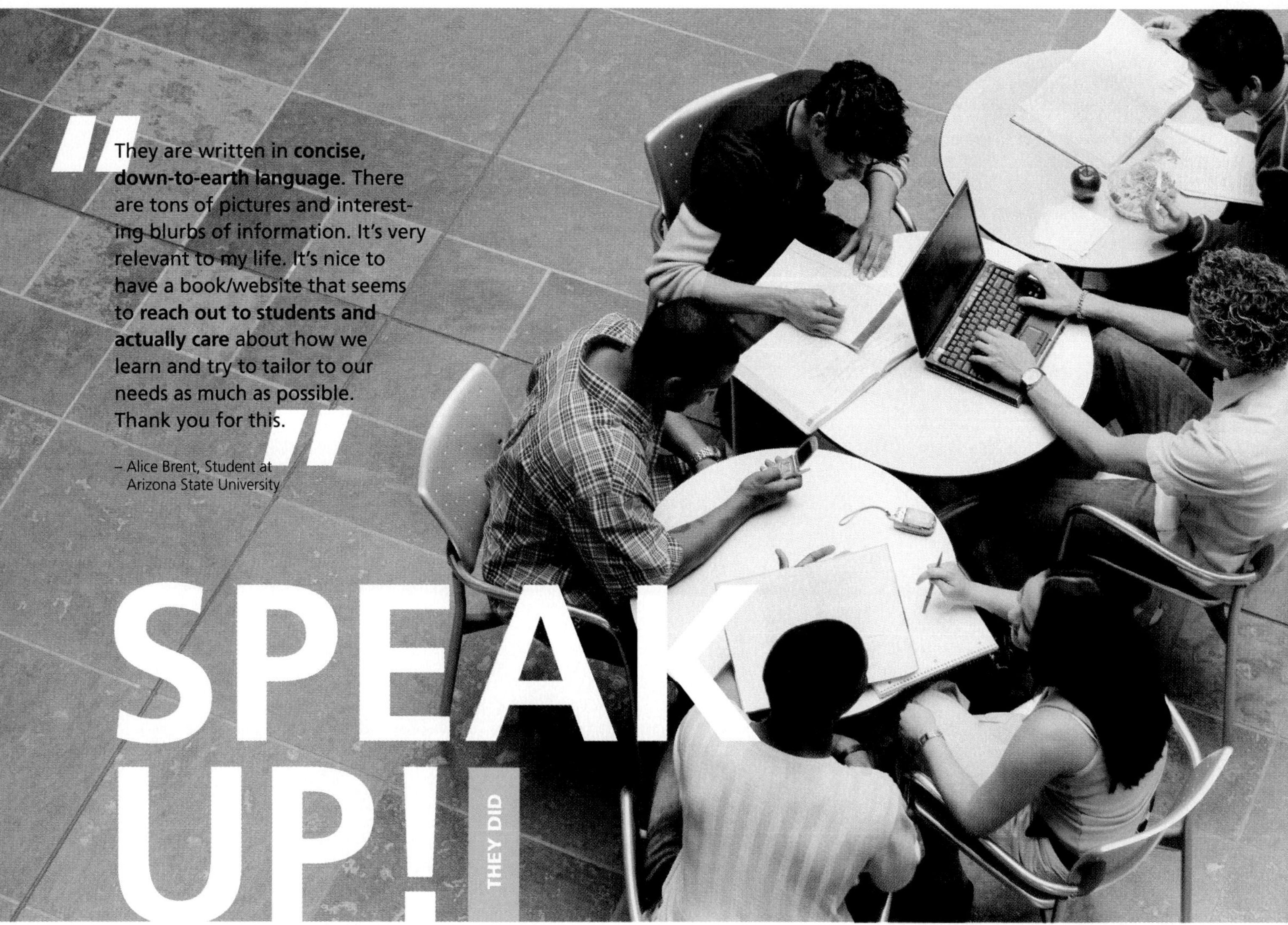

LIFE was built on a simple principle: to create a new teaching and learning solution that reflects the way today's faculty teach and the way you learn.

Through conversations, focus groups, surveys, and interviews, we collected data that drove the creation of the current version of LIFE that you are using today. But it doesn't stop there – in order to make LIFE an even better learning experience, we'd like you to SPEAK UP and tell us how LIFE worked for you.

What did you like about it?
What would you change?
Are there additional ideas you have that would help us build a better product for next semester's non-major biology students?

At **4ltrpress.cengage.com/life** you'll find all of the resources you need to succeed in introductory biology – **animations, visual reviews, flash cards, interactive quizzes,** and more!

Speak Up! Go to **4ltrpress.cengage.com/life**.

CHAPTER 2

Cells and the Chemistry of Life

Learning Outcomes

LO[1] Describe the general structure of HIV

LO[2] Differentiate among elements, atoms, molecules, isotopes, and ions

LO[3] Explain the biological importance of chemical bonds

LO[4] List the properties of water that make it so critical to life

LO[5] List and describe the four main types of biological molecules found in all living organisms

LO[6] Compare the characteristics of a virus and a cell

LO[7] List the organelles and structures found in cells, and discuss their functions

LO[8] List the seven steps in HIV's infective cycle

"It is remarkable that a simple, nonliving particle can wreak such havoc in the lives of infected people."

A Threat to Cells and Lives

In his years at the University of California at San Francisco, Dr. Jay Levy has watched acquired immune deficiency syndrome, or AIDS, grow from a seemingly isolated threat into a global epidemic affecting more than 33 million people. "I can't believe where the time has gone," he says, "and that today we still don't have a really long-lasting treatment." That, he says, is because the human immune deficiency virus (HIV) that causes AIDS keeps evolving drug-resistant forms. By now, some of Levy's research subjects have been able to stay relatively healthy for many years. But in other patients initially helped by various combinations of antiviral medicines, "the virus is coming back and there are no more drugs to use. I think these are going to be sad cases," he says, "and we're seeing up to 40 percent of the people at San Francisco General Hospital with this."

What do you know?

Could a single-celled, amoeba-like creature ever be as big as a truck? Why or why not?

The ongoing fight against HIV and AIDS make an ideal case study for this chapter. Our subjects here are the chemistry of life and the structure and function of cells, the fundamental units of life. By studying the human immune deficiency virus, you will see how virus particles contain different kinds of building blocks called biological molecules. We will compare viruses, which are nonliving, to bacterial, plant, and animal cells to reveal their basic differences and unique characteristics. By following the entry of an HIV particle into a human cell, as Levy and others have investigated in great detail, you will get an intimate tour of a functioning cell and its many internal organelles. From this, you will see for yourself why AIDS is such a deadly disease and come to understand the efforts now underway to control this global threat through drugs and vaccines.

As we go through the chapter, you'll find out why you can pick up a cold virus—but not HIV—from a doorknob. You'll see why health experts fear that AIDS could annihilate 40 percent of the population of Africa before a solution is found. You will also find the answers to these questions:

- ☑ How are atoms structured and how do they function?
- ☑ What are the special properties of water?
- ☑ What are the main kinds of biological molecules and their roles?
- ☑ How is a living cell different from a virus?
- ☑ How do the various cell parts function and how does HIV sabotage them?

AIDS (acquired immune deficiency syndrome) a partial or total loss of immune function based on infection by the human immune deficiency virus (HIV)

HIV (human immune deficiency virus) the causative agent in acquired immune deficiency syndrome (AIDS)

element
a pure substance that cannot be broken down into simpler substances by chemical means

atom
(Gr. *atomos,* indivisible) the smallest particle into which an element can be broken down and still retain the properties of that element

molecule
a cluster of two or more atoms held together by specific chemical bonds

proton
a positively charged subatomic particle found in the nucleus of an atom

neutron
(L. *neuter,* either) a subatomic particle without any electrical charge found in the nucleus of an atom

electron
a negatively charged subatomic particle that orbits the nucleus of an atom; the negative charge of an electron is equal in magnitude to the proton's positive charge, but the electron has a much smaller mass

LO[1] What Is HIV?

HIV is a nonliving particle that infects human cells. Biologists consider HIV to be *nonliving* because it has some but not all of the life characteristics we discussed in Chapter 1. A virus has internal order, based on the same groupings of atoms or biological molecules that make up living things. Viruses have genetic material that can change over time, and thus they can evolve. Finally, parts of the virus particle can move, so we can consider them to display the living trait of motility. However, a virus has no metabolism, it is unresponsive in the biological sense, and it lacks the ability to reproduce (without help from living cells). While virus particles do assemble themselves, this construction requires the machinery inside a host cell and so cannot be considered true growth or development. Virus particles are clearly not alive by our definitions in Chapter 1. Yet their structure and behavior inside the human body are still governed by the same set of chemical and physical laws that determine how all biological molecules form and act and how all living cells function.

It is remarkable that a simple, nonliving particle made up of just a few chemicals can wreak such havoc in the lives of infected people. Our goal in this chapter is to look more closely at our lethal yet nonliving enemy HIV, and at the atoms and molecules that make up all viruses, cells, and larger organisms.

LO[2] What Are Atoms?

All matter, including HIV particles and the cells they infect, are based on atoms of distinguishable types. That's where the elements come in and they take us back to some of history's earliest students of life and matter.

Elements

Ancient Greek philosophers realized that some materials, such as rocks, wood, and soil, are composed of more than one substance, while other materials, such as chunks of iron, gold, and sulfur appear to be pure materials. Chemists call pure substances like these that can't be broken down further into different constituent **elements**. Chemists also assign each known element a chemical symbol; for example, the symbols for the main elements found in an HIV particle are C (carbon), H (hydrogen), and O (oxygen).

Chemists have discovered 118 elements. Of these, 89 occur in nature, while scientists have created the rest in the laboratory. The properties of different elements vary widely. For example, carbon is a black solid, sulfur is a yellow solid, and helium is a colorless, odorless gas. Although the Earth contains dozens of elements, only seven elements, headed by oxygen, silicon, and aluminum, make up about 98 percent of Earth's surface layer. Researchers have found more than three dozen elements in living things, but most occur only in traces. Just three elements make up 98 percent of the body of a human or a fern—hydrogen, oxygen, and carbon. In our later discussion of water and carbon, we'll see why living tissue is a unique and special form of matter.

Atoms and Molecules

Never satisfied by superficial discoveries, early scientists wondered what makes each element distinct: How is gold, for example, fundamentally different from oxygen? (Knowing this might, among other things, have helped them turn iron or carbon into gold—or so they hoped!) In the 1800s, the English chemist John Dalton concluded that each element is composed of identical particles called **atoms** (Greek, *atomos* = indivisible). Atoms are the smallest particles of an element that still display that element's chemical properties. A **molecule** is the chemical combination of two or more atoms. In a molecule of water, for example, two hydrogen atoms are combined with one oxygen atom.

Atoms are extremely tiny. About a million carbon atoms could sit side by side on the period ending this sentence. A small gold nugget consists of billions of gold atoms. Tiny as they are, though, atoms themselves have an internal structure.

COURTESY OF NATIONAL INSTITUTE OF SCIENCE AND TECHNOLOGY

Structure of Atoms

Whether found in a lifeless rock or in a biological entity, such as a person, all atoms are composed of protons, neutrons, and electrons (Fig. 2.1). A **proton** is a subatomic particle with a positive electrical charge, and a **neutron** is a particle with no electrical charge. **Electrons** are much smaller (have less mass) and they have a negative electrical charge.

A Model of the Atom

Think of an atom as resembling a miniature solar system. The atomic nucleus at the center contains protons and neutrons and accounts for most of the atom's mass. A specific number of electrons, equal to the number of protons, orbit the nucleus at a relatively great distance. Figure 2.1a shows the simplest atom, hydrogen, with its single proton, single electron, and no neutron. If a somewhat larger atom like carbon (Fig. 2.1b) were the size of the Houston Astrodome, the nucleus would be a small marble on the 50-yard line.

What Gives Atoms Their Properties?

Why is a chunk of the element carbon black and solid, while the element oxygen is a clear, colorless gas? The answer is that each type of atom contains a unique number of protons in its nucleus. All carbon atoms have six protons, for example, and all oxygen atoms have eight protons (Fig. 2.1b, c). The number of protons affects the atom's mass and its attraction for electrons, and these two features, in turn, determine the atom's physical and chemical properties.

Electrons and Energy Levels

The attraction between the positively charged nucleus and the orbiting electrons, with their negative charge, sets up conflicting forces: The opposite charges pull the electrons toward the nucleus, but their rapid circling tends to throw them outward, away from the nucleus, the way a rock tied to a twirling string pulls outward.

Electrons are too small to be seen with the eye or most instruments, of course, but scientists picture them whizzing about in **energy shells** at specific distances from the nucleus, with higher energy levels the farther they orbit from the nucleus. Each shell can contain a certain maximum number of electrons, and the bonding of atoms into molecules depends on the order of shell-filling. The ring or electron shell nearest the nucleus can hold either one electron, as in hydrogen, or two, as in helium. The second shell can accommodate up to eight electrons, and will be filled before any electrons appear in the higher-energy third shell, which also holds up to eight electrons. Subsequent shells also become filled with set numbers of electrons, and tend to fill in order.

atomic nucleus
the central core of an atom, containing protons and neutrons

energy shells
energy levels occupied by electrons in orbit around an atomic nucleus; each shell can contain a maximum number of electrons, for example, two electrons for the first shell, eight for the second

isotopes
an alternative form of an element having the same atomic number but a different atomic mass due to the different number of neutrons present in the nucleus

Variations in Atomic Structure

Slight exceptions to the standard structure of atoms—either in the number of neutrons or the number of electrons—help explain phenomena as diverse as atomic bombs, acid rain, and the actions of your nerve cells.

An atom of a given kind contains a set number of protons, but the number of neutrons can vary. Atoms with the same number of protons but different numbers of neutrons are different **isotopes** of the same element. The most common carbon isotope is ^{12}C with six neutrons and six protons. Other carbon isotopes are ^{13}C and ^{14}C, with seven and eight neutrons, respectively. Both ^{12}C and ^{13}C are stable, nonradioactive forms, but ^{14}C is radioactive—it tends to break down and emit radiation. In 1991, hikers high on a ridge in the Swiss Alps found an unfortunate hiker's head and shoulders sticking out of a chunk of melting ice. Researchers used radioactive carbon 14 (^{14}C) to determine the age of the ice man. When an organism dies, it stops incorporating ^{14}C from the environment, and the isotope begins to decay into the isotope nitrogen 14 (^{14}N). (^{14}C has six protons

Figure 2.1
Models of an Atom

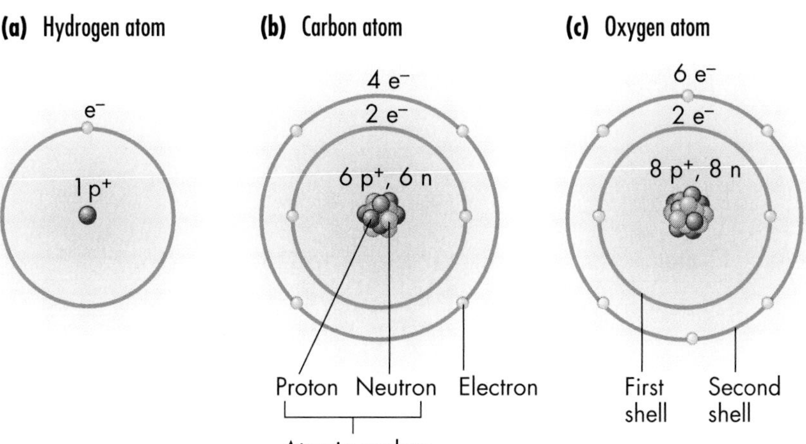

> “Atoms make bonds to fill their outer electron orbit.”

half-life
the length of time it takes for half of the total amount of radioactivity in an isotope to decay; as a result of such decay, for example, the concentration of carbon-14 relative to carbon-12 decreases; half of the carbon-14 will decay in 5,730 years; this is the half-life

ion
an atom that has gained or lost one or more electrons, thereby attaining a positive or negative electrical charge

chemical bond
an attractive force that keeps atoms together in a molecule

and eight neutrons, while ^{14}N has seven protons and seven neutrons.) As a result of this decay, the concentration of ^{14}C relative to ^{12}C decreases. One-half of the original amount of ^{14}C decays in 5,730 years; this is known as the **half-life** of ^{14}C. Through ^{14}C dating, scientists concluded that the ice man lived about 5,300 years ago, and that he was the oldest well-preserved human body yet found.

While in the case of isotopes, neutron numbers vary, in **ions**, electron numbers vary (Fig. 2.2). This means the entire atom has a specific positive or negative electrical charge; the number of electrons in an ion does *not* equal the number of protons. For example, the most common form of the hydrogen atom has one proton and one electron: Because the electrical charges cancel each other out, the atom has no net charge (Fig. 2.2a). A hydrogen ion is missing its electron; as a result, it has only one proton, and is positively charged (Fig. 2.2b). A chlorine ion (Cl^-), on the other hand, has a negative charge.

The properties of the elements emerge from both the structure of the atomic parts and the way those parts are arranged. As we'll see in the next sections, this idea of emergent properties also holds for the way atoms make up molecules and the way molecules make up living things.

LO³ How Do Atoms Form Molecules?

The atoms of life—carbon, hydrogen, oxygen, and the others—are joined in tens of thousands of combinations to form the molecules in your food, in other animals and plants, and in your body. How atoms combine to form different kinds of molecules helps determine their properties in living things.

In molecules, two or more atoms are linked by an attractive force called a **chemical bond**. The bonds that link atoms are not actual physical connections, like the couplings between railroad cars. Instead, they are links of energy acting like "energy glue," often based on shared or donated electrons. Bonds act like invisible springs; once a bond forms between

Figure 2.2
Ions

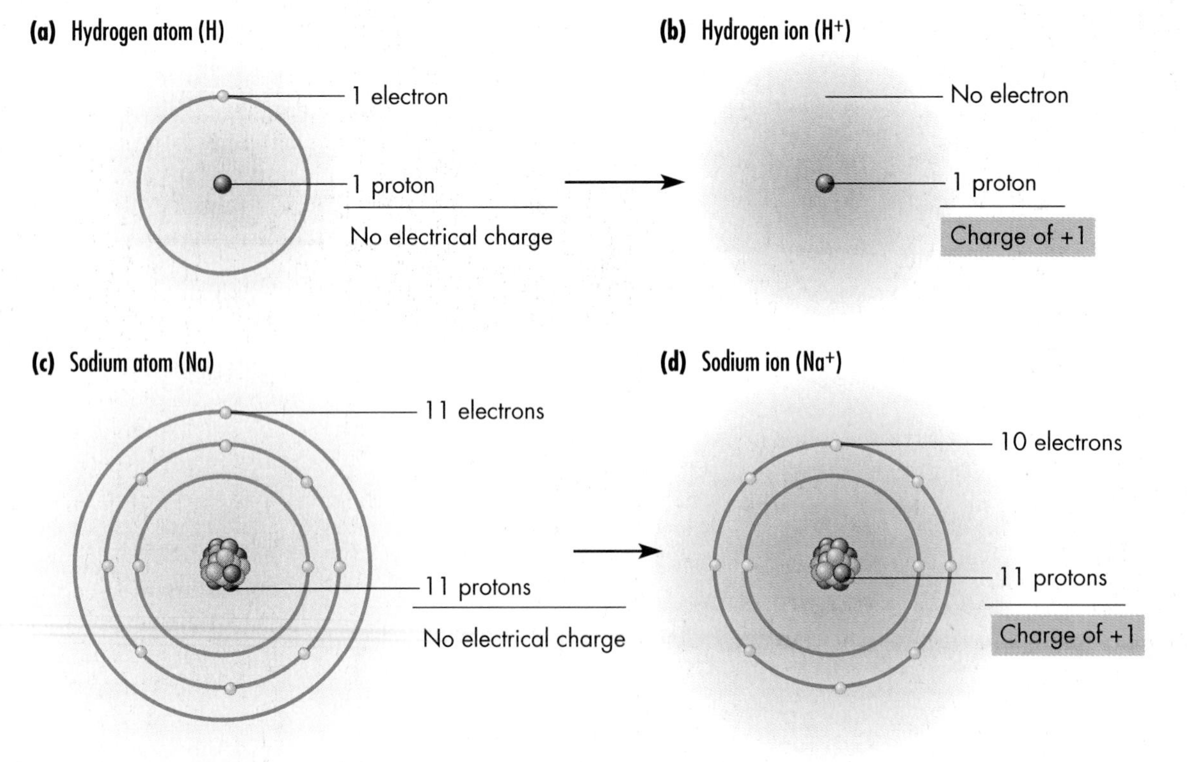

two atoms, it requires energy to pull the atoms apart or to push them closer together. We'll see three kinds of bonds in our exploration of biology: covalent bonds, hydrogen bonds, and ionic bonds.

Covalent Bonds

When two atoms share a pair of electrons, the most common type of chemical bond forms—a **covalent bond** (Fig. 2.3a on the next page). In a water molecule (chemical formula, H_2O), two hydrogen atoms each share a pair of electrons with one oxygen atom. As each hydrogen atom approaches the oxygen atom, its positively charged nucleus begins to attract electrons orbiting the other nucleus. Eventually, the electron orbits overlap and fuse, and the two atoms—the hydrogen atom and the oxygen atom—share a pair of electrons.

In some molecules, the electrons spend as much time orbiting one nucleus as the other, and the electrical charge is evenly distributed about both ends, or *poles*, of the molecule. A molecule with this equal sharing of charge is said to be **nonpolar**. In a molecule like water, however, the electrons spend more time orbiting the oxygen than the hydrogen. This leaves the oxygen pole of the molecule with a slightly negative charge, and the hydrogen pole of the molecule with a slightly positive charge, making H_2O a **polar** molecule.

Hydrogen Bonds

With their charged ends, some polar molecules can form another kind of chemical bond—a hydrogen bond. In liquid water, for example, a hydrogen atom from one water molecule can electrically attract an oxygen atom from an adjacent water molecule (Fig. 2.3b). The attraction of a hydrogen atom to an atom (usually oxygen or nitrogen) in another molecule is called a **hydrogen bond**.

Hydrogen bonds are much more easily broken and reformed than covalent bonds. Some of water's unusual properties (such as the tendency for ice to float) are based on hydrogen bonds, and some important biological molecules are held together by hydrogen bonds. Hydrogen bonds, for example, are involved in the interaction between the bead-like proteins on the surface of HIV particles and the surface of a human cell about to become infected. Hydrogen bonds also hold together the nucleic acids that give DNA molecules their "meaning" as the code of life. (We'll encounter this code in Chapter 6.)

Ionic Bonds

In the third type of chemical bond—an **ionic bond**—electrons from one atom are completely transferred to another atom rather than shared. Salt (NaCl) is a good example: a sodium ion is positively charged (Na^+), and a chlorine ion is negatively charged (Cl^-) (Fig. 2.3c). These oppositely charged ions can attract each other, rather like magnets, and an ionic bond forms between the Na^+ and Cl^- and holds the atoms together. Ionic bonds are much stronger than hydrogen bonds but still not as strong as covalent bonds. That explains why ionically bonded compounds dissociate (break down) into their component ions when dissolved in water, as when table salt dissolves in a pot of soup.

covalent bond
(L. *co*, together + *volere*, sharing) a form of molecular bonding characterized by the sharing of a pair of electrons between atoms

nonpolar
having a symmetrical distribution of electrical charge; i.e., a nonpolar molecule like most lipids will not dissolve readily in water

polar
having an asymmetrical distribution of electrical charge; i.e., a polar molecule like glucose will dissolve readily in water

hydrogen bond
a type of weak molecular bond in which a partially negatively charged atom (oxygen or nitrogen) bonds with the partial positive charge on a hydrogen atom when the hydrogen atom is already participating in a covalent bond

ionic bond
a type of molecular bond formed between ions of opposite charge

LO4 What Makes Water So Special for Life?

Let's look at the many properties of water molecules that make them so special for life.

Physical and Chemical Properties of Water

The tendency of water molecules to form hydrogen bonds gives water several of its important physical characteristics—all of which are important for living organisms. For example, hydrogen bonds make water molecules stick to each other and to soil, glass, and other substrates. This explains the capillary action and transport processes that draw water up into plants—even towering trees. Water's stickiness also creates surface tension or a "skin" on liquid water that some insects can glide across. Hydrogen bonds in frozen water make ice float. And because of hydrogen bonds, it takes a large amount of heat to increase the temperature of water, and this helps living organisms sustain steady internal temperatures.

Water also has chemical properties that explain why things dissolve and why some substances are acidic and some are basic (alkaline). In one sense, living things are forms of water moving about the planet: Our bodies, for instance, contain more than 60 percent H_2O, and so the movement and bonding

Figure 2.3
Automic Bonding

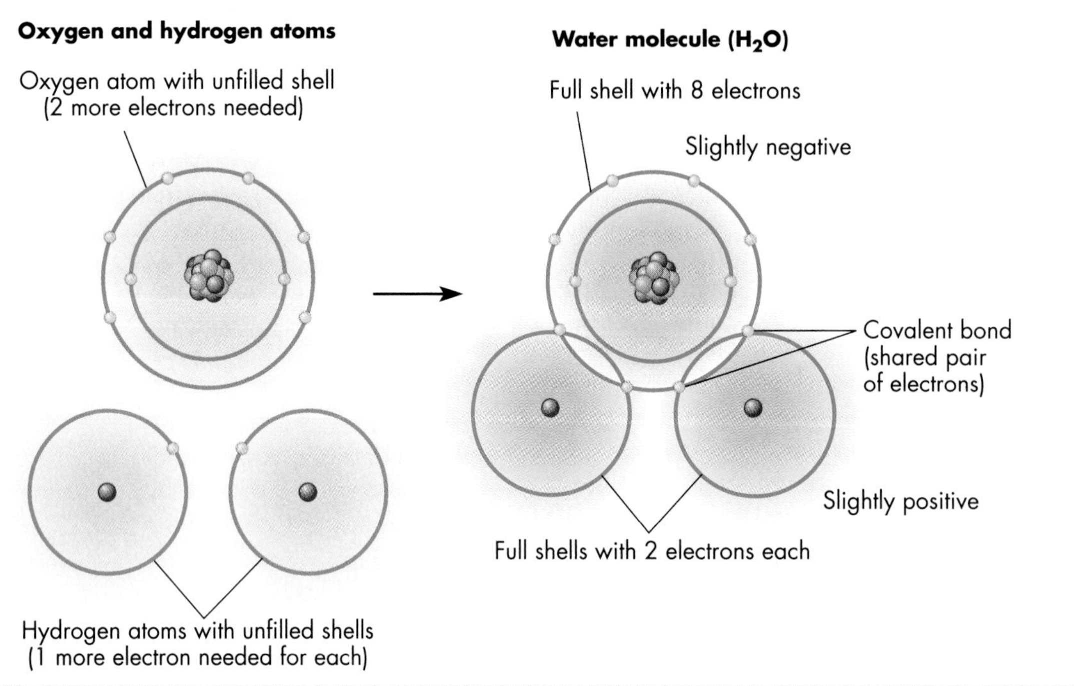

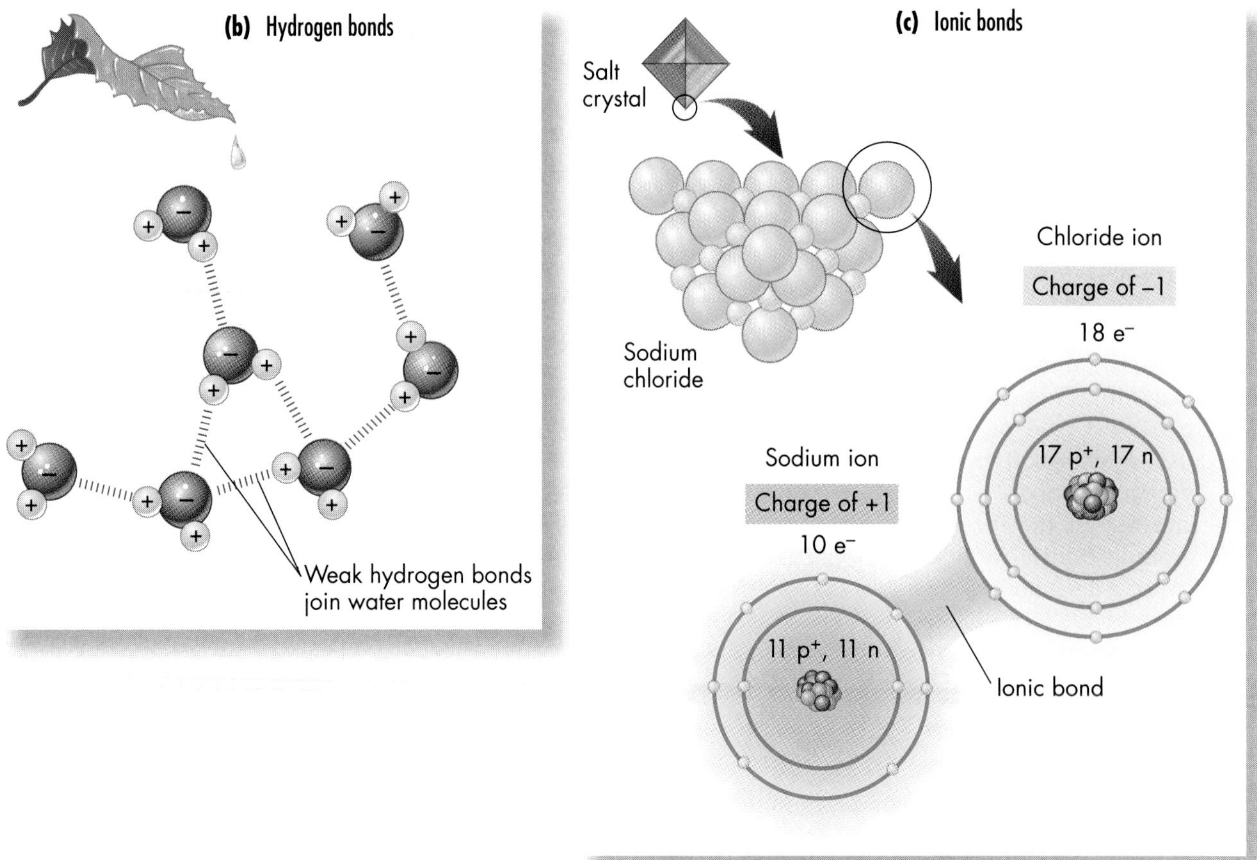

of water molecules is crucial to blood flow, food digestion, and so on. Life itself would be very different if water's chemical behavior was other than it is. Let's see why.

Why Water Dissolves Things

Washing dishes illustrates some of water's important chemical properties. Dishwater—in fact, all water—is a **solvent**, a substance capable of dissolving other molecules. Dissolved substances are called **solutes**. Water can dissolve polar compounds, such as table sugar, and most kinds of ionic compounds, such as table salt. When a polar solute such as sugar—say, syrup on dirty plates, or glucose molecules inside cells—becomes surrounded by water molecules, hydrogen bonds form. With an ionic solute like salt, the component ions dissociate, and each becomes surrounded by a cloud of water molecules (Fig. 2.4a).

Compounds such as sugar and salt that dissolve readily in water are called **hydrophilic**, or "water-loving," compounds. In contrast, nonpolar compounds, such as cooking oils and animal fats on dirty dinner dishes, do not dissolve readily, and are called **hydrophobic**, or "water-fearing," compounds. Instead of dissolving in water, hydrophobic compounds form a boundary (or interface) with the water (Fig. 2.4b). As we will see shortly, the membranes that surround all living cells are just such boundaries.

solvent
a substance capable of dissolving other molecules

solute
a substance that has been dissolved in a solvent

hydrophilic
compounds that dissolve readily in water, such as salt

hydrophobic
compounds that do not dissolve readily in water, such as oil

Acids and Bases

Water has another chemical property with significant implications for living things: Its molecules have a slight tendency to break down into a positively charged hydrogen ion (H^+) and a negatively charged hydroxide ion (OH^-):

$$H_2O \rightarrow H^+ + OH^-$$

Figure 2.4
Hydrophilic Substances Dissolve Well in Water, Hydrophobic Ones Don't

(a) Salt dissolves in water

(b) Oil and water don't mix

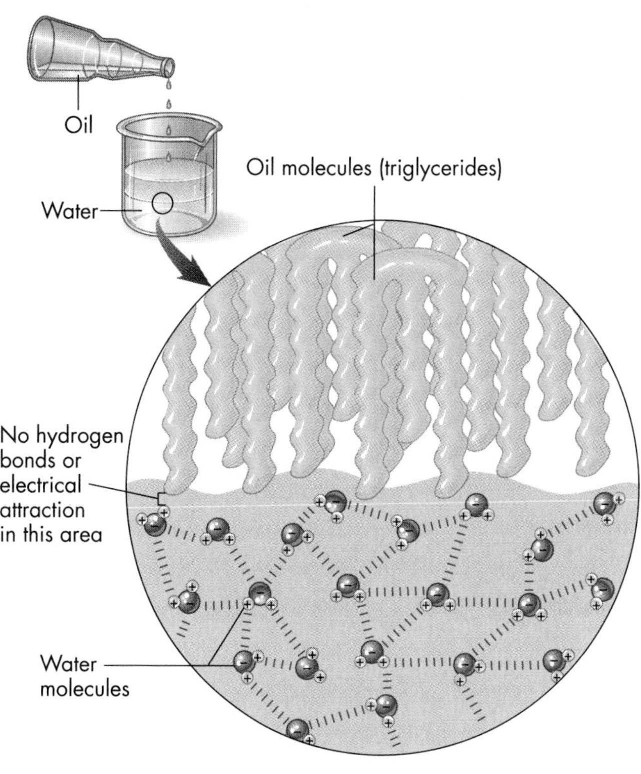

acid
a proton donor; any substance that gives off hydrogen ions when dissolved in water, causing an increase in the concentration of hydrogen ions; acidity is measured on a pH scale, with acids having a pH less than 7; the opposite of a base

base
any substance that accepts hydrogen ions when dissolved in water; basic solutions have a pH greater than 7

pH scale
a logarithmic scale that measures hydrogen ion concentration; acids range from pH 1 to 7, water is neutral at a pH of 7, and bases range from pH 7 to 14

buffer
a substance that resists changes in pH when acids or bases are added to a chemical solution

biological molecules
molecules derived from living systems; the four major types are carbohydrates, lipids, proteins, and nucleic acids

organic
molecules that are based on carbon and contain hydrogen

inorganic
molecules that are not based on carbon

This breakdown, however, is a relatively rare event in pure water. By definition, an **acid** is any substance that gives off hydrogen ions when dissolved in water, thereby increasing the H^+ concentration of the solution. A **base** is any substance that accepts hydrogen ions in water. This property allows a base to reduce the H^+ concentration of a solution.

The concentration of hydrogen ions is important to living cells because many of the chemical reactions that drive life's processes—the digestion of foods, for example—depend on specific concentrations of these ions. Biologists measure hydrogen ion concentration on the pH scale, which ranges from 1 to 14 (see Appendix A). On the pH scale, water has the neutral value of 7, in the middle of the scale. Acidic solutions like stomach acid or coffee have pH values between 0 and 7. Basic solutions like drain cleaners or baking soda in water have pH values between 7 and 14. The pH inside most cells stays fairly neutral, between about 6.5 and 7.5, and it is only within this narrow range that many vital cellular reactions take place at optimum speed. A subject related to acids and bases concerns buffers, agents that soak up or dole out hydrogen ions and help control pH level. Antacids buffer your stomach acid, for example, and cells contain natural buffering agents.

Water is clearly a key to life, but so is the element carbon and the biological compounds it forms.

LO5 What Are Biological Molecules?

All the living things on our planet contain four main types of **biological molecules**—carbohydrates, lipids, proteins, and nucleic acids—as well as millions of smaller molecules, all based on the special properties of the carbon atom. Why are carbon atoms so crucial for living things? And how do its bonding properties allow the raw materials of life to form?

Carbon Compounds

Our bodies may be mostly water, but a full 18 percent of our weight comes from carbon atoms. For a large tree, the figure can approach 50 percent. Carbon and its chemical bonds are so interesting that chemists divide all molecules into two broad types—those that contain carbon, **organic** molecules, and those not primarily based on carbon, or **inorganic** (lifeless) molecules.

Interestingly, there are far more organic than inorganic compounds. Why? Because carbon, with its unique structure, can form millions of different combinations with other atoms. This versatile bonding is carbon's key characteristic, and it explains why carbon is the "stuff of life."

Carbon Backbones

Carbon is "hungry" for electrons and can form covalent bonds with up to four atoms at a time (review Fig. 2.3a). A simple example of a "satisfied" carbon atom is methane gas (CH_4), sometimes called *swamp gas* (Fig. 2.5). Many biological compounds are much

Figure 2.5
The Versatility of Carbon

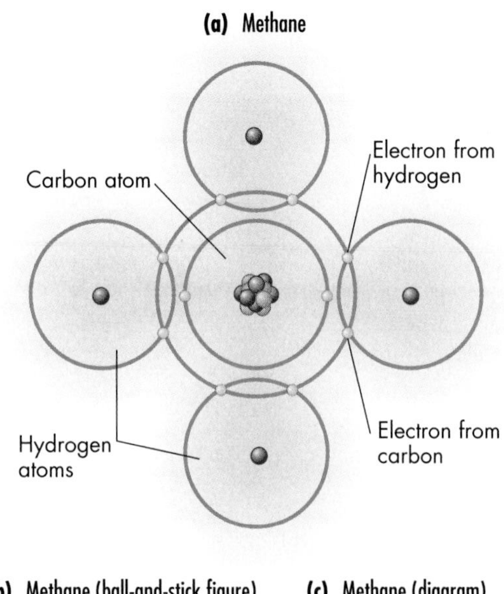

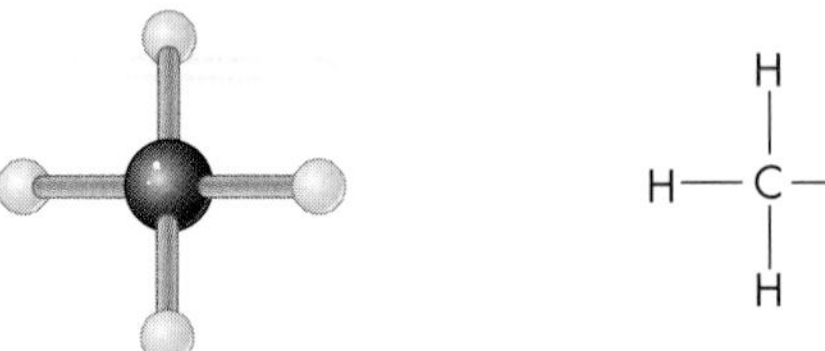

larger than methane and have a backbone of several carbon atoms bonded to each other in long, straight chains, branched chains, or rings—a sculptural property, again, based on forming up to four bonds per carbon atom.

Functional Groups

The chemical sculpturing of an organic molecule's shape contributes to its role in a cell, whether as water-tight seal, storage compound, messenger, protector, or reference library. But a molecule's specific activities often come from small clusters of atoms called **functional groups** that hang from the carbon backbone. Functional groups usually contain atoms other than carbon and hydrogen, and they give special properties to the molecules they are part of. The presence of a methyl group (CH_3), for example, prevents a molecule from quickly dissolving in water, and the hydroxyl group (OH) gives wood alcohol some of its properties like low boiling point and solvent activity.

The functional groups help explain why the four classes of large biological molecules—carbohydrates, lipids, proteins, and nucleic acids—form and act as they do. Together with a few other materials, these four types of compounds account for the diverse shapes, colors, and textures of organisms.

Figure 2.6
Carbohydrates

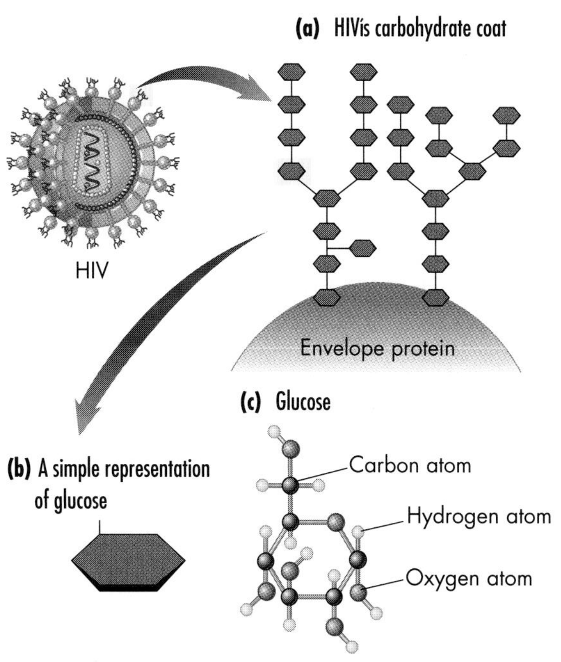

Carbohydrates

An HIV particle looks like a basketball studded with beads tipped with little threadlike branches (Fig. 2.6a). These branches are made of carbohydrates, and together they form a fuzzy coat that helps an HIV particle to recognize a target cell to infect. In addition to their role in recognition, carbohydrates are also crucial to cell structure and to energy storage. Carbohydrates include sugars and starches, and the term itself literally means "carbon-water." Indeed, carbohydrates usually contain carbon, hydrogen, and oxygen in a ratio of 1:2:1 (CH_2O). The formula for the sugar glucose, for example, is a multiple of that CH_2O subunit: $C_6H_{12}O_6$ (Fig. 2.6b,c). The same generally holds true for both simple and more complex carbohydrates.

functional group
a group of atoms that confers specific behavior to the (usually larger) molecules to which they are attached

monosaccharide
a simple sugar that cannot be decomposed into smaller sugar molecules; the most common forms are the hexoses, six-carbon sugars such as glucose, and the pentoses, five-carbon sugars such as ribose

disaccharide
a type of carbohydrate composed of two linked simple sugars

polysaccharide
a carbohydrate made up of many simple sugars linked together; glycogen and cellulose are examples of polysaccharides

starch
a polysaccharide composed of long chains of glucose subunits; the principal energy source of plants

glycogen
a polysaccharide made up of branched chains of glucose; an energy-storing molecule in animals, found mainly in the liver and muscles

Simple Carbohydrates: Mono- and Disaccharides

The simple sugars, including glucose and fructose, contain three to six carbon atoms. They share the same molecular formula, $C_6H_{12}O_6$, but have slightly different properties because their –OH functional groups are attached in different places. As a group they are called **monosaccharides** (*mono* = one), and their simple structures form the subunits of more complex carbohydrates.

A molecule with two simple sugars joined together is a **disaccharide** (*di* = two). For example, the joining of glucose and fructose makes the disaccharide sucrose. Sucrose is abundant in the saps of sugarcane, maple trees, and sugar beets—our major sources of sugar for refining.

Complex Carbohydrates: Polysaccharides

Many carbohydrates consist of thousands of simple sugar subunits joined into long chains or **polysaccharides** (*poly* = many). A polysaccharide molecule is like a string of beads, where each bead is a simple sugar, often glucose.

Starch is the polysaccharide stored as an energy reserve in plants. **Glycogen** is the polysaccharide

cellulose
(L. *cellula,* little cell) the chief constituent of the cell wall in green plants, some algae, and a few other organisms; cellulose is an insoluble polysaccharide composed of straight chains of glucose molecules

chitin
a complex nitrogen-containing polysaccharide that forms the cell walls of certain fungi, the major component of the exoskeleton of insects and some other arthropods, and the cuticle of some other invertebrates

alpha-helix
one possible shape of amino acids in a protein that resembles a spiral staircase

pleated sheet
secondary protein structure in which hydrogen bonds link two parts of a protein into a shape like a corrugated sheet

disordered loop
regions of a less specific but constant shape that link other parts of a protein

amino acid
a molecule consisting of an amino group (NH_2) and an acid group (COOH); the 20 amino acids are the basic building blocks of proteins

peptide bond
a bond joining two amino acids in a protein

polypeptide
(Gr. *polys,* many + *peptin,* to digest) amino acids joined together by peptide bonds into long chains; a protein consists of one or more polypeptides

storage molecule in animals. Cellulose and chitin (KYE-tin) are structural polysaccharides that give form and rigidity to plants and insects, respectively.

Proteins

The complexities of our bodies and the rich diversity of life in general—the millions of organisms of different textures, colors, and life styles—depend on different types of proteins in different types of cells. Proteins come in such a wide variety of forms (at least 10 to 100 million different kinds in the spectrum of the earth's organisms) that they can easily explain the myriad shapes and functions of specific cells and whole living things.

Proteins have many functions. They can:

- form structural parts of cells (e.g., the contractile machinery in a muscle cell)
- control cell processes (e.g., the steps involved in metabolism)
- act as messengers that move through fluids (e.g., hormones)
- carry other substances (e.g., the hemoglobin in red blood cells)
- protect animals from disease (e.g., antibodies)
- speed life processes (e.g., enzymes), and
- act as receptors on cell surfaces (e.g., the receptors for HIV).

Protein Structure: An Overview

The overall shape of a protein determines its function. A protein region shaped like a spiral staircase is called an alpha-helix and provides rigidity. A pleated sheet region (often represented by a broad flat arrow) gives proteins flat, box-like sides. The two types of regions within the protein ribbons are connected by disordered loops, which are usually gently curved. The helix, sheet, and loop regions occur in specific positions along the length of the protein and help provide each protein's unique overall shape.

Amino Acids: Building Blocks of Proteins

The long ribbon of a protein not only has coiled, looped, and pleated regions, but it can also be represented in finer detail as beads on a string, where each bead is an amino acid with the general structure shown in Figure 2.7a and b.

Proteins consist of hundreds of amino acids strung together. A single amino acid consists of an amino part, an acid part, and a side chain. The side chains of the 20 amino acids commonly found in proteins differ in chemical composition. Different side chains have different properties; for example, some attract water, some repel water, some are acidic, and some are basic.

Each amino acid "bead" links to the next one by means of covalent bonds called peptide bonds. The *order* of the beads is distinctive in each protein and is determined by the organism's genes.

Just as simple sugar units are joined into a polymer called a polysaccharide, amino acid subunits are joined into a polymer called a polypeptide (Fig. 2.7c).

The 20 types of amino acids with their different side chains function as subunits in a biological alphabet, forming complex proteins much as the 26 letters of our alphabet can form a nearly infinite array of words. And just as the order of letters in a word determines the word's meaning, the identity of a protein—its shape, properties, and functions—depends on the exact order of its amino acid "letters." Biologists refer to the order of these letters as the protein's primary structure. They call the specific helical portions, sheets, and loops we discussed earlier the protein's secondary structure. A protein's tertiary structure is the particular way these helices, sheets, loops, and intervening parts pack together into a three-dimensional ball like the HIV matrix protein or the long, fibrous shape of the keratin protein. Some proteins consist of more than one amino acid chain—the oxygen-carrying protein in your blood, hemoglobin, is an example—and the protein's quaternary structure is the way these chains pack together. Within the human body, there are over 50,000 different proteins, each with a unique shape. In the living world collectively, there are tens of millions of unique sequences of amino acids, giving rise to an enormous array of specific proteins that make possible life's amazing diversity.

To some, the chemistry of biological molecules and structures may seem detached from the living world they inhabit, but in fact, it's actually the basis for life.

1. The variable side chains (functional groups) on each of the 20 different amino acids found in proteins determine how those amino acid "beads" behave within their protein chains.
2. The order of the amino acid beads determines the shape of the overall protein molecule.
3. The overall shape of the protein molecule determines its function in the organism.
4. Collectively, the functions of an organism's proteins determine what the organism looks like and how it lives.

For an HIV particle, amino acid properties and sequences determine the shape of the matrix protein, which in turn helps strengthen the virus particle. Other proteins with their own sequences and shapes protect the virus's genetic instructions and help make new copies that will infect additional cells and eventually destroy a patient's immune system. The virus contains not just protein but another category of biological molecules—lipids. So let's look at that type next.

Lipids

The smooth surface of the HIV "basketball"—beneath the protein "beads" with their fuzzy carbohydrate extensions—is made of lipid molecules. Lipids are a class of biological molecules that tend not to dissolve in water. Thus they can keep water from rushing into cells and diluting their contents.

Lipids can serve as energy-storage molecules in plants and animals, just as carbohydrates do. Solid storage molecules are fats. Liquid storage molecules are oils. Waxes are semisolid types of lipids. Another important class of lipids, the steroids, includes certain vitamins, some hormones, and cholesterol.

fat
an energy storage molecule that contains a glycerol bonded to three fatty acids; fats in the liquid state are known as oils

wax
a sticky, solid, waterproof lipid that forms the comb of bees and waterproofing of plant leaves

steroid
(Gr. *stereos,* solid + L. *oi,* having the form of + *oleum,* oil) a major class of lipids based on a 4-carbon-atom ring system and often a hydrocarbon tail; cholesterol and sex hormones are steroids

nucleic acid
a polymer of nucleotides, e.g., DNA and RNA

Nucleic Acids and Nucleotides

The fourth major class of biological molecules, the nucleic acids, include DNA and RNA, which carry the chemical "code of life" and transmit genetic information from one generation to the next (Fig. 2.8a,b). The

Figure 2.7
The Overall Shape of a Protein and Amino Acids

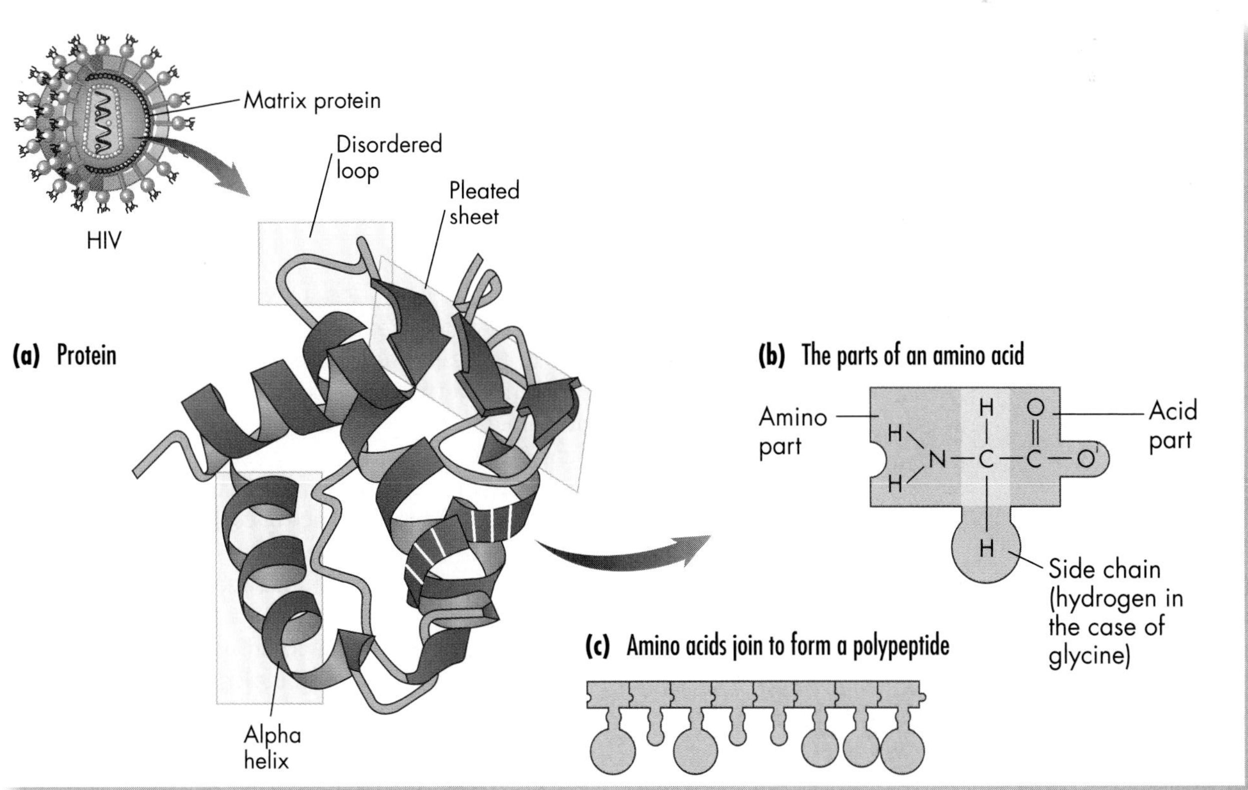

RNA (ribonucleic acid) a nucleic acid similar to DNA except that it is generally single-stranded and contains the sugar ribose and the base uracil replaces thymine

DNA (deoxyribonucleic acid) the twisting, ladderlike molecule that stores genetic information in cells and transmits it during reproduction; DNA consists of two very long strands of alternating sugar and phosphate groups that form a double helix; each "rung" of the ladder is a base (either adenine, guanine, cytosine, or thymine) that extends from the sugar and bonds with a base from the other strand; genes are made of DNA

reverse transcriptase the enzyme that facilitates "reverse" transcription of RNA to DNA

information necessary for building new HIV particles lies in the viral RNA molecule.

Nucleic Acids: Molecules of Information Storage and Processing

Ribonucleic acid, or RNA, is present in living cells and helps in the processing and use of the information stored in deoxyribonucleic acid, or DNA, which stores hereditary information in all living cells and most types of viruses. (Chapter 6 covers RNA and DNA in detail; here we'll just explore enough to understand the basic story of HIV, AIDS, and the chemistry of living cells.)

Nucleic Acids and HIV Infection

When a person is exposed to HIV, viral RNA molecules enter his blood cells and are copied into molecules of DNA (see Chapter 6). This so-called reverse transcription of RNA to DNA was facilitated by the enzyme reverse transcriptase. (The process is called "reverse" because in living cells, genetic information is copied from DNA to RNA, not RNA to DNA.)

The RNA molecule in an HIV particle carries three major genes, laid out one after the other. Each gene provides all the instructions for making a single type of protein. The three proteins initially made are then chopped into the smaller proteins that make up the virus. Chapter 6 explains how nucleic acids store the instructions for making proteins. For now, just remember that your DNA, by determining the shape of your proteins, has played a central role in shaping the way you look and the way your body functions, and the same is true for all plants, animals, and other living organisms.

You've seen how atoms join to form biological molecules and how an HIV particle is built of all four kinds. But why do we keep insisting that HIV is nonliving, while cells are alive? What makes a cell a cell and a virus . . . different?

Figure 2.8

Nucleic Acids and Nucleotides

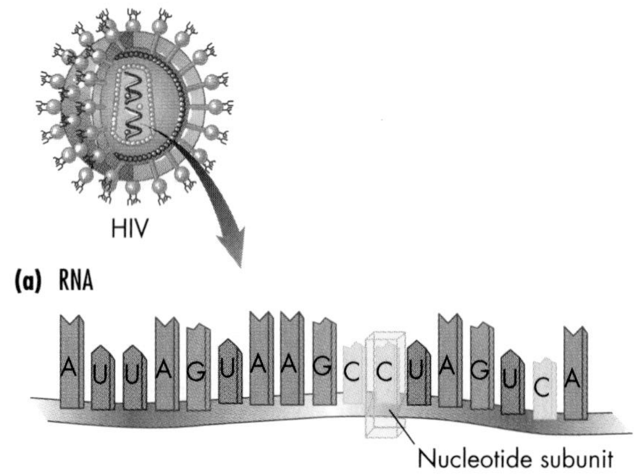

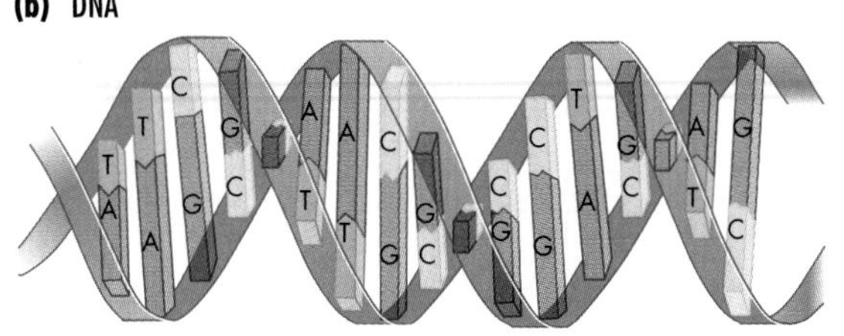

LO6 What Is a Living Cell, and What Makes HIV a Nonliving Enemy?

What makes cells alive and susceptible to being killed by HIV particles, which themselves are not alive and are so very difficult to fight? Answering this requires an explanation of cells—including those found in animals, plants, and single-celled organisms like bacteria and protozoa—as well as a discussion of viruses. How big are both entities? What are their parts? How do they function? Size is a good place to start, and that leads us to the story's 17th-century beginnings.

The Discovery of Cells

More than 300 years ago, English scientist Robert Hooke focused one of the very earliest microscopes on some everyday objects from his home: the point of a pin, the surface of a nettle leaf, and the body of a flea. Hooke was astonished by the fine detail he could make out in this new, previously unseen world. When Hooke looked at a thin slice of cork through his microscope, he saw what he called "cells," which reminded him of the small rooms inhabited by monks.

Hooke was apparently the first person to publicize seeing cells, but he could not fully define what he was observing. Modern biologists know that a *cell* is the smallest entity completely surrounded by a membrane and capable of reproducing itself independent of other cells. It is also the smallest unit displaying all the properties of life listed in Chapter 1, including the orderly chemical activities of metabolism, the capacity of self-propelled motion, the ability to reproduce and develop, and the potential to evolve over many generations.

Cells versus Viruses

Biologists have found that cells have three fundamental parts:

1. A surface envelope of lipid and protein (a *plasma membrane*) that controls the passage of materials into and out of the cell.
2. A central genetic region that controls all the cell's functions and stores DNA, the repository of the cell's hereditary information.
3. A gel-like substance (called *cytoplasm*) that fills the cell between the surface envelope and the genetic storage region, and surrounds small uniquely structured compartments called *organelles* (see Chapter 1) that carry out specialized functions.

Given these principles, how does a virus like HIV differ from a cell and why is it nonliving rather than living? HIV has a membrane made up of lipid and protein, and it has a central genetic region that contains RNA rather than DNA. Both of these features embody the life characteristic of order. (In many viruses, the genes are DNA.) HIV can also evolve or change and adapt over time; recall that HIV has evolved resistance to some drugs.

However, as we saw earlier, viruses lack metabolism, responsiveness, and independent reproduction. HIV, for example, is not filled with cytoplasm, and this material is integrally involved in a living cell's ability to carry out metabolic functions such as harvesting energy, generating self-powered movement, and building proteins. Virus particles are tiny compared to the living cell. There's simply not enough space inside a virus for the machinery of the cytoplasm that gives cells their capacity for independent life, including metabolism, motion, and reproduction. Instead, viruses have evolved to use a cell's machinery. Viruses in fact may once have been complete cells that lost elements and became parasites on true cells.

Cell Size: An Import-Export Problem

Even though a cell is thousands of times bigger than an HIV particle, cells are still minuscule. The average cell in your body is just one-fifth the thickness of the paper in this book. Why are cells so small? The answer is that cells have an import-export problem based on the physical relationship between surface area and volume.

cell theory
the biological doctrine stating that all living things are composed of cells; cells are the basic living units within organisms; the chemical reactions of life occur within cells, and all cells arise from pre-existing cells

A cell's active cytoplasm needs to take in materials to fuel activities and build cell parts, and it needs to get rid of wastes it produces as by-products—in general, the more cytoplasm, the more materials and wastes. A cell imports materials and exports wastes across its plasma membrane. The greater the surface area of this plasma membrane, the more rapidly the cell exchanges substances with its environment. When a cell increases in size, its volume increases more rapidly than its surface area, and its import-export needs outstrip its ability to exchange these items with the surroundings. If a cell got much larger than a certain typical size (for a bacterium, under 10 micrometers; for an animal cell, 5 to 30 micrometers; for a plant cell, 35 to 80 micrometers), it couldn't meet its material and waste needs quickly enough to survive. (Note that the abbreviation μm is frequently used for a micrometer, one millionth of a meter.) This is why an elephant's liver is hundreds of times bigger than a mouse's liver, but its *cells* are the same size. There are just millions more of them.

Viruses have one solution to this surface-to-volume problem and cells have others. Since viruses don't metabolize, particles like HIV don't have to "worry" about their surface-to-volume ratio; living cells import, export, and stockpile all their needed materials, and the viruses simply have to enter and use it. This strategy clearly wouldn't work for most living cells, and so one primary means of solving their surface-to-volume problem is through altered cell shape or contents. A long, thin cell, such as a nerve cell that reaches from a giraffe's spine down to its hoof, can have the same volume as a round or cube-shaped cell but a greatly expanded surface area.

The Cell Theory

Because cells are so small, large organisms such as people and trees consist of trillions of cells. About 160 years ago, biologists were trying to understand how cells could be so small and yet how billions of them could function in a coordinated way inside a plant or animal organ. Their efforts resulted in the cell theory, which illuminated the cell's significance to life for all of us who came later. According to the cell theory:

1. All living things are made up of one or more cells.

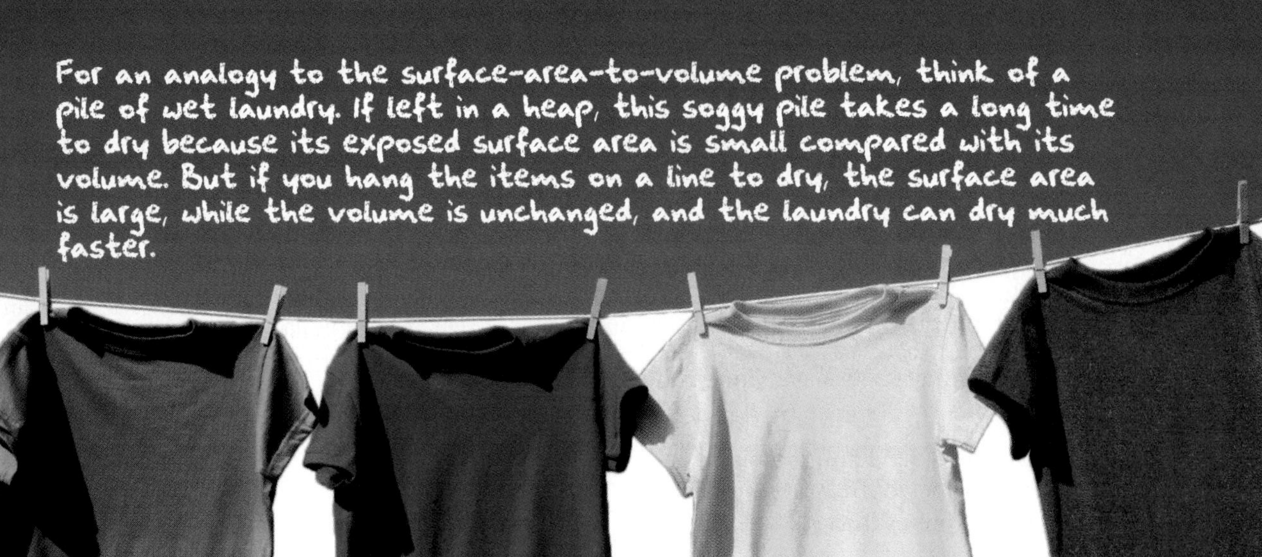

© MICHAEL FLIPPO/ISTOCKPHOTO.COM

2. Cells are the basic living units within organisms, and the chemical reactions of life take place within cells.
3. All cells arise from pre-existing cells.

What makes these simple statements so important? First, there are no living organisms made up of anything other than cells. Organisms such as bacteria consist of just one cell, while people and trees contain trillions, but the living subunits are always still cells. Viruses are considered particles not cells, because they don't carry on their own energy metabolism and can't reproduce independently.

Second, cells are the basic units of life because the individual components that make up cells lack the complete properties of life. For example, if you take the nucleus out of a cell, it can't carry out life functions or replicate on its own anymore.

Third, new cells arise today only from preexisting cells that divided into daughter cells. Each cell in your body can be traced back to a single fertilized egg cell generated when your mother's egg cell fused with your father's sperm cell. These sperm and egg, in turn, were produced by other cells in your parents' bodies; each of your parents arose from a single fertilized egg cell produced by your grandparents, and so on back in time. Virus particles do not arise from the division of a pre-existing particle into two new particles. The end of this chapter explores not only how new HIV particles are assembled from parts like the manufacture of cars on an assembly line, but also, why a living cell is required for the assembly.

eukaryotic cell
a cell whose DNA is enclosed in a nucleus and associated with proteins; contains membrane-bound organelles

prokaryotic cell
a cell in which the DNA is loose in the cell; eubacterial and archaebacterial cells are prokaryotic; prokaryotic cells generally have no internal membranous organelles and evolved earlier than eukaryotic cells

prokaryote
an organism made up of a prokaryotic cell

eukaryote
an organism made up of one or more eukaryotic cells

Cell Types in Life's Kingdoms and Domains

Biologists have never found a cell they can't assign to just one of two basic types: prokaryotic or eukaryotic.

A cell's most obvious distinguishing feature is the presence or absence of a cell nucleus. Eukaryotic cells (*eu* = true *karyo* = nucleus) contain a prominent, roughly spherical, membrane-enclosed body called the nucleus, which houses DNA, the cell's hereditary material. In contrast, in prokaryotic cells (*pro* = before *karyo* = nucleus) the DNA is loose in the cell's interior and not separated from the rest of the cell's contents by a membrane (Fig. 2.9). An organism made up of a prokaryotic cell is called a prokaryote; an organism made up of one or more eukaryotic cells is called a eukaryote.

Our own cells are eukaryotic and so we humans are eukaryotes, belonging to the domain Eucarya (see Chapter 1). So are all the members of the animal and plant kingdoms, the mushrooms and other fungi, and the algae and other protists. The two remaining domains of life, the Eubacteria (also called simply Bacteria) and Archaea, contain thousands of species in which each individual is made up of a single prokaryotic cell lacking a cell nucleus. Because HIV and other viruses are not cells, they're neither prokaryotic nor eukaryotic.

Prokaryotic Cells

Prokaryotes are the smallest cells, and some bacteria are no more than 0.2 µm in length, just twice as big as an HIV particle. (For comparison, the thickness of this page is about 100 µm.) Prokaryotes exploit environments not open to eukaryotes, with their more complex cells. Some prokaryotes, for example, thrive in the boiling waters of hot springs.

Eukaryotic Cells

Because they house nuclei and other compartments or organelles that perform specific tasks, eukaryotic cells are generally much larger than prokaryotic cells. An average-sized animal cell (Fig. 2.10a on the next page) is about 20 µm in length—about five animal cells could be lined up across the thickness of a sheet of paper. A typical plant cell (Fig. 2.10b) (also a eukaryote) is a bit larger at about 35 µm across.

LO7 What Are the Parts of the Cell, and How Does HIV Sabotage Them?

If HIV is not a cell and is not alive, how can it destroy white blood cells and cause the other devastating consequences of AIDS? More specifically, how can HIV take over the various organelles inside the cell and replicate more deadly particles of itself? This section gives the answers by following the way HIV infects a cell, step by step, cell part by cell part. This will serve as an introductory tour of the cell and how it functions, while at the same time explaining the dangers of HIV and AIDS.

HIV Infection and the Cell Surface

HIV enters a cell by injecting its contents through the cell surface and into the cell's interior. An animal cell contains many organelles, including those responsible for its metabolism and reproduction (Fig. 2.10a). But we'll start our tour of the eukaryotic cell at the cell's outer envelope, then see how HIV overcomes a component of that protective barrier to gain entry.

Crossing Plasma Membranes

As a flexible fatty boundary studded with proteins and carbohydrates, the cell's plasma membrane tends to keep the watery cell contents in and moisture,

Figure 2.9

HIV, Prokaryotic Cells, and Eukaryotic Cells Compared

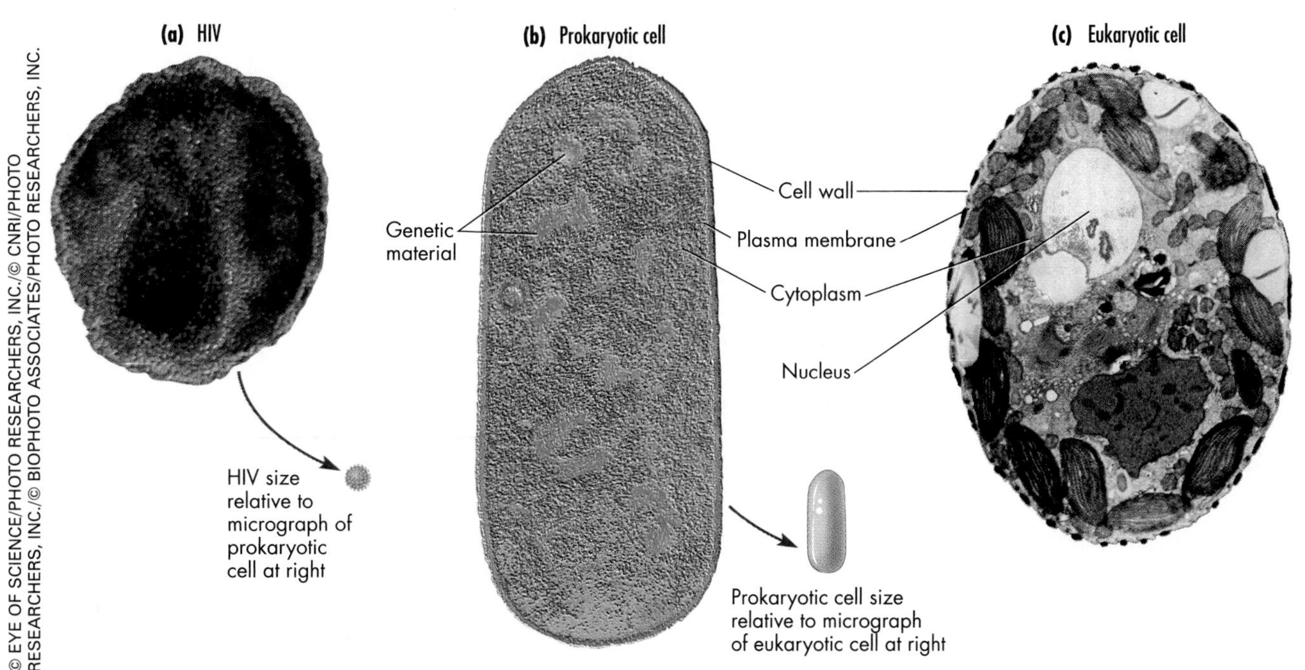

Figure 2.10
Generalized Animal and Plant Cells

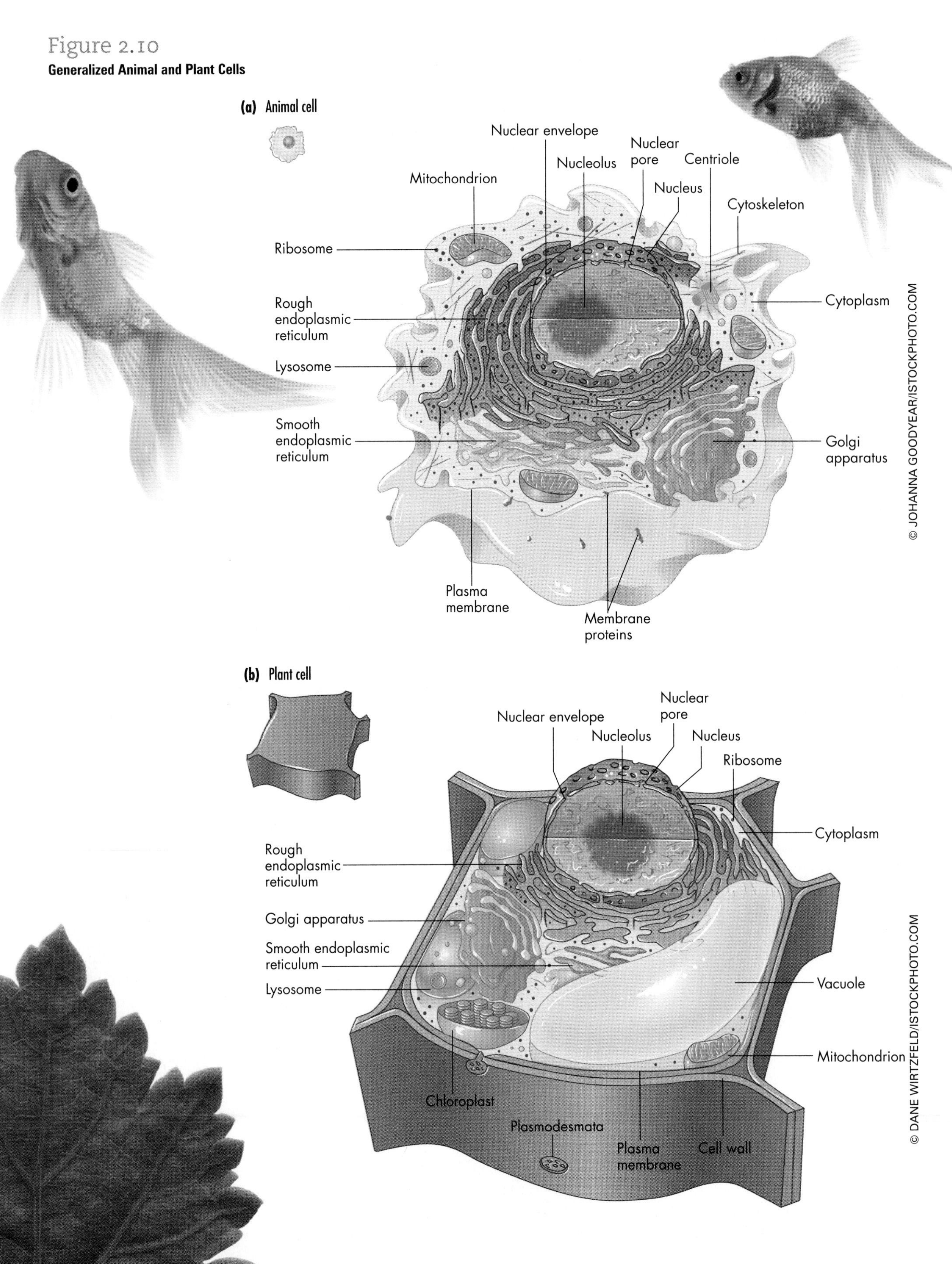

chemicals, and other elements of the external environment out (Fig. 2.11a on the next page). Recall, though, that nutrients must pass into cells and waste products must pass out. To be infective, HIV and other viruses must gain entry to the cell. The plasma membrane is *selectively permeable*, that is, permeable (penetrable) to certain substances but not all. So what accounts for the selectivity that allows nutrients, wastes, and viruses to pass through plasma membranes while most other substances are barred?

Fat-Soluble and Water-Soluble Molecules

Passage through plasma membranes depends first on size. Some very small inorganic molecules such as oxygen, carbon dioxide, and water pass through plasma membranes by simple **diffusion**—they move from a region of high concentration to one of low concentration (Fig. 2.11b).

Organic molecules tend to be larger, so instead of diffusing, they pass into cells in two other ways, depending on whether they dissolve in fat or water. Some organic substances are fat-soluble: They can dissolve in fats, and thus they can pass directly through the fatty plasma membrane.

Most organic molecules, however, are water-soluble, including nutrients such as sugars and amino acids, and cellular wastes such as urea. A plasma membrane made of pure lipid molecules (fat or oil) would act like a perfect raincoat, blocking out all water-soluble materials and eventually starving or poisoning the cell. Instead, small water-soluble organic compounds enter the cell by passing through proteins floating in the plasma membrane. Some membrane proteins allow only particular sugars, ions, or amino acids to pass, often through a channel in the protein. If the protein helps the substance pass down its concentration gradient without the expenditure of energy, we call the process **facilitated diffusion** (Fig. 2.11c). If a substance is too large to simply diffuse through the membrane or slip through a channel, the cell may have to expend energy to pump materials in or out by means of a process called **active transport** (Fig. 2.11d).

Large Materials

A cell's outer membrane obviously has lots of "gates" and lots of traffic in and out. Really large materials, though, such as virus particles and large protein molecules usually can't get through those gates. Often, soldiers of the body's immune defense system—for example, large, mobile blood cells—engulf debris left over from a dying cell or surround whole parasites such as bacteria or viruses within pockets of the cell membrane. The engulfment of solid material is called **phagocytosis** (FAJ-oh-sigh-toh-sis) (literally "cell eating"; see Fig. 2.12a on page 37). Jay Levy was the first scientist to show that these large cells or **phagocytes** can themselves be infected by HIV.

Cells can move substances in across the plasma membrane by the import process of **endocytosis** (of which phagocytosis is one type) or out across the plasma membrane via the export process of **exocytosis** (Fig. 2.12b). Some cells discharge wastes this way or secrete proteins, such as hormones or digestive enzymes, into the bloodstream or into a food-digesting organ like the stomach or small intestine.

How HIV Gets into a Cell

Recall that HIV looks like a basketball studded with beads and that the skin of the "basketball" is made of lipids while the beads are a type of envelope protein. To explain in a simplified way the lipid membrane surrounding the HIV particle is suddenly able to fuse with the lipid membrane of a white blood cell like two soap bubbles converging. The contents of the HIV particle can then spill into the cell and infect it.

Once the HIV membrane has fused with the cell membrane, the virus releases its genetic material, RNA. This gets copied into the cell's own genetic language, DNA, and this new viral DNA enters the cell's central repository of genetic information, the nucleus. We'll take a closer look at this process as we examine the nucleus.

diffusion
(L. *diffundere,* to pour out) the tendency of a substance to move from an area of high concentration to an area of low concentration

facilitated diffusion
a type of transport in which a protein helps a substance pass across a cell membrane down its concentration gradient without energy expenditure by the cell

active transport
movement of substances against a concentration gradient requiring the expenditure of energy by the cell

phagocytosis
the type of endocytosis through which a cell takes in food particles

phagocyte
a specialized scavenger cell that devours debris

endocytosis
the process by which a cell membrane invaginates and forms a pocket around a cluster of molecules; this pocket pinches off and forms a vesicle that transports the molecules into the cell

exocytosis
the process by which substances are moved out of a cell by cytoplasmic vesicles that merge with the plasma membrane

nucleus
the membrane-enclosed region of a eukaryotic cell that contains the cell's DNA

The Nucleus

When you look through a microscope at a typical animal or plant cell, often the most conspicuous organelle you will see is the **nucleus** (Fig. 2.10). This roughly spherical structure contains genetic information that controls most of the cell's activities. Just as an

Figure 2.11

The Plasma Membrane and Movement of Molecules Into and Out of Cells

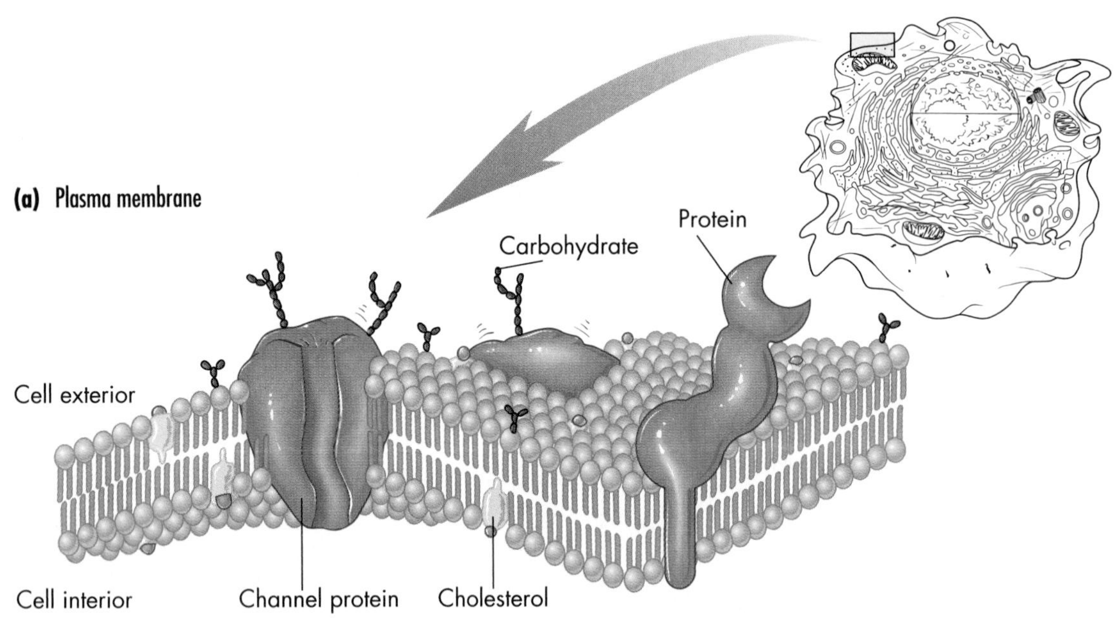

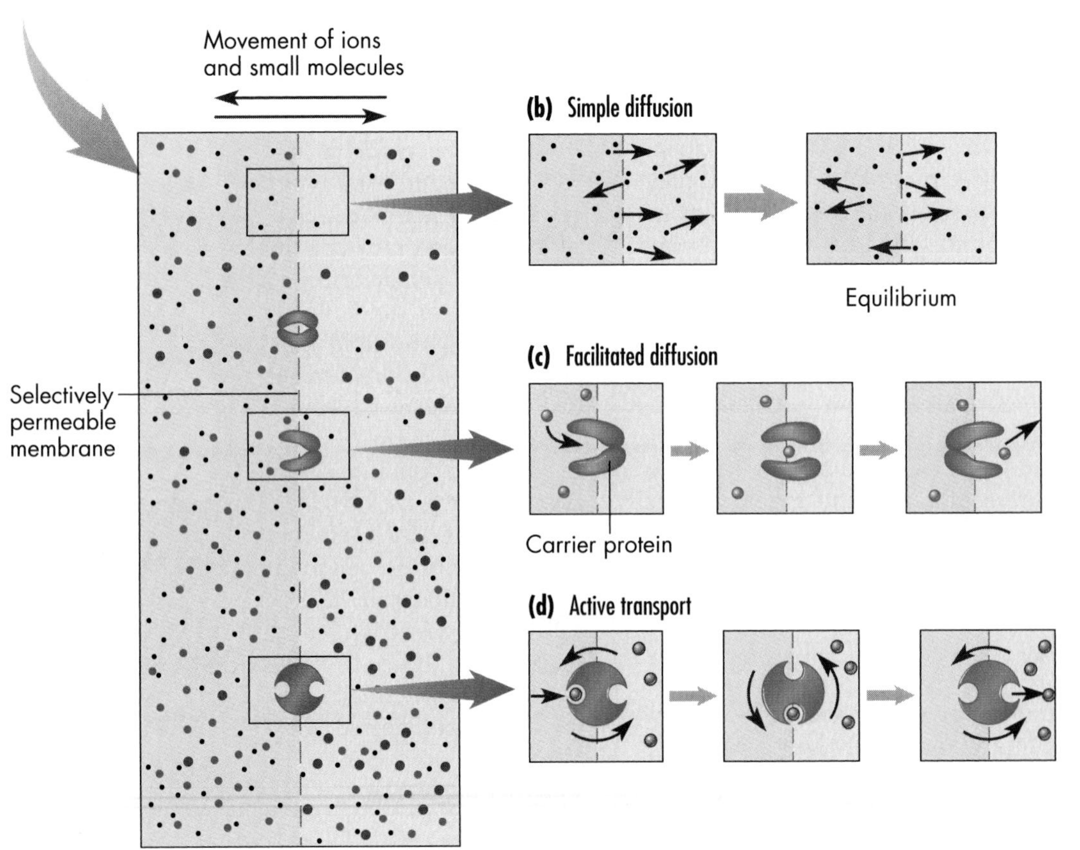

Figure 2.12
Cell Engulfing Large Materials

(a) A white blood cell engulfs a yeast cell

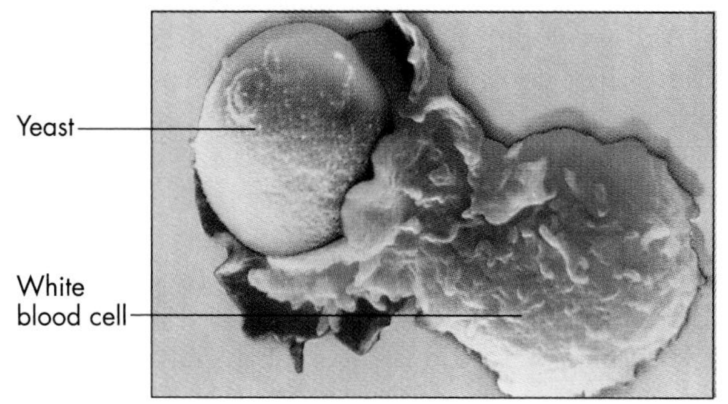

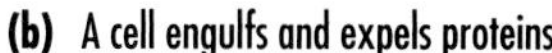

(b) A cell engulfs and expels proteins

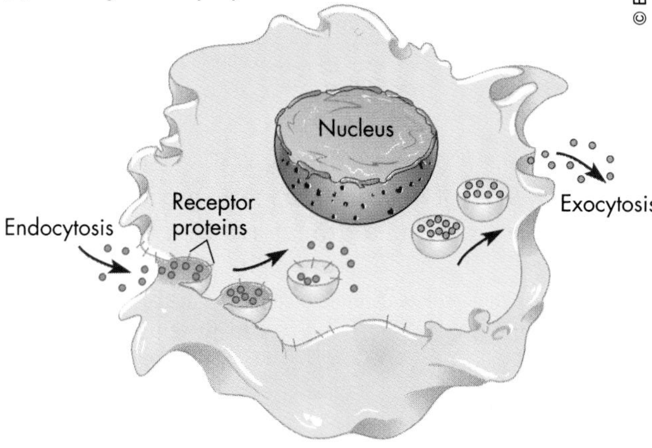

enemy tries to take over an opposing army's central command post, HIV must get into the nucleus if it's going to take over the cell's activities and commandeer them for its own purpose: making new virus particles.

Information Flow in the Nucleus

In the nucleus of all eukaryotic cells the information in DNA is copied into RNA, the RNA then moves out of the nucleus into the cytoplasm, and its information is used to make proteins, which carry out the work of the cell. This flow of information can be diagrammed:

DNA → RNA → Protein

We've said that when HIV enters the cell, its RNA is copied into DNA and this viral DNA enters the cell nucleus. Once inside, however, and once integrated into the cell's chromosomes, it follows the DNA→RNA information flow characteristic of healthy, uninfected cells. Thus new viral RNA is once again made and moves back out of the nucleus into the watery cytoplasm to begin diverting the cell to act as an HIV-manufacturing factory. This takeover involves a number of cell organelles.

ribosome
a structure in the cell that provides a site for protein synthesis; ribosomes may lie freely in the cell or attach to the membranes of the endoplasmic reticulum

endoplasmic reticulum (ER)
a system of membranous tubes, channels, and sacs that forms compartments within the cytoplasm of eukaryotic cells; functions in lipid synthesis and in the manufacture of proteins destined for secretion from the cell

Cytoplasm and Organelles

All living cells are filled with cytoplasm, a semifluid, highly organized pool of raw materials and fluid in which the cell's internal organelles are suspended. Cytoplasm is about 70 percent water, 20 percent protein molecules, and 10 percent carbohydrate, lipid, and other types of molecules. An average cell contains 10 billion or so protein molecules of about 10,000 different kinds! Many of these proteins are highly active *enzymes,* substances that speed biochemical reactions, while others are structural proteins that assemble into various cell parts. Suspended in the cytoplasm are numerous kinds of organelles—many of which HIV uses for its own ends.

Ribosomes

In order for new virus particles to form inside a cell, HIV needs new proteins. So it usurps the cell's protein-sythesizing machinery, the **ribosomes**. Thousands of ribosomes are embedded throughout the cell's cytoplasm. There are no functional ribosomes inside the nucleus, so protein building occurs only in the cytoplasm. We'll see more about how HIV takes over ribosomes shortly.

Rough Endoplasmic Reticulum

Most of a cell's ribosomes float freely in the cytoplasm, but some are attached to flattened membranous sacs within the cell—the **endoplasmic reticulum**, or **ER** (*endo* = inside; *plasmic* = cell; reticulum = network) (Fig. 2.13 on the next page). Part of the endoplasmic reticulum looks rough because it is so heavily studded with ribosomes; biologists call this the rough ER. This rough region makes many proteins that wind up being exported from the cell. For example, the rough ER helps produce the enzymes that digest most of the foods you eat. Since HIV takes over the rough ER, tracing how the membranous network makes and transports viral protein is a good way to see how the ER functions.

In a white blood cell infected with HIV, viral RNA made inside the nucleus attaches to floating ribosomes and then docks with the rough ER (Fig. 2.13, Step ①). Next, the ribosomes make some of HIV's bead-shaped proteins, which enter the interior

Golgi apparatus in eukaryotic cells, a collection of flat sacs that process proteins for export from the cell or for shunting to different parts of the cell

smooth ER the part of the endoplasmic reticulum folded into smooth sheets and tubules, containing no ribosomes; synthesizes lipids and detoxifies poisons

cavity of the rough ER (Step ②). Eventually, the proteins enter transport vesicles—little membranous bubbles that pinch off from the rough ER (Step ③). Most of these transport vesicles then fuse with membranes of another cell organelle, the Golgi apparatus (Step ④).

Golgi Apparatus

The **Golgi apparatus** (GOAL-gee) is a series of flattened membranous sacs resembling a pile of empty hot-water bottles. Some cells have just one Golgi apparatus, but others may have hundreds.

The Golgi apparatus acts like a traffic cop, directing different proteins to different parts of the cell, where they perform their functions. For example, it sends vesicles carrying the new HIV proteins to the cell surface, where they fuse with the cell's plasma membrane (Step ⑥). The HIV proteins then embed in the cell's membrane, so that the infected cell's surface now has the bead-studded appearance that usually characterizes an HIV particle (Step ⑦). Usually, when other proteins reach the cell surface, they just diffuse away and enter the bloodstream. The HIV proteins have their "tails" stuck into the cell membrane, so when they move to the surface, they remain embedded.

Smooth ER

Another part of the endoplasmic reticulum, the **smooth ER**, is folded into smooth tubes and small sacs. Smooth ER makes the lipids that line up side-by-side in cell membranes. When commandeered, they also make the HIV particle's lipid "basketball" or envelope. Smooth ER makes other familiar lipids, as well, including cholesterol, and the steroid sex

Figure 2.13

The Endoplasmic Reticulum and the Golgi Apparatus

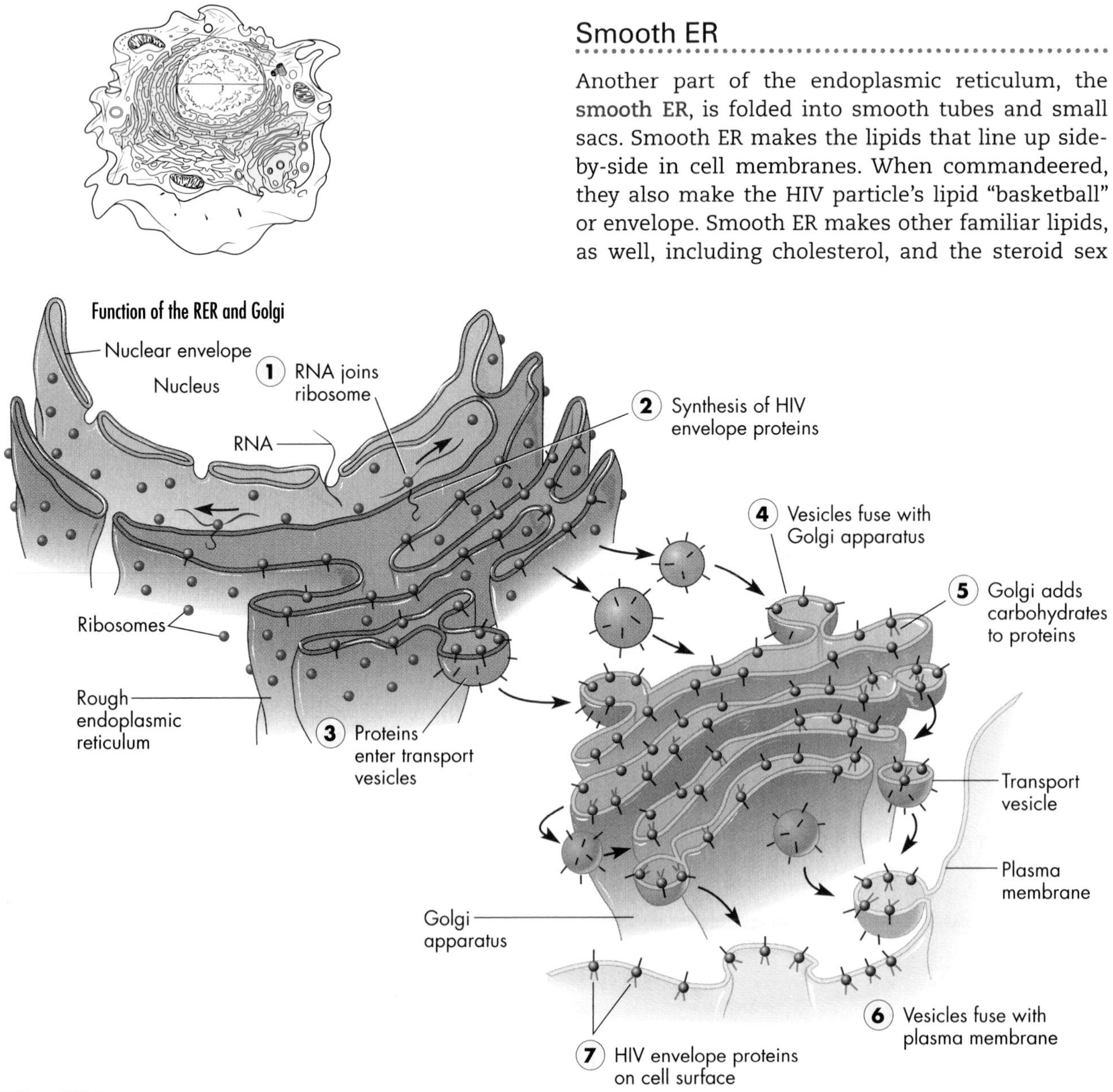

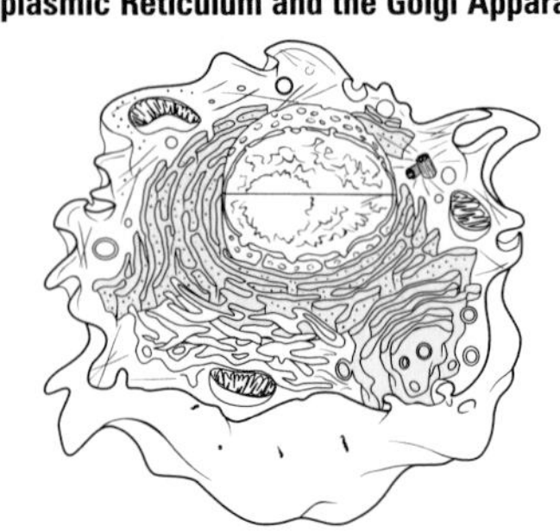

hormones estrogen and testosterone. Smooth ER in liver cells is especially active at detoxifying harmful lipids, making them more water soluble so that they can eventually be excreted in the urine.

Lysosomes: The Cell's Recyclers

Lysosomes are tiny spherical bags of powerful digestive enzymes that can digest invading bacteria or debris the cell has engulfed, or cell parts that have worn out internally. Digestive enzymes in the lysosome generally break down the refuse into smaller molecules, which then reenter the cytoplasm for reuse by the cell.

Cytoskeleton

The **cytoskeleton** is a structure of thin protein fibers forming a lattice throughout the cytoplasm, suspending the organelles and allowing cell parts to move. Protein fibers in the cytoskeleton also help maintain a cell's shape and transport cell organelles.

When cells divide in two, they use another cytoskeleton organelle called the **centriole** (see Fig. 2.10). Pairs of these short rod-shaped organelles organize certain cytoskeletal fibers that guide the separation and movement of chromosomes into the two new cells.

Mitochondria: Harvesters of Energy

Mitochondria (sing., mitochondrion) are cellular organelles that harvest energy from food by breaking down carbon-containing molecules and that release energy packets, or ATP. In this way mitochondria provide chemical fuel for cellular activities such as building proteins, copying DNA, and moving cells and cell parts.

Mitochondria have their own DNA and make some of their component proteins on their own ribosomes. The cell's DNA encodes the rest. They can also divide in half independently of the cell's normal division cycle.

The HIV drug AZT blocks the copying enzyme that translates HIV's RNA genes into DNA. Experiments show that AZT can also inhibit the enzyme that copies mitochondrial DNA before the organelle divides. This blockage slows HIV replication but also slows mitochondrial DNA. This, in turn, may interfere with the way the energy organelles function and lead to the weakness and fatigue of AIDS patients taking AZT.

Specialized Organelles

So far, all the cell structures we've talked about can be found in human cells and in most other eukaryotic cells, and are commandeered by HIV and contribute to AIDS. No exploration of cell biology would be complete, however, without examining a few specialized organelles for movement and storage found only in certain cell types. (These specialized organelles play no known role in HIV or AIDS.)

Organelles of Cell Movement. For some mobile cells throughout the kingdoms of life, movement is made possible by propelling extensions of the cytoskeleton. A whiplike organelle called a **flagellum** (Latin, small whip; pl., *flagella*) extends from the cell surface and pulls or pushes the cell through a liquid medium. Many kinds of sperm rely on propulsion by flagella.

Certain single-celled protists have thousands of projections called *cilia* (Latin, eyelashes; sing., *cilium*) that look and act much like flagella. Cilia beat in concert like the oars of a medieval galley ship, allowing the cell to swim quickly. In cells that line the human breathing passages, cilia sweep dust particles out toward the mouth and nose, where they are eventually expelled in mucus or swallowed.

Links between Cells. Most animal cells are attached to neighboring cells by links called *intercellular junctions,* which help weld cells together into functional tissues and organs (Fig. 2.14 on the next page). These junctions also allow free communication between cells and the coordination of cells in tissues and organs. Some linkages allow materials to flow between cells; others prevent leaks between cells and help organs like the urinary bladder hold fluid.

Junctions at the base of many cells attach fibrous proteins or an **extracellular matrix** that surround and support the cell and glue it to adjacent cells.

lysosome
spherical membrane-bound vesicles within the cell containing digestive (hydrolytic) enzymes that are released when the lysosome is ruptured; important in recycling worn-out mitochondria and other cell debris

cytoskeleton
found in the cells of eukaryotes, an internal framework of microtubules, microfilaments, and intermediate filaments that supports and moves the cell and its organelles

centriole
pairs of short, rod-shaped organelles that organize the cytoskeletal fibers called microtubules into scaffolds; these intracellular frameworks help maintain cell shape and move chromosomes during cell division

mitochondrion
(pl. mitochondria) organelle in eukaryotic cells that provides energy that fuels the cell's activities; mitochondria are the sites of oxidative respiration; almost all of the ATP of nonphotosynthetic eukaryotic cells is produced in the mitochondria

flagellum
(pl. flagella) long whiplike organelle protruding from the surface of the cell that either propels the cell, acting as a locomotory device, or moves fluids past the cell, becoming a feeding apparatus

extracellular matrix
a meshwork of secreted molecules that act as a scaffold and a glue that anchors cells within multicellular organisms

plastid
an organelle found in plants and some protozoa that harvests solar energy and produces or stores carbohydrates or pigments

vacuole
large fluid-filled sac inside cells surrounded by a single membrane; in plant cells, it is important for maintaining cell shape

cell wall
a fairly rigid structure that encloses some protists, and all prokaryotic, fungal, and plant cells

plasmodesmata
specialized junctions between plant cells that facilitate cell-to-cell communication

The most common of these fibrous proteins is collagen, which constitute 25 percent of all the protein in a typical mammal, including most of the material in tendons, the cables that enable muscles to move bones.

Specialized Organelles in Plant Cells. Some special organelles account for the successful stationary, light-harvesting lifestyle of plants.

Plastids are oval organelles surrounded by a double-layered membrane. They harvest solar energy, manufacture nutrient molecules, and store materials. *Chloroplasts* are one type of plastid that trap the energy of sunlight in a chemical form—generally, sugar molecules—in the process of photosynthesis (see details in Chapter 3). All animals and fungi, as well as most of Earth's other organisms, depend on photosynthesis—directly or indirectly—to supply the nutrients, materials, and energy compounds they need for survival.

Chloroplasts share several features with mitochondria, an organelle that plant cells also have. Both have an outer membrane and a convoluted inner membrane (see Fig. 2.10b). Both organelles house their own DNA molecules and make some of their own proteins. A major difference, though, is that the convoluted innermost set of chloroplast membranes contains the green, light-absorbing pigment *chlorophyll*.

Many plant cells have large fluid-filled sacs in the center called **vacuoles.** Plant vacuoles contain water and nutrients; they have a single surrounding membrane and can fill up to 95 percent of the total cell volume. That's a lot of space, and in fact, taking up space is the vacuole's main role. Essentially a bag of water that fills up a plant cell, the vacuole presses a small amount of cytoplasm and all the cell's organelles into a thin layer with a favorable surface-to-volume ratio (see Figure 2.10b). Full vacuoles keep plant cells plump and give firm shape to the leaves, stems, and other structures.

Cell walls surround most plant cells just outside the plant cell's plasma membrane. Cell walls are composed largely of cellulose. They remain stretchable and flexible until the cell has stopped growing and has begun to mature, then they harden. The porous wall allows water, gases, and some solid materials to pass through to the plasma membrane, and specialized junctions between plant cells called **plasmodesmata** facilitate cell-to-cell communication (Fig. 2.10b). If you are reading this book at a wooden desk, your work is being supported by thickened, dead cell walls—all that remains of a once-living tree.

Figure 2.14
Junctions between Cells

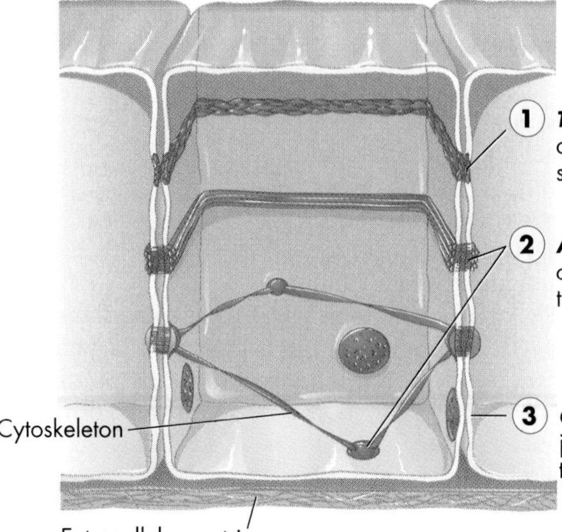

LO8 How Does HIV Complete Its Infective Cycle?

Our complete tour of cell parts puts us in position to return to HIV and understand its whole cycle of infection; how it involves the biological molecules and cell parts we discussed in this chapter; how it harms a patient; and how the drugs called protease inhibitors are helping many people with AIDS.

We've already seen:

- How HIV works its way into the cell (Fig. 2.15 Step 1).
- How the viral RNA is copied onto DNA (Step 2).
- How that DNA enters the nucleus and leads to new viral RNA being formed (Step 3).
- How the cell makes viral proteins on its own ribosomes (Step 4).
- How these proteins are transported to the cell surface (Step 5).

We haven't seen, however, how new viruses are assembled. This last step is crucial because it provides the target for the protease inhibitor drugs that help patients return to relative health. Here, then, is the final part of the HIV infection story.

Assembly of new HIV particles requires additional large viral proteins produced on cellular ribosomes floating free in the cytoplasm. After they're made, the active viral protein called protease cuts the large proteins into smaller proteins (Step ⑥). These proteins eventually surround the viral RNA, move toward the surface, and help assemble hundreds, even thousands of complete viral particles (Step ⑦). The role of protease in chopping large proteins into smaller ones explains why blocking that process with protease inhibitor drugs helps so many AIDS patients so dramatically.

Who knew?

As cells increase in size but maintain the same shape, their volume enlarges more than their surface. An amoeba-like creature the size of a truck would have too much internal volume to be supplied adequately by the area of its outer membrane. It couldn't support the intake of nutrients and excretion of wastes.

Figure 2.15
The Infective Cycle of HIV

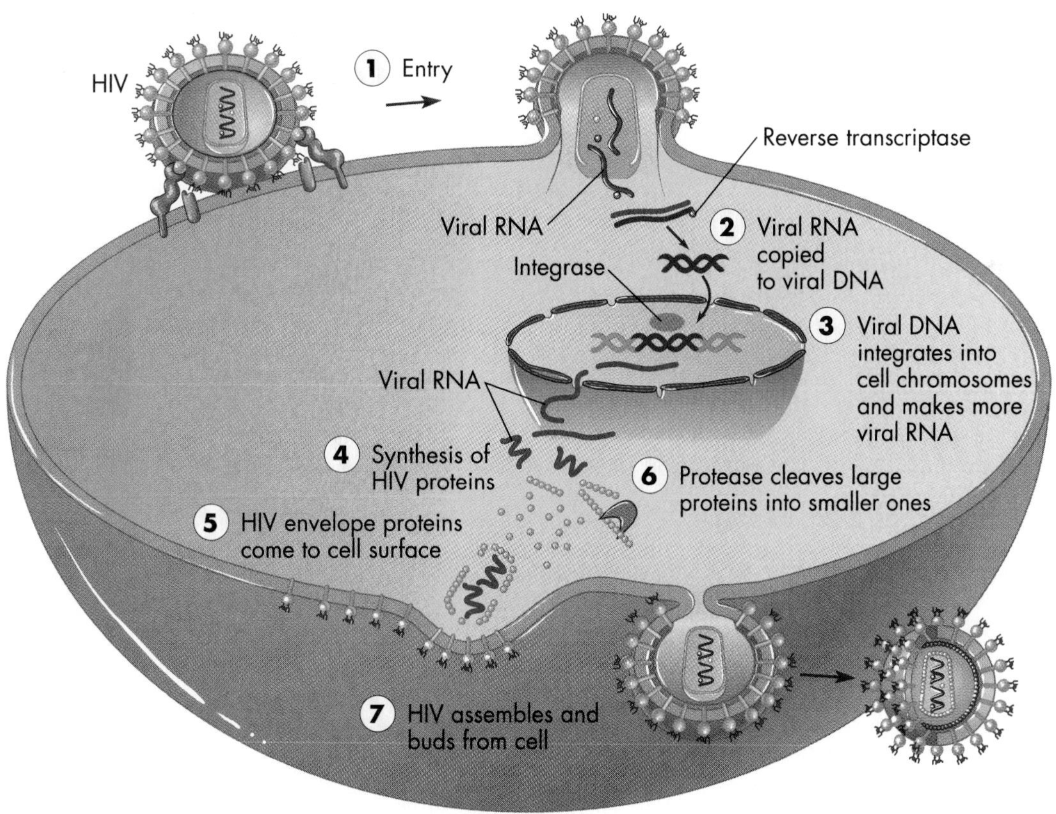

3

How Cells Take in and Use Energy

Learning Outcomes

LO 1 Identify the universal laws of energy conversions

LO 2 Explain the energy flow in living things

LO 3 Understand how living things harvest energy

LO 4 Identify the energy systems used for exercise

LO 5 Explain how plants trap and convert light energy

> "Energy flows from the nonliving physical world to the living world and back."

The Highly Improbable Hummingbird

Hummingbirds have a problem when it comes to energy use: Their bodies have to be very small in order to hover in front of delicate flowers and thus extract an energy resource not readily available to most other species. Their small size, however, means that they lose heat faster than bigger animals because hummingbirds have a larger surface area relative to their body mass. To maintain a high body temperature and still generate enough energy to hover, their rate of food use and calorie expenditure—for their size, anyway—far exceeds that of any other animal with a backbone. How do they do it? What mechanisms do they share with all other organisms for extracting energy from food? What special tricks do they have up their feathery sleeves that allow them to harvest energy fast enough so they can hover like nature's helicopters? These discoveries are related to, and require an understanding of, **energetics**, the study of energy intake, processing, and expenditure. So do each of the bird's other amazing adaptations. Energetics is part of our exploration of how animals, plants, bacteria, and other living things acquire energy and use it to fuel the ongoing enterprise of maintaining order and staying alive. Energy metabolism is a complex subject. But it helps us appreciate life's intricate beauty and its important biochemical symmetry and balance. It also allows us to explore other subjects later in the book, including how our cells build the proteins they need, how we digest the food we eat, how we burn that fuel during exercise, and—on a much bigger scale—what we must do to keep our global ecosystems healthy. In the course of this chapter, you'll find the answers to these questions:

What do you know?

Our planet can support a larger number of vegetarians than meat eaters. Why?

- What universal laws govern how the cells of living things gather and use energy?
- What are the energy routes and carriers in living things?
- How do oxygen-dependent organisms harvest energy?
- How do organisms that can survive without oxygen harvest energy through alternative pathways?
- How do we get the energy for exercise?
- How do plants trap solar energy and form sugars and other biological molecules?

energetics
the study of energy intake, processing, and expenditure

photosynthesis the metabolic process in which solar energy is trapped and converted to chemical energy (ATP), which in turn is used in the manufacture of sugars from carbon dioxide and water

cellular respiration the harvest of energy from sugar molecules in the presence of oxygen

energy the power to perform chemical, mechanical, electrical, or heat-related work

potential energy energy that is stored and available to do work

kinetic energy the energy possessed by a moving object

LO[1] What Universal Laws Govern How Cells Use Energy?

All cells need energy to live. That's true whether they are muscle cells driving a hummingbird's wing beat, nectar-producing cells in the red columbine flower the hummingbird sips from, or bacterial cells decomposing leaves that have fallen off the plant. Regardless of type, however, no cell can "make" energy; instead, it must get energy from some outside source in the environment. Let's follow the flow of energy through a mountain ecosystem that includes hummingbirds and the flowers on which they depend.

Sunlight arriving from space strikes the leaves of the red columbine plant, and they carry out **photosynthesis**, the trapping of energy in a series of metabolic steps that eventually stores energy in the chemical bonds of sugars. The sugar molecules are transported inside the plant, which uses some of the energy to build the delicate, tubular blossoms that attract the hummingbird. Other sugar molecules become concentrated in the flower's nectar. After a hummingbird sips and swallows the nectar, the sugars in the sweet droplets leave the bird's stomach and intestines, flow through the bloodstream, and enter muscle, brain, lung, and other tissue cells. The millions of cells then break down the sugar molecules in the presence of oxygen and release some of the energy in the process called **cellular respiration**. That energy is now available to power the hummingbird's breathing, heartbeat, hovering, and other activities. This represents an energy transformation from solar energy to chemical energy. Such transformations are never 100 percent efficient, however, and some energy is inevitably lost as heat. As a result, the hummingbird warms up as it hovers over one columbine flower then another. The bird's excess body heat dissipates in the cool mountain air around it. Eventually, when the bird dies, bacteria and fungi break down the little body and the rest of the energy still trapped in its tissues fuels the decomposer's activity or is lost as heat. We will explore the processes of photosynthesis and cellular respiration in more detail later in this chapter.

Our main message for now is that energy flows from the nonliving physical world to the living world and back, and it is essential for the activities and survival of all living cells. There are a few basic physical laws that underlie energy transactions in nature. Let's look at them now.

How can I transform this book's *potential energy* into *kinetic energy?*

The Laws of Energy Conversions

Lift this book and hold it parallel to the floor at head level. Now drop it. What seemed like nothing but a slam on the floor is really a series of energy conversions, **energy** being officially defined as the ability to perform work or to produce change.

States of Energy

The lifted book contained stored energy in the form of **potential energy**—energy that is available to do work. Likewise, water saved up behind a dam contains stored energy. This potential energy can be released and accomplish work, such as turning the turbines of a hydroelectric generator. When you dropped the book, it began falling and potential energy was transformed into **kinetic energy**, the energy of motion. These two forms of energy, potential and kinetic, are the two major states of energy in the universe, and each can be converted into the other.

The Law of Energy Conservation

As the book fell and hit the floor, energy changed in form but not in amount. This is because *energy is conserved in energy transformations;* it is neither created anew nor destroyed. When the falling textbook loudly met the floor and abruptly stopped moving, the amount of energy released as sound waves in the air, plus the amount of energy that warmed the floor, book, and surroundings, exactly equaled the kinetic energy in the moving book.

The Law of Inefficient Energy Changes

The amount of energy is the same before and after an energy conversion, but *the amount available to do useful work always decreases during the change.* For example, energy in the sound waves from the book slapping the floor is less than the total kinetic energy of the falling book. Energy conversions are never completely efficient because some energy is always lost as heat, the random movement of atoms and molecules.

Random movement is the opposite of order, so the inefficient conversion of energy from one form into another increases disorder in the system. Scientists use the term entropy (EN-tro-pee) to describe the disorder or randomness in a system. The more disorder, the greater the entropy.

Cells and Entropy

The law of inefficient energy transformations most directly affects an organism's cells. Cells remain healthy only if they obtain enough energy to fuel all their synthesizing and repair activities, over and above the loss of energy to inefficiency and heat. A living cell is just a temporary island of order supported by the cost of a constant flow of energy. If energy flow is impeded, order quickly fades, disorder reigns, and the cell dies.

LO2 How Does Energy Flow in Living Things?

If a cell is a temporary island of order, then hummingbirds are temporary archipelagoes with millions of islands, all maintained and orderly—at least so long as the animal is alive. What maintains that order? The answer is chemical energy, trapped and released through chemical reactions. Recall from Chapter 2 that chemical bonds are a type of "energy glue" that joins atoms together. In a chemical reaction, energy in chemical bonds shifts and atoms rearrange, forming new kinds of molecules. As a kid, did you ever mix vinegar with baking soda? The resulting chemical reaction is a vigorous release of energy in the form of fizzing and heat. The starting substances, or reactants, interact to form new substances, the products. The reactants are hydrogen ions (H^+), from the acetic acid in the vinegar, and bicarbonate ions (HCO_3^-), from the baking soda. Chemical bonds break, allowing the OH^- group to leave the bicarbonate ion, which becomes CO_2, or carbon dioxide gas, bubbling in the water. The OH^- combines with a hydrogen ion (H^+), producing H_2O, water. This reaction takes place spontaneously, and releases energy in the form of heat because the energy conversion is inefficient. The entropy (disorder) of the system increases as the carbon dioxide molecules bubble off into the air randomly.

The reaction of vinegar and baking soda satisfies the law of inefficient energy exchanges: The products contain less energy and are more disordered than the reactants. Some reactions release energy like the one we just discussed, but need some energy input before they will proceed, just as a rock poised at the top of a hill needs a shove before it will start rolling downhill and release potential energy. Even with this needed initial energy to get the reaction started, however, the overall result is a *release* of energy. Reactions of these two types that proceed spontaneously or need a starting "push" but eventually release energy are called energy-releasing reactions.

An entirely different set of reactions called energy-absorbing reactions will not proceed spontaneously and do not give off heat. Think about an egg cooking in a pan. The added energy (heat) causes the egg white proteins to form new chemical bonds with each other and to change into a white rubbery solid. Another example is the way red columbine or purple lupine leaves trap energy from sunlight and use it to power the building of carbohydrate molecules in flower nectar, cell walls, pollen, or other plant parts. Energy-absorbing reactions underlie most of the transformations that maintain order in the cell, such as the building of proteins and the replacement of worn-out cell parts.

So where, then, does the energy come from to fuel a cell's huge number of energy-absorbing reactions? The answer involves a beautiful symmetry: The power for *energy-absorbing reactions* comes from the cell's *energy-releasing reactions*. The two kinds of reactions are energetically *coupled,* so that the leftover energy of a releasing reaction provides the energy

heat
the random motion of atoms and molecules

entropy
a measure of disorder or randomness in a system

chemical reaction
the making or breaking of chemical bonds between atoms or molecules

reactant
the starting substance in a chemical reaction

product
a substance that results from a chemical reaction

energy-releasing reactions
reactions that proceed spontaneously or need a starting "push" but eventually release energy

energy-absorbing reactions
reactions that will not proceed spontaneously and do not give off heat

metabolic pathway the chain of enzyme-catalyzed chemical reactions that converts energy and constructs needed biological molecules in cells and in which the product of one reaction serves as the starting substance for the next

ATP (adenosine triphosphate) a molecule consisting of adenine, ribose sugar, and three phosphate groups; ATP can transfer energy from one molecule to another; ATP hydrolyzes to form ADP, releasing energy in the process

ADP (adenosine diphosphate) an energy molecule related to ATP but having only two phosphate groups instead of three

needs of an absorbing reaction, and so on down a chain of linked reactions called a metabolic pathway (more on this shortly).

ATP: The Cell's Main Energy Carrier

Most of the energy given off during an energy-releasing reaction is quickly trapped in the chemical bonds of a compound that can carry energy from one molecule to another. The most common energy-carrying molecule within a living cell is ATP (adenosine triphosphate). ATP is a *nucleotide*, a two-part molecule with a head composed of three molecular rings and a tail made up of three phosphate groups (PO_4^-, a phosphorus atom bonded to four oxygen atoms) (Fig. 3.1a). Living cells make the energy carrier ATP from ADP (adenosine diphosphate), a related molecule with only two phosphate groups (Fig. 3.1b). Cells can use the energy freed from an energy-releasing reaction to add a phosphate group to ADP and thus make ATP, with a higher level of stored energy in its chemical bonds.

Cells can then "spend" their ATP currency, because the release of a phosphate from ATP delivers energy in a controlled way. The cleavage of ATP releases ADP, a phosphate, and a small amount of energy (Fig. 3.1c). A cell can then use that energy to do work. A hummingbird flaps its wings 40 to 80 times per second and the muscle cells that contract to move those wings carry out millions of energy-absorbing reactions every second to keep up with that hard, fast work. An energetic phosphate released from ATP will often transfer to another molecule, energizing that molecule (Fig. 3.1d). When phosphate is split off an ATP, ADP forms; that ADP can itself lose another phosphate group and

Figure 3.1
ATP: The Cell's Main Energy Carrier

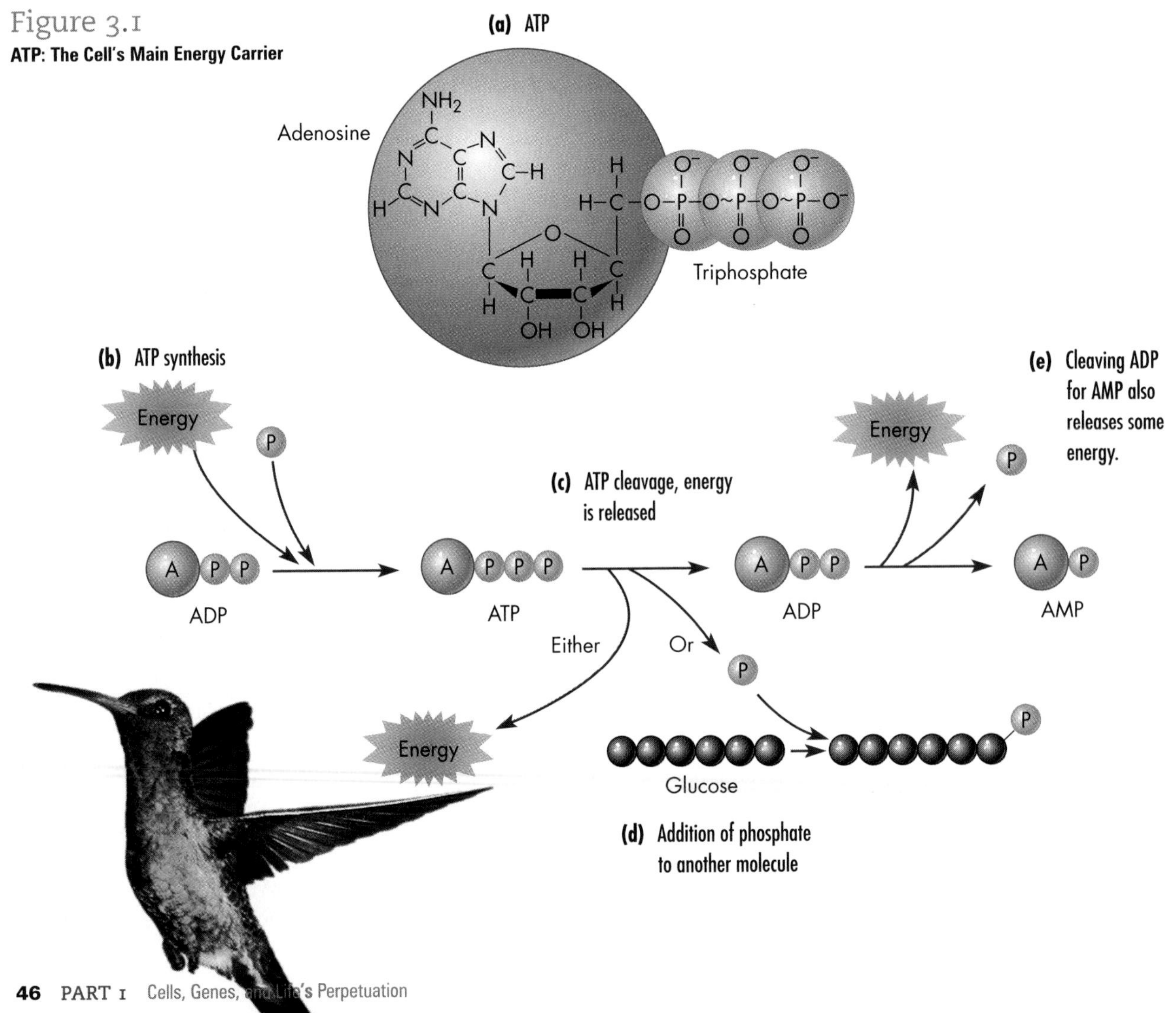

release more energy plus the molecule AMP (adenosine monophosphate) (Fig. 3.1e). Cells only use this second release and formation of AMP when energy demands are unusually high. The significance of ATP and its less energetic relatives ADP and AMP is that they function as links between energy exchanges in cells.

How Enzymes Speed Up Chemical Reactions

Chemical reactions rearrange chemical bonds and release or absorb energy. But how fast does this take place? For example, if you collected some flower nectar and allowed it to dry in a cool, arid place, the high-energy food (sucrose) would sit there intact for decades, like sugar in a jar, rearranging its chemical bonds very slowly and releasing energy at a nearly undetectable rate. When a hummingbird takes that same nectar into its digestive tract, however, special digestive proteins can break down the chemical bonds in the sucrose within seconds into simpler sugars. Clearly, these proteins can speed up chemical transformations. Let's look at how they do it.

Within living cells, most chemical transformations are facilitated by enzymes. Digestive enzymes are just one type. Enzymes are proteins that function as biological catalysts, agents that speed up specific chemical reactions without themselves being permanently changed by the reaction. Without dozens of kinds of enzymes to speed the breakdown of sugars, hummingbirds would not be able to maintain their high metabolic rates.

AMP (adenosine monophosphate) an energy molecule related to ATP but having only one phosphate group instead of two (ADP) or three (ATP)

enzyme a protein that facilitates chemical reactions by lowering the required activation energy but is not itself permanently altered in the process; also called a biological catalyst

catalyst a substance that speeds up the rate of a chemical reaction by lowering the activation energy without itself being permanently changed or used up during the reaction

transition state in a chemical reaction, the intermediate springlike state between reactant and product

Figure 3.2

Activation Energy and Enzymes

(a) Activation energy

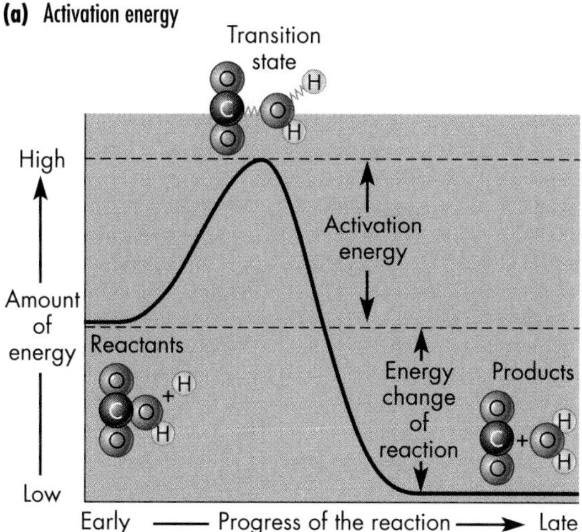

(b) Analogy for enzyme action

Enzymes Lower Activation Energy

In chemical reactions, there is an energy barrier separating the reactants and products. For a pair of reactants such as bicarbonate and hydrogen ions to be converted into products such as carbon dioxide and water, the reactants must collide with each other hard enough to break chemical bonds in the reactants and form new bonds in the products. For a fleeting instant during the conversion, chemical bonds in the reactants are distorted like springs being stretched, and this fleeting intermediate state cannot be reached without a very energetic collision between the molecules. We call the momentary, intermediate springlike condition a transition state. Because energy input is required to achieve it, the transition state is often characterized as an energy hill or barrier that separates reactants and products (Fig. 3.2).

Most molecules jostling about and colliding randomly lack the energy

activation energy the minimum amount of energy that molecules must have in order to undergo a chemical reaction

active site a groove or a pocket on an enzyme's surface to which reactants bind; this binding lowers the activation energy required for a particular chemical reaction; thus, the enzyme speeds the reaction

substrate a reactant in an enzyme-catalyzed reaction; fits into the active site of an enzyme

enzyme-substrate complex in an enzymatic reaction, the unit formed by the binding of the substrate to the active site on the enzyme

metabolism (Gr. *metabole,* to change) the sum of all the chemical reactions that take place within the body; includes photosynthesis, respiration, digestion, and the synthesis of organic molecules

of motion they need to overcome the energy barrier; they simply bounce off each other. Some individual molecules, however, do jostle about with enough kinetic energy so that they achieve a productive collision, an impact that generates the springlike transition state, from which reactants change into products. An energy of impact great enough to cause molecules to cross the energy barrier is called the activation energy (Fig. 3.2).

In a living cell, most molecules need help in overcoming the energy barrier that prevents the start of many reactions. Heat, of course, speeds up colliding molecules, but high temperatures would speed *all* the reactions in a cell, not just those special ones that help a cell function. This overall temperature increase would disorganize the cell and kill it. Luckily, enzymes can lower the level of activation energy enough so that biochemical reactions can take place without an increase of temperature. The action of an enzyme, in effect, bores a tunnel through the energy hill, allowing the reactants to become products without needing to roll like a heavy boulder up and over the high barrier represented by activation energy.

Enzymes not only allow reactions to proceed at the relatively low temperatures compatible with life processes, but they act only on specific reactions. For example, one specific enzyme speeds up a reaction that helps red blood cells transport CO_2 from tissues that produce it, such as active muscles, to the lungs, from which the waste gas can be exhaled. Each single molecule of this enzyme facilitates this change at the amazing speed of 600,000 times per second, nearly a million times faster than the reaction would occur without a catalyst. Biologists estimate that every second, a hummingbird's blood makes an entire circuit of its heart, lungs, and body and the animal breathes in and out four to eight times. Without this red blood cell enzyme working at near top speed, carbon dioxide produced in working cells could not possibly get transferred fast enough to the lungs to ensure the bird's survival.

Figure 3.3

How an Enzyme Works

(a) Enzyme binds substrate

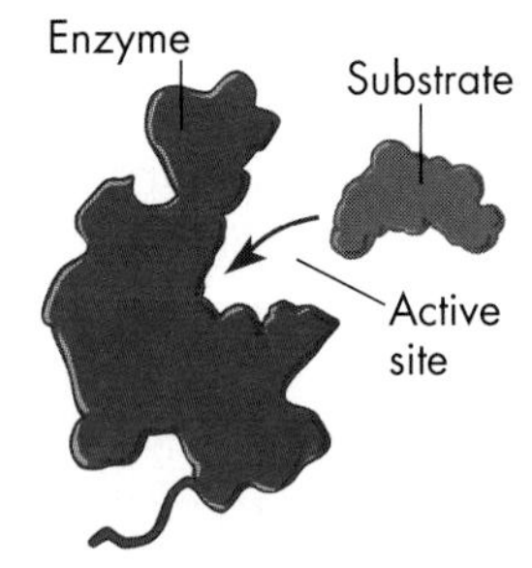

(b) Enzyme-substrate complex

(c) Active site's new shape induces a better "fit"

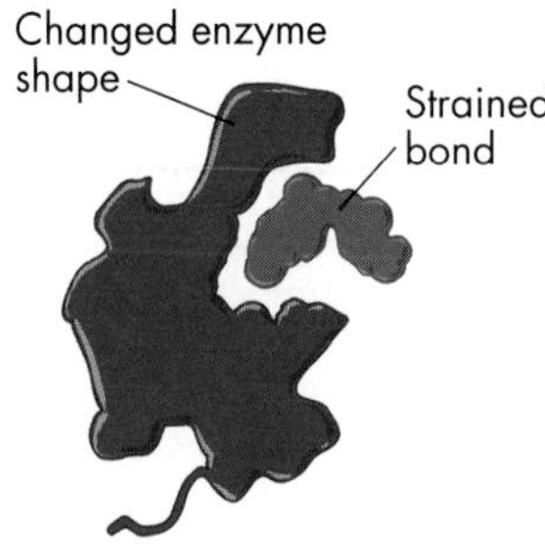

(d) Enzyme unchanged

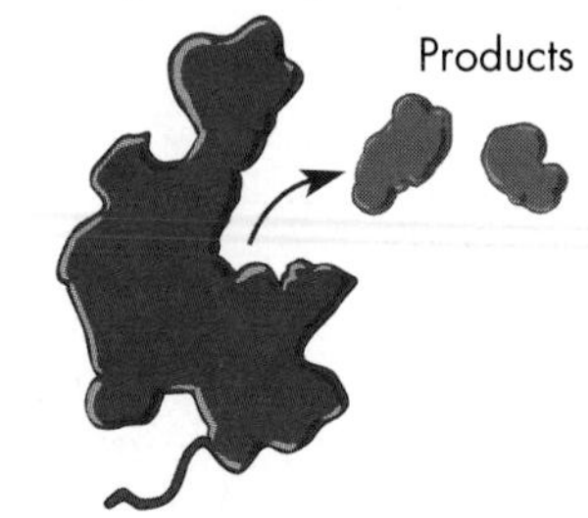

Induced Fit Model of Enzyme Action

Like other proteins, each enzyme has its own unique three-dimensional shape, but most share an important feature: a deep groove, or pocket, on the surface called the active site. The shape of the enzyme's active site fits a specific reactant, or substrate. The substrate fits into the active site forming an enzyme-substrate complex (Fig. 3.3a,b). The active site of the enzyme can change shape as the substrate binds to it, improving the "fit" (Fig. 3.3c). (For this reason, biologists sometimes call it the "induced fit" model of enzyme action.) In fitting to the active site, the substrate's bonds are strained, which makes it possible for the substrate to reach a transition state, which in turn lowers the activation energy that is needed for the substrate to react and form products (Fig. 3.3d).

Enzymes accelerate a huge number of processes in living cells. Among these processes are the ones that allow organisms to transform energy in food molecules, light, or other nonliving sources into forms of energy that can do work for the cell. These enzyme-catalyzed reactions, along with all the other reactions of the body that sustain life, constitute metabolism. Metabolic reactions can break down or build up molecules, and can give off or use energy. As we saw earlier, metabolic reactions are interlinked within cells in metabolic pathways. These pathways allow living cells to subdivide a big chemical change into a number of smaller steps. One such change might, for example, release a tremendous amount of heat. This, in turn, would liberate energy in packets small enough for the cell to use efficiently.

LO3 How Do Living Things Harvest Energy?

Some of the busiest molecules in the living world are the enzymes within hummingbird muscles that keep those flight "motors" whirring. They have to be amazingly active to keep up with the animals' constant energy needs for hovering, flying, defending territory, and staying warm. The energy comes primarily from sugars present in flower nectar. In this section, we'll examine the mechanisms that cells use to obtain energy from food molecules. In the next section, we'll go on to see how plants, algae, and some bacteria trap solar energy in energy-rich molecules. These are challenging subjects, but they're central to all that goes on in cells and organisms and crucial to understanding biology. The subject is as personal as how your body is currently utilizing what you ate for your previous meal. On a larger scale, the subject of cellular energy harvest is also a key to understanding ecology, including how energy flows in the environment as one organism feeds on another. We'll return to it in Chapter 15.

An Overview of Energy Harvest

A hummingbird lighter than a tea bag can migrate all the way across the Gulf of Mexico without stopping. Think about this animal's exertion as it migrates nonstop, and then contrast it with the rapid burst of effort the bird would make as it accelerated to escape a swooping predator. These exertion levels are similar to you strolling for miles through the fall leaves versus running to catch the bus. Where does the hummingbird's (or your) energy come from?

There are two pathways of energy harvest that take place in living cells: energy harvest with plenty of oxygen, the **aerobic pathway**, and energy harvest with minimal oxygen or none at all, the **anaerobic pathway**. Many living cells are committed to one type or the other. However, some cells, such as muscle cells, are somewhat special in their ability to obtain energy from either pathway, depending on oxygen availability at any given moment.

Consider a man hiking up a low hill. As he walks, his lungs and blood circulation deliver enough oxygen to his muscles so that the cells can use the aerobic pathway (the yellow arrows in Fig. 3.4). During a more strenuous climb, his heart, lungs, and blood vessels cannot provide oxygen as fast as muscle cells require it. For this reason, muscle cells switch to the anaerobic pathway (the orange arrows in Fig. 3.4).

The Aerobic Pathway

The aerobic energy harvesting pathway, also called *cellular respiration*, consists of three main parts (follow the yellow arrows in Fig. 3.4). The first part, **glycolysis**, begins the breakdown of glucose and produces just a few ATP molecules. The second part of the aerobic pathway, called the **Krebs cycle**, completely dismantles glucose to individual carbon atoms while capturing high-energy electrons. The third part of the aerobic pathway, the **electron transport chain**, is where the oxygen comes in: The high-energy electrons captured in the Krebs cycle join with oxygen and hydrogen. This produces water molecules and a large number of ATP molecules—a big energy payoff for the cell. Most cells in animals, plants, algae, and fungi, and most kinds of bacteria carry out a strictly aerobic energy harvest.

aerobic pathway energy harvest in the presence of oxygen; also called cellular respiration

anaerobic pathway the series of metabolic reactions that results in energy harvest in the absence of oxygen

glycolysis the initial splitting of a glucose molecule into two molecules of pyruvate, resulting in the release of energy in the form of two ATP molecules; the series of reactions does not require the presence of oxygen to occur

Krebs cycle the second stage of aerobic respiration, in which a two-carbon fragment is completely broken down into carbon dioxide and large amounts of energy are transferred to electron carriers; occurs in the mitochondrial matrix

electron transport chain the third stage in aerobic respiration, in which electrons are passed down a series of molecules, gradually releasing energy that is harvested in the form of ATP

fermentation extraction of energy from carbohydrates in the absence of oxygen, generally producing lactic acid or ethanol and CO_2 as byproducts

The Anaerobic Pathway

The anaerobic pathway consists of two main parts (follow the orange arrows in Fig. 3.4). The first, glycolysis, is the same as in the aerobic pathway. The second, **fermentation**, doesn't release any useful energy, but it does recycle materials necessary for glycolysis to continue. Without fermentation, a cell could not even use glycolysis to produce its meager amount of ATP. Animal muscle cells, yeasts in low-oxygen environments, and many kinds of bacteria harvest energy through this anaerobic pathway some or all of the time.

electron carriers
molecules that transfer electrons from one chemical reaction to another

How Do Cells Harvest Energy When Oxygen Is Present?

How can the tiny hummingbird store enough energy to make a migration of 800 km or more? What accounts for the hummingbird's tremendous feats of strength and endurance—or for that matter, your own? The answers can be summarized with a simple equation for cellular respiration, the process of energy harvest utilizing oxygen:

Glucose + Oxygen + ADP + Phosphate →
Carbon dioxide + Water + ATP

This equation basically says that in the presence of oxygen, the energy of sunlight trapped in glucose molecules is transferred to ADP along with a phosphate ion, thereby producing the more readily usable energy carrier ATP. Water and carbon dioxide are byproducts of this ATP formation. The arrow in the equation above represents the cellular mechanisms that bring about the phases of this energy harvest: glycolysis, the Krebs cycle, and the electron transport chain (yellow arrows in Fig. 3.4).

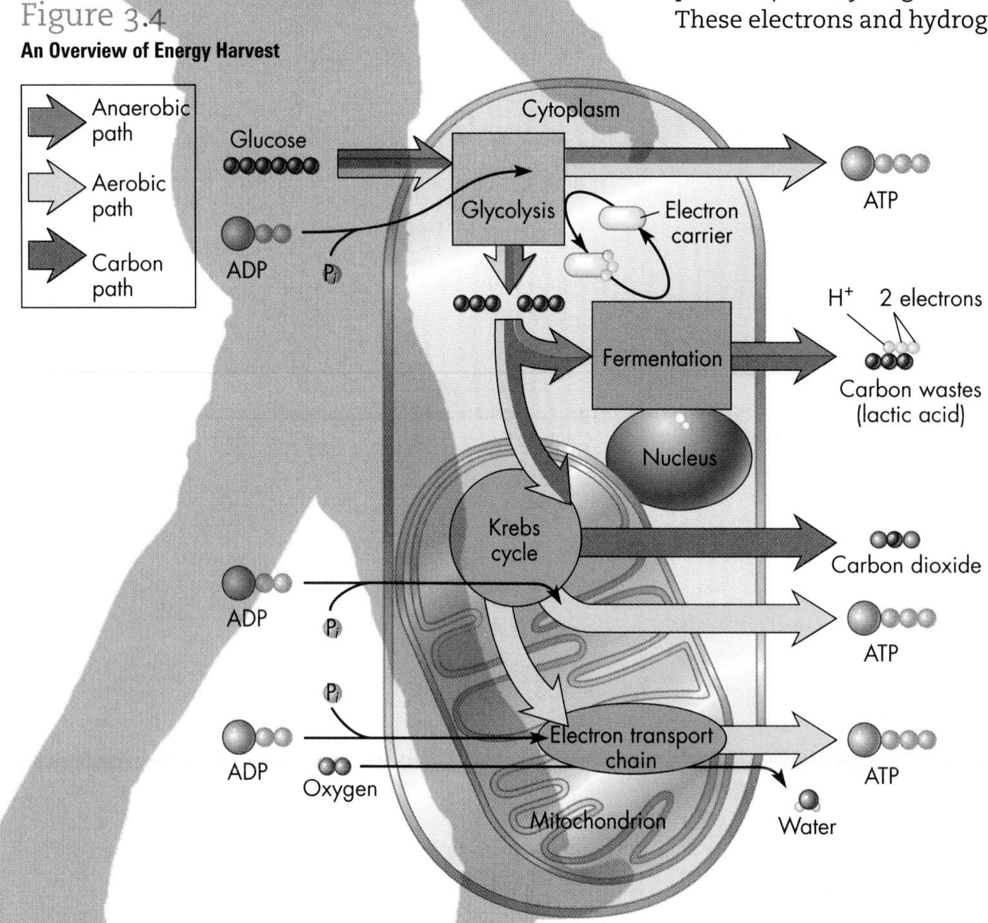

Figure 3.4
An Overview of Energy Harvest

Glycolysis

Energy harvest begins with the breakdown of the simple sugar glucose in a sequence of reaction steps called glycolysis (*glyco* = sugar + *lysis* = splitting). Cells can convert other kinds of sugar, such as the sucrose in flower nectar, into glucose and use it as a basic fuel. Energy stored in the chemical bonds of glucose molecules originally came from the sun, and it was trapped in molecular form by photosynthesis taking place in green plants as well as in algae and certain kinds of bacteria.

The reactions of glycolysis split the six-carbon sugar glucose into two molecules of the three-carbon compound, *pyruvate* (ionized pyruvic acid; Fig. 3.5). Pyruvate contains some stored chemical energy and acts as an *intermediate,* a compound that serves as a product for one reaction and a reactant for the next in a metabolic pathway. The splitting of glucose makes available energetic electrons (charged atomic particles) and hydrogen ions (H^-, also called protons). These electrons and hydrogen ions are transferred to a special **electron carrier** molecule—a type of biochemical "delivery van" we'll discuss in detail later. The steps of glycolysis take place in the cell's liquid cytoplasm, rather than in an organelle, and they are facilitated by enzymes dissolved in the watery cytoplasmic solution. For each glucose molecule split during glycolysis, there is a net gain of two ATPs and two pyruvate molecules. The ATPs can move through the cytoplasm to places in the cell where energy is immediately needed. The pyruvate molecules leave the cytoplasm and enter the cell organelle called the mitochondrion.

The Mitochondrion

Highly active cells—your own heart muscle cells, for example—contain large numbers of sausage-shaped organelles called *mitochondria*, which we encountered in Chapter 2 and which act as the cell's powerhouses, generating ATP.

Mitochondria consist of two membranes, like a large, uninflated balloon folded up inside a smaller, fully inflated balloon (see the micrograph in Fig. 3.7). A mitochondrion's *outer membrane* is directly bathed by the cell's cytoplasm. Large protein-bound pores perforate the membrane, and molecules up to the size of small proteins can pass through the openings. A mitochondrion's inner membrane is thrown into folds called **cristae**, which are studded with enzymes and pigment molecules (see Fig. 3.7). Some of these enzymes take part in an energy bucket brigade described below, and one of the enzymes synthesizes ATP.

In all eukaryotic organisms—even lethargic slugs and slow-growing lichens—the mitochondrion's inner membrane is much less permeable than the outer membrane; hence, the area enclosed by the inner membrane, the **matrix**, is a compartment well isolated from the rest of the cell. The matrix contains many enzyme molecules, including those that carry out the reactions of the Krebs cycle, which is so important in aerobic respiration. The matrix also contains several copies of the circular mitochondrial DNA molecule and hundreds of mitochondrial ribosomes.

Inside the mitochondrion, the pyruvate that formed after glycolysis gets broken down by means of cellular respiration, the breakdown of nutrients and the production of ATP energy using oxygen. Aerobic respiration in mitochondria consumes oxygen and yields carbon dioxide and water plus a large harvest of ATP molecules. That energy harvest, involving the Krebs cycle and the electron transport chain, accounts for the mitochondrion's reputation as a cellular powerhouse, regardless of a cell's metabolic speed.

crista (pl. cristae)
a fold or folds formed by the inner membrane of a mitochondrion

matrix
in HIV and many other viruses, a sphere of protein inside the envelope and outside the capsid; in a mitochondrion, the area enclosed by the inner membrane

The Krebs Cycle

The Krebs cycle is part of a series of chemical reactions taking place inside mitochondria that break down pyruvate completely into carbon dioxide and water (Fig. 3.6). In several steps, the carbons in pyruvate are cleaved off, one at a time, and released as carbon dioxide (CO_2). Your exhaled breath is the body's way of getting rid of carbon dioxide produced in the Krebs cycle.

The Krebs cycle starts with the end product of glycolysis, the three-carbon substance pyruvate

Figure 3.5
The Aerobic Pathway Part 1: Glycolysis

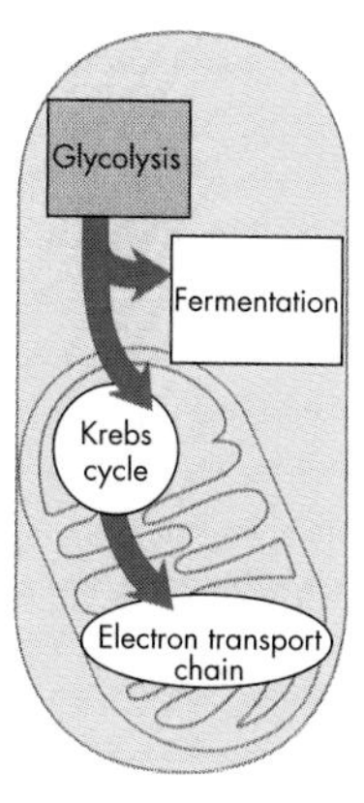

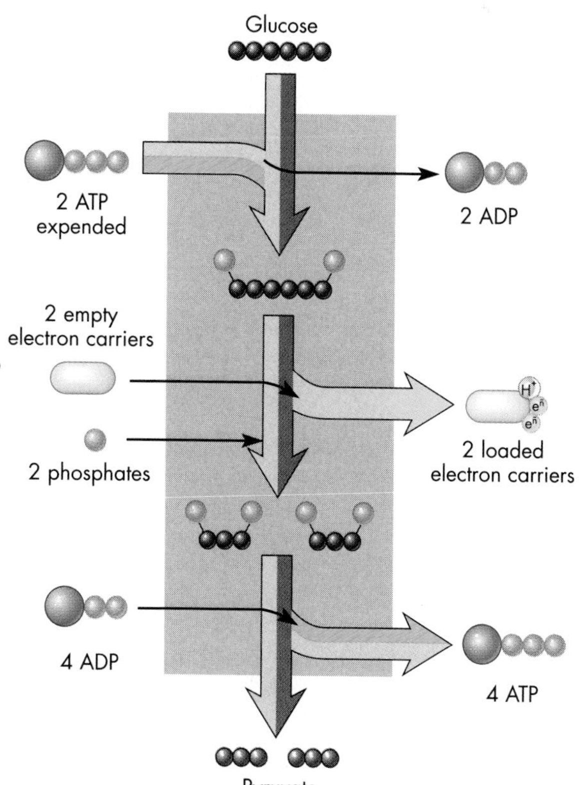

1. The enzymes of glycolysis ì superchargeî glucose by spending two molecules of ATP and adding their phosphates to the six-carbon sugar molecule.
2. Enzymes chop the six-carbon molecule in half, and then strip two electrons and a hydrogen from each three-carbon compound and add them to an electron carrier while adding another phosphate to each three-carbon compound.
3. In the final step, enzymes remove both phosphates from each three-carbon compound and use them to make 4 ATPs, giving a net payoff of 2 ATPs.

(review Fig. 3.5), which passes from the cell's watery cytoplasm into the mitochondrion. In the matrix of the mitochondrion, enzymes cleave pyruvate into a molecule of carbon dioxide plus a two-carbon portion (Step 4 of Fig. 3.6). Enzymes in the Krebs cycle join this two-carbon portion to a four-carbon compound to make a six-carbon molecule (Step 5). Other enzymes sequentially cleave two carbon dioxide molecules from this six-carbon molecule (a person exhales this in his or her breath) and produce four energized electron carriers. Notice the two electron carriers in Steps 6 and 7). In addition, enzymes in Step 7 regenerate the original four-carbon compound, completing the cycle. An additional step, not shown, results in one ATP. The end result is that the Krebs cycle converts all the carbons from the original glucose into carbon dioxide and stores the energy in electron carriers.

The Krebs Cycle as a Metabolic Clearinghouse

The various compounds taking part in the Krebs cycle are intermediates, reactants in one metabolic reaction and products in the next. Cells can use these intermediates in several ways. The Krebs cycle can help a cell use lipids or proteins for energy in addition to carbohydrates. When an animal or plant cell's supply of sugar falls, the cell

{Give Me Some Sugar}

An important exception to the clearinghouse principle is the human brain cell. For complex reasons, the brain can use only glucose as fuel, and this fact has major implications:

- First, it explains why sugary foods are temporary mood elevators. Soon after you eat a candy bar or drink a sugary soda, glucose molecules are cleaved from sucrose (table sugar) and enter the bloodstream and brain. Infused with their favorite fuel, the brain cells function at peak efficiency, leading one to feel happier, smarter and livelier—at least for a time.
- Second, this property of the brain cell explains why dieters are warned to consume at least 500 calories in carbohydrates per day and not to restrict themselves to liquid or powdered protein diets—the brain alone needs at least 500 calories of glucose for normal functioning, and without it, a person can grow faint and even lapse into unconsciousness.

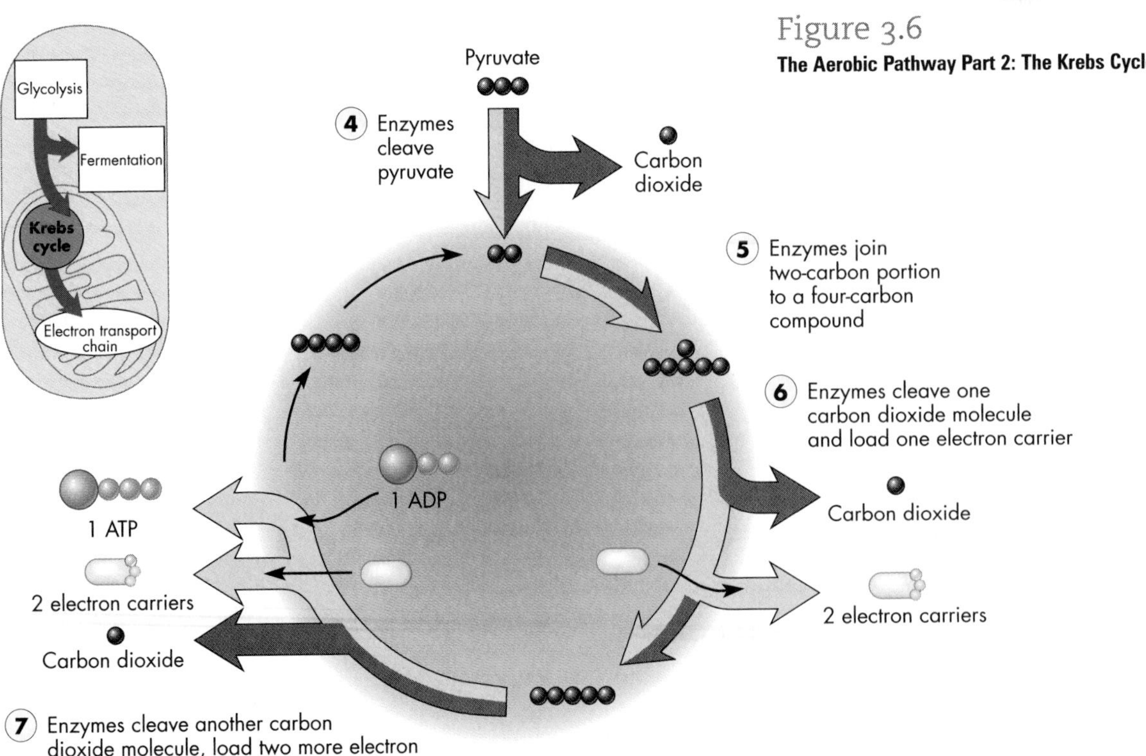

Figure 3.6
The Aerobic Pathway Part 2: The Krebs Cycle

begins to break down lipids or proteins. Some of the subunits from these reactions can then be converted into pyruvate or into Krebs cycle intermediates. These compounds can then enter the Krebs cycle pathway at the appropriate stage and be dismantled for energy harvest. If an organism is starving, this process can provide enough energy to keep the body alive, but the body is literally digesting itself to provide that energy, with the Krebs cycle making it possible. When food once more becomes available, intermediates from the Krebs cycle can become the raw materials for growth: They can serve as skeletons for the synthesis of new fats, proteins, and carbohydrates to restore those broken down during lean times or that wear out and need replacing.

The Electron Transport Chain

Still inside the mitochondrion, eight electron carrier molecules per initial glucose molecule are loaded with electrons from the Krebs cycle. These carriers move to the *electron transport chain,* a group of enzymes and pigment molecules embedded in mitochondrial membranes (Fig. 3.7). Each member of the

Figure 3.7

The Aerobic Pathway Part 3: The Electron Transport Chain

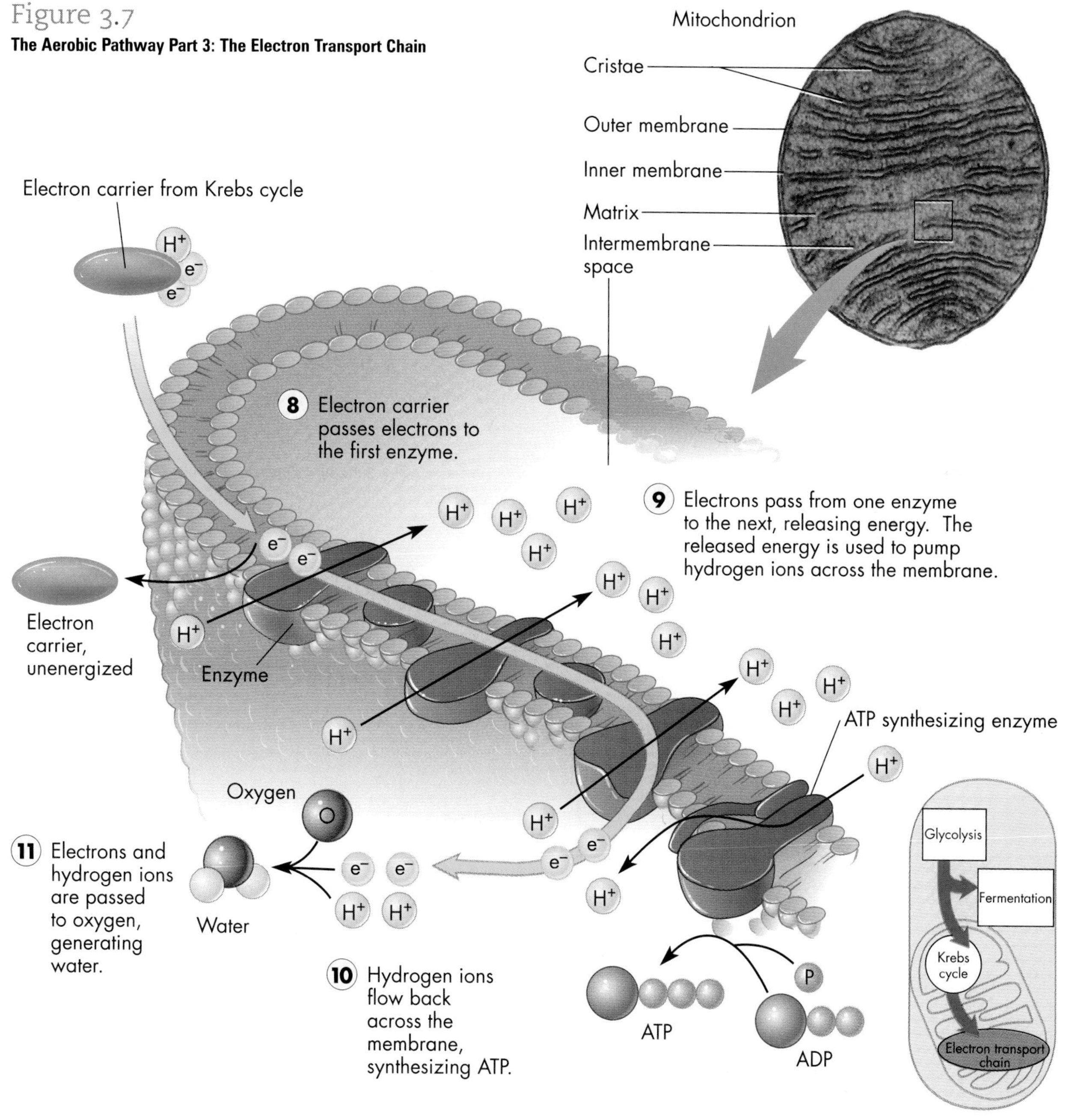

chain passes electrons to the next member, like a bucket brigade. During each electron transfer, the electron loses a bit of energy (like a splash of water spilling from the bucket). Particular mitochondrial enzymes use this released energy to synthesize ATP from ADP. For each initial molecule of glucose that entered glycolysis, the yield from the electron transport chain is a whopping 32 ATPs.

The electron transport chain is the stage of the aerobic pathway that actually uses the oxygen because the last electron acceptor in the chain is oxygen. As electrons are added to oxygen atoms, and hydrogen ions follow along, the hydrogen and oxygen combine to form water, H_2O. Thus, the oxygen you are breathing in as you read these sentences will be converted to water in your mitochondria as oxygen accepts electrons and aerobic respiration takes place. The legend to Figure 3.7 describes this step by step.

If oxygen is not present to accept the electrons and hydrogen ions once they have passed down the electron transport chain, the entire process quickly stops. The organism or cell, if strictly oxygen-requiring, then dies.

How Do Cells Harvest Energy Without Oxygen?

A hummingbird has a greater blood supply to its wing muscle cells than a mammal does to its wings or legs. Sometimes, though, the bird must exert itself so strenuously that its furiously beating heart still cannot deliver oxygen to the muscles as fast as they are using the gas. As a result, the hummingbird's muscle cells have to rely for short periods of time on anaerobic energy harvest with its two phases, glycolysis and fermentation (the orange arrows in Fig. 3.4).

Glycolysis

Anaerobic energy harvest or the *anaerobic pathway* begins with glycolysis, the same process that starts the aerobic pathway (Fig. 3.8). Because glycolysis doesn't vary from one pathway to the other, *all* organisms possess the enzymes of glycolysis. This universal prelude to energy metabolism leads biologists to think that glycolysis probably evolved in earth's earliest cells over 3 billion years ago, and was

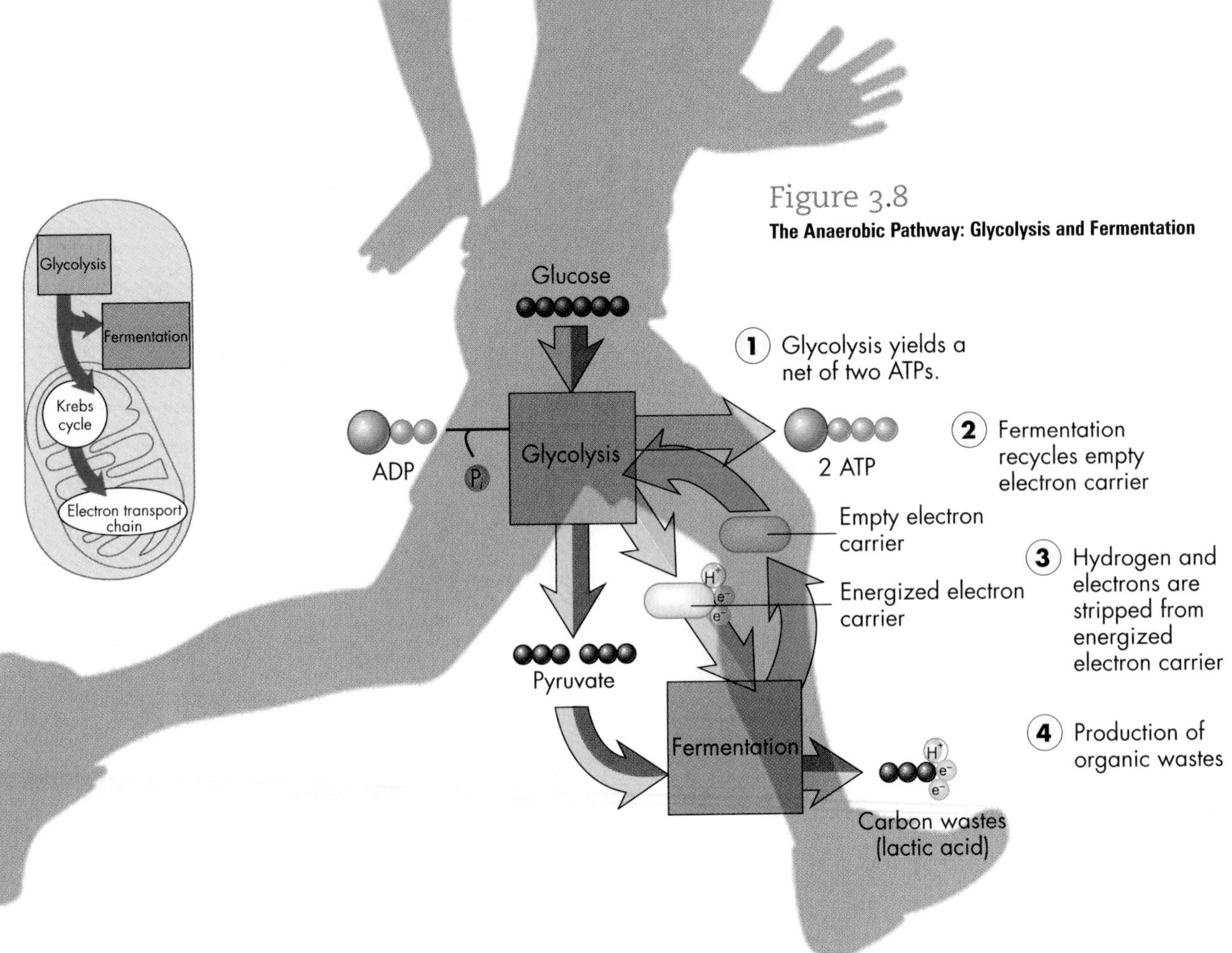

Figure 3.8
The Anaerobic Pathway: Glycolysis and Fermentation

passed down to all surviving organisms. This common biochemical thread underscores the evolutionary relatedness of all life forms.

Fermentation

Recall that energized electron carriers and the 3-carbon compound pyruvate are end products of glycolysis. During the second phase of the anaerobic pathway, called *fermentation*, enzymes modify pyruvate in the absence of oxygen. The enzymes that speed the reactions of fermentation lie in the cell's cytoplasm, just like those that facilitate glycolysis. Depending on the organism, however, fermentation converts the pyruvate into various end products, such as ethanol and carbon dioxide, or lactic acid (Fig. 3.8). The intoxicating effect of wine and beer and the fragrant aroma of baking bread come from ethanol, the alcohol produced during fermentation in yeast cells. Likewise, fermentation carried on by certain bacteria growing in milk releases the lactic acid that gives some cheeses their sharp taste. Your muscle cells also produce lactic acid (or the electrically charged form called *lactate*) when you exercise anaerobically. The "burn" that you feel when you work a muscle very hard comes from the liberation of lactate.

Ironically, fermentation itself yields no ATP. The wastes—the ethanol or lactic acid— squander the energy that would be captured in ATP if the cell were using the aerobic pathway. You might wonder, then, why cells would carry out fermentation and produce toxic waste products such as ethanol and lactate? The answer is that fermentation reactions recycle the electron carrier molecule needed for glycolysis, stripping away the electrons and hydrogens from the carrier and making the carrier available for a new round of glycolysis (blue arrows, Fig. 3.8). In fermentation, it is as if the electron carrier "delivery van" dumps its load unused just so it can return and pick up a new load. Without the recycling of the energy carrier, the cell would run out of this necessary "delivery" molecule, and glycolysis would cease.

The energy yield of the anaerobic pathway may seem meager, just 2 ATPs per glucose compared to 36 for the aerobic pathway. Anaerobic metabolism, however, is crucial to the global recycling of carbon and the stability of the environment. Organic matter from dead leaves, dead microorganisms, and other sources often sinks into an environment devoid of oxygen, such as the soft layers at the bottom of lakes or oceans. If it weren't for anaerobic decomposers—organisms capable of breaking down organic matter via anaerobic metabolism—most of the world's carbon would eventually be locked up in undecomposed organic material in these oxygen-poor environments. As a result, there would be too little carbon dioxide available as a raw material for photosynthesis, plants would be unable to generate new glucose molecules, and neither plants nor animals would survive.

feedback inhibition
the buildup of a metabolic product that in turn inhibits the activity of an enzyme; since this enzyme is involved in making the original product, the accumulation of the product turns off its own production

Control of Metabolism

Hummingbirds can burn carbohydrates by either the aerobic or the anaerobic pathway if their activity demands it temporarily. As we saw, they can burn fats when they are migrating nonstop, and, thanks to the Krebs cycle clearinghouse, they can also use proteins for energy. These intricacies of metabolism bring up new questions: What makes a cell start to burn its stores of lipids or proteins when glucose runs out? Then, when the supplies of glucose return, what stops the cell from burning all its lipids or proteins, literally eating itself up from the inside and destroying its cellular structures? Finally, what triggers a cell to build molecules only when they are needed?

When levels of ATP build up in a cell so high that the cell requires no more of the molecular fuel, the ATP binds to a special regulatory site on a specific enzyme that facilitates an early step in glycolysis. This binding shuts down enzyme activity. This, in turn, switches off the entire glycolytic pathway. So, when the cell already has high levels of ATP, the presence of the ATP molecule itself serves as a control to turn off its own production. When ATP levels drop, glycolysis resumes once more. This kind of metabolic regulation is called **feedback inhibition**.

Through feedback inhibition and other forms of control, the activity of a cell's metabolic enzymes is turned on or off so that the cell burns glucose when that sugar is available; it burns lipids or proteins when glucose is lacking; and it builds the appropriate biological raw materials just when they are needed for growth or maintenance activities. These control mechanisms ensure that order rather than metabolic chaos reigns within living organisms.

LO4 How Do We Get Energy for Exercise?

Energy for different forms of exercise comes from different parts of the metabolic pathway. Exercise physiologists have determined that three energy systems—the immediate system, the glycolytic system, and the oxidative system—supply energy to a person's muscles during exercise. The duration of physical activity and availability of oxygen dictate which system the body uses. The immediate energy system is instantly available for a brief explosive action, such as one bench press, one tennis serve, or one ballet leap, and the system has two components. One component is the small amount of ATP stored in muscle cells, immediately useful like the few coins you carry around in your purse or pocket. This stored ATP, however, runs out after only half a second—barely enough time to heave a shot or return a tennis serve, let alone to trudge up a long hill. The second component of the immediate energy system is a high-energy compound called *creatine phosphate*, an amino acid–like molecule that has an energetic phosphate, like ATP. Muscle cells store creatine phosphate in larger amounts than ATP. Creatine phosphate is more like a handful of dollar bills than a few coins. When the ATP in a muscle cell is depleted, creatine phosphate transfers its phosphate to ADP; this regenerates ATP, and ATP can then fuel the muscle cell to contract and move the body. Even the cell's store of creatine phosphate, however, becomes depleted after only about a minute of strenuous work; thus, muscles must rely on more robust systems than the immediate energy system to power longer-term activities.

immediate energy system
energy in the body instantly available for a brief explosive action, such as one heave by a shot-putter

glycolytic energy system
energy system based on the splitting of glucose by glycolysis in the muscles; the glycolytic system can sustain heavy exercise for a few minutes, as in a 200-m swim

oxidative energy system
the longest-sustaining energy system of the body, which relies on the Krebs cycle and electron transport chain in mitochondria and its ability to use fats as fuel; typically fuels aerobic activity

The glycolytic energy system, which depends on splitting glucose by glycolysis in the muscles, fuels activities lasting from about 1 to 3 minutes, such as an 800-m run or a 200-m swim. This storage form is more like an account at the bank than like dollars in your wallet, and it "purchases" more activity. Glycolysis going on in the muscle cell cytoplasm can cleave glucose in the absence of oxygen and generate a few ATPs. Fermentation takes the product of glycolysis a step further and makes the waste product lactic acid. Recall that there is a net yield of only two ATP molecules for each molecule of glucose from glycolysis. For this reason, the glycolytic energy system can sustain heavy exercise for only about 3 minutes, and all of it would be considered anaerobic exercise, powered by the cells' anaerobic pathway. It's also why lactic acid begins to build up in muscles in just that short amount of time and can lead to a muscle cramp or "stitch" in your side.

Activities lasting longer than about 3 minutes—a jog around the neighborhood, an aerobic dance session, or a long uphill hike—require oxygen and employ the oxidative energy system. This system can supply energy for activity of moderate intensity and long duration (aerobic exercise), and is like money invested in stocks and bonds that give long-term steady income. The oxidative energy system is based on cellular respiration and includes the Krebs cycle and the electron transport chain. It uses oxygen as the final electron acceptor and generates many ATP molecules per molecule of glucose burned. Clearly, anyone interested in melting away body fat should engage in aerobic (oxygen-utilizing) activities like jogging, swimming, bicycling, or strenuous hiking which rely primarily on the oxidative energy supply system and its ability to use fats as fuel.

The glycolytic energy system can make ATP only from glucose or glycogen (cleaved to release glucose), but the oxidative energy system can produce energy by breaking down carbohydrates, fatty acids, and amino acids mobilized from other parts of the body and transported to the muscle cells by way of the bloodstream. The oxidative energy supply system can provide many more ATP molecules than creatine phosphate or glycolysis, but its supply rate is slower.

LO5 How Do Plants Trap Energy in Biological Molecules?

We've been talking so far only about how cells extract energy from biological molecules. But how did the energy get trapped in those biological molecules in the first place? It's less obvious, perhaps, but no less amazing that plants use photosynthesis to convert sunlight into chemical energy, store it in the bonds of organic molecules, and use those same molecules to build leaves, stems, flowers, and nectar as well as to fuel the plants' own energy needs and the steps of aerobic respiration taking place in all the plants' cells.

Compare, for a moment, the way you gather energy with the way a plant does it. The plant obtains its energy from the nonliving environment in the form of sunlight and inorganic raw materials, while you must get yours by eating plants or by eating other animals that ate plants. Biologists classify organisms that take in preformed nutrient molecules from the environment as heterotrophs (*hetero* = other + *troph* = feeder). Heterotrophs include many prokaryotes; most protists (such as protozoa); mushrooms, yeasts, and other fungi; and all animals. In contrast to heterotrophs, autotrophs (*auto* = self) are organisms that take energy directly from the nonliving environment and use it to synthesize their own nutrient molecules. Autotrophs include photosynthetic organisms—green plants and certain protists and prokaryotes that obtain energy from sunlight. Autotrophs also include a small group of *chemosynthetic* organisms—a few kinds of prokaryotes that extract energy from inorganic chemicals such as hydrogen gas and hydrogen sulfide with its rotten-egg smell. We can see that, ultimately, autotrophs are the source of all energy that flows through living systems, since heterotrophs obtain nutrients either from autotrophs or from heterotrophs that once consumed autotrophs.

Physical Characteristics of Light

Visible light is just a small part of the electromagnetic spectrum, which is the full range of electromagnetic radiation in the universe, from highly energetic gamma rays to very low–energy radio waves. Such radiation travels through space behaving both as particles called photons and as waves. The amount of energy in a photon determines its wavelength, the distance it travels during one complete vibration.

Photons of visible light have wavelengths in a narrow range: If the entire electromagnetic spectrum wrapped once around Earth, the visible part would be the length of your little finger. Living things can absorb and use light within the restricted wavelength range of visible light. Gamma rays are so energetic that they disrupt and destroy biological molecules they strike, while radio waves are so low in energy that they do not excite biological molecules.

heterotroph
(Gr. *heteros*, different + *trophos*, feeder) an organism, such as an animal, fungus, and most prokaryotes and protists, that takes in preformed nutrients from external sources

autotroph
(Gr. *auto*, self + *trophos*, feeder) an organism, such as a plant, that can manufacture its own food

electromagnetic spectrum
the full range of electromagnetic radiation in the universe, from highly energetic gamma rays to very low-energy radio waves

photons
a vibrating particle of light radiation that contains a specific quantity of energy

absorption spectrum
the wavelengths of light absorbed by a pigment

chlorophyll
(Gr. *chloros*, green + *phyllon*, leaf) light-trapping pigment molecules that act as electron donors during photosynthesis

carotenoids
plant pigments that absorb green, blue, and violet wavelengths and reflect red, yellow, and orange light

Chlorophyll and Other Pigments Absorb Light

Various things absorb certain colors of light and reflect others. For instance, a flower's petals are red because pigment molecules in the petal absorb light from various parts of the visible spectrum and reflect only red light. This range of absorbed light is called the absorption spectrum, and is unique for each pigment. The green pigment in leaves is called chlorophyll, and it takes part in photosynthesis as well as giving a green leaf its color.

Chlorophyll is often accompanied by colorful carotenoids, pigments which absorb green, blue, and violet wavelengths and reflect red, yellow, and orange light. Carotenoids are generally masked by chlorophyll and thus tend to be unnoticed in green leaves. However, they give bright and obvious color to many

chloroplast
an organelle present in algae and plant cells that contains chlorophyll and is involved in photosynthesis

stroma
in chloroplasts, the space between the inner membrane and the thylakoid membranes

thylakoid
(Gr. *thylakos,* sac + *oides,* like) a stack of flattened membranous disks containing chlorophyll and found in the chloroplasts of eukaryotic cells

nonphotosynthetic plant structures, such as roots (carrots), flowers (daffodils), fruits (tomatoes), and seeds (corn kernels). Chlorophyll absorbs all but green wavelengths of light, and carotenoids all but the red, orange, and yellow wavelengths. Functioning together in pigment complexes, chlorophylls and carotenoids can absorb most of the available energy in visible light. The importance of these pigments, of course, is not just that they absorb light, but what becomes of that captured energy.

The Chloroplast: Solar Cell and Sugar Factory

What gives a plant the ability to gather sunlight in its leaves? The photosynthetic pigments in green leaves are concentrated in layers of green cells that carry out photosynthesis (Fig. 3.9a–c). These cells contain **chloroplasts**, green organelles in which both the energy-trapping and carbon-fixing reactions of photosynthesis take place (Fig. 3.9c,d). Each leaf cell may contain about 50 chloroplasts, and each square millimeter of leaf surface may contain more than half a million of the green organelles. Chloroplasts are similar to mitochondria in several ways: Both are elongated organelles with an inner and outer membrane and interior flattened sacs (see Fig. 3.7 and Fig. 3.9d); both carry out energy-related tasks in the cell; and both have their own DNA in circular chromosomes. However, while mitochondria are "powerhouses" that generate ATP, chloroplasts are both solar cells and sugar factories that capture sunlight and generate sugar-phosphates and other carbohydrates.

Chloroplast Membranes

Chloroplasts have an outer membrane and inner membrane that lie side by side and that collectively enclose a space filled with a watery solution, the **stroma** (see Fig. 3.9d). A third membrane system lies within the inner membrane and forms the **thylakoids**, a complicated network of stacked, disklike sacs, interconnected by flattened channels. Chlorophyll and other colored pigments are embedded in the thylakoid membrane and make this membrane the only part of an entire plant that is truly green. The fact that we see most leaves as green shows how incredibly abundant thylakoids are in nature. In addition to pigments, the thylakoid membrane contains members of an electron transport chain and, in some areas, many copies of an ATP-synthesizing enzyme (Fig. 3.9f).

An Overview of Photosynthesis

There is a beautiful symmetry to the metabolic processes of respiration and photosynthesis that is revealed by their nearly opposite overall equations.

Figure 3.9
Leaves and Photosynthesis

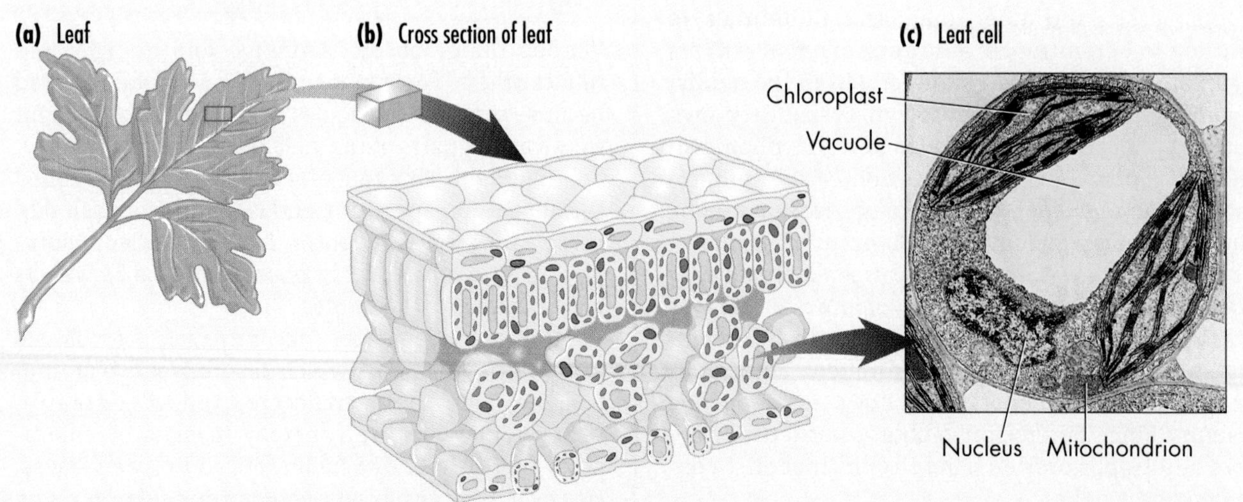

Earlier, we saw that the aerobic cellular respiration could be summarized like this:

Glucose + Oxygen + ADP + Phosphate →
Carbon dioxide + Water + ATP

The process of photosynthesis is a nearly opposite equation:

Carbon dioxide + Water + Light energy →
Glucose + Oxygen

Recall that when oxygen is present, aerobic respiration in mitochondria breaks down glucose into carbon dioxide and water and releases chemical energy that becomes stored in ATP. Nearly the reverse takes place in photosynthesis: the chloroplast traps light energy, transforms it into chemical energy, and then uses that chemical energy to convert carbon dioxide and water into sugars, releasing *oxygen* as a waste product.

Clearly, living things must have both a source of energy and a means of releasing it, and for green plants and most other autotrophs, the direct energy source is sunlight. To understand photosynthesis, we'll follow the path of light striking a leaf and track the electrons whose energy level is boosted by sunlight. That pathway has two phases: a light-trapping phase and a carbon-fixing phase.

The Light-Trapping Phase of Photosynthesis

When sunlight reaches the leaf of a plant, some of the solar energy strikes chlorophyll or other colored pigment molecules in the chloroplasts and becomes trapped as it boosts electrons in the pigments to higher energy levels (Fig. 3.10a, Step ①). The electrons leave the chlorophyll and pass down an electron transport chain (Step ②). As the electrons travel down the electron transport chain, they release their energy bit by bit, and this energy is then stored in the chemical bonds of ATP and the electron carriers (Step ③). The "hole" in the chlorophyll left by the energized electrons is filled by electrons stripped from a water molecule (H_2O) (Step ④). The hydrogens from the water molecule stay in the chloroplast, but the oxygen is released to the atmosphere, where plants and animals can use it in aerobic respiration. These events make up the first phase of photosynthesis, the **light-dependent reactions**, also known as *energy-trapping reactions*. The reactions are driven by light energy and can take place only when light is available, and they produce oxygen, ATP, and energized electron carriers.

light-dependent reactions
(also known as *energy-trapping reactions*) the first phase of photosynthesis, driven by light energy; electrons that trap the sun's energy pass the energy to high-energy carriers such as ATP where it is stored in chemical bonds

carbon-fixing reactions
(also known as *light-independent reactions* or the *Calvin-Benson cycle*) the second stage of photosynthesis in which the energy trapped and converted during the light-dependent reactions is used to combine carbon molecules into sugars

The Carbon-Fixing Phase of Photosynthesis

Now we move to the second part of the story. The ATP and electron carriers produced by the energy-trapping reactions supply the energy needed for the second phase of photosynthesis, a biochemical cycle called the **carbon-fixing reactions** (Fig. 3.10b). (These

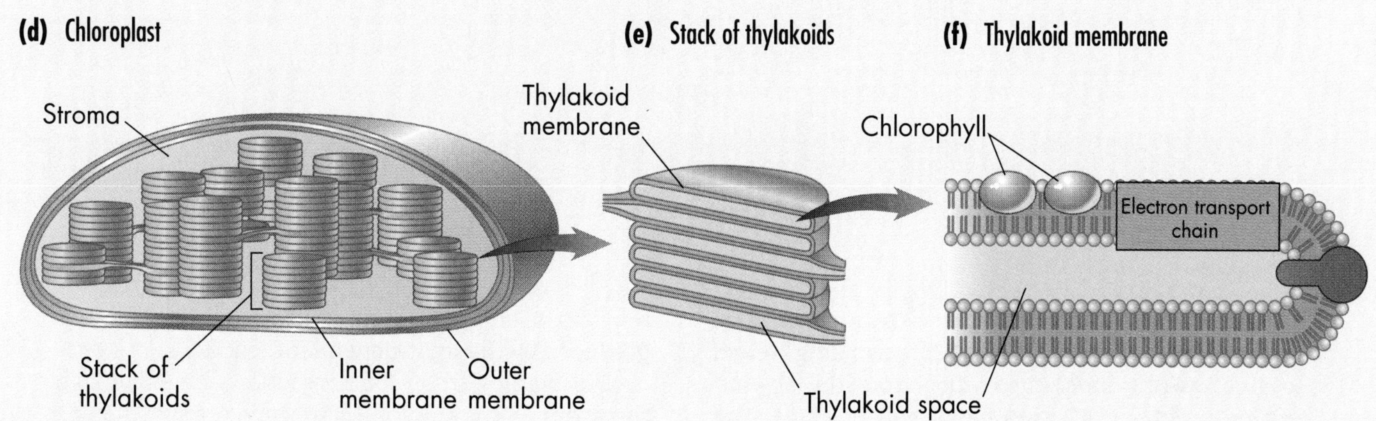

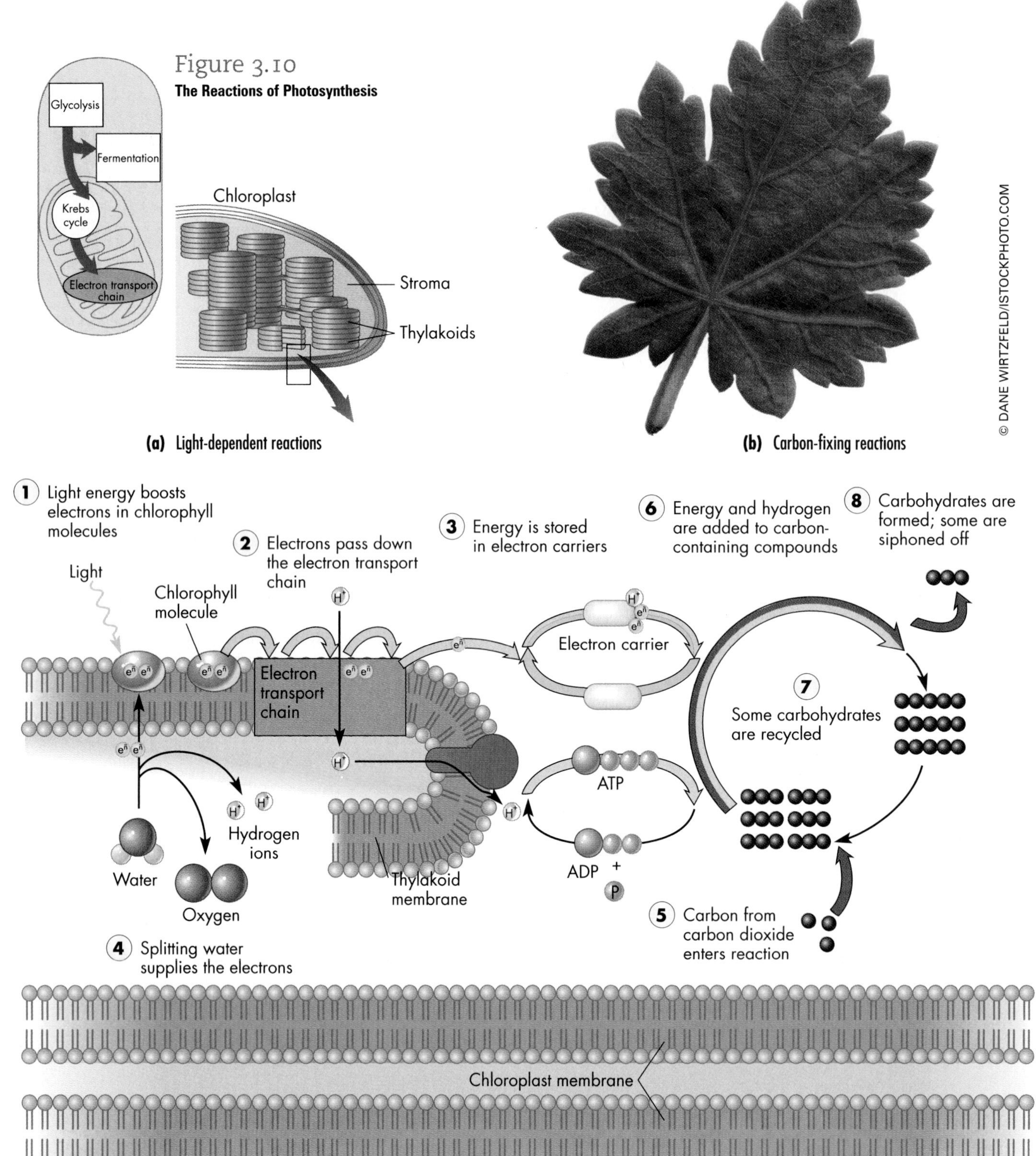

Figure 3.10

The Reactions of Photosynthesis

(a) Light-dependent reactions

(b) Carbon-fixing reactions

are also called the *Calvin-Benson cycle* or sometimes the *light-independent reactions*.) "Carbon fixing" refers to a cell taking inorganic carbon from the air and joining it ("fixing it") to a biological molecule. The carbon-fixing reactions can go on day or night because they do not directly require light energy; they only require the energy carriers ATP and electron carriers produced by the light-dependent reactions.

During the carbon-fixing reactions, an enzyme in the stroma of the chloroplast first adds carbon dioxide from the air to a previously formed five-carbon compound, making a six-carbon compound that immediately breaks into two three-carbon compounds (Fig. 3.10b, Step ⑤). Then the chloroplasts transfer the energy stored in the bonds of ATP and electron carriers, and the hydrogens from the electron carriers, to the newly made three-carbon compounds (Step ⑥). Some of the newly formed three-carbon molecules are joined together and rearranged to regenerate the original starting molecules of the cycle (Step ⑦), while others can be siphoned off in energy-storing carbohydrate molecules (Step ⑧). Chloroplasts have to run this cycle with three carbon dioxide molecules to get out one three-carbon carbohydrate molecule. Cells can use the newly available carbohydrate to fuel the plant cell's own survival activities (and those of nonphotosynthetic plant parts such as roots) via the energy-harvesting steps of glycolysis and aerobic respiration in the plant's mitochondria. Or the plant cell can make sugar, as in flower nectar; cellulose, a structural material in cells walls, stems, leaves, and other plant parts; or starch, a form of long-term energy storage. Thanks to the formation of cellulose, we have the paper and wood we use daily. And the plant starch stored in rice, wheat, oats, potatoes, corn, and other crops are staples of the human diet.

Who knew?

Eating "low on the food chain" means eating more vegetables and less meat. Because energy is lost as heat every time it is converted to a different form, the amount of energy represented by a vegetarian meal is less than one composed of meat, since the loss of heat energy (entropy) must be included. The amount of energy from the sun is fixed; losing less food energy to heat by eating veggies means more energy is conserved, less land is required to feed one individual, and more vegetarians can survive than meat-eaters.

The Global Carbon Cycle

It's hard to imagine how the metabolic pathways inside microscopic cells could possibly help perpetuate a cycle of planetary proportions. But a global carbon cycle moves vast amounts of carbon-containing compounds through the atmosphere, soil, water, and living organisms based on the carbon-fixing activities of autotrophs, the release of carbon dioxide by heterotrophs, and on geological phenomena such as erosion. You'll read more about the global carbon cycle in Chapter 15.

It is also hard to imagine that people could be altering that vast carbon cycle through their activities. But we are releasing millions of tons of extra carbon dioxide into the atmosphere each year by burning rain forests to clear new agricultural land and burning carbon-containing fossil fuels like coal and oil in factories and cars. This extra carbon dioxide causes extra heat to be trapped in the atmosphere. Scientists call this process the *greenhouse effect*. The greenhouse effect appears to be causing a slow but steady increase in average air and water temperatures called *global climate change* (also discussed in Chapter 15).

Aside from the changes in global air and water temperatures, wouldn't an increase in carbon dioxide benefit plants, since they take in the gas as a carbon source? Plant researchers have shown that they can add extra carbon dioxide to the plants growing in a controlled environment and measure an increase in the carbon-fixing reactions of photosynthesis and plants can grow faster. In an equivalent way, supplying more gasoline will speed up a car's engine. The problem is, some kinds of plants might respond better than others to the increased carbon dioxide, leading to imbalances as one plant species becomes overpopulated at the expense of others. If the benefit went mostly to desirable crop plants and not to the weeds that tend to choke them out, this might be a welcome outcome. Unfortunately, our limited understanding of natural ecosystems does not allow us to predict which plants would respond to heightened carbon dioxide levels and take over certain ecosystems.

Associated with the rapid increase in carbon dioxide in the atmosphere, we are experiencing an increase in global temperature as well. In fact, all of the century's warmest years have occurred in the past decade. This warming trend appears to be affecting all kinds of plant and animal populations. As Chapter 15 explains in detail, the distribution of animals is changing, along with certain disease-causing organisms. What this next century of warming will mean to populations of hummingbirds and to our own burgeoning numbers—we can only guess.

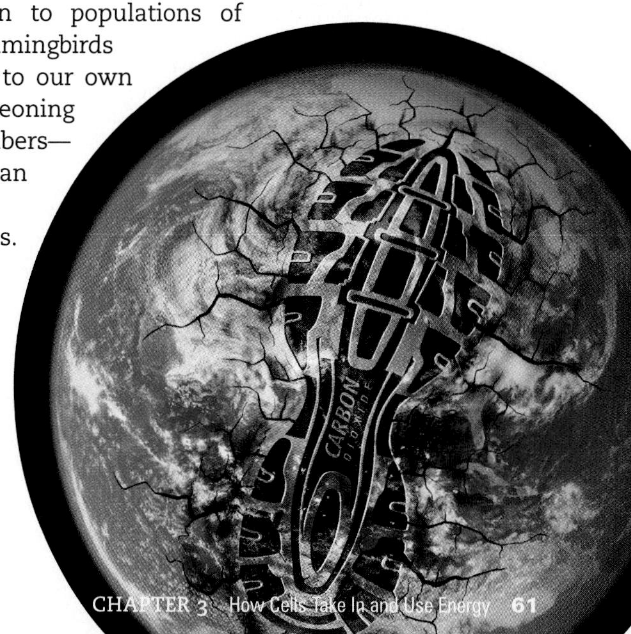

4

The Cell Cycle

Learning Outcomes

LO[1] Understand the patterns of cell growth and cell division

LO[2] Explain mitosis and the mechanisms of cell division

LO[3] Identify what controls cell division

LO[4] Describe meiosis and the cell divisions that precede sexual reproduction

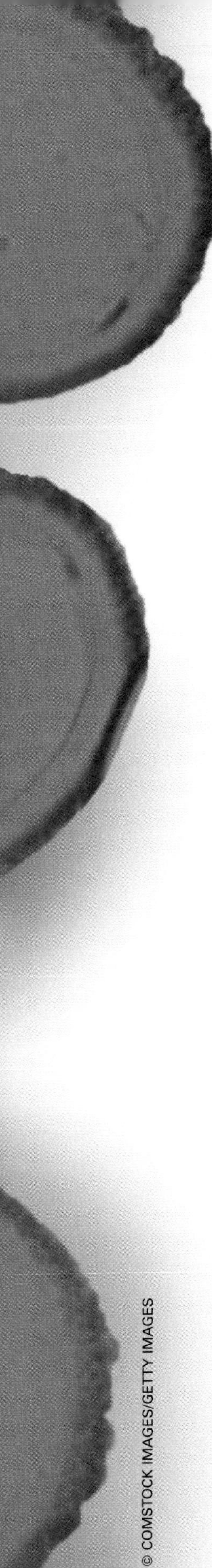

"There are two basic types of cell division in humans and other eukaryotic organisms: mitosis and meiosis. Of the two types, cancers involve the cell divisions of mitosis."

An Unavoidable Loss of Control

During the summer of 1972, 17-year-old Patricia's dermatologist found a small, pearly pink bump on her big toe and diagnosed it as a *basal cell carcinoma*. This patchy overgrowth of cells is the most common form of skin cancer—in fact, it is the most frequently diagnosed human cancer. The typical skin cancer patient, however, is middle aged or older. So finding a lesion in a teenager was somewhat unusual—fair-skinned though Patricia was and susceptible to sunburns. The discovery of a second basal cell carcinoma on her temple a few weeks later was more ominous. It inspired a detailed search of her entire skin surface—and the doctor found yet more tiny, pearly cancers.

What do you know?

Where in your body are cells dividing by mitosis? By meiosis?

Their multiple occurrences in a girl of her age was diagnostic of a disease called *Basal Cell Nevus Syndrome (BCNS)*—a rare inherited condition that starts in adolescence and is characterized by numerous skin cancers throughout life. Today, at age 44, Patricia still sees a dermatologist every few weeks and, she says, she "can easily have 10 new carcinomas every visit." But she's accepted her situation with remarkable grace. Patricia remains hopeful that "still in my lifetime, some treatment will come up" to alleviate the disease she's been fighting for so long.

That treatment could come through the study of BCNS patients at the University of California, San Francisco—a study Patricia has been part of for years. The study's principal investigator, Ervin Epstein, is trying to find answers for the millions of people who will develop one or more basal cell carcinomas during their later years. If caught early enough, these cancers are "essentially 100 percent curable," Epstein explains. But his goal is to thoroughly understand how cells grow and divide at the level of genes, proteins, and other biological molecules. With that detailed knowledge of growth and division, he and his colleagues could possibly devise much more effective treatments and even preventative measures for skin cancers.

Patricia's long battle with skin cancer and Ervin Epstein's study of basal cell carcinoma are fitting subjects for this chapter. Here we explore the cell cycle—the stages of growth, duplication, and division in living cells. There are two basic types of cell division in humans and other eukaryotic organisms: mitosis and meiosis. *Mitosis* takes place in somatic (non-sex) cells, and *meiosis* occurs in gametes (sex cells). Of the two types, cancers involve the cell divisions of mitosis. As you explore the cell cycle, you'll see that a cancer starts when the normal controls over mitosis go haywire within a single cell. Instead of starting and stopping at appropriate times, cancerous cell division goes on and on until a mass of cells called a *tumor* results and displaces or invades other tissues.

cell cycle the events that take place within the cell between one cell division and the next

epidermis
the outer layer of cells of an organism

dermis
the skin layer just below the epidermis or outer layer; the dermis contains tiny blood vessels, sweat glands, hair roots, and nerve endings

subcutaneous
literally below the skin; often refers to fat (adipose) cells that bind the skin to underlying organs found below the dermis layer

basal layer
the deepest layer of the epidermis, the outer layer of the skin; the basal layer contains cells that divide and replace dead skin cells; also called *germ layer*

basal cells
cells that lie in a layer just below the epidermis

The key to understanding both cancer and the normal growth and reproduction of cells and organisms lies in the intricacies of the cell cycle. That makes it one of the more important and fascinating topics in modern biology.

As we move through this chapter on the cell cycle, you'll find the answers to these questions:

- What are the patterns of normal cell growth and cell division?
-  What is mitosis and what happens during cell division?
-  What controls the timing and location of cell division?
-  What is meiosis, the special cell divisions that precede sexual reproduction, and what happens during this process?

LO1 Patterns of Cell Growth and Cell Division

To understand how skin cancers arise, we first need to understand normal cell growth and cell division in skin. Then we can compare these typical patterns to the rare case of genetically determined BCNS in which cell growth and division patterns are abnormal and the cell cycle goes on uncontrolled.

Where and When Do Skin Cells Divide?

Your skin is your body's largest organ, and it forms a wonderful protective barrier around the muscles and other tissues. The outer skin surface layer is continually abraded away by the rubbing of clothes or the scrubbing of a washcloth in the shower. How does the skin normally replace all this lost exterior material? The answer is a controlled replacement process of the outermost layers by deeper, dividing layers.

Skin consists of two main zones, a thin outer zone, or epidermis, and a thicker zone underneath, or dermis. Below the dermis is a subcutaneous layer, mostly fat (adipose) cells that bind the skin to underlying organs. The dermis contains tiny blood vessels, sweat glands, hair roots, and nerve endings sensitive to heat and touch. None of these structures plays a direct role in the tumors of BCNS.

The epidermis is the skin's protective zone and it consists of several layers (Fig. 4.1). The layer nearest the dermis is a basal layer or dividing layer (also called the *germ layer*). Because this dividing region lies at the base of the epidermis, the cells there are sometimes called basal cells. This dividing layer is close to the blood in the dermis and is the only part of the epidermis that contains reproducing cells.

Figure 4.1
Where Do Cells Divide in Skin?

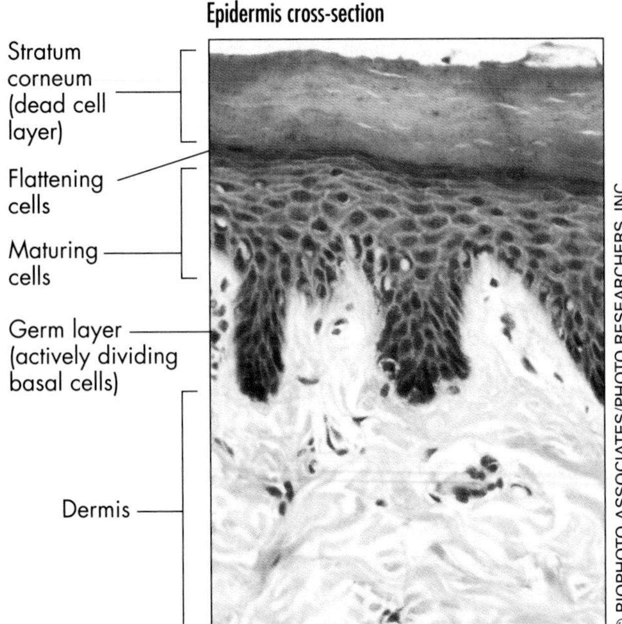

Figure 4.2
Where Do Basal Cell Skin Cancers Arise?

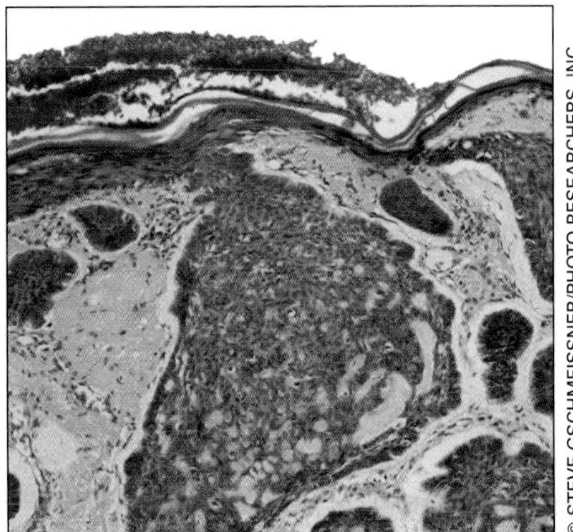

As these cells divide, some of them get pushed up into a layer of cell differentiation where they take on their mature characteristics. As they mature, the cells become tightly joined by cell junctions (review Fig. 2.14), they accumulate large quantities of the cytoskeletal protein keratin, and they become very flat, like miniature cookie sheets or pizza pans. As they move up further toward the body surface, their supply of nutrients becomes limited, and the cells die. The outermost portion of the epidermis is a dry layer consisting of sheet after sheet of flat, dead cells that are easily rubbed away.

To determine which cells give rise to a skin tumor, a surgeon can remove a basal cell carcinoma—say, on the palm of the hand—with a sharp instrument and then slice the cluster of cells very thinly to observe sections of it (Fig. 4.2). Comparing Figure 4.2 to the normal skin section in Figure 4.1, what do you see? You should notice a clump of cells in the dermis with the characteristics of cells that belong in the epidermis. What has happened is that basal cells from the dividing germ layer of the epidermis have divided too many times and made a mass that extends downward into the underlying dermis, an "overgrowth" process biologists call *proliferation* (an increase in cell numbers).

The Cell Cycle: Action in the Dividing Layer

The cells in the skin's dividing layer share a general strategy of cell reproduction with other living cells: they take in nutrients, increase in size, and then divide into two daughter cells. These two daughter cells may then, themselves, go through a period of growth followed by division. This alternation of growth and division is called the *cell cycle* (Fig. 4.3). During the division phase of the cell cycle, there is a partitioning of the cells' internal organelles, and then a dividing of the cell in two. The growth period between two division phases is called interphase.

As we saw in Chapter 2 (Fig. 2.10), the material inside of a cell is not a homogeneous mass like a loaf of bread that you can just slice to yield halves containing very similar materials. Cells contain various organelles in differing amounts and places. Some of these organelles, like the ribosomes, mitochondria, Golgi apparatus, and so on, are present by the thousands or millions in each cell. If the cell is just cleaved in two, each half will have sufficient numbers of these organelles to ensure smooth functioning. Some organelles, however, are present in a single copy and may be offset into one part of the cell. These organelles include the chromosomes inside the nucleus with its genetic contents and the centrioles in the cytoplasm. Special mechanisms have evolved that distribute these "singleton" organelles correctly to daughter cells.

division phase
during the cell cycle, the partitioning of the cells' internal organelles, and the dividing of the cell in two

interphase
the period between cell divisions in a cell; during this period, the cell conducts its normal activities and DNA replication takes place in preparation for the next cell division; interphase is divided into three periods: G_1, S, and G_2

chromosome
a self-duplicating body in the cell nucleus made up of DNA and proteins and containing genetic information; a human cell contains 23 pairs of chromosomes; in a prokaryote, the DNA circle that contains the cell's genetic information

Figure 4.3
Overview of the Cell Cycle

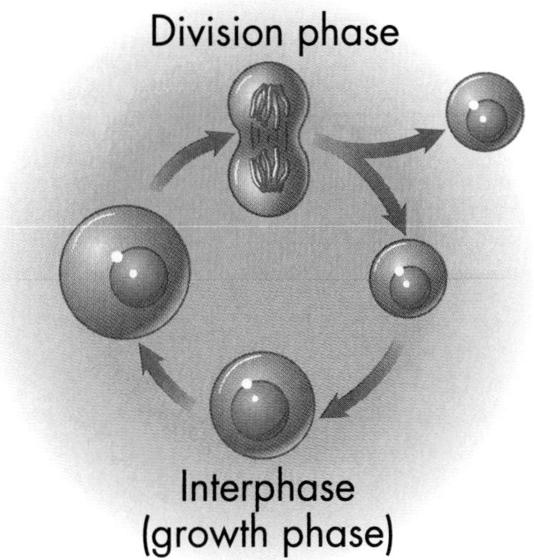

Chromosomes

Patricia inherited her BCNS; this means her cells include the instructions for the disease in their genetic information. As we saw in Chapter 2, chromosomes contain a cell's hereditary information—the instructions the cell uses to construct itself—and thus the organism of which it is a part. We'll look at chromosomes in more detail in Chapter 6. The important thing, here, is that cells need an orderly distribution mechanism so that as they divide, each of the two daughter cells receives the same hereditary information and both have the same information as did the parent cell.

All organisms contain chromosomes, although the size, shape, and number of these hereditary structures differ from species to species. Eukaryotic cells have 2 or more chromosomes: a cell in a roundworm contains 4 chromosomes; the nucleus of a cell

chromatid a daughter strand of a duplicated chromosome; duplication of a chromosome gives rise to two chromatids joined together at the centromere

replication the copying of one DNA molecule into two identical DNA molecules

alignment the positioning of chromosomes on the mitotic or meiotic spindle

separation the movement of chromatids or chromosomes away from each other toward opposite poles of the dividing cell during mitosis or meiosis

sister chromatids the two rods of a replicated chromosome

from a giant sequoia leaf contains 22; a goldfish nucleus contains 104; and the nucleus in a human skin cell has 46.

Chromosomes in the Cell Cycle

When a cell divides, each new offspring cell receives its own set of chromosomes containing an identical copy of the hereditary material. For this reason, the structure, duplication, and distribution of the chromosomes are central to the cell cycle. Chromosomes consist of protein and DNA, the long molecule that bears hereditary information. You can see chromosomes as distinct bodies by looking through a microscope, but they are visible only during a certain portion of the cell cycle, the division phase (review Fig. 4.3). At this time, the DNA is wound up into tight bundles we can see—bundles that resemble a kite string wound around a spool. The rest of the time, in interphase, the DNA is unwound and spread out, like a loose pile of kite string on the ground. It's impossible to visualize individual chromosomes in this state.

At the end of the division phase, each chromosome consists of a single long rod called a **chromatid** (Fig. 4.4b, Step ①). During interphase, the DNA in the chromosome undergoes **replication**, the copying of one DNA molecule into two identical DNA molecules (Step ②). When chromosomes become visible once again at the beginning of the division phase (Step ③), each chromosome consists of two rods (two chromatids), with each rod containing just one double-stranded DNA molecule. (Note that replication by itself does not change the number of chromosomes in a cell. Each chromosome simply doubles from one chromatid to two.) At this stage, chromosomes undergo **alignment**, aligning themselves in the middle of the cell as we will investigate shortly. Finally, in Step ④, **separation** takes place; the replicated chromosome with its two chromatids splits into two chromosomes, each with one chromatid made up of a double-stranded DNA molecule.

To summarize, the chromosome cycle involves one rod replicating to two rods during interphase, the replicated chromosome aligning in the cell, and then the two rods separating to different cells during the division phase.

Chromosome Structure

Now, let's look a bit closer at the anatomy of a chromosome. Figure 4.4a shows a sketch of a two-chromatid chromosome in the division phase of the cell cycle. You can see that each individual rod is formed by great loops of DNA complexed with protein that extend out from the axis of the condensed chromosome. In these loops, the DNA winds around proteins like string wrapped around thousands of tiny spools. Because each individual rod, or chromatid, in a chromosome contains a single extraordinarily long double-stranded DNA molecule, this packaging decreases the tangling of DNA during chromosome alignment and separation in the division phase.

The chromosomes we have just seen consist of two identical rods, called **sister chromatids**, held

Figure 4.4
Chromosome Structure and the Chromosome Cycle

(a) Sketch of a chromosome

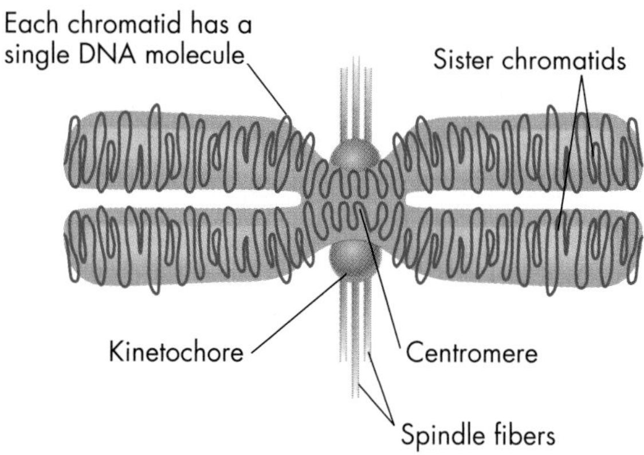

(b) The chromosome cycle

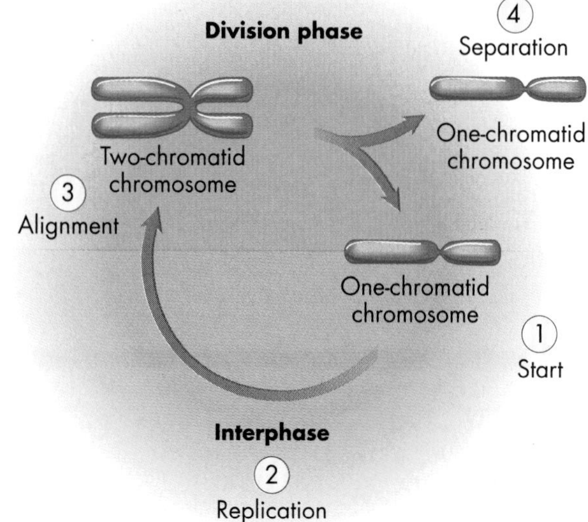

together at a single point, the centromere (Fig. 4.4a). Located at the centromere is a group of proteins called the kinetochore, and this structure attaches to long fibers in the cell called spindle fibers (Fig. 4.4a). These spindle fibers move chromosomes around, for reasons we'll see shortly. Each chromatid has just one double-stranded DNA molecule (Fig. 4.4a), and the DNA molecules in two sister chromatids are identical. As cell division progresses, the two sister chromatids separate from each other (Fig. 4.4b, Step 4). One of the sister chromatids ends up in one of the daughter cells, and the other, genetically identical chromatid ends up in the other daughter cell. Thus, immediately after division, each chromosome consists of a single chromatid with a single DNA molecule.

Interphase: The Growth Period in the Cell Cycle

Cells proliferate unchecked in basal cell carcinomas like the ones Patricia Hughes develops. Because the alternating periods of growth and division are so central to understanding this proliferation, we must continue exploring the stages of the cell cycle. Recall that the cell cycle has a division phase and a period of growth, the interphase. Biologists divide interphase into three portions called G_1, S, and G_2. G_1 and G_2 stand for *Gap 1* and *Gap 2*, because they are gaps in the cell cycle that come between the time of chromosome division and the time of chromosome replication. The S stands for *synthesis* of DNA. And biologists usually call the division phase M, for *mitosis* (Fig. 4.5).

Different cells require different amounts of time to pass through the entire cell cycle. Certain cells, like mature nerve cells of the brain, arise in the embryo and never divide again. In contrast, cells in your bone marrow replace worn-out blood cells by dividing every 18 hours or so. Cells that line your stomach divide in a cycle about 24 hours long. Most skin cells, even in the dividing basal layer, replicate only every week or so. A basal cell carcinoma, however, can undergo new rounds of the cell cycle every 67 hours. Clearly, the cell cycle can accelerate (we'll discuss the reasons later).

G_1: Active Growth

G_1 and G_2 are important growth phases. During the G_1 phase, cells manufacture new proteins, ribosomes, mitochondria, and other cell components in preparation for DNA synthesis and cell division. The length of the G_1 phase determines the length of the entire cell cycle. G_1 can be quite short or very long, depending on the type of cell, its role in the organism, and conditions in its environment. For example, skin cells normally have a long cell cycle and a long G_1 (a few days). That can change, however. If a wound removes some of those cells, the G_1 phase may shorten in the skin cells at the edge of the wound, speeding up growth and division and enabling the wound to heal rapidly. (Skin cancer cells have an accelerated cycle a bit like a wound that never heals.) Plants can have an analogous process. In an aspen tree, if a deer or beaver nibbles away the bark, the bark-forming cells below can enter a shortened G_1 phase, rapidly producing new bark, which protects the damaged area. Unlike in cancer cells, however, in both normal skin and tree bark the faster cell cycling stops when the wound is healed. Following G_1, eukaryotic cells enter the synthesis phase, S.

centromere the point on the chromosome where the spindle attaches and also where the two chromatids are joined

kinetochore a group of proteins located at the centromere that attaches to long fibers called spindle fibers

spindle fibers the cytoskeletal rods (microtubules) that move chromosomes toward the cell poles and that cause cell poles to separate during cell division

G_1 phase the portion of the cell cycle that follows mitosis but precedes DNA synthesis

S phase the portion of the cell cycle during which the cell synthesizes DNA

Figure 4.5
Phases of the Cell Cycle

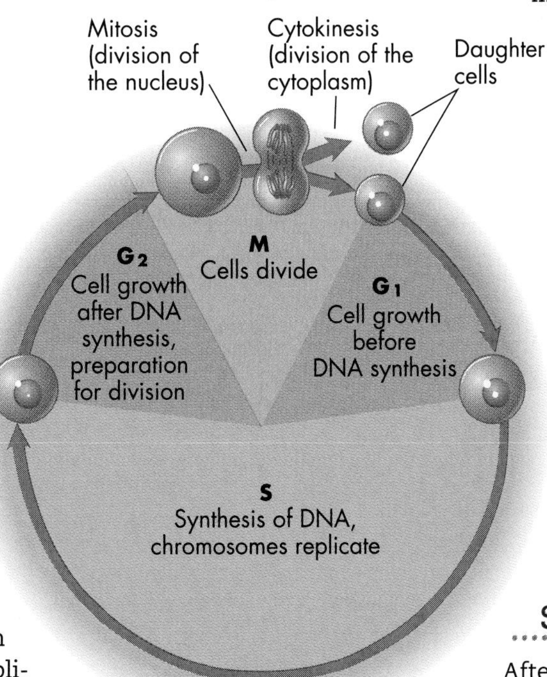

S: Synthesis of DNA

After the G_1 phase, cells enter the S phase, during which enzymes replicate the double-stranded DNA

G_2 phase
the portion of the cell cycle that follows DNA synthesis but precedes mitosis

M phase
the portion of the cell cycle during which the nucleus divides by mitosis and the cytoplasm divides by cytokinesis

mitosis
the process of nuclear division in which replicated chromosomes separate and form two daughter nuclei genetically identical to each other and the parent nucleus; mitosis is usually accompanied by cytokinesis (division of the cytoplasm)

cytokinesis
the process of cytoplasmic division following nuclear division

prophase
the first phase of nuclear division in mitosis or meiosis, when the chromosomes condense, the nucleolus disperses, and the spindle forms

metaphase
the period during nuclear division (mitosis) when the spindle microtubules cause the chromosomes to line up at the center of the cell

molecule in each chromosome (see Chapter 6 for details). Cells also synthesize certain proteins necessary for maintaining chromosome structure as the S phase proceeds. When the S phase ends, each chromosome consists of two identical and parallel double-stranded DNA molecules packaged into two chromatids. Every stretch of DNA in a chromosome is copied once and only once during each S phase. After copying is complete, the cell enters G_2. As you might imagine, processes occurring during S phase provide physicians with important targets for cancer therapies. Nondividing cells don't go through S, so agents that block the events of the S phase affect only dividing cells, such as Patricia's skin tumors. Such therapeutic agents include substances that mimic DNA subunits and compounds that block DNA synthesis.

G_2: Preparation for Division

During the **G_2 phase**, the cell continues to synthesize many proteins. If a researcher artificially blocks this synthesis, the cell fails to divide, suggesting that some proteins synthesized during G_2 promote mitosis, division of the nucleus. When all the necessary proteins have been synthesized, the cell leaves the final growth phase and begins to divide.

M: The Cell Divides

The **M phase** generally consists of two main events: **mitosis**, the division of the nuclear material, and **cytokinesis**, the division of the cytoplasm (Fig. 4.5). Both mitosis and cytokinesis are vitally important for the equitable distribution of genetic material and other cell components, so we'll talk about each one separately. But keep in mind what's at work: mechanisms to insure that a cell gives rise to identical daughter cells with the same genetic information and role in the organism. Without these carefully controlled and timed divisions, nothing would keep a toe cell from becoming a liver cell, or a skin cell from becoming a tumor.

LO2 Mitosis and the Mechanisms of Cell Division

In the previous section, we've seen that cells in the lower, basal layer of the epidermis pass through the cell cycle, thereby generating new skin. If their growth is uncontrolled, the cells instead grow into basal cell carcinomas. The division phase, or M phase, is therefore a crucial part of the cell cycle. So let's see how a normal cell divides.

The Phases of Mitosis

Whether in the skin or other tissue, the mitotic dance of the chromosomes goes on continuously in a dividing cell. Biologists have named several prominent phases of mitosis to simplify its description. The events of these phases are summarized here and illustrated and explained in more detail in Figure 4.7 on pages 70–71. Recall from Figure 4.4 that chromosome replication has already occurred during the S phase, before mitosis begins.

Prophase

In **prophase** (*pro* = before), the chromosomes condense and become visible, the nucleolus (a dense organelle within the nucleus; review Fig. 2.10) disappears, and a mitotic spindle forms (Fig. 4.7a–c). The mitotic spindle is a bundle of certain filaments of the cytoskeleton (microtubules) that suspends and moves the chromosomes. In late prophase, a stage cell biologists call prometaphase (*meta* = middle), the nuclear envelope disappears, the spindle enters the nuclear region, and the spindle attaches to the chromosomes at the centromere (Fig. 4.7c). Individual chromosomes jostle back and forth, as if they were involved in a tug-of-war between the two poles.

Metaphase

In **metaphase**, the spindle microtubules align the chromosomes in the middle of the spindle, each chromosome lined up independently of the others in a single plane, called the metaphase plate, in the middle of the spindle (Fig. 4.7d). The metaphase plate is a bit like the flat surface of a grapefruit that has been sliced in half.

Anaphase

In **anaphase** (*ana* = apart, opposed), the centromeres split and the spindle microtubules separate the chromatids (now called chromosomes) and pull them toward opposite poles (Fig. 4.7e). Figure 4.8 on page 72 describes in more detail the role of the spindle in separating the chromatids during anaphase.

Telophase

In **telophase** (*telo* = goal), the chromosomes arrive at opposite poles of the cell, and the preparatory events are reversed: the nuclear envelope reappears, the spindle dissolves, and so on (Fig. 4.7f,g). Once telophase is over, the division of the cell nucleus (mitosis) is complete. Now the cell has two nuclei carrying identical sets of chromosomes. The M phase continues, however, with cytokinesis, the division of the cytoplasm (Fig. 4.7h).

> ### "Bye Bye, Parent Cell"
>
> There is a key difference between the replication of a cell involving mitosis and the reproduction of a person or a plant involving meiosis. When these large, complex organisms reproduce, the parent and offspring both generally continue to exist. In contrast, during cell reproduction, the parent cell (Fig. 4.7a) ceases to exist as an entity and its parts are distributed to the two offspring cells (Fig. 4.7h).

The Mitotic Spindle

Figure 4.7 shows where the chromosomes move during mitosis. It doesn't, however, reveal the mechanism that causes the chromosomes to move. This mechanism involves the **mitotic spindle**, a microscopic scaffolding of fibers made of microtubules that suspends and moves the chromosomes (Fig. 4.8). *Spindle fibers* extend from the **centrioles**, organelles consisting of two short cylinders that organize the cell's network of microtubules. Some spindle fibers attach directly to special structures (kinetochores) at the centromeres of chromosomes, and pull the two chromatids toward the poles (Fig. 4.8). Other spindle fibers overlap at the cell's center and cause the two poles to move apart from each other. Knowing how the spindle works is also important in designing anticancer drugs. The physician wants to block mitosis in tumor cells. But how?

Cytokinesis: The Cytoplasm Divides In order for two cells to appear where there was once one, the original cell must cleave in two. Toward the end of mitosis, the cytoplasm of most plant and animal cells begins to divide by means of cytokinesis (literally, "cell movement") (Fig. 4.7f–h). In both animal and plant cells, new plasma membranes form at or near the place once occupied by the chromosomes during metaphase and separate the two nuclei into the two new cells. The details of cytokinesis vary because animal cells have a pliable outer surface, while plant cells have a rigid cell wall.

Cytokinesis in Animal Cells Animal cells divide from the outside in, as a circle of microfilaments called a **contractile ring** pinches each cell in two. Late in mitosis, a ring of filaments containing the contractile protein, *actin*, creates a furrow in the cell surface in much the same way that a purse string tightens around the neck of a purse (Fig. 4.7g–h). The furrow deepens, and eventually squeezes the cell in two.

Cytokinesis in Plant Cells Plant cells, with their rigid cell walls, retain their shape throughout the cell cycle, dividing from the inside out (Fig. 4.6). At the end of

anaphase
a period during nuclear division when the chromosomes move toward the poles of the cell

telophase
the final phase of nuclear division when the chromosomes are at opposite poles of the cell, the nuclear membrane and nucleolus reappear, and the spindle disappears

mitotic spindle
a weblike structure of microtubules that suspends and moves the chromosomes; formed during prophase in mitosis

centriole
pairs of short, rod-shaped organelles that organize the cytoskeletal fibers called microtubules into scaffolds; these intracellular frameworks help maintain cell shape and move chromosomes during cell division

contractile ring
the ring of cytoskeletal elements (actin filaments) that separates one cell into two during the division of the cytoplasm

Figure 4.6

Cytokinesis in Plant Cells

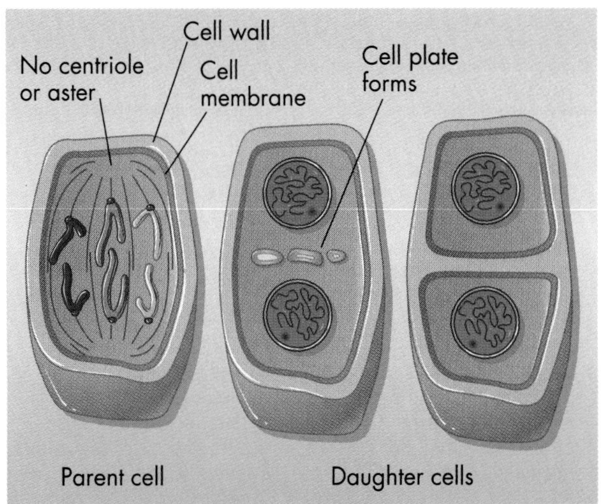

cell plate
in plant cells, a partition that arises during late cell division from vesicles at the center of the cell and that eventually separates the two daughter cells

mitosis, vesicles filled with cell wall precursors collect in the center of the cell. The separate vesicles gradually fuse, forming a central partition, or cell plate, made of cell-wall material sandwiched between plasma membranes. This fusion completes the central partition and divides the plant cell into two identical daughter cells, which remain connected. Each cell now has its own nucleus and is ready to begin interphase.

Applying Our Knowledge of Mitosis to Cancer Therapy

Patricia's multiple skin cancers result from an overgrowth of cells. In her cancer and in all cancers, a cell starts growing and dividing by means of the phases we've just explored, but then continues unchecked, forming a tumor that can displace or invade other cells and tissues. The real key to cancerous cell division is to understand how the normal control of the cell cycle is lost, and we'll return to that shortly. In the meantime, some aspects of mitosis apply to the treatment of cancer once it arises.

The best way to treat most tumors, including BCNS lesions, is to surgically remove them. This makes sense for basal cell carcinomas because, being at the surface, they are readily accessible. Furthermore, while the masses they form crowd other cells, they do not grow especially aggressively, nor do they tend to invade and take over adjacent tissues. Once the skin tumor is removed or zapped through freezing with liquid nitrogen, it seldom returns. Certain other tumors, however, such as melanomas (pigmented skin tumors), and cancers of the breast, ovaries, and pancreas are more aggressive and invasive and often do require other methods in addition to surgery, such as chemotherapy and radiation.

Chemotherapy and the Spindle

We see in Figure 4.8 that the spindle is crucial for moving chromosomes around in mitosis. Does this open an avenue for cancer therapy? Researchers thought

Figure 4.7
Chromosome Choreography: The Stages of Mitosis in an Animal Cell

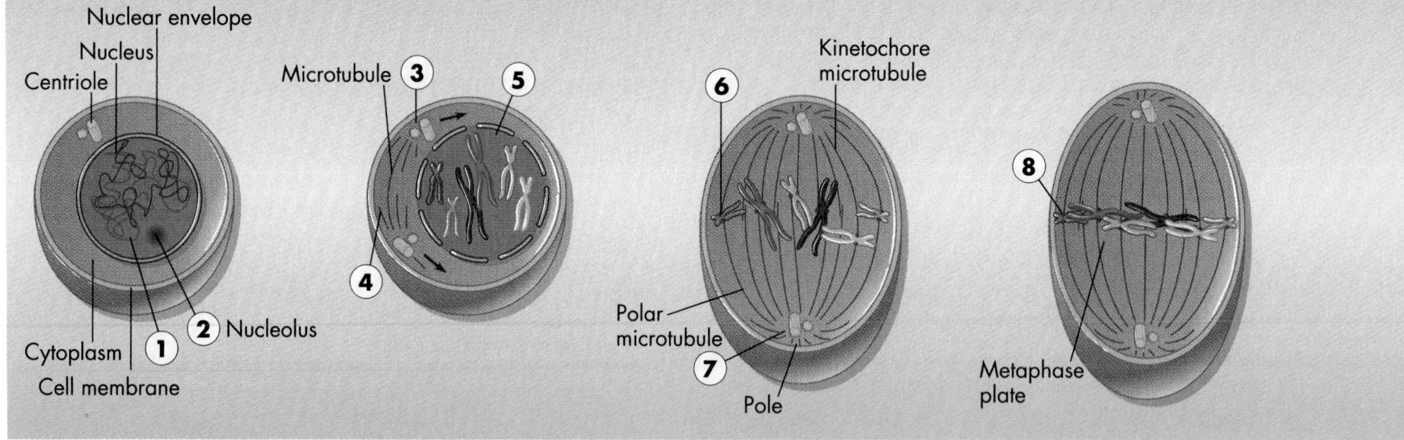

(a) Late interphase

The cell's DNA has already replicated during S in the previous interphase. As the cell enters the first part of mitosis, called *prophase,* the DNA changes from its diffuse and tangled state in interphase (1) to become more tightly packaged. Also, the nucleolus, a dense organelle within the nucleus (2), disperses.

(b) Middle prophase

In animal cells, centrioles, cylindrical groups of short microtubules, duplicate and organize spindle microtubules (3). Plant cells lack centrioles. Centrioles separate and move toward opposite ends, or poles, of the cell, spinning out the mitotic spindle (4). As the nuclear envelope disperses (5), the spindle invades the nuclear region.

(c) Late prophase

Microtubules attach to chromosomes by kinetochores (6). The chromosomes then jostle back and forth as the polar microtubules, which suspend and move the chromosomes, interact with the kinetochore microtubules, and the centrioles complete their migration to the poles (7).

(d) Metaphase

During metaphase the chromosomes become aligned on the metaphase plate (8), a plane lying halfway between each pole.

so. They reasoned that if they could block the action of the spindle, they could block cell division and hence slow cancer growth. In fact, several chemotherapy drugs now used to treat cancer attack the spindle. Taxol and related compounds, extracted from the bark of the Pacific yew tree, have proven useful in treating carcinomas of the cervix and ovary. Taxol binds to microtubules and blocks their breakdown into protein subunits. Thus, a cancerous cell gets stuck part way through mitosis and, with the spindle still in place, can't divide into new malignant daughter cells. Other drugs, like vinblastine from the periwinkle plant, stop the division of cancer cells by blocking the formation of the spindle in the first place. These chemotherapeutic drugs have powerful side effects, however, and so pharmaceutical researchers are always trying to develop new and better treatments.

Radiation Therapy

Have you known someone with cancer who received radiation therapy? During this treatment, technicians aim a radiation beam precisely at the patient's tumor and shield other parts of the patient's body from the beam. How does radiation therapy work? And why is it necessary to shield the patient's body?

Radiation can break chromosomes. If such breakage occurs in a cell that does not divide, the cell will still be able to carry on in a normal fashion because most body cells contain two copies of each chromosome. If chromosome breakage takes place in a cell that later divides—a skin cell, for example, or a tumor cell—the consequences are often quite different, and your knowledge of mitosis can help explain why. Let's say radiation breaks a chromosome and leaves a fragment that is no longer attached to a centromere. What would happen to the chromosome fragment during cell division? It might not be distributed normally to the daughter cells. This is because the centromere is the only part of the chromosome that is directly attached to spindle microtubules, which pull each chromatid to one of the cell's poles during mitosis (review Fig. 4.8b). In a case of chromosome breakage like this, one cell might end up with an extra chromosome part, while the sister cell ends up missing this chromosome part. The resulting genetic

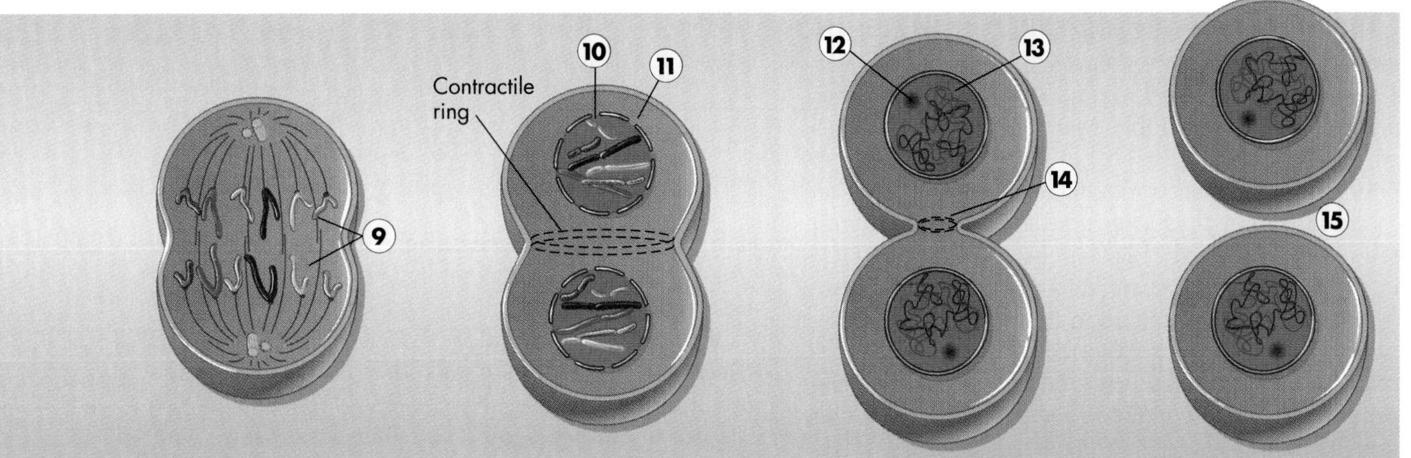

(e) Anaphase

Next, during anaphase the centromeres split, and microtubules pull sister chromatids apart, toward opposite poles (9).

(f) Early telophase

Early telophase marks the beginning of cytokinesis, the division of the cytoplasm. The daughter chromatids (now independent chromosomes) arrive at each pole (10), and the nuclear membrane re-forms around the chromosomes (11).

(g) Late telophase

As telophase progresses, the nucleolus reappears (12), the spindle dissolves, and the chromosomes reel out again into a tangled mass of DNA and protein (13). In addition, in late telophase in animal cells, a contractile ring tightens around the cell's midline where the metaphase plate had been, creating a furrow (14).

(h) Completion of cytokinesis

After the completion of nuclear division, daughter cells finally separate in cytokinesis (15).

Figure 4.8
What Moves the Chromosomes?

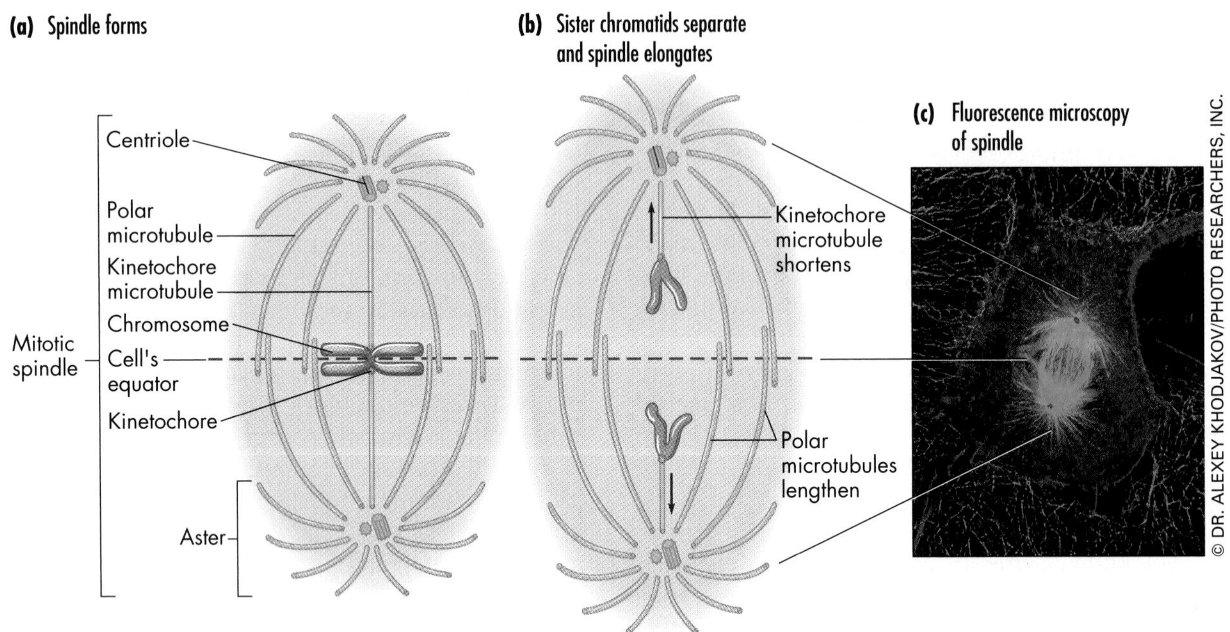

imbalance can alter the cell's information and lead to its death.

The most rapid cell division taking place in a cancer patient often involves the cancer cells themselves. Therefore, well-aimed radiation therapy damages and kills those cells more than it hurts the patient's surrounding cells. Like all humans, however, cancer patients have other rapidly dividing cells in addition to the cancerous cells. These include the precursors of red and white blood cells, skin cells, and cells of the intestinal lining. Not surprisingly, these cell types can be damaged during radiation therapy, and their disruption explains some of the short-term side effects of radiation therapy: anemia (too few red blood cells), susceptibility to infection (too few white blood cells), hair loss (damaged skin cells), and nausea (damaged intestinal cells). Ironically, long-term side effects include an increased risk of cancer, since, as we say, radiation can cause genetic mutations and these can lead to the uncontrolled divisions in a tumor. The disruption of rapidly dividing cell types also explains the destructive, often lethal effects of exposure to high levels of radiation from a hydrogen bomb or an accident at a nuclear power plant. So-called radiation sickness is like a severe set of symptoms from radiation therapy. These facts about mitosis underscore the devastating consequences nuclear war or nuclear accidents could have for people and most other life forms on Earth. They also highlight the restorative powers of mitosis, which keep most of us healthy most of the time.

LO3 What Controls Cell Division?

Let's return, now, to Patricia's Basal Cell Nevus Syndrome (BCNS) and the frequent formation of basal cell carcinomas. The growth of these tumors is evidence that some mechanism is missing that would normally stop her skin cells from dividing so rapidly and continuously. Some mechanism may be broken that normally acts like a car's brake to reduce the speed of the cell cycle. Or perhaps her cells have an "accelerator" that is working overtime. Normal cells in the skin and other tissues have some means of regulating the cell cycle. So what is the nature of that control? And what goes wrong with it in BCNS?

Stem Cells and Growth Control

Think, for a minute, about cell division in the skin. Normally after a basal cell in the dividing layer of the skin undergoes mitosis, then one of the daughter cells stops cycling and matures into a cell of the outermost layer of the skin (Fig. 4.9a). Although the process is entirely healthy, this cell is, in effect, on a suicide mission—it's "born to die," since it will never divide again and will eventually expire and flake off. The other daughter, however, can continue to divide. Cells with the ability to continue dividing are called

stem cells and their progeny can behave in two ways, with one daughter maturing into a differentiated cell type and the other retaining the ability to divide.

How do stem cells in the dividing cell layer "know" when they should divide? Here's a simple thought experiment: If you were to scrape off the upper layer of the epidermis in a small patch of skin, the dividing layer below it would somehow "recognize" the thinness. More cells in that dividing basal layer (or germ layer) would then leave the G_1 phase of the cell cycle, enter S, and eventually complete a division. When the epidermis becomes thick enough once more, the basal layer "senses" this, too, and slows its rate of cell division.

Now, in a basal cell carcinoma, the signal to divide is somehow turned on but the signal to slow or cease dividing never comes, or if it does, it goes undetected. One possibility is that cells produce too much of the signal to divide; as a result, at any given time, more of the stem cells would undergo division than is needed to replace sloughed off skin. Here's another possibility: The daughter cell that should mature into an epidermal cell doesn't and instead, it too continues to divide (Fig. 4.9b). This is part of the overall question: What tells a cell to divide or stop dividing? Let's look at some possibilities for what is most likely happening.

The Molecular Basis of Basal Cell Carcinoma

People like Patricia have inherited genetic factors that predispose skin cells to form basal cell carcinomas. If she were to have children, about half of them would be likely to be affected with the condition (and Chapter 5 explains why). By investigating the inheritance of the disease gene in many families, Dr. Erwin Epstein and colleagues at the University of California at San Francisco were able, in 1996, to isolate the specific gene that is disrupted in BCNS. They were astonished at what they found. The gene turned out to be closely related to a gene already isolated from a fruit fly, *Drosophila melanogaster.* The gene is called *patched,* because fly embryos with the defective gene have abnormal little patches of hair-like structures in their skin. Epstein's group found that human families with BCNS had mutations in the human *patched* gene in all of their cells. They also found that people with individual, spontaneous basal cell carcinomas due to sun exposure and not inheritance often had mutations in the *patched* gene, but only in the cells of their skin tumor, not in *all* of their skin cells. It was clear that the *patched* gene normally functions in some way to prevent the growth of basal cell carcinomas and that mutations of the gene allow skin tumors to form.

Figure 4.9
Stem Cells and Skin Cancer

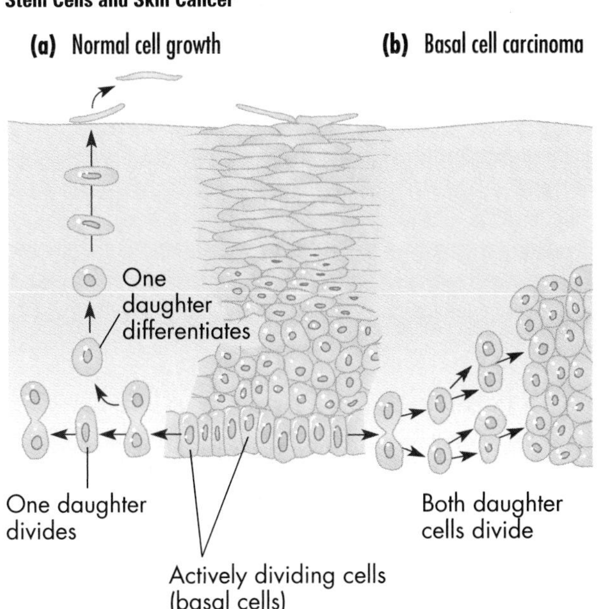

stem cells
a normal body (somatic) cell that can continue to divide, replacing cells that die during an animal's life

growth factors
proteins that can enhance the growth and proliferation of specific cell types

receptor
a protein of a specific shape that binds to a particular chemical

By studying flies, workers had shown that the *patched* gene causes a certain protein to appear on the surface of cells. That protein, in turn, is necessary for the cell to interpret signals coming from the outside of the cell that stimulate cell division. The basic idea is that a substance in the cell's environment, often a protein called a **growth factor** made by a nearby cell, can act as a signaling protein telling other cells to divide. In flies, researchers showed that *patched* is the receptor for the signaling protein. In humans, the specific signaling protein is called sonic hedgehog because it is made by the *sonic hedgehog* gene. (This odd name comes from a mutant fly that, in this case, has prickly skin, making it look like a hedgehog .)

How Growth Factors Act

Still looking, then, at what instructs a cell to divide or stop dividing, let's see how growth factors might work. At the site of a cut or other wound, growth factors such as the human sonic hedgehog protein, may be released from dying cells (Fig. 4.10a, Step ①). The factors could diffuse locally, and bind to a **receptor**, a protein that chemically recognizes the specific growth factor. Receptors may be embedded in the membranes of nearby cells (Step ②). The binding

meiosis
the type of cell division that occurs during gamete formation; the diploid parent cell divides twice, giving rise to four cells, each of which is haploid

sexual reproduction
a type of reproduction in which new individuals arise from the mating of two parents

gamete
(Gr., wife) a specialized sex cell, such as an ovum (egg) or sperm, that is haploid; a male gamete (sperm) and a female gamete (ovum) fuse and give rise to a diploid zygote, which develops into a new individual

of a growth factor from the wounded cell to the receptor of a nearby cell then stimulates an interior cascade of signals (Step ③). Some of those internal signals include *cyclins*, proteins whose levels rise and fall at different parts of the cell cycle and cause the cell to move from one phase to the next. The net result of this signaling (Step ④) is that the cells lining the wound are stimulated to divide. After enough cell division in the neighboring cell, the tear in the skin eventually fills in with new cells, growth factor levels fall, and the rate of cell division slows down.

Dr. Epstein's work showed that BCNS cells have a defective *patched* gene (Fig. 4.10b). This mutated gene makes an altered receptor that acts as if it is bound to the sonic hedgehog even when that signaling molecule is absent. Even without the sonic hedgehog "signal," the mutant patched protein stimulates the cascade of internal signals telling the cell to divide. This causes the cells in BCNS sites like those on Patricia's face and arms to enter S phase and divide inappropriately. Further investigations of this pathway in BCNS cells may eventually allow Dr. Epstein to understand exactly why and how basal cells start dividing in a skin cancer but never stop. This, in turn, could lead to treatments for BCNS as well as for the more common basal cell carcinomas that some people get later in life from too much exposure to sunlight.

Sonic hedgehog protein signal →
Patched receptor → Turned-on cell division

Cancer as a Disease of Altered DNA

Recent evidence indicates that most cancers are related to changes in a cell's DNA. Numerous studies have shown that many substances in the environment, including ultraviolet light, industrial chemicals, radiation, the tar in cigarettes, and certain viruses, cause such changes in DNA structure. Some DNA changes can affect either growth factors or receptors for growth factors on the cell's plasma membrane, as Dr. Epstein found for BCNS. Surprisingly, some cancer cells may produce *both* the growth factor to turn on cell division *as well as* its specific receptor; the result is cells that are constantly stimulating themselves to divide. There is still much uncertainty about the causes of cancer, but it is clear that the answers will be found in the regulation of the cell cycle, and the stages of growth, division, and rest. In the meantime, it's wise to avoid the most common environmental hazards associated with cancers, such as ultraviolet light and cigarette smoke.

LO4 Meiosis: The Special Cell Divisions That Precede Sexual Reproduction

If a BCNS patient like Patricia has children, there is a 50–50 chance that each child will inherit the disease of multiple skin cancers. Yet consider a teenager playing in the sun who experiences an ultraviolet light-induced change in the DNA of the *patched* gene in a cell on, say, the nose. This change or mutation could then lead to basal cell carcinoma arising on that person's nose. But this individual would *not* pass on to his or her children an increased likelihood of having either a single basal cell carcinoma or the multiple reoccurrences of BCNS. How can this be? What's the difference in their response? The answer revolves around the special cell population from which sex cells arise and the special type of cell division that produces them: **meiosis**.

Sexual Reproduction: Offspring from Fused Gametes

In **sexual reproduction**, parents (usually two, but sometimes one) generate specialized sex cells called **gametes** (Fig. 4.11). When gametes from two individuals (usually one male and one female) fuse during

Figure 4.10

Growth Factor Control, Normal Cell Division, and Cancer

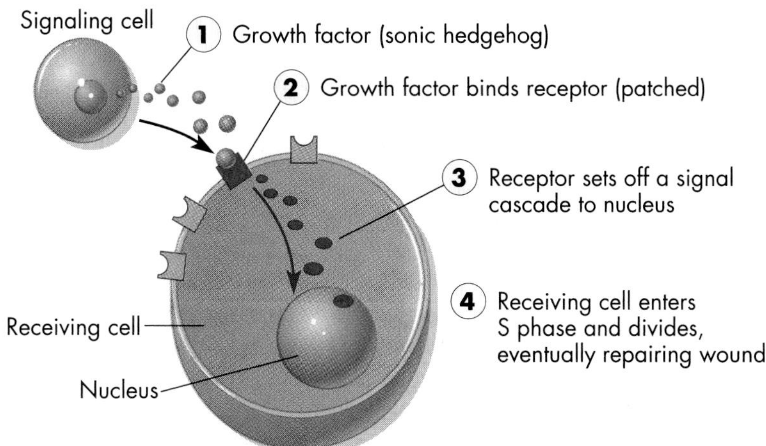

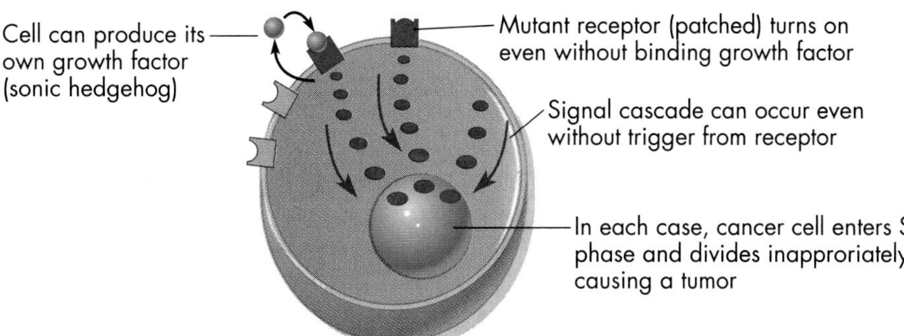

fertilization the fusion of two haploid gamete nuclei (egg and sperm), which forms a diploid zygote

egg the haploid female gamete

sperm the haploid male gamete

gonad an animal reproductive organ that generates gametes; testes and ovaries

ovary egg-producing organ

testis (pl. testes) the male reproductive organ that produces sperm and sex hormones

zygote (Gr. *zygotos,* paired together) the diploid cell that results from the fusion of an egg and a sperm cell; a zygote may either form a line of diploid cells by a series of mitotic cell divisions or undergo meiosis and develop into haploid cells

germ cell a sperm cell or ovum (egg cell) or their precursors; the haploid gametes produced by individuals that fuse to form a new individual

fertilization, they set into motion the life of a new individual.

Most of us are familiar with sexual reproduction and the human life cycle. As with most plants and animals, the human female's gametes are large cells that are incapable of spontaneous movement called **eggs**, while the male gametes are small motile cells called **sperm**, which can move or be carried from the male to the egg (Fig. 4.11a). Eggs and sperm are usually produced in specialized organs. In flowering plants, they are produced by structures in the flowers; in animals, gametes are made by special organs called **gonads**. The female gonad is the **ovary**, which produces eggs, and the male gonad is the **testis**, which produces sperm.

In many complex organisms, including human beings and ginkgo trees, each individual produces just one kind of gamete, either egg or sperm. However, in pear trees, earthworms, and a number of other species, each adult individual can produce both types of gametes.

The fusion of egg and sperm, or fertilization, results in a single cell called the **zygote** (Fig. 4.11b). In the zygote, the hereditary information from both parents unites, creating a genetically unique combination of genes and chromosomes. The single-celled zygote undergoes *development*, usually a period of rapid mitosis and cellular specialization during which the new cells emerge and take on their specific roles in the organism (Fig. 4.11c). As a result of development, an immature form emerges, continues to grow, and eventually changes into a mature adult.

Germ Cells and Somatic Cells

Early in animal development, a group of cells called **germ cells** is set aside. Germ cells are like the stem cells we discussed earlier; they retain the potential to divide to produce more germ cells, or to differentiate into gametes (eggs or sperm), and to migrate to the developing gonad during development (Fig. 4.11d). Animal eggs have a particular group of proteins that become separated into certain cells as the egg divides by mitosis (Fig. 4.11c). Whichever cells

Fertilization *doubles* the chromosome number, while meiosis divides it in *half*. ”

somatic cell
(Gr. *soma,* body) A cell in an animal that is not a germ cell

receive these proteins develop into germ cells. The cells that don't receive these proteins become the rest of the body's cells in the muscles, brain, nose, and so on. These special proteins are called germ cell determinants, because they determine whether embryonic cells will become germ cells.

Figure 4.11
A Human Life Cycle

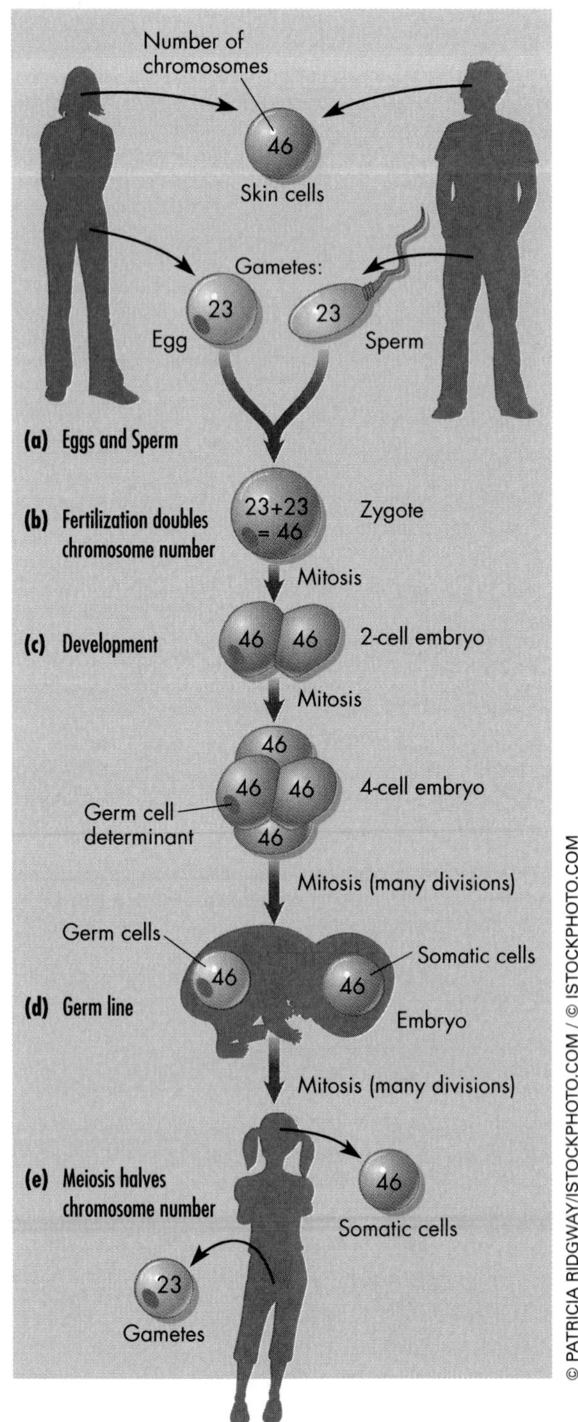

© PATRICIA RIDGWAY/ISTOCKPHOTO.COM / © ISTOCKPHOTO.COM

Germ cell determinants form a continuous cell lineage extending back to your mother, her mother, and so on back in time. The normal body cells, or somatic cells, of a multicellular animal can't form egg and sperm, and these somatic cells eventually die. If, however, an organism's egg or sperm cells unite with those of another individual of the same species, then in a sense, that germ cell lineage lives on in the offspring even though the somatic cells die.

Now let's return to the question of how a BCNS patient might pass on the condition to offspring. In the case of BCNS, the disease gene is in every body cell, including the germ cells (Fig. 4.11). It is therefore passed into the eggs or sperm and can be inherited. In a sunbathing teenager who develops a basal cell carcinoma, however, the diseased copy of the *patched* gene brought about by solar radiation occurs only in one or more somatic cells (in this case, skin cells on the nose) and not in the germ cells, because the teenager did not inherit the genes for skin cancer. Therefore, the mutation could be passed on to the progeny of that one skin cell, but not to the person's future offspring.

An organism's life cycle comes full circle when its germ cells undergo the special type of cell division called meiosis. This division decreases the number of chromosomes present in germ cells by half, from 46 to 23 in the case of humans (Fig. 4.11e). The resulting cells can then become egg or sperm and lead to a new generation.

Meiosis: Halving the Chromosome Number

Why would an organism need a special type of cell division before producing gametes? Each of your somatic cells has 46 chromosomes. Let's say you produced gametes that had 46 chromosomes. How many chromosomes would your children's somatic cells have? Without a special mechanism, your child would get 46 chromosomes from you and another 46 from your mate, giving a total of 92. The next generation would have 184 and so on. Clearly this doesn't happen because each human baby has 46 chromosomes just like each parent. So what prevents the doubling of chromosomes in each generation?

The special type of cell division, meiosis, prohibits this runaway increase in chromosome number. Meiosis ensures that gametes contain half as many chromosomes as normal body cells (Fig. 4.12). In other words, each sperm or egg cell you make has 23 chromosomes, not 46. And the fusion of gametes at fertilization restores the original parental chromosome number, 46.

Figure 4.12
Haploid and Diploid Cells

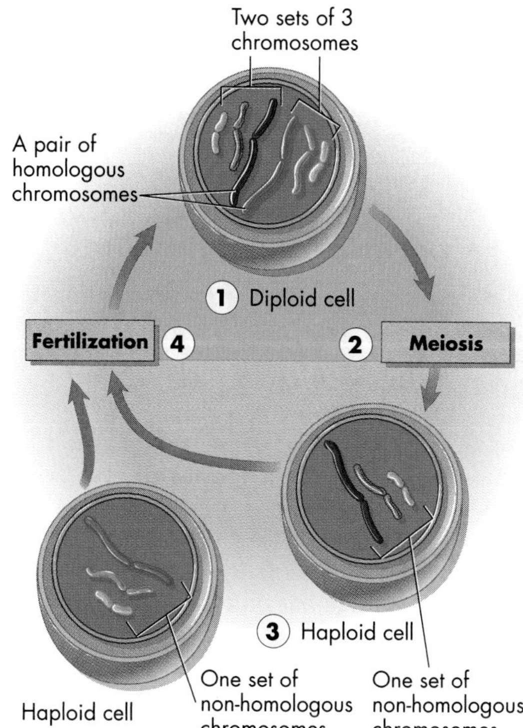

Meiosis (literally, to make smaller) produces gametes or, in some species, such as mushrooms and ferns, other specialized reproductive cells called *spores*, whose chromosome number is half that of other body cells (see Fig. 4.13i on page 79).

In terms of chromosome number, fertilization and meiosis play opposite roles in a life cycle involving sexual reproduction. Fertilization *doubles* the chromosome number, while meiosis divides it in *half*. As we will see later in the chapter, meiosis also increases genetic variation, which is a precondition for evolution.

Chromosome Sets

Before we can investigate how meiosis reduces the number of chromosomes, we must understand how many sets of chromosomes each cell contains at different phases of the life cycle. As we have seen, the body cells of each species have a characteristic number of chromosomes: human beings have 46, chimpanzees 48, houseflies 12, onion plants 16, roundworms 4. What do all these chromosome numbers have in common? They are all even numbers. This is because in the body cells of most eukaryotes, *chromosomes are present in pairs*. A pair of chromosomes that have similar size, shape, and usually gene order are called homologous chromosomes (*hom* = alike).

A human skin cell contains 23 pairs of homologous chromosomes, for a total of 46 chromosomes. One member of each pair came from the individual's mother, the other member of each pair came from the individual's father. Thus, a skin cell has two *sets* of chromosomes: a maternal set of 23 and a paternal set of 23. A cell such as this that has two sets of chromosomes is called a diploid cell (*di* = two). A cell that has just one set of chromosomes, one copy of each homologous pair, is called a haploid cell. Human gametes are examples of haploid cells.

Follow through the steps of Figure 4.12 to see these principles at work in an organism with just three pairs of chromosomes in each diploid cell.

homologous chromosomes
chromosomes that pair up and separate during meiosis and generally have the same size, shape, and genetic information; one member of each pair of homologous chromosomes comes from the mother and the other comes from the father

diploid
a cell that contains two copies of each type of chromosome in its nucleus (except, perhaps, sex chromosomes)

haploid
having only one copy of a chromosome set; a human haploid cell has 23 chromosomes

meiosis I
the first division of meiosis, during which the number of chromosomes in a diploid cell is reduced from a diploid set of duplicated chromosomes to a haploid set of duplicated chromosomes

meiosis II
the second division of meiosis, during which a haploid cell with duplicated chromosomes divides to form two haploid cells with unduplicated chromosomes

The Cell Divisions of Meiosis

Because you've already explored the divisions of mitosis, you are in a good position to understand the variations that give rise to meiosis. Keep in mind that:

- Only somatic cells such as skin cells undergo mitosis and that the process leads to cell replacement and to growth and repair.
- Only germ cells like those in the gonads undergo meiosis and lead to the production of gametes and the possibility of future generations.
- The result of a meiotic division is to reduce the number of chromosomes by half, changing a diploid stem cell into haploid cells that can become gametes.

The first point to notice about meiosis is that the change from diploid to haploid chromosome number involves two sequential cell divisions called meiosis I and meiosis II. The chromosomes are duplicated before meiosis I, and then they divide. No duplication occurs before the second division, so when the cells divide in meiosis II, four haploid cells are produced. As in mitosis, chromosome *replication, alignment,* and *separation* are central concepts. Figure 4.13 charts chromosome movements during meiosis.

polar bodies haploid cells that form during meiosis in the female but do not become the egg

genetic recombination the reshuffling of maternal and paternal chromosomes during meiosis, resulting in new genetic combinations

Replication: The Interphase Before Meiosis I

Let's return to our organism with three chromosomes in a haploid set, one long, one medium sized, and one short. As Figure 4.12 showed, the germ cell from which this haploid cell arose is initially diploid, and so it has two long, two medium, and two short chromosomes. Each of the two chromosomes make a homologous pair (Fig. 4.13a).

We start with the cell just after its parent cell has divided (for example, the lower cell in Fig. 4.7f, with two sets of one-chromatid chromosomes). At this stage of meiosis, our cell is in the G_I phase of the cell cycle (Fig. 4.13a). This germ cell leaves G_I and enters S phase, and replicates each chromatid. Now, after replication, each chromosome has two chromatids (Fig. 4.13b).

Alignment: Homologous Chromosomes Pair in Meiosis I

As meiosis I begins, homologous chromosomes pair, lining up very close together: the two long chromosomes next to each other, the two short ones next to each other, and so on (Fig. 4.13c). In fact, homologous chromosomes pair so closely together that they exchange genetic material, as we'll see later.

As meiosis I progresses, the paired homologous chromosomes align in the center of the cell, moved by the spindle fibers (Fig. 4.13d). The order of nonhomologous chromosomes on the spindle, however, is random.

Separation: Homologous Chromosomes Separate from Each Other in Meiosis I

As meiosis I continues, homologous chromosomes separate from each other (Fig. 4.13e). When the nucleus divides and cytokinesis is completed after meiosis I, the two resulting cells now have the haploid content of chromosomes (Fig. 4.13f). Look at the figure to confirm that. Note that each cell in Figure 4.13f has three chromosomes: one long, one medium, and one short; these two cells, therefore, are haploid: Meiosis I has reduced the chromosome number from diploid to haploid. Notice, however, that each chromosome is a two-chromatid chromosome, not a one-chromatid chromosome as at the end of a mitotic division. Meiosis II takes care of that problem.

Replication: Chromosomes Do Not Replicate Between Meiosis I and II

After meiosis I, an interphase follows that is special in that it involves no DNA synthesis or chromosome replication. At the beginning of meiosis II, then, each cell still has a haploid set of two-chromatid chromosomes (Fig. 4.13f).

Alignment: Chromosomes Align Independently in Meiosis II

As meiosis II progresses, the chromosomes line up on the spindle independently of each other (Fig. 4.13g).

Separation: Sister Chromatids Separate in Meiosis II

Meiosis II then follows like a normal mitotic division, and the sister chromatids separate to opposite poles of the cell (Fig. 4.13h). The result, after cytokinesis (Fig. 4.13i), is now four cells, each of which is haploid and each of which has one set of one-chromatid chromosomes.

Compare, now, the parental cell and the daughter cells of one complete meiotic division. While the parental cell had two sets of one-chromatid chromosomes (Fig. 4.13a), the daughter cells—four of them from the two successive divisions of meiosis—each have one set of one-chromatid chromosomes (Fig. 4.13i). Depending on the adult's sex, those cells can develop into either sperm or eggs.

The Fate of the Haploid Products of Meiosis

The four haploid products of meiosis can have different fates in different species or in males and females of the same species. For example in the human female, only one of the four products becomes the egg; that cell keeps nearly all of the cytoplasm. The other three cells are called polar bodies and eventually die. Geneticists sometimes use polar bodies during genetic tests because from them, they can determine the genes in the egg cell. In the human male, each of the four haploid products will become a sperm.

Meiosis Contributes to the Origin of Genetic Variation

Meiosis plays a large role in the genetic differences that exist in a group of brothers and sisters because during meiosis, there is a reshuffling of the maternal and paternal chromosomes. This reshuffling is called genetic recombination, and it occurs in two ways: crossing over and independent assortment.

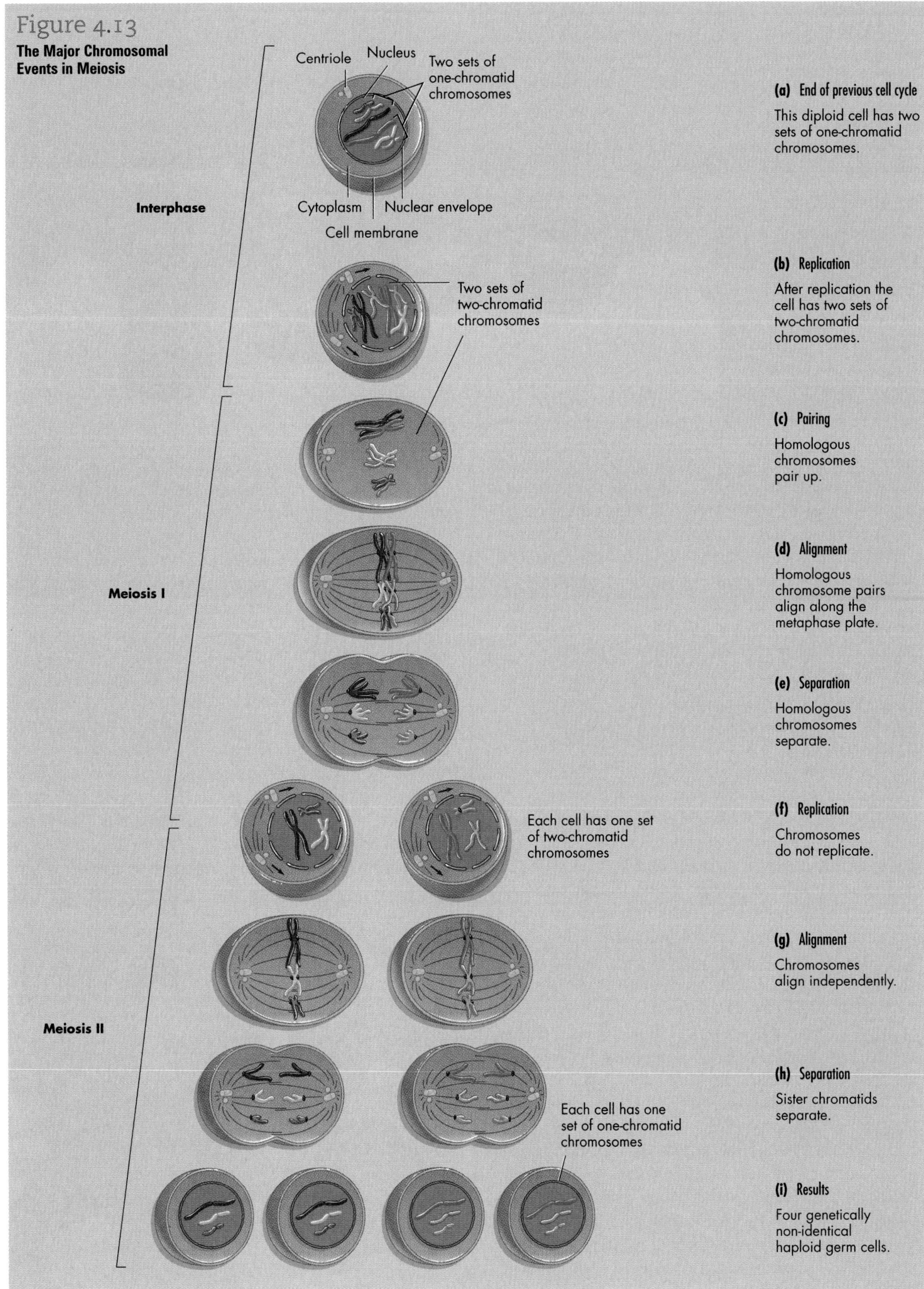

Figure 4.13
The Major Chromosomal Events in Meiosis

independent assortment the random distribution of genes located on different chromosomes to the gametes; Mendel's second law, the principle of independent assortment

Figure 4.14

Genetic Recombination through Crossing Over

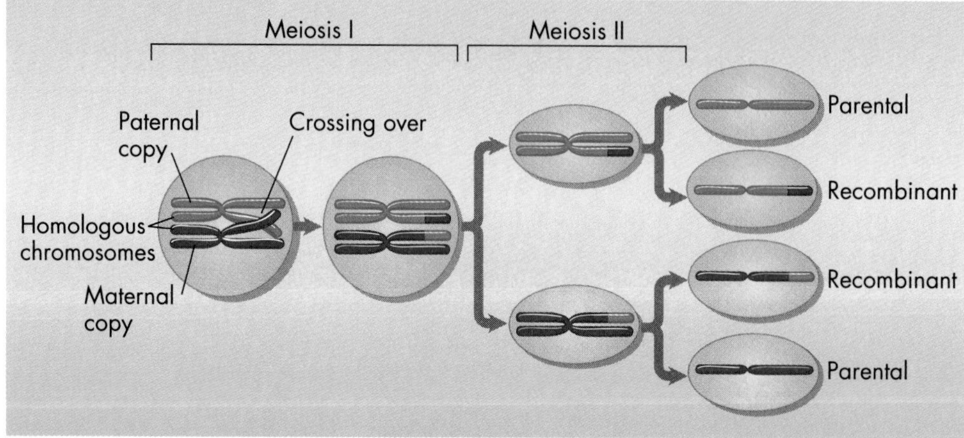

Crossing Over: Homologous Chromosomes Exchange Parts

Crossing over occurs during meiosis I, when homologous chromosomes pair (Fig. 4.14a). While the chromosomes are paired, enzymes can break the DNA molecule in each homologue, switch corresponding regions of each chromosome (the actual crossing over), and then attach them to the new chromosome (Fig. 4.14b). After meiosis II, some of the individual haploid cells will contain recombinants; that is, they will contain chromosomes of mixed ancestry, as the red and blue colors in Figure 4.14 indicate. Note that in Fig. 4.14c, just one crossover event makes each haploid cell genetically unique.

Independent Assortment: Chromosome Pairs Align Randomly

A second mechanism also contributes to genetic variety after meiosis and requires no chromosome breaks. It is called independent assortment, a property based on the fact that nonhomologous chromosomes align independently during meiosis I (review Fig. 4.13d). The chromosomes in Figure 4.15a are arranged with all of the paternal copies of the chromosomes oriented toward the same pole of the cell. Because of this alignment, each of the haploid products of meiosis will contain either all paternal copies or all maternal copies. Another possible arrangement is shown in Figure 4.15b, with one of the paternal chromosome copies oriented toward one pole and the other toward the opposite pole. With this arrangement, each meiotic product will be recombinant, with one paternal and one maternal chromosome.

Take a moment to look at the set of eight gametes in Figure 4.15a and b. Count how many different types of gametes have formed. Each gamete pictured has a haploid set of chromosomes (one short and one long). But there are four combinations present based on the parental origin: (1) both paternal, (2) both maternal, (3) long maternal and short paternal, and (4) long paternal and short maternal. These four combinations differ not only from each other, but also from the diploid parent cells.

The independent assortment of chromosomes during meiosis can be compared to choosing from the menu of a restaurant where there are two choices for each course. Each course represents a chromosome, and the two choices—chicken or fish, rice or potatoes, apple pie or chocolate cake, and so on—represent the two homologous copies of each chromosome. Just as the potential number of different meals depends on the number of courses, the potential number of unique genetic combinations in the gametes depends on the number of chromosomes. From a menu with 3 courses and 2 choices per course, you could make up eight (= $2 \times 2 \times 2$) different meals. For each chromosome added, you'd have to multiply again by 2. So for a human being, that number would be 2^{23}, or

Figure 4.15
Genetic Recombination through Independent Assortment

(a) One possible chromosome arrangement

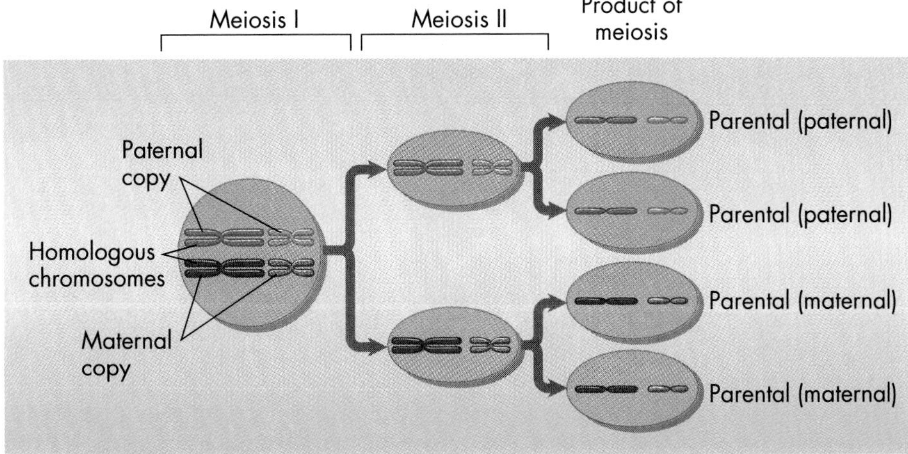

(b) Another possible chromosome arrangement

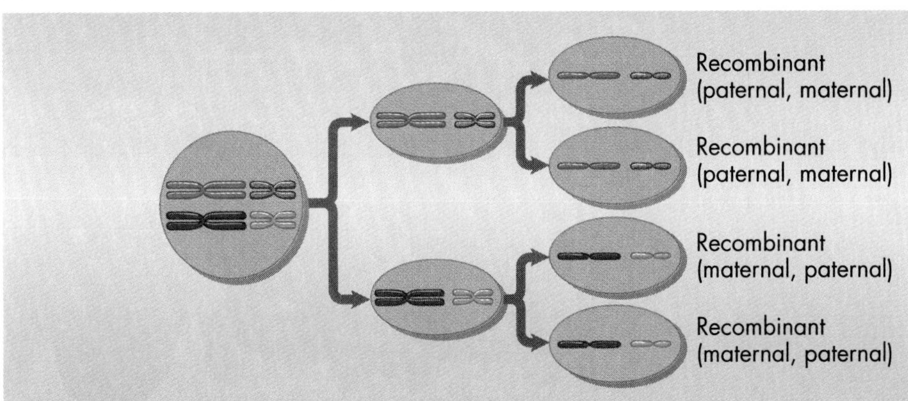

a potential of 8 million different chromosome combinations of eggs or sperm in the nuclei. Because crossing over (review Figure 4.14) adds even more new combinations, the number of actual possible types of gametes made by a single person is astronomical. Finally, the random combination of maternal and paternal chromosomes in the zygote at fertilization further increases the number of genetic combinations.

The incredible genetic diversity that results from genetic recombination helps explain why an organism is highly unlikely to produce two genetically identical gametes, and why, in turn, all other organisms resulting from sexual reproduction may resemble their parents but are never exactly like either parent.

Mitosis and Meiosis Compared

The details of mitosis and meiosis are relatively easy to confuse, so it's worth one last comparison of the two (see Figure 4.16).

Mitosis occurs only in somatic cells and allows for the growth of the organism and repair of its parts. Mitosis accomplishes two things:

1. Reproduction of cells.
2. Equal distribution of DNA to each new daughter cell.

Meiosis occurs only in germ cells and can result in the production of gametes that can take part in sexual reproduction. Meiosis accomplishes three things:

1. By reducing the chromosome number from diploid to haploid, meiosis prevents an increase in chromosome number that otherwise would occur at fertilization.
2. Crossing over during meiosis permits new combinations of maternal and paternal hereditary traits.
3. Independent assortment allows for the further random combination of maternal and paternal chromosomes.

Figure 4.16
Mitosis and Meiosis Compared

(a) Mitosis

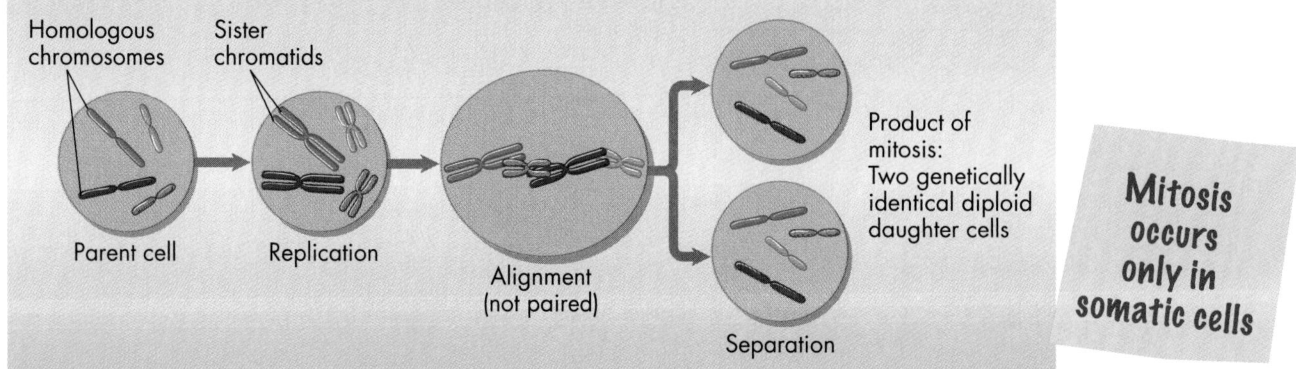

Mitosis occurs only in somatic cells

(b) Meiosis

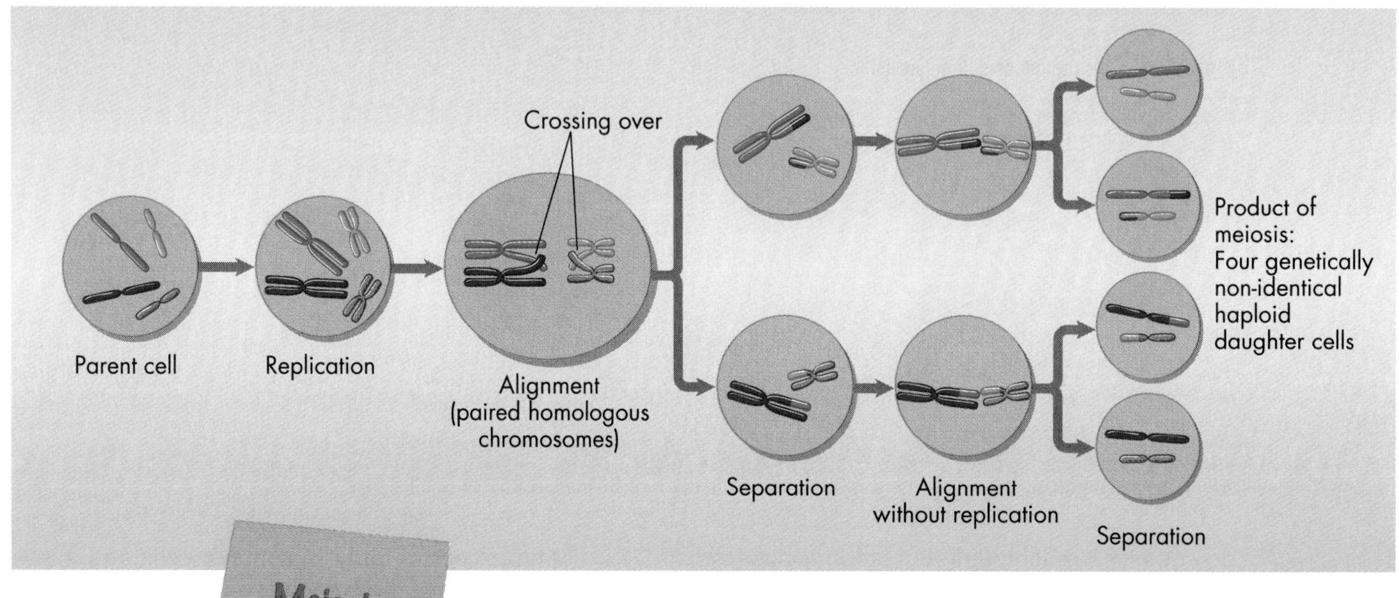

Meiosis occurs only in germ cells

Who knew?

Somatic cells divide by *mitosis* for growth and replacement throughout the body. Cells divide by *meiosis* only for the production of gametes (eggs and sperm); these cells are found only in the gonads (ovaries and testes).

82% **of students surveyed find LIFE is designed more closely to match the way they naturally study.**

LIFE was designed for students just like you – busy people who want choices, flexibility, and multiple learning options.

LIFE delivers concise, focused information in a fresh and contemporary format. And… **LIFE** gives you a variety of online learning materials designed with you in mind.

At **4ltrpress.cengage.com/life,** you'll find electronic resources such as **animations, visual reviews, flash cards,** and **interactive quizzes** for each chapter.

These resources will help supplement your understanding of core biology concepts in a format that fits your busy lifestyle. Visit **4ltrpress.cengage.com/life** to learn more about the multiple resources available to help you succeed!

5

Patterns *of* Inheritance

Learning Outcomes

LO[1] Understand the universal principles of heredity

LO[2] List the rules that govern the inheritance of a single trait

LO[3] Explain how geneticists analyze human inheritance patterns

LO[4] List the rules governing how organisms inherit multiple traits

LO[5] Explain how sex influences the inheritance of traits

LO[6] Identify how genetics is changing our world

© FHF GREENMEDIA/GAP PHOTOS/GETTY IMAGES

> "140 years ago, most observers thought each individual's traits resulted from a blending of their parents' traits."

A Devastatingly Common Illness

With one affected child in every 3,000 births, cystic fibrosis (CF) is the most common lethal inherited illness among Caucasians. It causes a constellation of breathing, digestion, and other medical problems, and until recently an affected child born was unlikely to survive past his or her teens.

What do you know?

Can you tell an organism's genotype by its phenotype?

Cystic fibrosis is essentially a disease of clogged ducts. A child who inherits one mutated cystic fibrosis gene from each parent will produce a faulty version of a protein. The protein is involved in salt and water movement across cell membranes, and being defective, it prevents normal fluid transport. The walls of the ducts and the protective coatings they secrete tend to dry out, creating a thick, sticky mucus layer. This, in turn, clogs narrow passageways and ducts in the lungs, stomach, pancreas, sweat glands, and reproductive organs. Because of this, an individual who inherits cystic fibrosis usually has difficulty breathing. He or she repeatedly contracts dangerous bacterial infections in the lungs and suffers stomachaches and a diarrhea-like condition due to poor absorption of fats in the diet. Most also exude a salty secretion on the skin. And affected males are almost always sterile because of blocked ducts leading from the testes.

Our goal in this chapter is to help you understand inheritance patterns like these—patterns that determine who will display lethal disease symptoms and who won't. Inheritance patterns underlie each of our thousands of traits—eye color, hair color, height, and so on—and those of all other living organisms. The science of **genetics** explores the nature of genes and how they are organized on chromosomes; how genes govern our appearance, physical functioning, and even behavior; and how medical researchers can manipulate genes to treat diseases like cystic fibrosis.

As you explore Chapter 5, you'll find the answers to these questions:

- ☑ How did a 19th-century monk discover the universal principles of heredity?
- ☑ What rules govern the way organisms inherit individual traits?
- ☑ How do geneticists analyze inheritance patterns in people?
- ☑ What rules govern the way organisms inherit several traits at the same time?
- ☑ How does our sex influence how we inherit traits?
- ☑ How is the study of genetics changing the way we predict and treat diseases?

genetics
the study of genes and inheritance

hybrid
an offspring resulting from the mating between individuals of two different genetic constitutions

blending model of heredity
the idea that maternal and paternal characteristics *blend* to produce the characteristics found in the offspring; disproved by Mendelian genetics

particulate model of heredity
Mendel's idea that heredity could be governed by "particles" that retain their identity from generation to generation

self-fertilization
the ability of a plant or animal to fertilize its own eggs

LO[1] How Did Scientists Discover the Universal Principles of Heredity?

To help potential parents calculate the risk of having a child with cystic fibrosis, geneticists apply the universal laws of heredity, the principles that govern how traits are passed from parents to offspring. These rules have a long and interesting history. They were discovered by a European monk named Gregor Mendel, who first made his discoveries public in 1865. Even today, in the 21st century, it is easiest to understand these rules of heredity if we learn how Mendel himself discovered them.

In Mendel's day, 140 years ago, most observers thought each individual's traits resulted from a blending of their parents' traits. Looking at organisms in nature, it is not hard to see why people believed that offspring were intermediates between their parents. Consider, for example, two monkey flower plants whose flowers have petals of vastly different sizes. Let's say we mated plants with these different flower shapes—one with long petals to one with short petals. The result would be a **hybrid**: the offspring of two individuals with differing forms of a given trait. In this case, the hybrid's flowers had petals intermediate in length between those of the two parents. Such observations made people think that the hereditary "stuff" of a mother and father was liquid and would *blend* to produce the characteristics found in the offspring, just as cream mixes with dark-brown coffee to produce the beige-colored café au lait. This idea became known as the **blending model of heredity**.

While Mendel was at the University of Vienna, he learned that all matter is made up of discrete atoms and molecules. He wondered if heredity could also be governed by "particles" that retain their identity from generation to generation. He put his new **particulate model of heredity** to the test in a long-term study involving pea plants, controlled matings between them, and careful tabulations of the kinds of offspring each cross produced.

© DAVID CROCKETT/ISTOCKPHOTO.COM

Genetics in the Abbey

The blending hypothesis predicted that, like café au lait, each hereditary factor would be permanently diluted in the hybrid. Mendel's particulate model, however, predicted that each hereditary factor would remain unchanged in a hybrid, like dark-brown and cream-colored marbles mixed in a bag. Mendel's key insight was that he could disprove one of these two models not by looking at the hybrid itself—the first generation of the mating—but by checking the *offspring* of hybrids—the second generation. If the original parental forms reappeared in the second generation, this would show that the hereditary factors had passed through the hybrids unchanged and remained as some kind of intact particles. If, however, the original forms *failed* to reappear in the hybrid's offspring, then the factors would appear to have been blended.

Mendel chose common garden pea plants as his test subject because peas have several advantages. From seed stores he could purchase strains of pea plants that showed clear alternative forms for single traits, such as stem length or flower color. For example, long-stem plants versus short-stem, or purple flowers versus white. By selecting strains that differed in only one trait such as height or flower color, he could study inheritance of one feature unconfused by all other variations. In addition, Mendel could also easily control which pea plant mated with which other pea plant. A pea flower normally **self-fertilizes** or mates

Figure 5.1

Mendel's Evidence for a Nonblending (Particulate) Model of Heredity

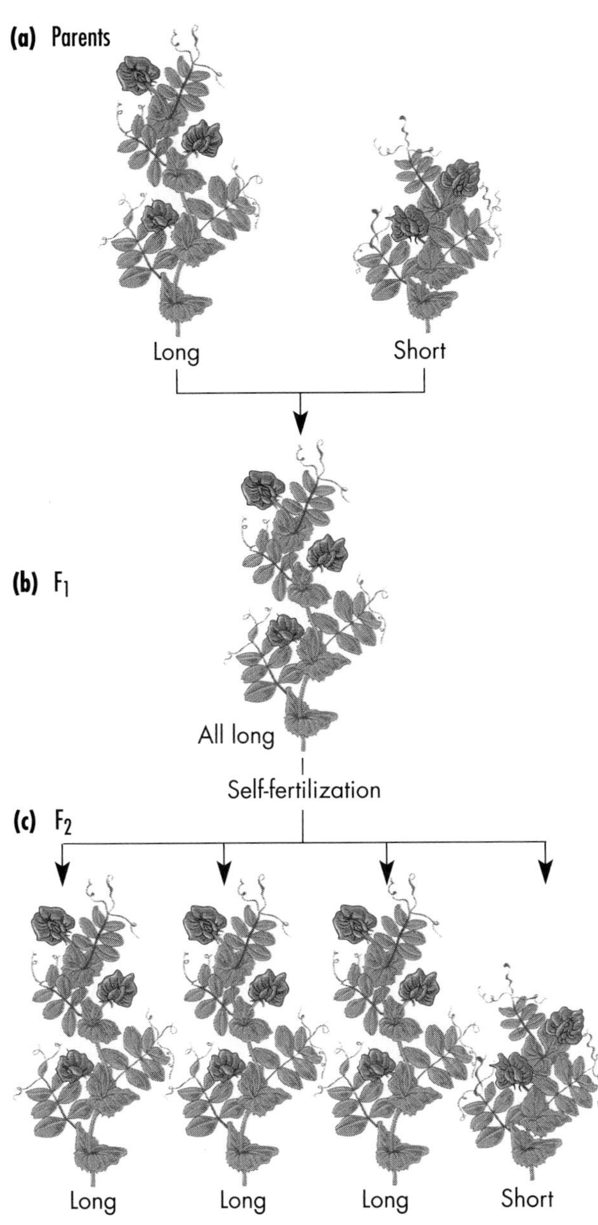

with itself. But Mendel cross-fertilized plants. From a purple flower, for example, he could simply clip off the organ that produces pollen, the sources of the sperm, and dust the egg-containing organ of that flower with pollen from another plant (for example, a white flower). From the seeds of this cross fertilization or "cross," Mendel could grow a new generation of pea plants and watch to see which traits were expressed.

With his clearly stated hypotheses and well-chosen experimental system, Mendel was now ready to perform the scientific tests that would lead to the rules of heredity, rules that medical doctors can still apply more than a century later to patients.

Mendel Disproves the Blending Model

In one of Mendel's first crosses, he planted seeds from long-stem and short-stem plants early one spring and let them grow into the parental (P_1) generation (Fig. 5.1a). Later that spring, when the parental plants had flowered, Mendel cross-fertilized long-stemmed plants with pollen from the short-stemmed plants. In the summer, when the pods became swollen with plump peas, he collected the seeds. These seeds would produce the next generation, called the first filial (F_1) generation, meaning the first generation in the line of descent. Planted in the spring of the second year, the F_1 seeds of the long-stem/short-stem cross all grew into plants with stems just as long as the original long-stemmed parent (Fig. 5.1b). Mendel repeated this type of experiment for other traits—flower color (purple vs. white), seed shape (round vs. wrinkled), and so on—and he found that in each case, only one alternative of each trait appeared in the F_1 hybrid generation. It was as if one of the traits had totally disappeared. The trait that appears in the F_1 hybrid (such as long stems in peas) is said to be dominant, while the trait that does not show in the hybrid (such as short stems in peas) is referred to as recessive.

Now came the crucial part of the experiment. What happened to the recessive characteristic in the hybrid? Did it blend with the dominant characteristic? Did it disappear completely and forever? Or did it remain intact but hidden in the F_1 generation? To find out, Mendel allowed the long-stemmed F_1 hybrid plants to self-fertilize, and the next spring he planted the seeds of the second filial (F_2) generation. When the second generation of pea plants grew up,

cross-fertilize
to deliberately cross two organisms; in plants, to transfer pollen from one self-fertilizing flower to another

parental (P_1) generation
in Mendelian genetics, the individuals that give rise to the first filial (F_1) generation

first filial (F_1) generation
in Mendelian genetics, the first generation in the line of descent

dominant
in genetics, an allele or corresponding phenotypic trait that is expressed in the heterozygote (in other words, that shows in the hybrid)

recessive
an allele or corresponding phenotypic trait that is hidden by a dominant allele in a heterozygote

second filial (F_2) generation
in Mendelian genetics, the second generation in the line of descent

allele
one of the alternative forms of a gene

most of them had long stems, but significantly, some plants had short stems. Again, there were no stems of intermediate length (Fig. 5.1c). The reappearance of plants with stems just as short as the stems of the original short-stemmed parents, and the absence of any intermediates were the results predicted by the particulate model of heredity and dramatic disproof of the blending model of heredity.

LO² What Rules Govern the Inheritance of a Single Trait?

Being a careful and inquisitive person, Mendel was not satisfied with just saying that "some" of the F_2 plants had short stems and therefore the blending hypothesis was wrong. He wanted to understand what he saw. So he counted the plants and by analyzing the numbers, was able to infer the mechanisms that hid the short-stemmed trait in the F_1 and its reappearance in the F_2.

The good monk found that 787 of the F_2 plants he counted had long stems and 277 had short stems. These numbers showed a 787:277 ratio or approximately 3:1 ratio of long-stemmed to short-stemmed plants in the F_2 generation. (A perfect 3:1 ratio for 1064 plants would be 798:266, not much different from the 787:277 he actually observed.) It turns out that the results of Mendel's observations for pea stem length apply to many traits in eukaryotic organisms. The general finding is that with two clear alternative traits such as long versus short stems, purple versus yellow seeds, or presence of cystic fibrosis versus absence of the disease, the hybrid (the F_1 generation) shows only one trait, the dominant one. The mating of two hybrids (the F_2 generation) produces offspring in which three quarters show the trait that appears in the hybrid (the dominant trait), while one quarter show the trait that is hidden in the hybrid (the recessive trait). How can we understand the mechanism that causes such a result to occur?

Genes and Alleles

Mendel reasoned that because short stems reappeared in the F_2 plants, the hereditary factor that causes short stems had to be an individual unit, like a particle, and not like a liquid that could be mixed with another liquid of a different color. Modern geneticists call this particulate factor a *gene* (see Chapter 1). While Mendel did not use that term, we will use it in the following discussion for clarity.

A gene influences a specific trait in an organism, such as the length of a pea stem, the color of a corn kernel, or the presence or absence of a hereditary disease like cystic fibrosis. The gene is not the trait itself. Instead it is a factor that causes the organism to develop a specific trait.

Mendel's insight was remarkable. Even though he had no knowledge of DNA or genes, he reasoned that hereditary "particles" must come in different forms. Nearly a century later, molecular researchers would show that genes do, in fact, have different forms, which are now called alleles. An allele (AL-eel) is an alternative form of a gene. In pea plants, the gene for stem length has two alleles, one causing long stems and one causing short stems. Likewise, modern geneticists know that one allele of the cystic fibrosis gene causes the disease, while another allele (alternative form) of the same gene is necessary for the normal functioning of the airways and other ducts.

Geneticists have also known for a half-century that a gene is a portion of a DNA molecule in a chromosome (Fig. 5.2 and Table 5.1). Although an individual chromosome may contain thousands of genes controlling hundreds of different traits, each chromosome will have just one allele for any individual gene. Because eukaryotic cells contain pairs of homologous chromosomes (review Fig. 4.13), each individual pea plant or person generally has two alleles for each gene, which may be the same or different.

Dominant and Recessive Alleles

Mendel realized that the reappearance of short-stemmed plants in the F_2 generation meant that the short-stem allele was present but invisible in the F_1 hybrids. If the short-stem allele had not been present

Figure 5.2
Genes, Alleles, and Chromosomes

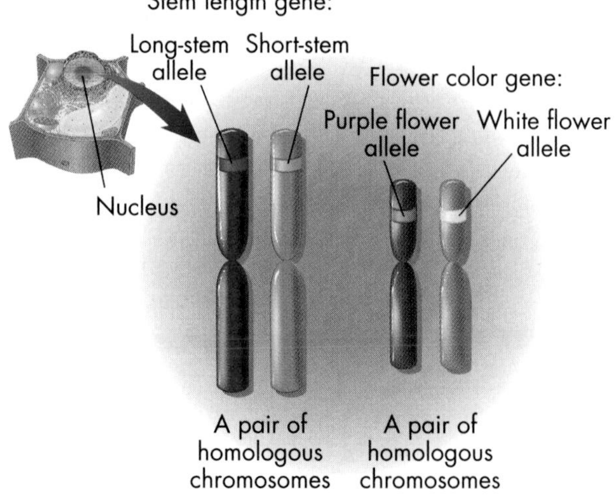

Table 5.1

Principles of Heredity

1. A hereditary trait is governed by a gene.
2. Genes reside on chromosomes and are specific sequences of DNA in all cells, but they are RNA in some viruses.
3. A gene for each trait can exist in two or more alternative forms called alleles. An individual's alleles, interacting with the environment, determine its external appearance, biochemical functioning, and behavior.
4. Most higher organisms have two copies of each gene in body cells (they are diploid). Gametes (eggs or sperm), however, have only one copy of each gene (they are haploid).
5. Homologous chromosomes are two chromosomes that are similar in size, shape, and genetic content.
6. A homozygote has two identical alleles of a gene; a heterozygote has two different alleles of a gene.
7. An individual's physical makeup (the way it looks and functions) is its phenotype; an organism's genetic makeup is its genotype.
8. In a heterozygote, generally only one of the two alleles shows in the phenotype, while the other allele is hidden. The allele that shows is the dominant allele, and the hidden allele is the recessive allele.
9. Pairs of alleles separate, or segregate, before egg and sperm formation, so each gamete has a single copy of each gene. At fertilization, sperm and egg combine randomly with respect to the alleles they contain, and the resulting zygote in general has two copies of all genes.
10. Genes on different chromosomes assort independently of each other into gametes.
11. Linked genes lie on the same chromosome and tend to be packaged into gametes together.

in the hybrid, it could not have been passed on to the F_2 offspring. Because the hybrids showed the long-stem trait, Mendel knew that the long-stem allele was also present in the hybrid. So Mendel concluded that a hybrid plant contains two copies, or alleles, of each gene, one visible and one invisible.

The allele whose trait shows in a hybrid is said to be *dominant*. The allele that is overshadowed each time it is paired with a dominant allele is said to be *recessive*. The long-stem allele of the stem length gene in peas was dominant to the recessive short-stem allele. Can you guess which allele is dominant and which is recessive for cystic fibrosis?

In his work with pea plants, Mendel reasoned that because each hybrid plant has two alleles of each gene, each pure-breeding parent plant must also have two copies of each gene. (A pure-breeding organism always produces offspring with traits identical to its own.) In the case of the hybrid plant, the two alleles are different, one dominant and one recessive. But in the case of the pure-breeding parents, both alleles are identical, either both dominant or both recessive.

phenotype the physical appearance of an organism controlled by its genes interacting with the environment

genotype the genetic makeup of an individual

heterozygous (Gr. *heteros,* different + *zygotos,* pair) Having two different alleles for a specific trait

homozygous having two identical alleles for a specific trait

heterozygote an organism with two different alleles for a given trait

homozygote an organism with two identical alleles for a given trait

Genotype and Phenotype

Although the long-stemmed F_I hybrid plants Mendel studied looked just like the long-stemmed pure-breeding plants of the parental generation, they were genetically different. Today, we refer to an organism's physical characteristics—stem length or airway functioning, for example—as its **phenotype**. We call the organism's specific alleles or genetic makeup its **genotype**. In the case of a long-stemmed phenotype, there are two possible genotypes. Some long-stemmed plants could have two identical dominant long-stem alleles, but other long-stemmed plants could have two different alleles, the visible dominant long-stem allele and the hidden recessive short-stem allele (Fig. 5.3a,b). Geneticists often indicate dominant alleles with uppercase (capital) letters and recessive alleles with lowercase letters. We can represent with *L* the dominant long-stem allele for stem length, and with *l* the recessive short-stem allele. In the case of long-stemmed plants, then, the genotype would be either *LL* or *Ll*. Organisms with two different types of alleles for a given trait are said to be **heterozygous** for that trait (Fig. 5.3b). Pure-breeding organisms, with a pair of identical alleles for a given trait, are **homozygous** for that trait (Fig. 5.3a,c). A heterozygous individual is called a **heterozygote**, and a homozygous individual is called a **homozygote**. Again, geneticists following in Mendel's footsteps learned that pure-breeding long-stemmed and short-stemmed parents are homozygotes, while their hybrid offspring are heterozygotes.

Mendel's Segregation Principle

Mendel concluded that each individual has two copies of each factor (each gene)—two copies of the stem length gene and two copies of the flower color gene. Where did these two copies come from? Mendel suggested that each individual receives one allele from its mother and the other from its father for each of its many traits. Thus, the two alleles possessed by a

law of segregation Mendel's first law, or principle, states that sexually reproducing diploid organisms have two alleles for each gene, and that during the gamete formation these two alleles separate from each other so that the resulting gametes have only one allele of each gene

parent must separate, or *segregate,* from each other so that only one allele of each gene goes into each egg and only one allele of each gene goes into each sperm. Recall from Chapter 4 that a cell with just one copy (allele) of each gene is said to be haploid and a cell with two copies (alleles) of each gene is diploid. Eggs and sperm are haploid, but all other cells in a human being and nearly all other cells in a pea plant are diploid.

If we generalize from Mendel's pea experiments, we can define his law of segregation this way: Sexually reproducing diploid organisms have two copies of each gene, which segregate from each other during meiosis without blending or being altered. When gametes form, they each contain only one copy of each gene.

> "A cell with just one copy (allele) of each gene is said to be haploid, and a cell with two copies (alleles) of each gene is diploid."

Genetic Symbols and Punnett Squares

The segregation principle is probably easiest to understand using the upper and lowercase symbols *L* and *l* for the alleles of the stem length gene (Fig. 5.4). The heterozygous F_1 generation is then designated as *Ll*.

During meiosis in the *Ll* heterozygote, the alleles separate. As a result, half the gametes end up with the capital *L* allele and the other half with the lowercase *l* allele (Fig. 5.4b). Mendel pointed out that to get the 3:1 phenotypic ratio, eggs and sperm can come together totally at random with

Figure 5.3
Phenotype Versus Genotype

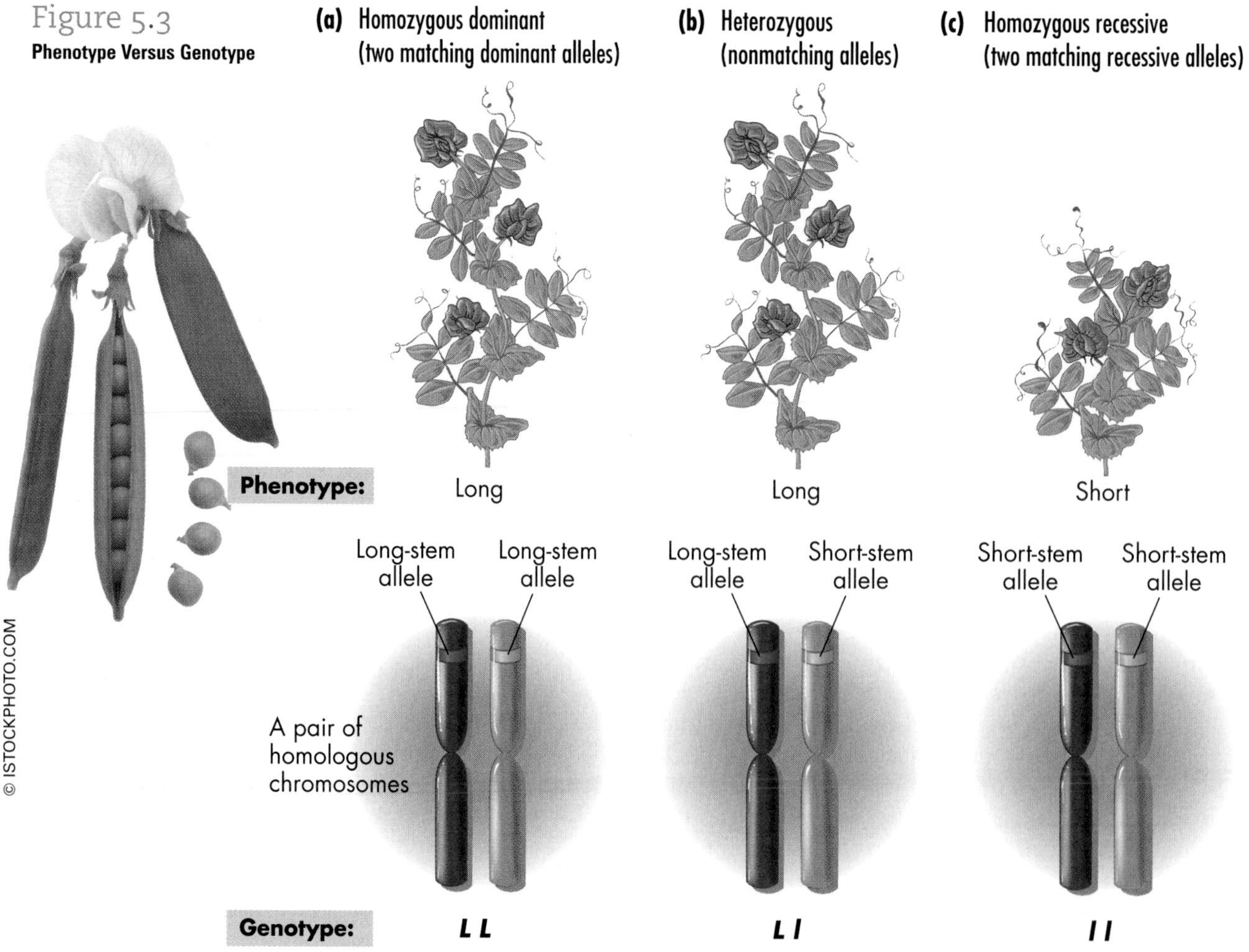

respect to the allele they carry (Fig. 5.4c). In other words, an egg cell with an *l* allele is just as likely to be fertilized by a sperm cell with an *l* allele as it is to be fertilized by a sperm carrying an *L* allele.

Figure 5.4
Meiosis and Mendel's Principle of Segregation

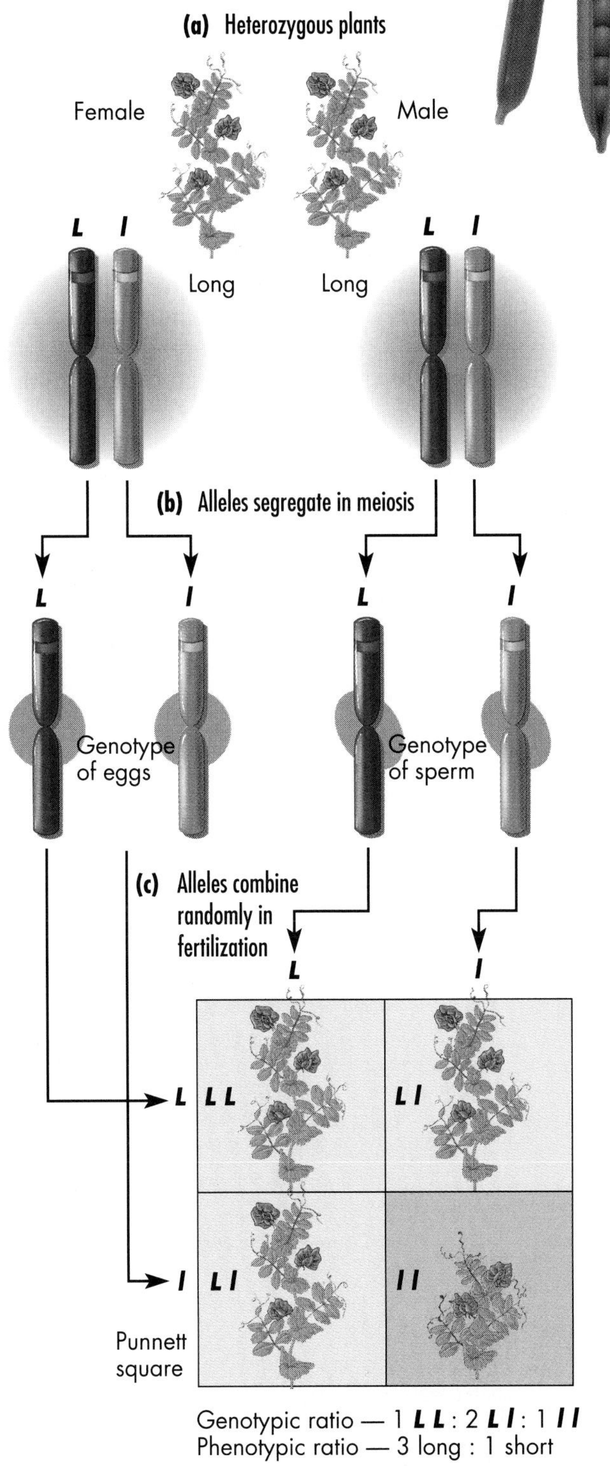

Punnett square
in genetics, a diagrammatic way of presenting the results of random fertilization from a mating

A good way to visualize the consequences of random fertilization is to draw an organized diagram called a **Punnett square**, as shown in Figure 5.4c. To construct a Punnett square for the mating of two heterozygous pea plants, draw a large square made up of four smaller squares. Along the top of the large square, write the two possible genotypes of the pollen (*L* and *l*) and along the left side of the square, write the two possible genotypes of the eggs (also in this case *L* and *l*). Then in the four empty boxes, fill in the genotypes of the offspring that result from the fertilization of each egg type with each pollen type.

Fig. 5.4c shows four F_2 genotypes: *LL*, *Ll*, *lL*, and *ll*. Because the order of alleles is not important, *Ll* and *lL* are equivalent; thus, there are really only 3 genotypes, found in the ratio 1 *LL* to 2 *Ll* to 1 *ll*. If we look at the physical characteristics of the plants themselves, however, we find that the 1:2:1 genotypic ratio produces a 3:1 phenotypic ratio (3 long stem to 1 short stem). The reason is that the single *LL* genotype and both *Ll* genotypes have the same long-stem phenotype, because *L* is dominant to *l*. Recall that this 3:1 ratio is very close to what Mendel observed in his experiment (787:277).

Genetics and Probabilities

Mendel's segregation principle predicts a 3:1 phenotypic ratio in the offspring from a mating of two heterozygotes. But he actually found 2.84:1 for the mating we discussed, and 3.15:1 and 2.96:1 for other similar matings involving different traits. Why don't the figures come out to exactly 3:1? The answer is that the principles of genetics rely on the laws of chance and probability.

You can demonstrate the probability of obtaining the 3:1 relationship by tossing two different coins simultaneously. Let a penny represent sperm from a pollen grain, and let a nickel represent an egg. The head of each coin represents the dominant allele, and the tail represents the recessive allele. Note that each coin has an equal number of dominant and recessive alleles, just like the population of gametes from a heterozygote.

To model fertilization, flip both coins at the same time and record whether they land heads up or tails up. If both are heads, the genotype is homozygous dominant; if both are tails, the genotype is homozygous recessive; and if one coin is heads and the other is tails, the "offspring" will be heterozygous. Flip the

pedigree
an orderly diagram of a family's relevant genetic history

pair of coins 20 times. How many times would you expect each of the three possible outcomes? Did you obtain exactly what you would expect? If not, how can you explain the discrepancy?

What would be the probable result if you tossed the coins many more times than 20, say, 1,064 times, as Mendel did when he was experimenting with pea stem length? Like the toss of a coin, the combination of alleles in fertilization is governed by the laws of chance. In a low number of trials, as you conducted, the results may differ substantially from those predicted for random tossing, but as the number of trials increases, the results will come closer to the mathematically predicted values. This principle is especially important when doing human genetics because of small family size, as we will see in the next section.

LO3 How Do Geneticists Analyze Human Inheritance Patterns?

How can we apply Mendel's principles of heredity to the problem of how likely a set of parents is to have children with cystic fibrosis?

Homo sapiens: An Inconvenient Experimental Animal

Even with Mendel's principles in mind, humans are uniquely difficult subjects for a geneticist to study. First of all, geneticists aren't matchmakers and can't convince people to choose mates and produce offspring just to satisfy their curiosity. Investigators must search for existing subjects and matings that happen to express traits of interest. In addition, there is never a true F_2 generation available for study because brothers and sisters rarely mate. Beyond that, individual human families are too small for statistical analysis; couples rarely produce more than ten children, and usually produce fewer than three. Finally, the human life cycle is too long. It could take an entire career to follow the traits in two human generations. So scientists and physicians rely heavily on collecting and analyzing family histories.

Pedigrees: Family Genetic Histories

A major method in human genetics is to follow the inheritance of a trait through all the members of a family. Geneticists search out families with particular genetic traits, and then interview family members, check their medical records, and collect samples of blood or other tissues from as many family members as possible. From such records, the investigator draws up **pedigrees**, orderly diagrams that show family relationships, birth order, gender, phenotype, and, when possible, the genotype of each family member.

To see how a family pedigree works, let's consider the family tree of a family with the cystic fibrosis trait (Fig. 5.5). In a pedigree, each generation occupies a separate horizontal row, with the ancestors at the top and more recent generations below. Males are indicated by squares and females by circles. Symbols for people affected with the trait are filled in (red, in this case). Geneticists designate each generation with a Roman numeral (I, II, etc.) and each individual with an Arabic number (1, 2, etc.) from left to right. For example, the matriarch of the family, a woman born in 1859, is I1, and her two daughters are II2 and II3. The boy and girl VI1 and VI2 are the only family members with cystic fibrosis. As in many pedigrees, this one arranges a group of brothers and sisters in order from oldest (left) to youngest (right).

Another convention is that a horizontal line joins two parents, and the offspring are attached to the line below. Parents II1 and II2, for example, produced two daughters and two sons, individuals III2 to III5. Geneticists sometimes omit from a pedigree parents who are unrelated and unaffected. They also tend to show consanguineous marriages (unions between blood relatives) with double horizontal lines like the ones in Figure 5.5.

Figure 5.5

Pedigree for a Family with Cystic Fibrosis

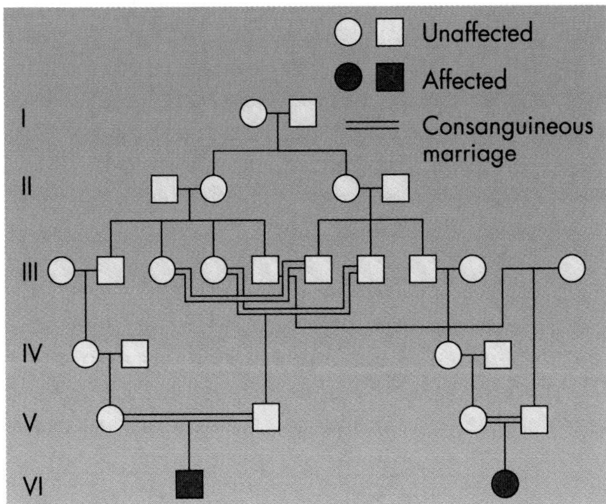

Is Cystic Fibrosis Inherited as a Recessive or Dominant Trait?

A pedigree can look rather formidable, with its marching rows of grandparents, aunts, brothers, and sisters. Nevertheless, the rules for analyzing a pedigree follow Mendel's principles. Let's analyze our hypothetical family's pedigree more closely. By doing this, we can determine whether cystic fibrosis shows a pattern of dominant or recessive inheritance.

The pedigree (Fig. 5.5) shows that several of the siblings in the family lacked the cystic fibrosis trait; neither parent showed the trait, either. Let's represent the cystic fibrosis gene by *CF*, with CF^- representing the disease allele and CF^+ the healthy allele. Because our affected female VI2 has cystic fibrosis, she must have at least one CF^- allele and must have inherited it from one of her parents. That shows that at least one of the parents must carry at least one CF^- allele. Since neither parent shows the trait, we can conclude that each parent has at least one CF^+ allele, and that at least one parent is a heterozygote with one CF^+ allele and one CF^- allele. Because both parents are healthy, we must conclude that a person with the heterozygous genotype has a healthy phenotype. We saw earlier that in a heterozygote, the dominant allele shows. Therefore, the CF^+ (healthy) allele must be dominant to the CF^{--} (disease) allele. In other words, the allele that causes cystic fibrosis must be recessive and the genotype of VI2 is $CF^- CF^-$. To generalize this argument: If an offspring inherits a condition but neither parent shows the condition, the trait is usually recessive.

Because the disease allele of the cystic fibrosis gene is recessive, affected people must have two copies of the disease allele. One of those copies must have come from the dad, and the other from the mom. Since both parents passed on the disease allele but do *not* show its effects, they must be heterozygotes, or carriers. In carriers, the dominant normal allele masks the recessive allele, which is mutant (the result of a mutation).

It turns out that many human genetic diseases are inherited as recessive traits like cystic fibrosis. Some, however, are inherited as dominants. Let's look at some of each.

carrier
[1] in genetics, a heterozygous individual not expressing a recessive trait but capable of passing it on to her or his offspring; [2] in biochemistry, a substance, often a protein, that transports another substance

mutant
the allele that results from a mutation; also used to refer to the organism containing such a mutation

sickle-cell anemia
a genetic condition inherited as a recessive mutation in a hemoglobin gene and characterized by pains in joints and abdomen, chronic fatigue, and shortness of breath

How Alleles Interact

Experimenting in his quiet abbey garden, Mendel showed that each gene—for plant height, flower color, and so on—has two alleles, which are either dominant or recessive. Life, however, is not always so simple, as later geneticists found with more sophisticated experiments. The alleles of some genes fail to fall clearly into either the dominant or recessive category, and some genes have many more than two alleles.

Incomplete Dominance

In 1905, a young African American experiencing pains in his joints and abdomen, chronic fatigue, and shortness of breath consulted a Chicago physician. A blood test showed that the man had too few red blood cells (a condition called *anemia*) and that many of his blood cells were shaped like crescents, or sickles, instead of the normal disks (Fig. 5.6a). Studies revealed sickle-shaped blood cells to be fragile and easily destroyed. This condition is called *sickle-cell anemia*, and it's the most commonly inherited lethal disease among African Americans.

A condition related to sickle-cell anemia is called *sickle-cell trait.* In people with sickle-cell trait, red blood cells form a sickle shape when deprived of oxygen in a test tube—a condition that fails to induce sickling in normal red blood cells. People with sickle-cell trait are normal except when exposed to extreme conditions, such as high altitude or severe physical exertion. For example, several men with sickle-cell trait living in low-altitude cities suffered severe spleen pain within two days of arrival in a part of Colorado with high altitudes. Sickle-cell trait is

incomplete dominance the genetic situation in which the phenotype of the heterozygote is intermediate between the phenotypes of two homozygotes

codominance the genetic situation in which both alleles in a heterozygous individual are fully expressed in the phenotype; this is a characteristic of human blood types

thus intermediate in severity between full-blown sickle-cell anemia and normal health. What is its genetic basis?

Figure 5.6b shows a pedigree for sickle-cell anemia in a family from Jamaica. Notice that each person with sickle-cell anemia has two parents who both display sickle-cell trait. By examining a large number of families with sickle-cell anemia, geneticists have found that the mating of two people with sickle-cell trait produces offspring in a 1:2:1 ratio with 1/4 showing full blown sickle-cell anemia, 1/2 showing sickle-cell trait, and 1/4 showing neither condition. The Punnett square in Figure 5.6c reveals the origin of this 1:2:1 phenotypic ratio. Sickle-cell anemia displays incomplete dominance: the phenotype of heterozygotes—in this case individuals with sickle-cell trait—is intermediate between the homozygous dominant and the homozygous recessive conditions. In incomplete dominance, the phenotypic and genotypic ratios are the same. The principle of incomplete dominance can help explain why early observers devised the incorrect blending hypothesis of inheritance (review LO1). Several genes act together to control petal length in monkey flowers, for example, and these are inherited in typical Mendelian fashion—except that their alleles display incomplete dominance.

Codominance

We just discussed incomplete dominance, in which the phenotype of the heterozygote is intermediate between the two homozygotes. Another variation is called codominance, in which the phenotype of the heterozygote simultaneously shows *both* phenotypes. A familiar example of codominance is the blood-type gene called *ABO* (Fig. 5.7a). For this gene, allele A causes a certain carbohydrate molecule called type A to appear on the surface of red blood cells, and a

Figure 5.6
Sickle-Cell Anemia: A Pattern of Incomplete Dominance

(a) Sickle cells

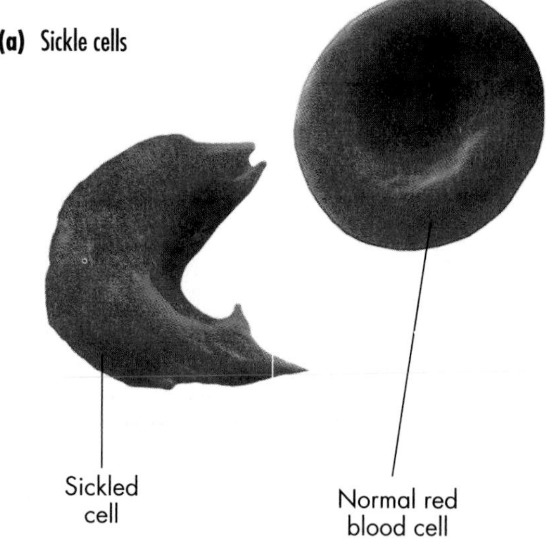

(c) Punnett square

Male with sickle-cell trait (columns); Female with sickle-cell trait (rows)

	HBB +	HBB Ò
HBB +	HBB + HBB + Non-sickling	HBB + HBB Ò Sickle-cell trait
HBB Ò	HBB + HBB Ò Sickle-cell trait	HBB Ò HBB Ò Sickle-cell anemia

(b) Pedigree

I
II
III
IV

Unaffected
Sickle-cell trait
Sickle-cell anemia
Deceased

Figure 5.7

Blood Groups, Incomplete Dominance, and Multiple Alleles

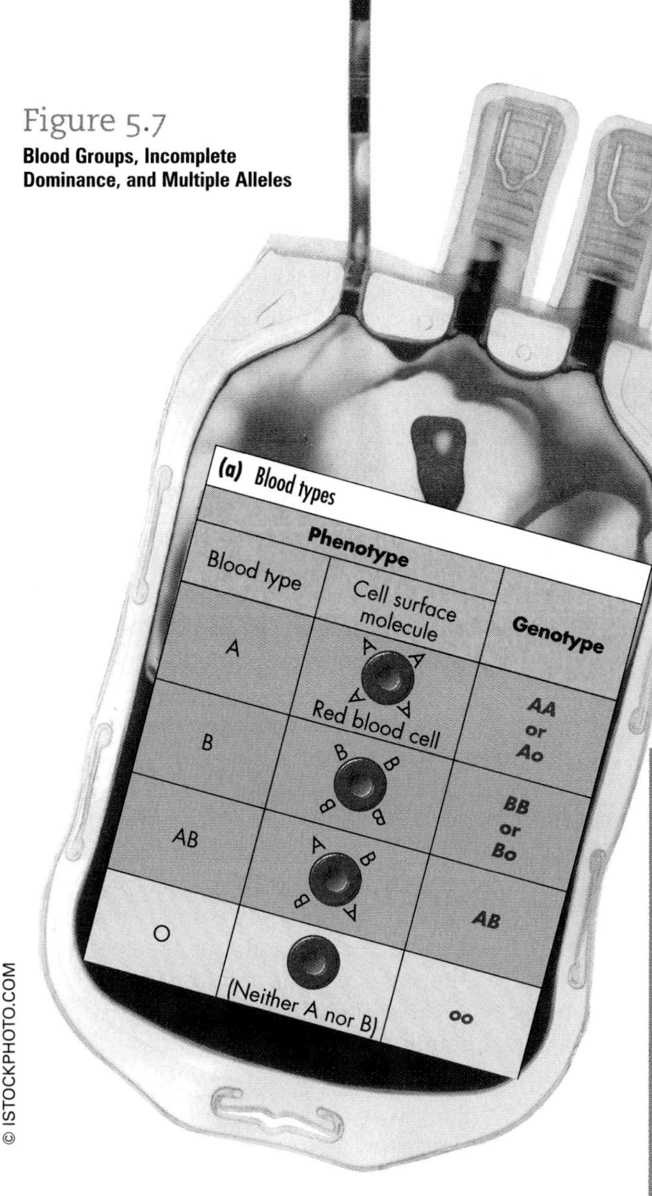

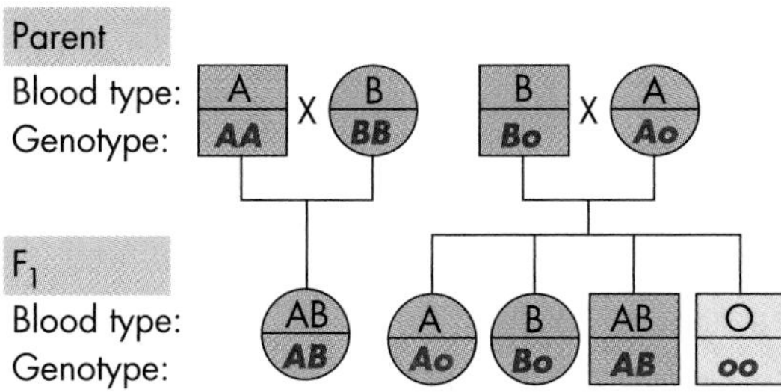

person with this allele may have blood type A. Allele B causes a different carbohydrate molecule called type B, to appear on the surface of red blood cells, and produces blood type B. Someone who has two A alleles has only the A molecule, and a person with two B alleles has only the B molecule. But a heterozygote with one A allele and one B allele has both A and B molecules on the surface of red blood cells. That person has blood type AB, a codominant phenotype.

Students are sometimes confused by the difference between codominance and incomplete dominance. In codominance, *both* alleles are fully expressed in the heterozygote, while in incomplete dominance, the phenotype is intermediate. AB blood type is not intermediate between A and B: it is fully A *and* B. But people with sickle-cell trait are nearly normal in phenotype, becoming ill only under extreme circumstances, and thus incomplete dominance is at work.

major histocompatibility complex (MHC) a complex of proteins that are specific for each individual and are the factors that cause the body to reject transplanted tissues; their main function is to aid in communication among immune cells

Multiple Alleles

If your blood type is O, not A, B, or AB, then you are a good example of another genetic concept: some genes have more than two alleles, and the human *ABO* blood group gene is an example. In addition to the two codominant alleles *A* and *B*, the *ABO* gene has a third allele that is fully recessive to both A and B. This recessive allele is called *o*. A person with two doses of *o* has neither the A nor B molecular marker and has blood type O. Because *o* is recessive, an *Ao* heterozygote has blood type A, and a *Bo* heterozygote has blood type B. Although there are three alleles of the *ABO* gene found in the human population, no one person can have all three at once, because each child gets only *one* allele of each gene from each parent, for a total of two copies of each gene. Figure 5.7b shows some pedigrees for the *ABO* gene.

The *ABO* gene has three alleles, but some genes have even more than that, and this is important to patients with severe cystic fibrosis who have received transplanted organs. Because the effects of cystic fibrosis on the lungs and airways can be life-threatening, hundreds of cystic fibrosis patients have received transplanted lungs as a treatment of last resort. Unfortunately, as you have probably read in newspapers or magazines, tissue transplants from unrelated people are likely to be rejected, and the reason hinges on multiple alleles.

The major histocompatibility complex (MHC) is a group of genes that encode certain proteins on cell surfaces. These substances serve as identification markers that help the body distinguish its own

> In codominance, *both* alleles are *fully* expressed in the heterozygote, while in incomplete dominance, the phenotype is intermediate.

parental type
an offspring having the characteristics of one of the parents

recombinant type
an offspring in which characteristics of the parents are combined in new ways

independent assortment
Mendel's second law, or principle; the random distribution of genes located on different chromosomes to the gametes

cells from foreign substances like bacteria, viruses, or parasites that might otherwise successfully invade the body and cause disease. Because there are several genes in the MHC, and because each gene has many alleles, it is highly unlikely that two unrelated persons will have precisely the same combination of alleles. That is why a person with kidney or liver disease or lung damage from cystic fibrosis must often wait a long time before being matched with a suitable donor. If there are too many allelic differences between the tissues of donor and recipient, the immune system cells of the recipient will kill the cells of the donated organ. Even with immunosuppressant drugs that help stop tissue rejection, the multiple alleles for tissue types constitute a major obstacle to many life-saving transplants.

LO4 What Are the Rules Governing How Organisms Inherit Multiple Traits?

So far, we have discussed single genes and their alleles, whether for flower color or cystic fibrosis. But organisms have thousands of genes. A single-celled yeast has about 6,000 genes, a soil-living roundworm has about 19,000, and you have about 40,000. What happens when a geneticist studying the cross between two individuals focuses on more than one gene at the same time in the same cross?

Formation of Gametes

Suppose a female is homozygous for the cystic fibrosis trait and has two recessive alleles: CF^-CF^-. She also has blood type A and is homozygous dominant for this blood type, with the genotype *AA*. Suppose a male is heterozygous CF^+ CF^-, and he is homozygous recessive for blood type O, with genotype *oo*. Suppose they decided to have a child and produced a son who is simultaneously heterozygous for cystic fibrosis CF^+ CF^- and has blood type A with the genotype *Ao* (one dominant *A* allele, one recessive *o* allele of the *ABO* blood group gene). This son's genotype for the two traits would then be CF^+ CF^- *Ao*.

When their son grows up and begins to generate gametes, what types would he make? Recall that Mendel's principle of segregation says that each gamete gets one copy of each gene, and so each gamete must have one copy of the cystic fibrosis gene and one copy of the blood type gene. Mendel's principle further suggests that half of the sperm cells would get a CF^- allele and the other half a CF^+ allele. Likewise, half of the sperm would get an *A* blood type allele and the other half an *o* allele. So the son's sperm could be of four types:

But in what proportions will these four types of gametes actually form? Would the sperm possess only the parent's original genotypes, CF^- A and CF^+ o? Geneticists call these **parental types** because they are like the original parents. Or would some of the sperm also possess the new combinations CF^- o and CF^+ A? Geneticists call these types the **recombinant types** (see Chapter 4), because they are "recombined" and not present in the original parents.

In working with peas, Mendel found that alleles of different genes move independently into the gametes, a process he called **independent assortment**. What it means is that the segregation of a particular allele pair into separate gametes is independent of other allele pairs. As a result, all four types of gametes (such as those we just showed for cystic fibrosis and blood type) are equally likely, each occurring one quarter of the time.

One way to visualize this is to use the following diagram:

Cystic fibrosis alleles		ABO alleles		gametes
1/2 CF^-	→	1/2 *A*	→	1/4 CF^- *A*
	→	1/2 *o*	→	1/4 CF^- *o*
1/2 CF^+	→	1/2 *A*	→	1/4 CF^+ *A*
	→	1/2 *o*	→	1/4 CF^+ *o*

This diagram shows the essential features of Mendel's principle of independent assortment. For genes that are inherited independently of each other, half of the gametes from an individual that is doubly heterozygous are of the parental type (CF^-A and CF^+o) and half are of the recombinant type (CF^-o and CF^+A).

Recall that Mendel always took his experiments through an F_2 generation. So let's consider what would happen if the theoretical son we've been discussing married a woman with exactly his same genotype for cystic fibrosis and blood type (CF^+ CF^- *Ao*). Furthermore, since this is hypothetical, let's say that this couple had hundreds of children so we can

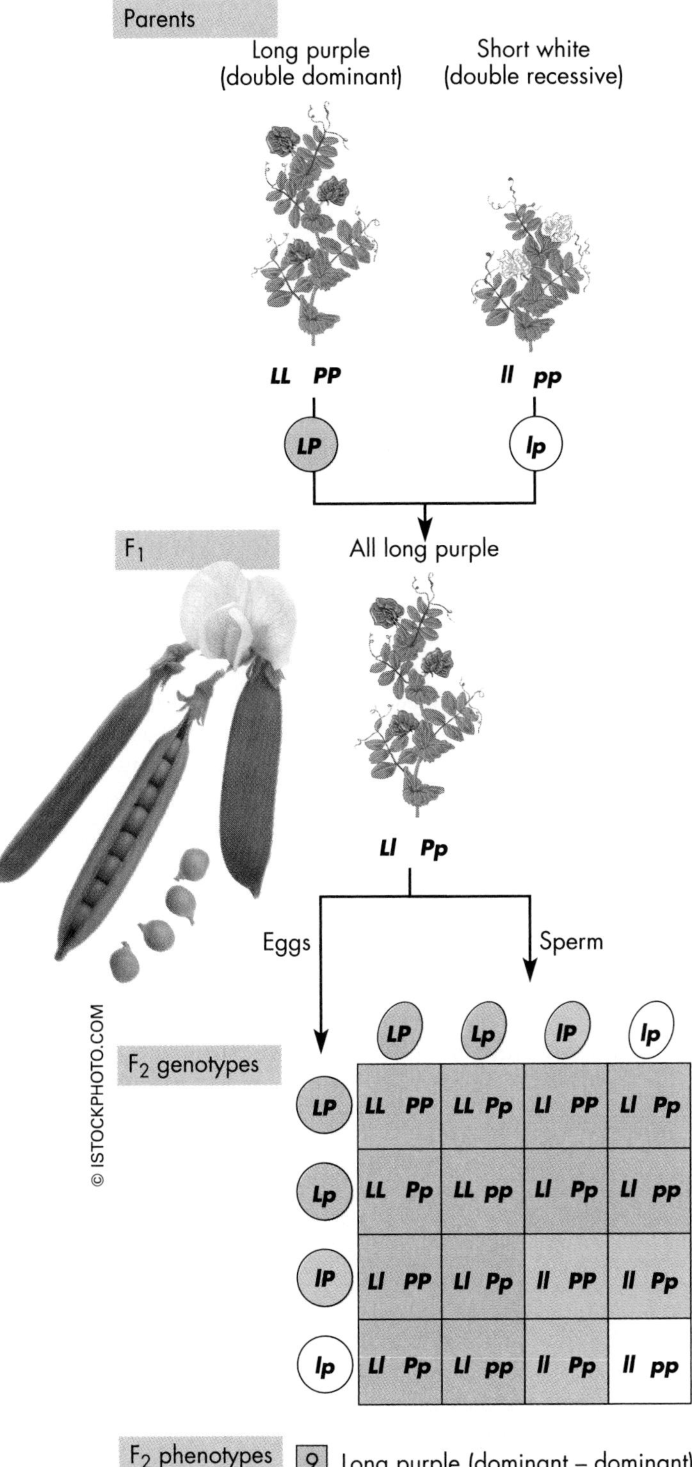

Figure 5.8

Independent Assortment in Pea Plants and People

obtain statistically meaningful results. What genotypes would the children have and in what proportions? Mendel carried out similar crosses, which he called dihybrid crosses (that is, crosses following two traits), with pea plants. He followed both long and short stems and purple and white flowers (represented by these alleles *LL*, *Ll*, *ll* and *PP*, *Pp*, *pp*). Figure 5.8 shows the results in the offspring, and we can see that the phenotypic ratio is 9:3:3:1. We can apply Mendel's results to the human situation as well and get the same 9:3:3:1 ratio. What we would see is 9 double dominant showing both the healthy (noncystic fibrosis) phenotype and the A blood type, 3 with the recessive cystic fibrosis phenotype but the dominant A blood type, 3 with the dominant noncystic fibrosis phenotype but the recessive O blood type, and 1 double recessive with both the cystic fibrosis phenotype and the O blood type. (Note that $9 + 3 + 3 + 1$ adds up to 16, which is the number of squares in the 4×4 Punnett Square in Figure 5.8.)

dihybrid cross a mating between two individuals in which the investigation follows the inheritance of only two traits

9:3:3:1 ratio the ratio of phenotypes found in the offspring of two individuals, both of whom are heterozygous for two traits whose alleles assort independently

In summary, Mendel's second principle shows that different hereditary factors segregate into gametes independently of each other. As a consequence of independent assortment, we see the 9:3:3:1 ratio in the F_2 generation.

Genes Are Located on Chromosomes

There are several parallels between the inheritance of genes and the distribution of chromosomes during meiosis:

1. Two copies of each gene and two copies of each chromosome exist in each body cell.
2. Pairs of alleles and pairs of homologous chromosomes both segregate during gamete formation.
3. Genes for different traits and nonhomologous chromosomes both assort independently when egg and sperm are formed.

These facts suggest that genes are physically linked to chromosomes. To test that possibility, early investigators had to locate individual chromosomes and show that when an organism inherits that chromosome, a specific trait is always transmitted with it. That became possible by investigating sex chromosomes.

autosome
a chromosome other than a sex chromosome

sex chromosomes
pairs of chromosomes where the members of the pair are dissimilar in different sexes and are involved in sex determination, such as the X and Y chromosomes

X chromosome
the sex chromosome found in two doses in female mammals, fruit flies, and many other species

Y chromosome
the sex chromosome found in a single dose in male mammals, fruit flies, and many other species

mutation
any heritable change in the base sequence of an organism's DNA

LO[5] How Does Sex Influence the Inheritance of Traits?

The pedigree we looked at earlier for cystic fibrosis show about the same number of affected males as affected females. Many other genetic conditions, however, such as *color blindness* (the inability to see specific colors) and *hemophilia* (inability to form a blood clot) are much more prevalent in males than in females. By investigating traits influenced by sex, early 20th-century geneticists were able to show that genes are indeed located on chromosomes.

Sex Chromosomes

Have you ever wondered why there are roughly as many boy babies as girl babies (the actual ratio is about 106 boys to 100 girls)? A *karyotype* or arrayed set of chromosome photographs reveals why (Fig. 5.9). For 22 of our 23 chromosome pairs, both members are identical in size and shape. For the 23rd chromosome pair, however, males and females differ. Chromosome pairs in which both chromosomes look the same in both sexes are **autosomes**, while chromosome pairs with dissimilar members in males and females are **sex chromosomes**. Humans have 22 pairs of autosomes and one pair of sex chromosomes. Females have two identical sex chromosomes, called **X chromosomes**, and males have one X chromosome and another, often smaller chromosome called a **Y chromosome**. Although sex chromosomes are common in animals, they are rarely found in plants, fungi, or protists.

The way sex chromosomes become distributed during meiosis explains the appearance of about equal numbers of males and females. An XY male is like the heterozygous parent, and an XX female is like the homozygous parent (see Fig. 5.10). In the male, the X and Y segregate during meiosis, and as a result, one half of the sperm contain a Y chromosome and one half an X chromosome. In the female, the two X chromosomes segregate during meiosis; as a result, each egg contains one X chromosome. If the X and Y sperm randomly fertilize a group of eggs, then half of the zygotes formed will be male (XY) and half female (XX). Note that a male's single X chromosome must be inherited from his mother. Because males and females have different chromosomes, we know that at least one trait—sex—is regulated by chromosomes. But are there any others?

Figure 5.9
A Karyotype: A Set of Human Chromosomes

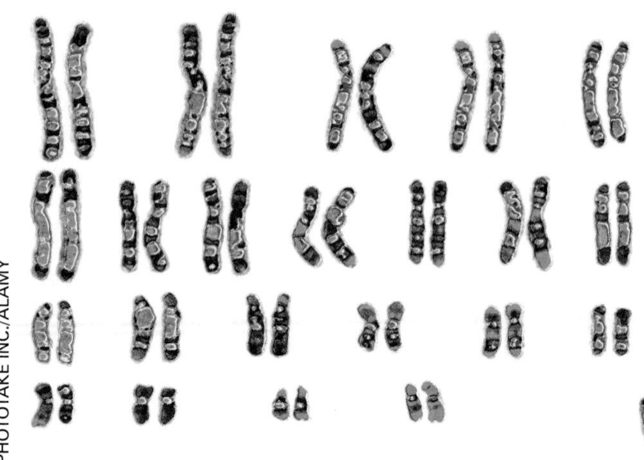

Sex-Linked Traits

In 1910, Thomas Hunt Morgan and his associates at Columbia University began a series of experiments that would change genetics forever. Morgan wanted to find out about genes and chromosomes but did not have Mendel's monastic patience, so he chose the fast-breeding fruit fly, *Drosophila melanogaster.* No bigger than an "l" in this sentence, fruit flies are easy to raise and breed, and in just 12 days, an egg becomes a reproductive adult ready to produce hundreds of offspring.

One day, as Morgan observed fruit flies under the microscope, he noticed a fly with white eyes instead of the usual red. A **mutation**—a permanent change in the genetic material—had altered a gene for eye color from the normal red-eye allele (symbolized by fly geneticists as w^+) to the mutant white-eye allele (w).

From a series of crosses like the one shown in Figure 5.10, Morgan realized that the gene for eye color is carried on the X chromosome. That gene is therefore a sex-linked gene, or more specifically, an **X-linked** gene.

Other genes were found on the fly X chromosome, including *yellow* body and *singed* bristles. After studying such genes, Morgan and his coworkers drew an important conclusion: The Y chromosome, being considerably smaller, carries no allele of the gene for eye color or for most of the other X-linked genes. From this series of experiments, Morgan drew three important conclusions:

1. genes are located on chromosomes,
2. each chromosome carries many different genes, and
3. genes on the X chromosome have a distinct pattern of inheritance.

X-linked
characteristic of a heritable trait that occurs on the X chromosome

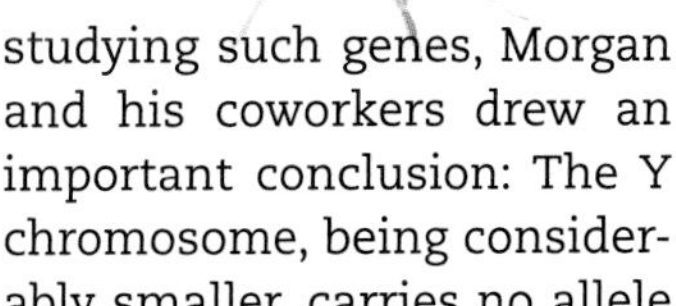

Figure 5.10
Fly Eyes and Sex-Linked Inheritance

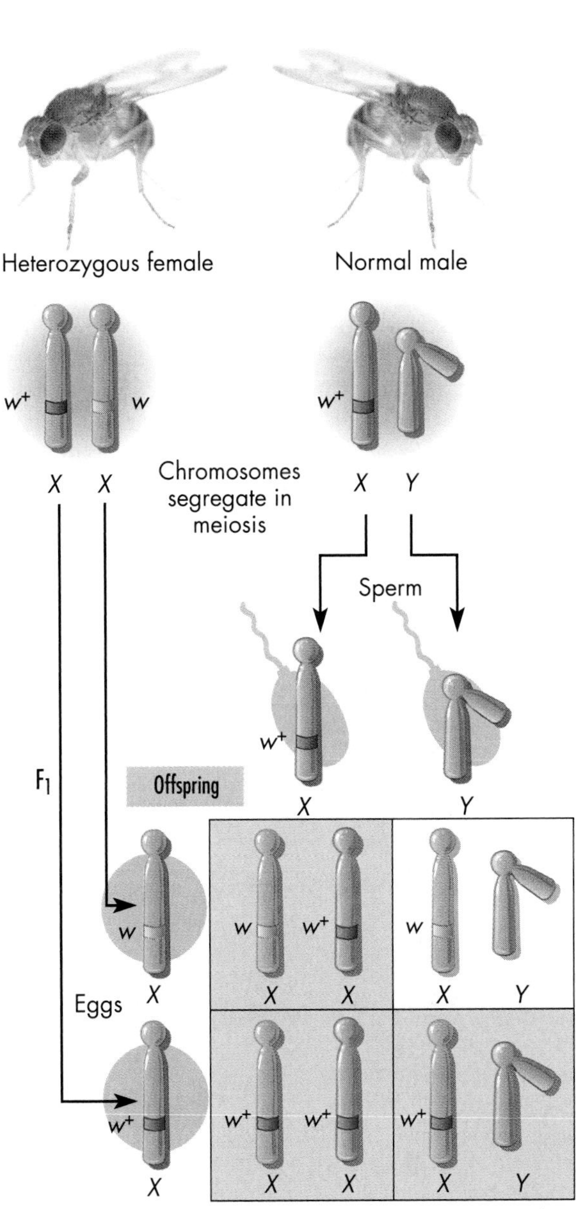

Chromosomes and Sex Determination

The historic experiments with sex chromosomes in fruit flies showed that the expression of male and female characteristics depends on chromosomes. They did not show, however, how sex determination works. Consider this: In flies as well as in people, XX individuals are females and XY individuals are male. But in both cases, males differ from females in two factors: the presence or absence of a Y, and the number of X chromosomes. Which is more important? To answer, geneticists have studied individuals with unusual numbers of sex chromosomes. Before reading on, look at the data in Table 5.2 to see if Y or X chromosomes are more important for flies. Then do the same for humans.

By analyzing the data in the table, you can see that any fly with at least two X chromosomes is a female, regardless of how many Y chromosomes she has. What about humans? Did your analysis of the table show that people with a Y chromosome

Table 5.2
Chromosomes and Sex Differentiation

CHROMOSOME CONSTITUTION	FRUIT FLY SEX	HUMAN SEX
X	Male	Female (Turner syndrome)
Y	Lethal	Lethal
XY	Male	Male
XX	Female	Female
XXY	Female	Male (Klinefelter syndrome)
XYY	Male	Male
XXX	Female	Female
XXXY	Female	Male

SRY specific gene on the Y chromosome that has been isolated as the "sex-determining region" and necessary for causing male development in mammals

develop as males, regardless of the number of X chromosomes? Good! This fact indicates that there must be a genetic factor on the human Y chromosome that is essential for producing the male phenotype. Geneticists have isolated that gene, and named it SRY, for "sex-determining region, Y chromosome." They still don't know yet exactly how it works, but in some way, SRY turns on some or all of the genes that stimulate the male phenotype and suppresses those leading to the female phenotype.

More evidence of human sex determination comes from the Turner and Klinefelter syndromes. A person with one X and no Y chromosome (XO) is a sterile female with Turner syndrome. About 1 in 2,200 newborns show this condition, characterized by folds of skin along the neck, a low hairline at the nape of the neck, a shield-shaped chest, and later in life, failure to develop adult sexual characteristics at puberty. About 1 newborn male in 1,000 has two X chromosomes and one Y chromosome (XXY), a condition called Klinefelter syndrome. Affected people develop as sterile males with small testes, long legs and arms, and somewhat diminished verbal skills, although their IQ scores are near normal. Most men with Klinefelter syndrome manage well in society, and many are unaware of their chromosomal abnormality until they marry and are unable to father a child.

Y-Linked Genes

We said earlier that the human Y chromosome was small, and you can see that on Figure 5.9. The Y actually contains only about 20 genes, in contrast to about 1000 on the X chromosome. A few Y chromosome genes have copies on the X chromosome, but most do not. Most of those 20 Y-chromosome genes are expressed only in the testes, where they are probably responsible for male fertility. Many sterile men who come to fertility clinics have mutations in a Y-linked gene. These mutations probably arose in a single sperm cell in the sterile man's dad, because a man whose Y chromosome carries a sterile mutation will have no offspring. If a Y-linked gene had a phenotype other than sterility, then an affected man would pass the trait to all of his sons and none of his daughters.

X Chromosome Inactivation

A person can get along quite well with only one X chromosome even though the absence of two copies of any other chromosome causes death before or shortly after birth. What's so special about the X? The answer is, no matter how many X chromosomes are present, *both sexes have only a single functional copy of it*. At the stage when a female human embryo (or other mammal) consists of only about a thousand cells, one of the X chromosomes in each of her cells becomes genetically inactive—it no longer reads out any genetic information. Hence, the genes on that inactivated chromosome can have no effect on the phenotype. After one of the X chromosomes in a female embryonic cell becomes inactive, all of the millions of daughter cells derived from it will have the same inactive X. Thus, a female mammal is a mosaic of cells containing active X chromosomes of maternal or paternal origin. You can see this mosaicism in the patches of black and orange fur in a calico cat.

In a female with two Xs (the typical number), a geneticist can see the inactive X chromosome in cells scraped from the inside of the mouth as a small, dark spot on the edge of a nucleus. XY males, who have no inactive X, lack this dark spot in the nucleus. For several Olympic games before 1992, officials relied on this procedure as a test to certify the "femininity" of female athletes, regardless of the other sex chromosomes.

LO6 How Is Genetics Changing Our World?

The majority of cystic fibrosis patients are now surviving into midlife, although there has been a leveling off of life expectancy over the last few years. To move beyond the environmental remedies now available (diet, airway clearing therapies, and antibiotics), new therapies are needed to extend patients' lives further. To this end, researchers are mapping genes, learning what the genes do in healthy and diseased cells, and working on molecular solutions. The history of gene mapping helps explain the pivotal importance of this technique.

Gene Mapping

Gene mapping, the assignment of genes to specific locations or loci (singular *locus*) along a chromosome, was developed in fruit flies, like so many other genetic principles. Recall that Thomas Hunt Morgan used flies to show that a single chromosome can carry many genes—that is, to show genetic linkage. In 1913, an enterprising undergraduate in Morgan's genetics laboratory, Alfred Sturtevant, showed that genes lie in a straight line along a chromosome and that simple mating experiments can map the genes—that is, reveal their order and relative distance from each other.

Gene Mapping and the Human Genome Project

In 1985, Dr. Francis Collins of the University of Michigan used gene mapping information to identify all the base pairs in the cystic fibrosis gene *CFTR*, and how it works. Today Dr. Collins heads the Human Genome Project at the National Institutes of Health. The goal of the project was the complete sequencing of the human genome, that is, the precise order of As, Ts, Gs and Cs for all 23 human chromosomes. This would be the ultimate gene map, revealing the position of every human gene.

By the year 2000, the sequencing of the entire human genome, all 3 billion nucleotides, was completed. Biologists were amazed to learn that humans have far fewer genes than they had previously thought. For instance, chromosome 22 has just 545 genes and chromosome 21 has just 225 genes. This suggests that we humans have just 40,000 genes, rather than the 100,000 to 140,000 genes they once predicted. This is truly amazing when you consider that a nematode worm has 19,000 genes and a fruit fly has 14,000 genes. How can humans have only two to three times as many genes as worms and flies? The answer may lie with the escalating ways genes can interact as their numbers increase, but a full understanding awaits further research.

Sequencing the human genome is important because researchers will have a far easier time now identifying the structure and activity of disease-related genes. As a result, they'll be better able to design new drugs to treat debilitating conditions. Knowing the sequence may also help geneticists determine why different people respond differently to therapies for serious diseases.

Isolating Disease Genes

Knowing a gene's nucleotide sequence can help reveal its function. By isolating and then analyzing the cystic fibrosis gene, for example, researchers showed that it causes cell membranes to make a particular protein, which causes aberrant ion transport in the ducts of the lungs, pancreas, sweat glands, and other organs leading to sticky mucus build up. This discovery gave patients and their doctors hope that we would someday have a cure for this fatal genetic disease.

The mapping and detection of diseases has another important application: revealing unaffected carriers (heterozygotes) of the disease. Mutations change DNA structure, and this fact can be used to help detect carrier status. If two people who are carriers know their status as heterozygotes for a disease gene, they can learn (often with the help of a genetic counselor) what the chance would be of passing the disease to their own child. After conception, physicians test early-stage fetuses for genetic diseases, including the most common forms of cystic fibrosis.

gene mapping
the assignment of genes to specific locations along a chromosome

locus (pl. loci)
the location of a gene on a chromosome

linkage
alleles of two genes located so close to each other on the same chromosome that they fail to assort independently

Human Genome Project
the research effort to sequence the entire set of human genes and to understand their functions

Treating Genetic Disease

To researchers, doctors, and patients, developing adequate treatments for genetic diseases is an urgent problem; living with a disease and making reproductive and other decisions are very real, day-to-day issues. The future of genetic research, based on Mendel's principles of heredity, is sure to be fascinating, powerful, and with any luck, life-extending.

Who knew?

You can't always tell genotype by looking at phenotype. For dominant characteristics, the homozygous genotype and heterozygous genotype are phenotypically indistinguishable. For example, if B (black fur) is dominant to b (white fur), both BB and Bb will exhibit black fur.

6

DNA: The Thread of Life

Learning Outcomes

LO[1] Understand how scientists identified the nature of hereditary material LO[2] Describe the structure of the DNA molecule LO[3] Identify the chemical structure of DNA and how it carries information LO[4] Explain how DNA's structure allows it to copy itself

"We knew that you can get only about one-tenth as much DNA from an animal hair root as from a human hair."

A Cat, A Crime, An Identity

One day in October 1994, a 32-year-old mother of five named Shirley went out on an errand and failed to return to her home on Prince Edward Island, Canada. Within days, the Canadian police found her car abandoned in the woods and splattered with blood, soon identified as Shirley's. Toward the end of October, soldiers on a military maneuver found a man's leather jacket hidden in underbrush and also spotted with Shirley's blood. In May 1995, a fisherman spotted an earthen mound in the woods and dug up Shirley's decomposing body.

What do you know?

Why was the structure of DNA so important that researchers raced to find a model?

The detective in charge of what was now a murder investigation began amassing evidence. The chief suspect was Douglas, Shirley's estranged husband, who had been released from prison not long before Shirley disappeared. Did the discarded jacket belong to Douglas? The crime lab found no blood, human hairs, saliva, or other traces that could have linked the clothing to the suspect. They did, however, find 27 white cat hairs in the jacket's lining, and since his release from prison, Douglas had been living with his parents and their white cat Snowball! The detective arrested the suspect and contacted the U.S. National Institutes of Health in Frederick, Maryland. Through the Internet, the detective had learned that a laboratory group at NIH had for years been studying the DNA of felines—cheetahs, tigers, pumas, ocelots, and domestic cats. If anyone could tie the cat hairs to Douglas it was this team of scientists.

The analysis fell to Dr. Marilyn Menotti. Among the cat hairs, she found four with roots, each one made up of several skin cells and each cell containing DNA in its cell nucleus. "We knew," says Menotti, that you can get "only about one-tenth as much DNA from an animal hair root as from a human hair." So the task was to carefully extract the minute quantities of DNA, amplify it with a special copying procedure, and then identify its unique "fingerprint."

After months of testing and retesting, the team confirmed that the DNA from the hair root in the jacket's bloody lining matched Snowball's DNA exactly and would be unlikely to turn up by chance even in millions of domestic cats of the genotypes on the island. Dr. Menotti testified before the Supreme Court of Prince Edward Island in 1996. The lab work was careful and irrefutable. And the jury, convinced that the cat hairs were Snowball's and the jacket had been worn by Douglas, convicted the defendant of murder.

In this chapter, you'll find a story within a story. Inside the case history of cat hairs and a murder trial, you'll see the race to discover the famous **double-helix** shape of DNA.

double helix
the term used to describe the physical structure of DNA, which resembles a ladder twisted along its long axis

You'll also see how that shape explained the molecule's information coding and copying. Along the way, you'll find answers to these specific questions:

- How did scientists identify the nature of hereditary material?
- How is DNA like a twisted ladder?
- What is the exact chemical structure of DNA and how does it carry information?
- How does that structure allow DNA to copy itself?

LO1 Identifying the Hereditary Material

The DNA that Dr. Menotti extracted from one of Snowball's hairs contained the animal's hereditary instructions. Along with environmental factors, it determined Snowball's unique combination of body size; hair color, texture, and length; eye color; disposition; and hundreds of other traits both obvious and subtle, morphological and biochemical. As in all living organisms, the cat's DNA contained its *genes,* the very same hereditary "factors" Mendel discovered over 150 years ago (see Chapter 5). At that time, however, Mendel had no idea what these hereditary factors might be made of or how they might work. The first step toward our present knowledge of genes was the insight that genes occur in chromosomes. Biologists were able to determine in the 1930s that chromosomes in eukaryotes contain mainly protein, RNA, and DNA. They presumed that one of those substances had to make up the genes carrying genetic information. But which one?

In the 1940s, some geneticists argued that only proteins were versatile enough molecules to carry the complex information in genes. These biologists pointed out that proteins are constructed from 20 different subunits (amino acids), while DNA has only four subunits (nucleotides): the nucleotide bases adenine (A), thymine (T), guanine (G), and cytosine (C). Clearly, 20 amino acids can form many more combinations—and hence carry much more information—than 4 bases, just as you can form more words from an alphabet of 20 letters than from an alphabet of 4 letters. In the mid-20th century, geneticists devised tests to determine whether genes are made of protein or nucleic acid, and these experiments led to a biological revolution.

Clearly, 20 amino acids can form many more combinations—and hence carry much more information—than 4 bases, just as you can form more words from an alphabet of 20 letters than from an alphabet of 4 letters.

Evidence for DNA: Bacterial Transformation

In 1928, British researcher Frederick Griffith had two strains of pneumococcus bacteria, one that grew into smooth colonies in the lab and could cause a lethal infection if injected into a mouse, and another that grew into rough colonies and did not cause mice to die when injected. Griffith injected mice simultaneously with live cells from the nonlethal bacterial strain *and* with cells from of the disease-causing strain that had been killed with heat. These dually injected mice died. The formerly nonlethal bacteria had somehow acquired the ability to cause lethal pneumonia from the heat-killed cells. As a control, Griffith injected mice with only the heat-killed lethal strain and found that the mice survived. (Why is that control procedure important?) The "transformed" cells—formerly nonlethal bacteria exposed to the lethal strain—could pass the lethal trait to their daughter cells during cell division.

In 1944, Oswald Avery in New York sought to determine which chemical component was carrying hereditary information from the lethal bacteria to the nonlethal strain. To do so, he extracted the carbohydrates, proteins, and DNA from the lethal strain (Fig. 6.1a,c), incubated the nonlethal bacteria (Fig. 6.1b,d) with the various chemicals from the lethal strain, and checked to see which substance would transform the nonlethal cells (Fig. 6.1e). They observed that the nonlethal bacteria treated with either carbohydrate or protein remained nonlethal (Fig. 6.1c–f). In contrast, the nonlethal bacteria treated with DNA from the lethal cells picked up the ability to cause pneumonia in mice and passed that trait to their progeny (Fig. 6.1c–f). They concluded that DNA contains the genes for physical and biochemical traits, in this case, the genetic information for the ability to cause fatal pneumonia in mice. Researchers have a word for the transfer of an inherited trait by the uptake of DNA: transformation. This process is responsible for the transfer of genes for resistance to antibiotics from one bacterium to another and is a serious problem in modern medicine (see Chapter 9).

Confirmation That Genes Are Made of DNA

The transformation experiments showed that DNA can carry the information for one trait in one species. But what about other traits in other organisms? In 1952, Alfred D. Hershey and Martha Chase got an answer by experimenting with viruses that infect and reproduce in bacteria (Fig. 6.2). We'll refer to them as bacterial viruses, but their official biological name is bacteriophages (literally "bacteria eaters"; also called simply phages).

transformation the process of transferring an inherited trait by incorporating a piece of foreign DNA into a prokaryotic or eukaryotic cell

bacteriophage (or phage) "bacteria eater"; a virus that infects bacterial cells

They chose to use bacterial viruses for several reasons. First, these viruses are easy and inexpensive to maintain. Second, they can produce new viruses rapidly by injecting a part of themselves into a host cell, which then spawns 100 or so exact copies of the original virus about 25 minutes later. Third, and most important, bacterial viruses consist simply of a core of DNA surrounded by a protein coat. This

Figure 6.1
The Chemical Composition of Genes

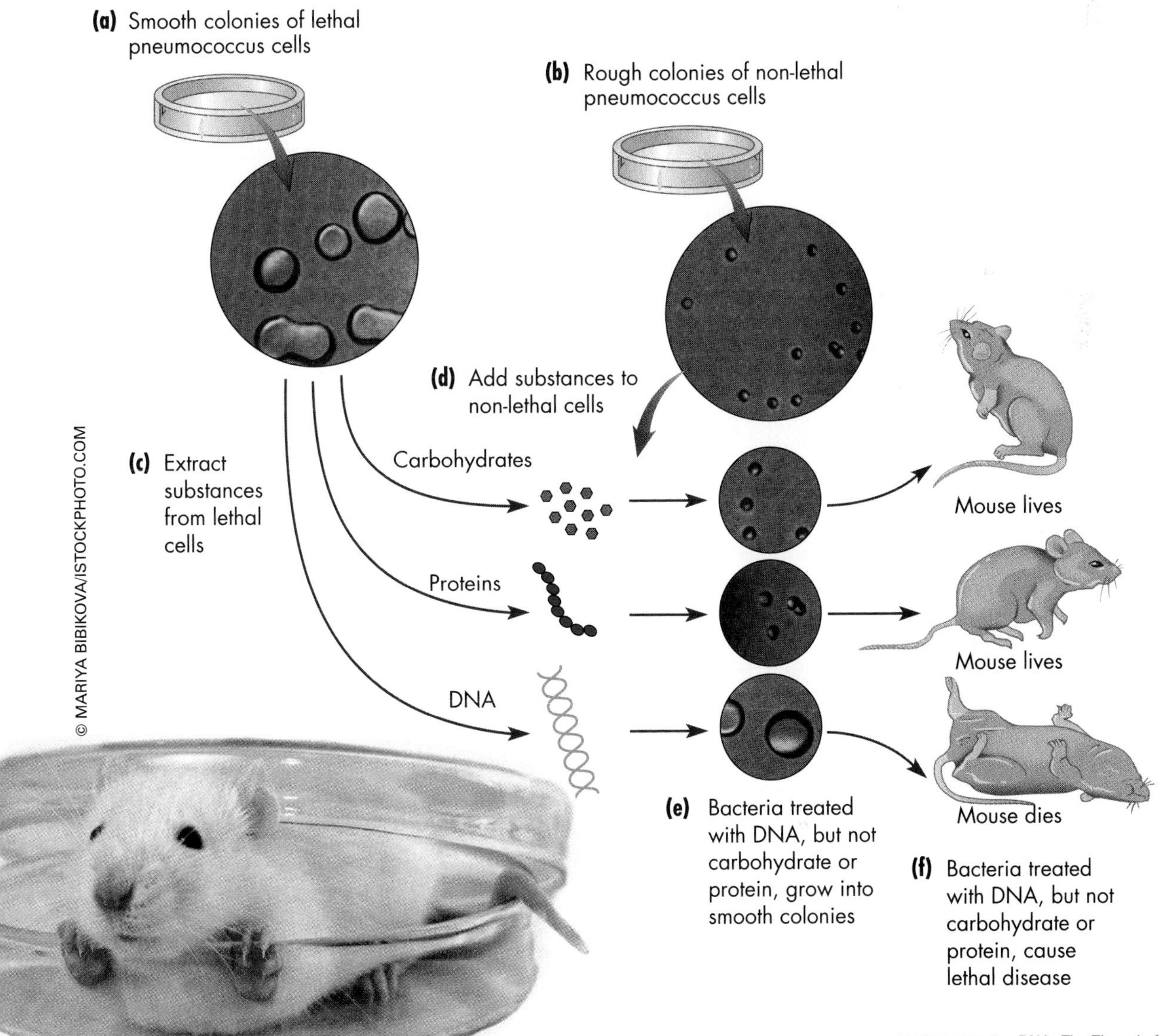

Figure 6.2

Life Cycle of a Bacterial Virus

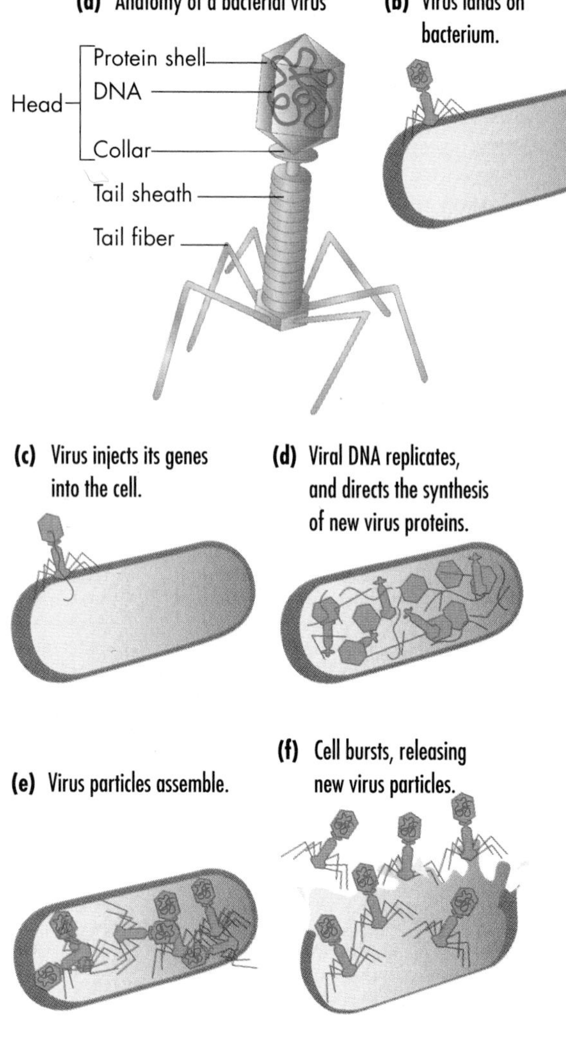

gave Hershey and Chase a clear shot at showing which component contains genes and is responsible for heredity. If only protein from the virus entered the host cell, then protein must be the hereditary material. But if only DNA from the virus entered the bacterial cell, DNA must be the molecule containing the genes. (What would you conclude if both substances entered the host cell? What if neither entered?) Hershey and Chase knew that:

1. proteins contain sulfur but do not contain phosphorus, and
2. DNA contains phosphorus but no sulfur. (Review protein and nucleic acid structure in Chapter 2.)

They labeled the proteins and DNA with radioactive isotopes of sulfur and phosphorus, respectively, which give off detectible signals. By tracing these chemicals, Hersey and Chase were able to determine whether protein (bearing radioactive sulfur) or DNA (bearing radioactive phosphorus) entered the host cell and altered that recipient's genetic activity. The results showed that after infection, "hot" phosphorus was on the inside of the infected cell, and "hot" sulfur was on the outside (Fig. 6.2g). This demonstrated that DNA had entered the cell while protein did not, and that the DNA encoded new phage particles that eventually killed the cell. The researchers concluded that DNA, not protein, was responsible for directing the genetic activity of the virus.

With these experiments, the researchers showed that DNA is definitely the hereditary material in a virus and a bacterium. But what about in more complex species? Research by others confirmed that even in complicated organisms, the answer is still DNA. (In a few viruses, including the virus that causes AIDS [see Chapter 2], genes can be made of RNA.)

These historic experiments changed our concept of the gene from Mendel's abstract hereditary "particle" to a tangible chemical that biologists could see and manipulate. Nowadays, molecular biologists such as Marilyn Menotti break open cells, separate the DNA from the proteins, add ice-cold ethanol to the DNA, and put it in the freezer. A few hours later, a stringy white precipitate a bit like cotton fibers sits at the bottom of a test tube (see photo p. 111)—pure DNA, pure genes.

Learning that genes are made of DNA was a huge step. But to be able to use DNA for sophisticated analyses such as identifying an individual cat from a couple dozen hairs, a researcher needs to know the precise structure of the hereditary molecule.

LO2 DNA: The Twisted Ladder

Both a cat and the mice it preys upon have genes encoded in molecules of DNA. DNA can replicate so perfectly that the offspring of two Siamese cats will have the same creamy fur and dark colored "points" on the nose, ears, feet, and tail. DNA can still contain enough variation, however, so that a Siamese cat looks and acts different from Snowball, a tiger, or a mouse. The variability inherent in DNA is the basis of natural selection—the key to how organisms evolve—as well as the explanation for life's wonderful variety. This leaves us with an interesting puzzle, however. How does the structure of DNA account for both the *unity* of life—the shared traits and common descent of living things—and also its stunning *diversity*?

In the early 1950s, James D. Watson, an American postdoctoral fellow, and Francis H.C. Crick, a British researcher, met in Cambridge, England, and began their quest to understand the structure of DNA. They knew that American biochemist Linus Pauling, who had already done his Nobel Prize–winning work on the nature of the chemical bond, was also researching DNA structure, and they wanted to beat this outstanding scientist to the finish line. Watson and Crick also knew from published experiments that DNA is a linear molecule; that is, it is a very long thread.

Watson and Crick knew two additional facts: that DNA is a chain of nucleotides (see Fig. 6.3b on the next page), and what those nucleotides are. Recall that each nucleotide consists of three parts—a sugar, a phosphate, and a portion called a base. Because of the shape of the sugar portion of the molecule, the nucleotide has a front and a rear end, like a car. The carbon atoms in the sugar are labeled 1′ (pronounced "one prime"), 2′, 3′, 4′, and 5′. At the rear end of the "car" is a "trailer hitch," the phosphate group, attached to the 5′ carbon of the sugar (labeled in Fig. 6.3b). The sugar portion acts like the car's chassis, and the base like the car's interior compartment. At the front is the "grill," representing the 3′ carbon. Watson and Crick also knew that DNA has nucleotides of four types (see Fig. 6.3b), identical except for the bases they contain. These are the **adenine (A)**, **cytosine (C)**, **guanine (G)**, or **thymine (T)** we encountered in Chapters 2 and 5.

Geneticists often describe a section of DNA by the *sequence* of bases in the chain. For example, a 22-nucleotide portion of a single-stranded DNA chain might have bases in the sequence AGGAAAA TGAAGTCAAGAAAATGG. As we will see later, this specific 22-base sequence played a role in helping to convict Douglas of killing Shirley.

But first, back to Watson and Crick. From the work of Erwin Chargaff, the team knew another significant detail about DNA: in any molecule of a cell's DNA, the bases A and T appear in equal amounts, and the bases C and G also exist in equal amounts. Watson and Crick were anxious to build a model of the DNA molecule that accounted for this peculiar regularity. But they couldn't do it until a last piece of information came from the British biophysicist Rosalind Franklin, who was working in the laboratory at King's College in London. Franklin took some particularly good images of DNA in an attempt to see how the four subunits—A, T, G, and C—were arranged. She made the images using **x-ray diffraction**, a process in which a beam of x-rays passes through a crystalline fiber made of many parallel strands of pure DNA. Within the DNA fiber, similar structures repeat, like the pattern of tiles on a floor. When x-rays pass through these repeated structures, the x-rays bend, just as light rays bend when they pass from air into water, causing a twig sticking out of a pond surface to appear bent. This bending results in spots cast on photographic film (Fig. 6.3d). By analyzing patterns of spots, biophysicists can map the relative positions of atoms in a molecule.

A, C, G, T (adenine, cytosine, guanine, thymine)
the four types of nucleotides contained in DNA; they are identical except for the bases they contain

x-ray diffraction
a process in which a beam of x-rays is passed through a crystalline material to help determine its three-dimensional structure

helix
a structure similar to a spiral staircase; DNA is a double helix

When Watson and Crick saw the x-ray photos, they noticed a certain symmetry, suggesting that the molecule might consist of *two* connected strands of DNA. The pictures further suggested that DNA was most likely a **helix**, a structure similar to a spiral staircase. Each loop of the helix consisted of ten nucleotides. Now the young researchers had to find out how the bases and sugar-phosphate backbones were arranged in three-dimensional space. They hoped that knowing the structure would also reveal how the molecule replicates itself and stores genetic information.

Excited by Rosalind Franklin's stunning data, Watson and Crick chose a model-building approach to testing their hypotheses about the helical structure. They arranged and rearranged pieces of Tinker Toy–like sticks and balls into various combinations to see which arrangement of basic components might best reproduce the evidence from Franklin's x-ray data. After building one incorrect structure after another, Watson and Crick were finally able to visualize the possible molecular structure. The structure was a *double helix,* which resembles a twisted ladder. By combining their intuition, tinkering, and knowledge from pre-existing research, Watson and Crick arrived at the structure of DNA before any other researchers. In 1953, Watson and Crick published their findings on the structure of DNA in a two-page report in the international journal *Nature.* In a classic understatement, they wrote that it had "not escaped our notice" that the model immediately suggested ways in which DNA could fulfill its two major biological functions: replicating itself and storing information.

The model's simplicity, plus its enormous power to explain observations of nature, led to its rapid acceptance by the scientific world.

In 1962, Watson, Crick, and Wilkins were awarded the Nobel Prize for their assessment of the structure of DNA. Franklin had died four years earlier of breast cancer at the age of 37, and most observers agree that she had played a crucial, but underappreciated, role in the elucidation of the structure of DNA.

LO3 The Structure of DNA

What are the detailed features of the structure of DNA that explain its function?

1. A DNA molecule is composed of two nucleotide chains (Fig. 6.4a).
2. These two chains are oriented in opposite directions, like the northbound and southbound

Figure 6.3
Elucidating the Structure of DNA

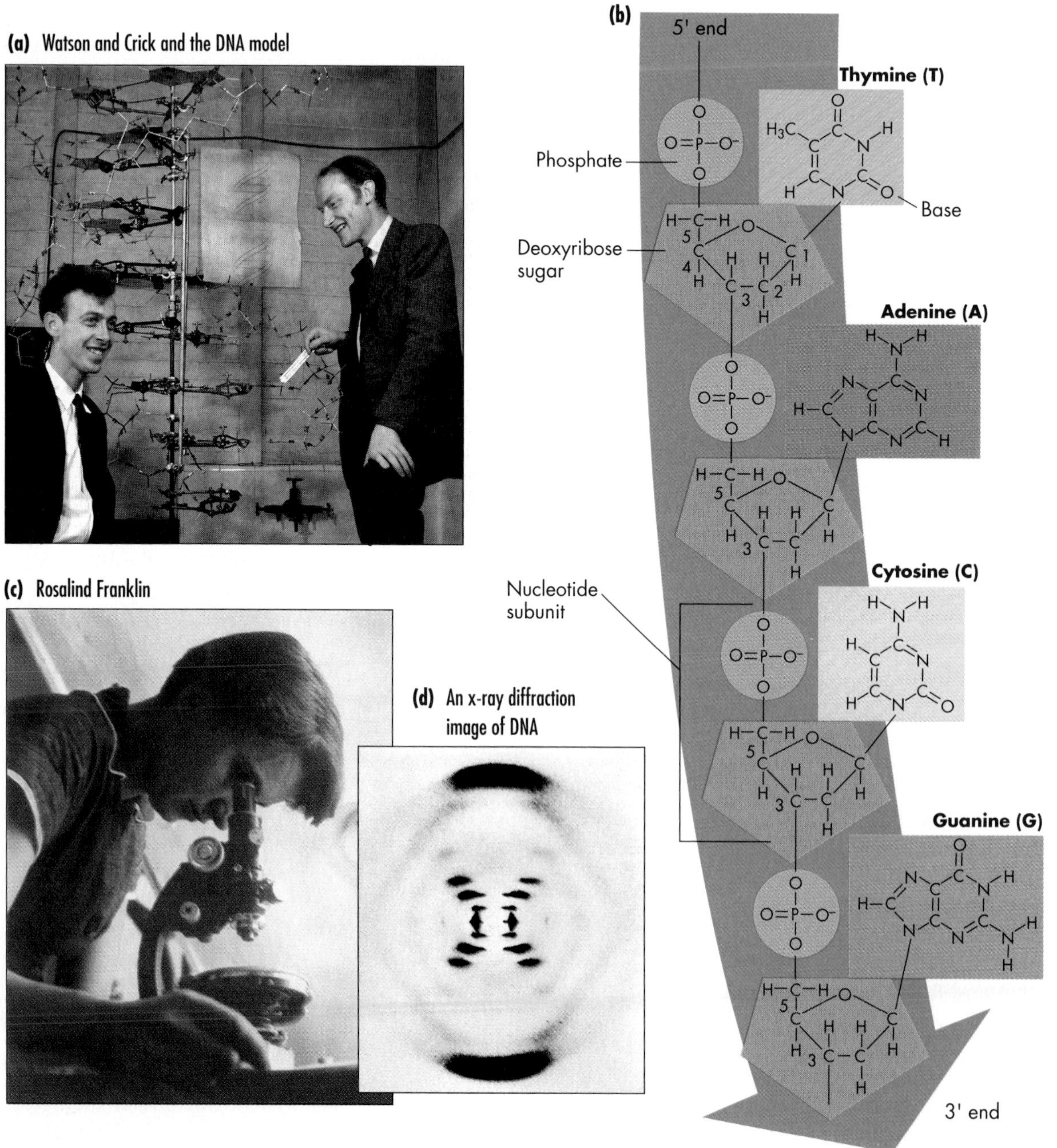

lanes of a highway (Fig. 6.4b). All the nucleotide "cars" in each opposing lane of the highway face in the same direction, with the 3′ sugar leading forward and the 5′ phosphate trailer hitch behind (review Fig. 6.3b). To emphasize the opposite orientation of the two strands, geneticists refer to them as "antiparallel."

3. The two sugar-phosphate chains form the outside of the molecule and are the uprights of the twisted ladder (Fig. 6.4c). In contrast, the bases attached to the backbones face inward, connecting in the middle like the rungs of a ladder (Fig. 6.4d).
4. The bases A and T pair with each other, and the C and G bases also pair; that is, A is complementary to T, and C is complementary to G—the two fit together like adjacent puzzle pieces (Fig. 6.4e). Bases are held together by hydrogen bonds (Fig. 6.4e), hydrogen atoms shared between an A and a T base or a C and a G base.

complementary base pairing in nucleic acids, the hydrogen bonding of adenine with thymine or uracil, and guanine with cytosine; it holds two strands of DNA together and holds different parts of RNA molecules in specific shapes, and is fundamental to genetic replication, expression, and recombination

This complementary base pairing by hydrogen bonds is significant in three ways:

- First, it provides the force that holds two single strands of DNA together into a double-stranded molecule. Because hydrogen bonds are relatively weak, however, they are easily broken by temperatures of 60°C or so.
- Second, complementary base pairing by hydrogen bonds explains Erwin Chargaff's finding (see p. 107) that cellular DNA always has equal amounts of A and T and equal amounts of C and G: Whenever one strand has a T, the other has an A, and so on. Knowing the rules of base pairing, you should be able to finish writing out the sequence complementary to the 22 bases shown here:

5′ A G G A A A A T G A A G T C A A G A A A A T G G 3′
3′ T C C T _ _ _ _ _ _ _ _ _ _ _ _ _ _ _ _ _ _ 5′

- Third, complementary base pairing is important for DNA's major biological activities: replication and information storage.

DNA's major biological activities: replication and information storage

5. A final feature of DNA structure is that the two strands of the DNA molecule twist together to form the double helix (Fig. 6.4f). This twist is important to at least two of DNA's features. It's crucial to the regulation of DNA expression, since specific proteins bind in the grooves of the double helix and control which genes a cell will use. And it's also crucial to DNA replication since a cell's DNA must unwind like the fibers of a rope and allow matching up of complementary base pairs. Figure 6.4g points out the parts of a nucleotide: a sugar, a base, and a phosphate.

Figure 6.4
The Structure of Double-Stranded DNA

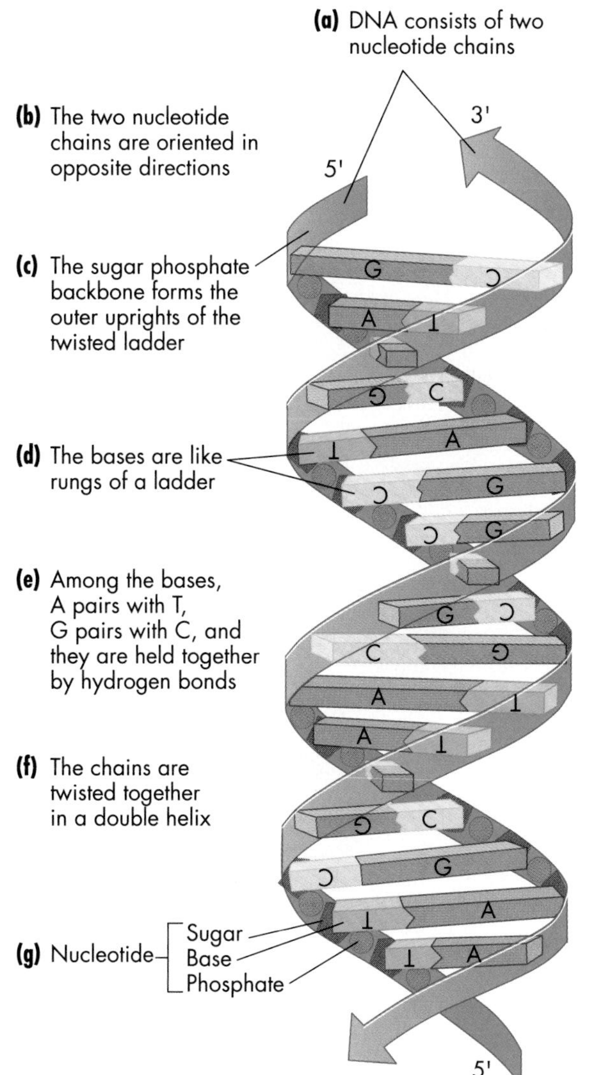

Packaging DNA in Chromosomes

There are very few cells in the follicle of a cat hair. As a result, Dr. Menotti had to use micromethods to purify the DNA from those few cells and separate it from lipids, carbohydrates, and proteins in the hair root. Prying DNA away from proteins is especially difficult because the two kinds of biological molecules are intimately associated in chromosomes.

histone
protein in the nucleus around which DNA molecules of the chromosomes wind, allowing extremely long DNA molecules to be packed into a cell's nucleus

nucleosome
the basic packaging unit of eukaryotic chromosomes; a histone wrapped with two loops of DNA

chromatin
the substance of a chromosome

In the nucleus of a cat's cells and those of most other eukaryotes, each chromosome consists of a single, long, tightly wound DNA molecule. In the fruit fly, for example, the actual length of the DNA molecule from the largest of the fly's 4 chromosomes is more than an inch long—about 12 times as long as the fly itself! Despite the molecule's length, each of the fly's millions of cells contains two copies of this chromosome, as well as pairs of the other three chromosomes. Likewise, a cat has 19 chromosomes, each even longer than a fruit fly's. How do cells package such a huge genetic molecule into a structure as small as a chromosome?

The enormous length of DNA in a eukaryotic cell can't just be wadded up haphazardly. If it were, the separation of DNA molecules during cell division would be as difficult as unraveling two tangled kite strings. What happens instead is that DNA, like a proper kite string, is wound in an orderly way (Fig. 6.5). Specifically, DNA is wound around spools of proteins called **histones** (see Fig. 6.5d). A single spool consisting of several histone molecules wrapped with two loops of DNA (140 base pairs long) is called a **nucleosome** (Fig. 6.5d). Adjacent nucleosomes pack closely together to form a larger coil, somewhat like a coiled telephone cord. This cord, in turn, is looped and packaged with scaffolding proteins into **chromatin**, the combined proteins and genetic material that constitute the substance of chromosomes.

Compare this understanding about DNA packing in chromosomes with what you already know about chromosome activity in mitosis. Recall that during the interphase portion of mitosis, individual chromosomes are invisible (review Figs. 4.5 and 4.7). This is because in interphase, the spools of DNA and protein are not packed closely together, allowing the DNA to spread diffusely in the nucleus. In contrast, during prophase (review Fig. 4.7), the spools compact together as shown in Fig. 6.5d, leading ultimately to a visible chromosome (Fig. 6.5b).

Figure 6.5
DNA Is Packaged into Chromosomes

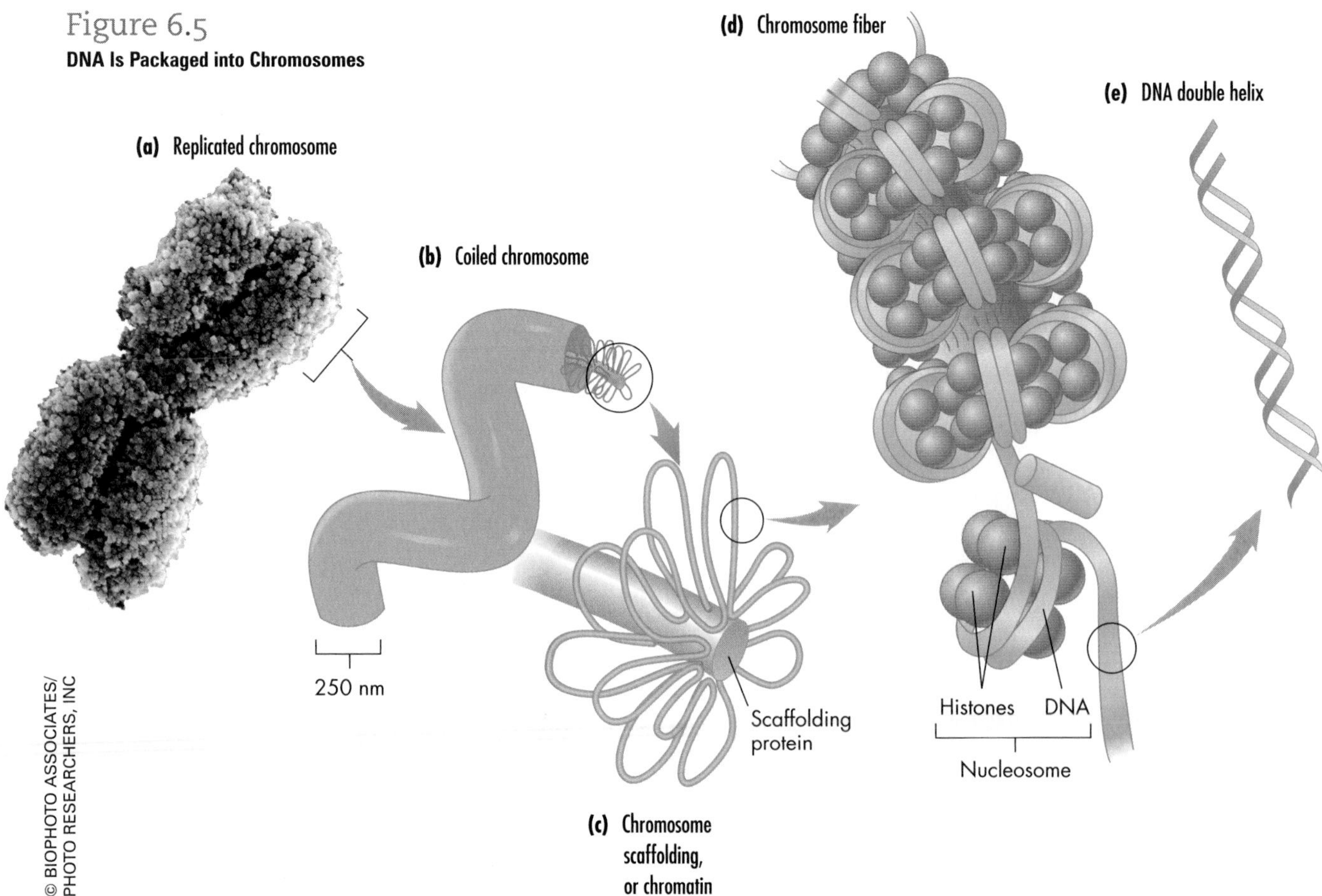

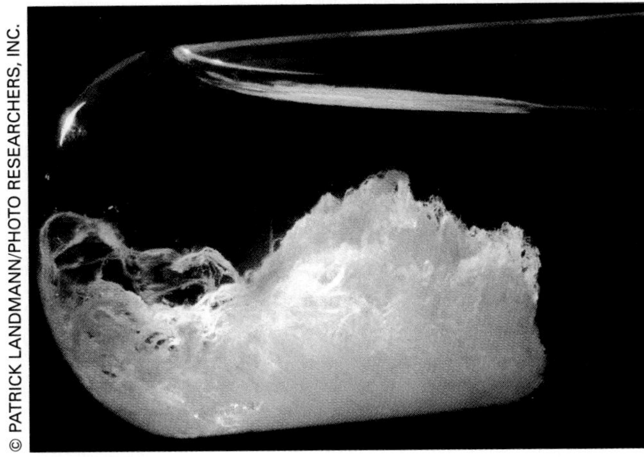
© PATRICK LANDMANN/PHOTO RESEARCHERS, INC.

The orderly packaging of DNA around proteins prevents massive DNA tangles during cell division. A molecular biologist like Marilyn Menotti studying DNA separates the genetic molecule from these protein spools by using special solvents that destroy the structure of proteins but leave DNA intact. After this, Menotti's next step was to make millions of copies of certain regions of Snowball's chromosomes so she could investigate their genetic properties, identify the cat hairs, and eventually help solve the mystery of who killed Shirley.

LO4 DNA Replication

As we've seen, the amount of DNA in a single hair follicle is very small. Because there is a single DNA molecule in each chromosome, any individual cat, for example, will have only two copies of any gene. (Recall from Chapter 5 that a gene is a unit of inheritance and is specifically a portion of a DNA molecule encoding an RNA and usually a polypeptide chain.) One copy of the gene on the DNA is inherited from the mother and the other copy on the DNA is inherited from the father. Recall that only four cat hairs in Douglas's coat had root cells attached. To get enough DNA to work with, therefore, Menotti had to make many copies of specific regions of DNA by using a test tube version of the normal method of DNA replication. Let's first examine the ways in which cells copy DNA and why that's important. Then we can see how Menotti harnessed the process of DNA replication to help solve the murder case.

Let's start by placing the biochemical process of DNA replication into the familiar context of the cell cycle. We saw in Chapter 4 that cells cycle through a period of growth (interphase) and a period of division (M phase) (see Fig. 4.5). Interphase, remember, has three parts: In G_1, cells have one double-stranded copy of each nuclear DNA molecule; in the S phase, DNA replicates; as a result, in G_2, there are two double-stranded copies of each DNA molecule. Finally, in mitosis, those two DNA copies separate into the two nuclei of the daughter cells. Let's now focus on the events of the S phase, the copying of DNA.

polymerization the joining together of newly paired bases, creating a DNA strand identical to the original double helix strand of DNA

DNA polymerase the enzyme that catalyzes the polymerization of DNA strands

Steps in Replication

The replication of DNA occurs in the cell nucleus (see Chapter 4) and follows directly from the principle of complementary base pairing. It can be divided into three steps: (1) strand separation, (2) complementary base pairing, and (3) joining. Keep in mind that before the S (synthesis) phase of the cell cycle, DNA is present in the double helix form (Fig. 6.6).

1. *Separation.* For replication to begin, the two strands of the double helix must first unwind and then strands must separate from each other (Fig. 6.6, Step 1). In cells, the unwinding and separation of the strands are catalyzed by enzymes that help break the "rungs" of the ladder. Those rungs, remember, are the bases on each strand that are bound together by weak hydrogen bonds. In the test tube, Marilyn Menotti and her colleagues used heat to "melt" the hydrogen bonds. After that, for a short time, the separated base pairs are unpaired.
2. *Complementary base pairing.* The unpaired bases form new hydrogen bonds with free nucleotides (As, Ts, Gs, Cs) that happen to diffuse into the area (Fig. 6.6, Step 2). An A base on one DNA strand pairs only with a free T base (complete with its sugar-phosphate backbone), and an attached C pairs only with a free G. Likewise, attached Ts bond only to free As, and attached Gs bond only to free Cs. Thus, the sequence of bases in the original strand specifies the same sequence in the new strand according to the rules of complementary pairing.
3. *Joining.* The joining together of the newly paired bases creates a new strand that is complementary to the parent strand, forming two new double helices that are identical to the original double helix (Fig. 6.6, Step 3). The joining, or **polymerization**, of the new double helices is catalyzed by an enzyme called **DNA polymerase**. This enzyme joins the phosphate group (the trailer hitch in our car analogy) of one nucleotide to the "grill" of the previous nucleotide.

semiconservative replication the universal mode of DNA replication, in which only one of the two strands of DNA of the parent molecule is passed to each daughter molecule; this inherited strand serves as the template for the synthesis of the other strand of the double helix

Semiconservative Replication

The three steps of DNA replication—strand separation, base pairing, and joining—occur over and over again along the length of the DNA molecule and produce two double-stranded DNA molecules identical to the parental molecule. Each new DNA has a base sequence identical to the base sequence of the original. You can see in Figure 6.6 that each of the two daughter DNA molecules has one strand intact from the original parent, while the other strand is completely new. Because only one of the two strands in the daughter molecule is inherited intact—or conserved—from the parent molecule, this type of replication is called semiconservative replication. Apparently, all living creatures share this mode of DNA replication. Semiconservative copying of DNA is very different from, say, copying a piece of paper on a photocopy machine. The machine produces a totally new copy but conserves the original fully intact. In contrast, during DNA replication, the original molecule ceases to exist, but one half of it becomes part of one offspring molecule and the other half becomes part of the other offspring molecule.

Accuracy of DNA Replication

It is a wonder that DNA molecules are ever copied correctly, considering the immense length of DNA

Figure 6.6

An Overview of DNA Replication

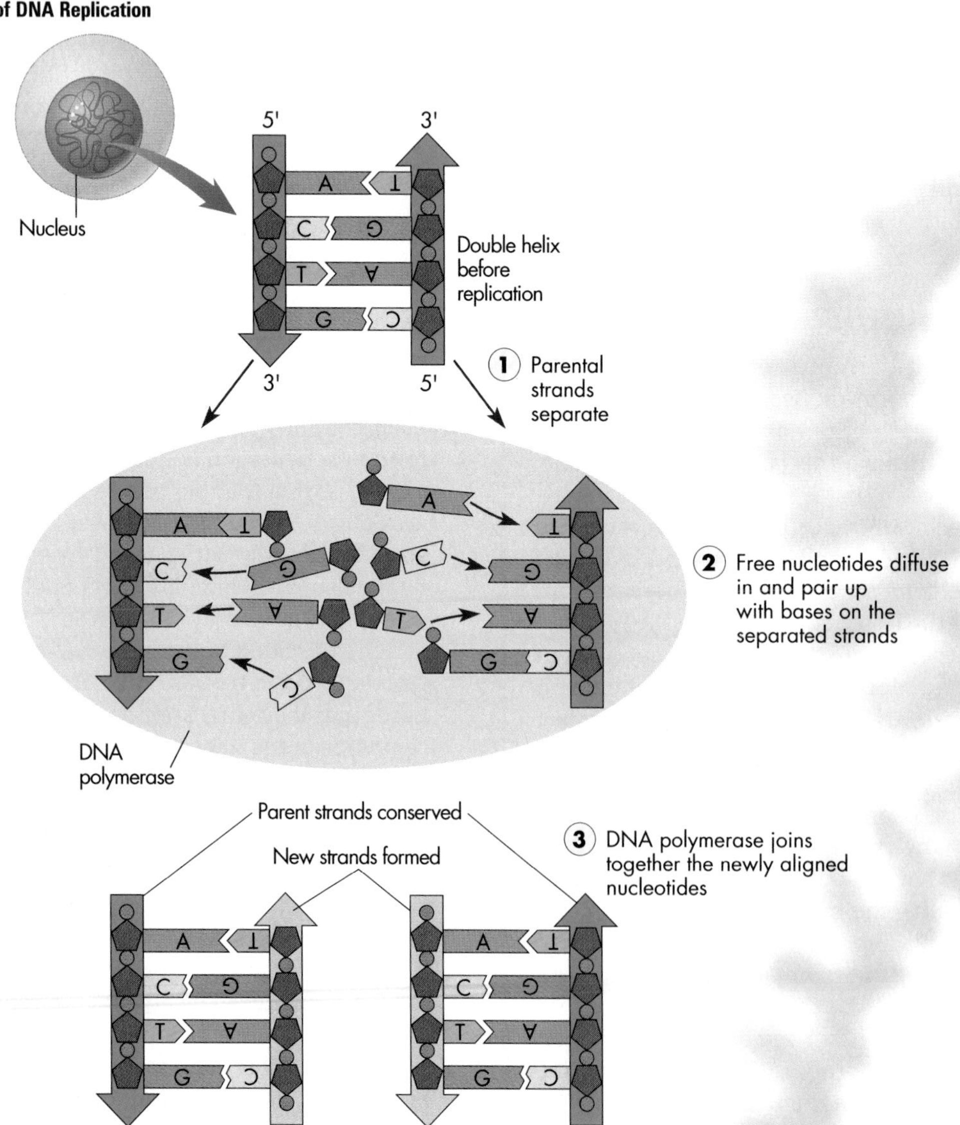

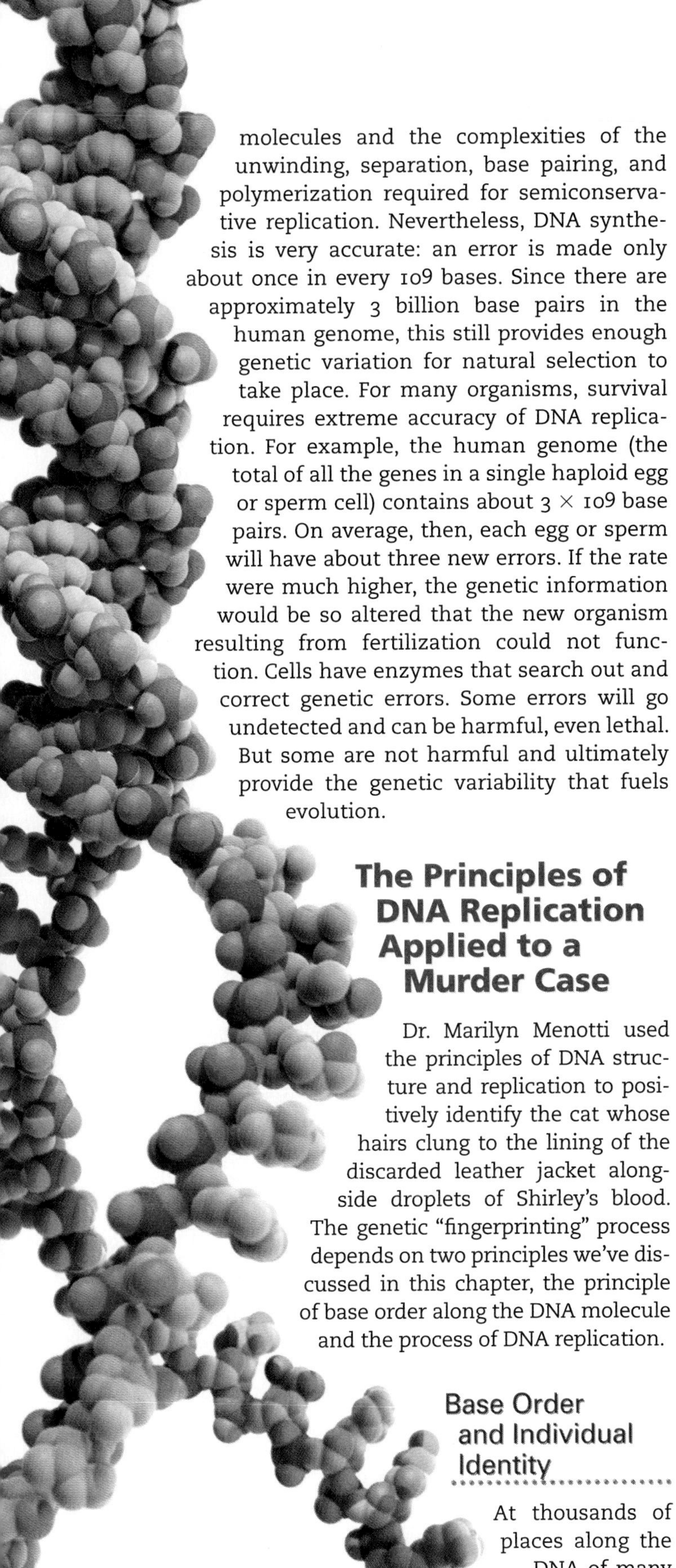

molecules and the complexities of the unwinding, separation, base pairing, and polymerization required for semiconservative replication. Nevertheless, DNA synthesis is very accurate: an error is made only about once in every 109 bases. Since there are approximately 3 billion base pairs in the human genome, this still provides enough genetic variation for natural selection to take place. For many organisms, survival requires extreme accuracy of DNA replication. For example, the human genome (the total of all the genes in a single haploid egg or sperm cell) contains about 3 × 109 base pairs. On average, then, each egg or sperm will have about three new errors. If the rate were much higher, the genetic information would be so altered that the new organism resulting from fertilization could not function. Cells have enzymes that search out and correct genetic errors. Some errors will go undetected and can be harmful, even lethal. But some are not harmful and ultimately provide the genetic variability that fuels evolution.

The Principles of DNA Replication Applied to a Murder Case

Dr. Marilyn Menotti used the principles of DNA structure and replication to positively identify the cat whose hairs clung to the lining of the discarded leather jacket alongside droplets of Shirley's blood. The genetic "fingerprinting" process depends on two principles we've discussed in this chapter, the principle of base order along the DNA molecule and the process of DNA replication.

Base Order and Individual Identity

At thousands of places along the DNA of many species, the two nucleotides C and A are repeated over and over: CACACACACA. The opposite DNA strand, of course, has the sequence TGTGTGTGTG, which is the complementary sequence in the opposite orientation. For convenience, however, we'll talk about the sequence of just one strand, because given that ordering of base pairs, one can figure out the sequence of the other strand by applying the rules of base pairing.

genetic marker any genetic difference between individuals that can be followed in a genetic cross; for example, areas of repeated base sequences along a DNA strand

Cats have 19 pairs of chromosomes. In comparison, we humans have 23 pairs, fruit flies have 4 pairs, and corn plants have 10 pairs. Studying Snowball the cat's DNA, Menotti found an important spot along the DNA from one of the two copies of chromosome B3. She called this locus, this region of 115 base pairs, "FCA88" and it had the following sequence of bases: AGG AAAATGAAGTCAAGAAAATGGCTTAATCCAAAGTCA CACAGTACTTAATGTGTGTGTGTGTGTTTGTGTG TGTGTGTGTGTGTGTGTATGTGTGTAACGGGAA AAAGAAAA.

Note that the sequence "TG" appears over and over again in this portion of the chromosome. Also notice that we've printed the sequence of just one of the two strands of the DNA double helix. What would the repeated portions "read" on the other strand? If you said, "CA" repeated over and over, you're right!

Dr. Menotti carefully studied the repeated sequences in this FCA88 position on the chromosome. Areas of repeated base sequences are also called **genetic markers** because different numbers of repeats tend to occur in different individuals and this can serve as a specific recognizable section of the DNA that adds to the unique identity. Next, she found 28 copies of the TG repeat on one copy of chromosome number B3 and 25 copies of the TG repeat on the other copy of chromosome B3 from the hair root cell DNA left on the bloody jacket. (The differences are understandable if you consider that each individual has two copies of each chromosome, one from the mother and one from the father.) Once she saw this "fingerprint" in the DNA from the cat hairs on the bloody jacket, Menotti could determine that it matched the number of TG repeats at this locus for the cat Snowball. She also had to test other cats on Prince Edward Island to see if this fingerprint was common or truly unique.

Menotti examined DNA from 19 cats who lived on Prince Edward Island but were unrelated to Snowball, as well as another 9 cats from around the United States. As we just saw, Snowball's DNA had precisely 28 copies of the TG repeat at the FCA88 locus on copy 1 of chromosome B3 and 25 copies of the TG repeat at the FCA88 locus on copy 2 of chromosome B3. The researcher found that an unrelated cat—say, a Siamese cat living on Prince Edward Island—might have as few as 12 or as many as 29 TGs in a row at position FCA88 on chromosome B3. Only a few cats on the island had the same combination of 28 and 25 TG repeats at locus FCA88 as found in the DNA from the cat hairs taken into evidence. By studying repeated base sequences at nine more loci along the DNA retrieved from the cats as well as from the leather jacket, Menotti came up with a set of sequences absolutely individual to one single cat. With it, she could rule out 219,999,999 of every 220 million cats living on Prince Edward Island. (This is a theoretical number, of course, since only a few thousand cats actually live there). She could also rule out 69,999,999 out of every 70 million cats on the U.S./Canadian mainland. This virtually guaranteed that the hairs in the discarded jacket came from the one cat, Snowball, who lived with Douglas and his parents.

Our cat hair case study has demonstrated DNA's capacity for identifying individuals with unique "fingerprints" of repeated sequences at different loci along the chromosomes. But the significance of DNA's elegant double helix structure goes far beyond genetic fingerprints, important as they are to the medical and legal systems. DNA's structure—two twisted strands of nucleotides, oriented in opposite directions—explains its ability to replicate with amazing accuracy as the strands separate and enzymes synthesize two daughter molecules, each a complementary copy of the parent strand. DNA structure also explains the way the "thread of life" controls an organism's physical make up and abilities. Living things, of course, are made up of cells, and it is the collective form and function of those cells that determines the entire organism's appearance, its physical and chemical properties, and its day-to-day behaviors. In our next chapter, we'll see how the blueprints for form and function inscribed in DNA are actually decoded into proteins and cellular activities, and how molecular geneticists have learned to manipulate these blueprints for medicine, agriculture, and industry.

Who knew?

Once DNA was known to be the carrier of hereditary information, knowledge of its structure was needed to understand how it could account for both the unity of life (the heritability) and the diversity of life (the enormous amount of variation). Only by understanding the structure of DNA could scientists explain both how DNA is replicated and how it directs the activities of cells that lead to the formation of differences between and within species. To be known for making such an important and fundamental discovery was a valuable "prize" for many researchers.

81% **of students surveyed find that LIFE helps them prepare better for their exams**

LIFE puts a multitude of study aids at your fingertips. After reading the chapters, check out these resources for further help:

- **Chapter in Review cards**, found in the back of your book, include all learning outcomes, definitions, and visual summaries for each chapter.
- **Online printable flash cards** give you three additional ways to check your comprehension of key biology concepts.

Other great ways to help you study include **animations, visual reviews,** and **interactive quizzes.**.

You can find it all at **4ltrpress.cengage.com/life**.

7

Genes, Proteins, *and* Genetic Engineering

Learning Outcomes

LO[1] Explain how genes specify the amino acid sequence of a protein

LO[2] Understand the processes of transcription: copying DNA into RNA

LO[3] Understand the processes of translation and the synthesis of polypeptide chains

LO[4] Explain how genes are controlled

LO[5] Explain how geneticists create recombinant DNA molecules

“Protein synthesis is one of the most important topics in biology.”

Milking and a Midnight Brainstorm

Harry Meade was raised on a dairy farm and still goes home sometimes to milk the cows. As a genetic engineer, Meade grew interested in the possibility of transferring human genes into bacteria and producing quantities of desirable human proteins. At the time, Meade was working on a human protein that can dissolve blood clots. He found, however, that it was next to impossible to produce the protein in bacteria. So instead, he started searching for other ways to engineer it. Naturally, Harry thought about using milk.

First, he found the "promoter" for the most common protein in milk, casein. A **promoter** is a sequence of nucleotide base pairs that dictates when and where a gene will be turned on—in this case, the gene for casein. Then he had to attach the promoter to a human gene—in this case a gene for a protein that prevents blood clots, AT3—and insert the combination into an animal host. His goal was to produce large quantities of human proteins in animal milk, and for that, goats were a practical choice. For the new business of "drug farming," goats mature to milking age much earlier than cows.

If Meade's group could make a large supply of AT3 available for use in a commercial drug, physicians could transfuse the protein into patients during heart surgery to prevent them from forming dangerous clots.

Today Meade and his colleagues have over 1,800 goats; dozens are already producing large amounts of AT3 in their milk, and dozens more are churning out various types of antibodies and other desirable human proteins. Meade proved that "drug farming," or making pharmaceuticals in barnyard animals (also called "pharming") is a particularly economical way to make large amounts of useful proteins.

This chapter explores two of the most important topics in modern biology: First, the elegant way in which living cells use the information inscribed in genes to direct the production of proteins for cell survival, growth, and division. Second, how biologists have learned to use these basic processes to splice foreign genes into organisms to make useful proteins in new ways. You'll learn in this chapter how cells make proteins, how the cell controls this vital process, and about the tools and techniques of genetic engineering, including pharming.

Along the way, you'll find answers to these questions:

- How is DNA copied into RNA?
- How is RNA translated into proteins?
- How are genes controlled?
- What is recombinant DNA technology?

What do you know?

If carcinogens are mutagens, are mutagens carcinogens?

promoter
a series of nucleotides to which RNA polymerase binds and initiates transcription of the adjacent gene

transcription (L. *trans,* across + *scribere,* to write) the transfer of information from a portion of a DNA molecule into an RNA molecule; the process is catalyzed by the enzyme RNA polymerase

translation the conversion of the information on a strand of RNA into a sequence of amino acids in a protein; occurs on ribosomes

LO[1] How Does a Gene Specify the Amino Acid Sequence of a Protein?

To *transcribe* means to *copy.*

Genes specify protein structure. This concept is so essential to understanding biology that in the second half of the 20th century, geneticists dubbed it the "central dogma." So let's take a look at that well-accepted truth—that the information stored in genes is played out in protein structure.

An Overview of Information Flow in Cells

Here's a shorthand way of understanding biology's central dogma: in cells, *genetic information generally flows from DNA to RNA and then from RNA to protein.* In the first step, called transcription, DNA codes for RNA. In the second step, called translation, RNA codes for the structure of a specific protein (Fig. 7.1). (In a few instances, information can flow backwards, from RNA to DNA. This happens, for example, when the AIDS-causing virus HIV infects a human cell; see Fig. 2.15.)

DNA to RNA: Transcription

In the first step of information flow, **transcription**, enzymes copy a portion of a DNA molecule into an RNA molecule. RNA is an information-storing molecule with a structure somewhat similar to DNA, as we'll see.

RNA to Protein: Translation

In the second step of information flow, **translation**, the informational content of one type of RNA molecule (called messenger RNA) is translated by ribosomes into specific sequences of amino acids that make up polypeptide chains.

Because transcription copies DNA to RNA, the copying takes place where DNA resides in the cell: in a eukaryote, the nucleus (Fig. 7.1). Likewise, because translation occurs on ribosomes, translation takes place where the ribosomes are located—the cell's cytoplasm. After an RNA is transcribed in the nucleus, it moves through the pores in the nuclear envelope to the cytoplasm. There the RNA binds to ribosomes, which translate the RNA into protein molecules. Cells then transport proteins to their place of action in the cell via the routes we've discussed (Fig. 2.13). Let's take just one example of this: human liver cells making the AT3 protein. After this protein is synthesized on ribosomes, tiny vesicles carry the newly made molecules to the cell surface and release them. The AT3 molecules then travel in the blood to various body sites where AT3 regulates blood clotting.

Now that we've seen an overview of the DNA → RNA → protein process, let's take a closer look at each step. We'll start with transcription, which allows the hereditary information stored in DNA to leave the nucleus and reach the cytoplasm, where protein synthesis occurs.

LO[2] Transcription: Copying DNA into RNA

The structures of RNAs and DNA are clearly important to understanding what genes *do.* So let's compare their architectures.

Figure 7.1
Information Flows from DNA to RNA to Protein

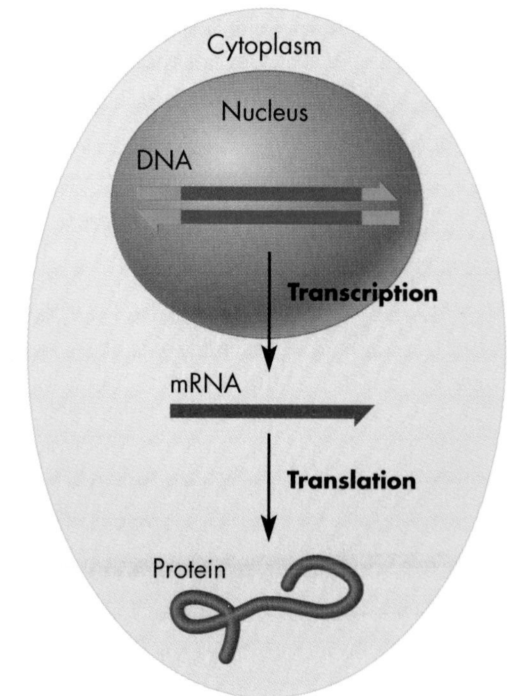

(For eukaryotes...) DNA is transcribed into RNA in the nucleus; RNA is translated in the cytoplasm.

Comparing DNA and RNA

Like DNA, RNA consists of a long string of nucleotides linked by sugar-phosphate backbones (Fig. 7.2). Unlike DNA, RNA can move from the nucleus to the cytoplasm for its role in protein synthesis. RNA also differs from DNA in four other ways.

- RNA nucleotides contain the sugar ribose instead of deoxyribose, the sugar in DNA. Deoxyribose has one less oxygen atom than ribose (Fig. 7.2b). This is the only structural difference between the two sugars, but it leads to important differences in function.
- RNA contains the base uracil (U) instead of the base thymine (T), which is found in DNA (Fig. 7.2b). Just as thymine can pair with adenine (A–T) in DNA, uracil can pair with adenine (A–U) in RNA.
- RNA usually consists of a single strand of nucleotides (Fig. 7.2a,b), whereas DNA usually consists of two strands. In some kinds of RNA, however, a single RNA molecule folds back on itself and forms short, double-stranded regions connected by complementary base pairs.
- RNA molecules are much shorter than the DNA molecules that make up chromosomes. Each DNA molecule carries hundreds or thousands of genes, but an RNA molecule usually contains information from only one gene.

The Transcription Process

The transcription of a portion of DNA into RNA involves three basic steps: DNA strand separation, complementary base pairing, and nucleotide joining. Let's take the *AT3* gene as the portion of DNA and go through those three steps to see how it's transcribed into RNA in a person's liver cell.

1. *Strand separation.* As transcription of the *AT3* gene begins in the nucleus of a liver cell, enzymes unwind and separate portions of the DNA double helix near one end of the gene (Fig. 7.3, Step ①). For any given gene, only one of the two separated DNA strands serves as a template (or model) for making a complementary strand of RNA. The other DNA strand simply stays out of the way. As we saw in the chapter introduction, a certain sequence of nucleotides called the promoter defines where the RNA copying enzyme (RNA polymerase) binds to DNA and hence where a new RNA copy begins.

Figure 7.2
The Structure of RNA

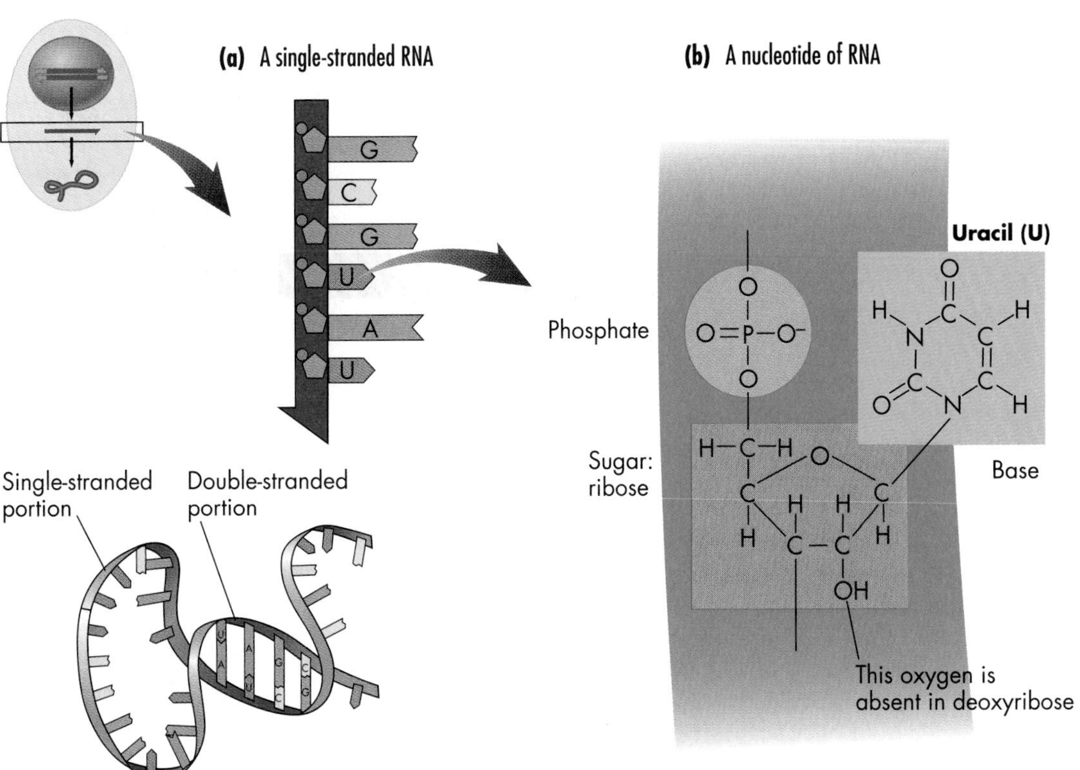

Figure 7.3

How DNA is Transcribed into a Messenger RNA.

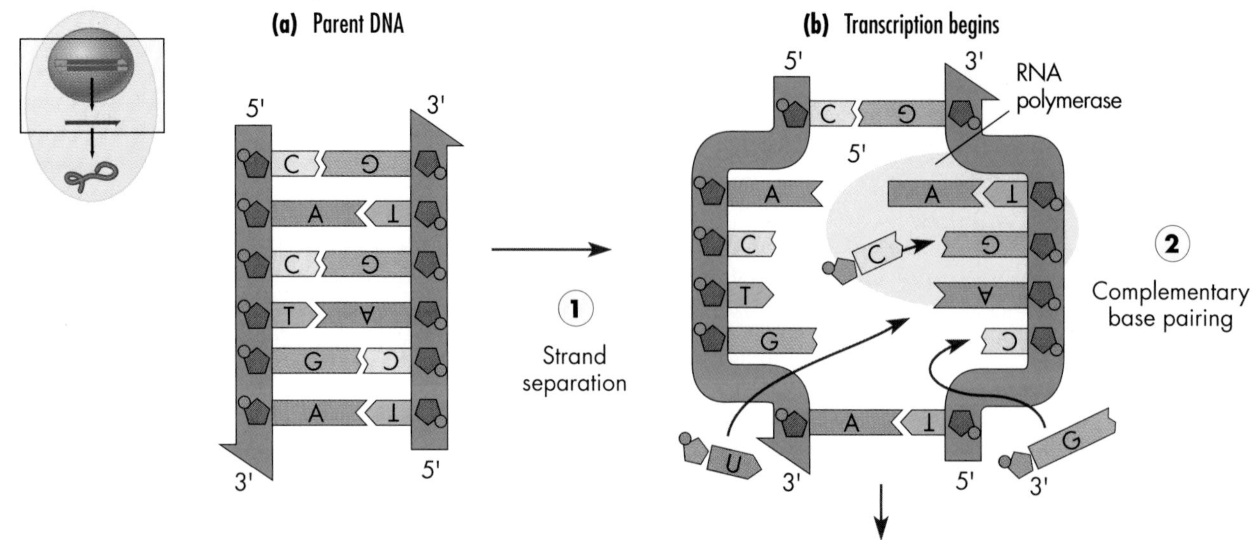

2. *Complementary base pairing.* Nucleotides containing the sugar ribose diffuse around the separated DNA and pair up with complementary nucleotides on that strand (Fig. 7.3, Step 2). (This is much like the replication of DNA we saw in Chapter 6.) In this case, however, the base A on the DNA pairs with a U nucleotide for RNA. As in DNA replication, T on the DNA pairs with an A nucleotide for RNA and C pairs with G.

3. *Nucleotide joining.* The enzyme RNA polymerase joins two adjacent RNA nucleotides together (Fig. 7.3, Step 3). Other RNA nucleotides diffuse in and pair with their complementary bases in the DNA strand, and the enzyme joins them together one by one, forming a single strand of RNA. Although the figure shows only a few nucleotides in the RNA molecule, a completed RNA molecule for the AT3 gene actually will have a total of 1,389 nucleotides. The enzyme machinery for joining new nucleotides to an RNA molecule adds about 30 nucleotides per second. So it would take a liver cell about 46 seconds to transcribe the AT3 gene onto a single RNA molecule. The RNA polymerase finally stops making an RNA copy of the gene when it hits a "stop signal," a specific nucleotide sequence in DNA. As transcription is completed, the parental DNA rewinds. The DNA is now available to produce more copies of the same RNA. Some chemical modification may be made to the new RNA molecule (lower part of Fig. 7.3), but once completed, it passes out of the nucleus into the cytoplasm.

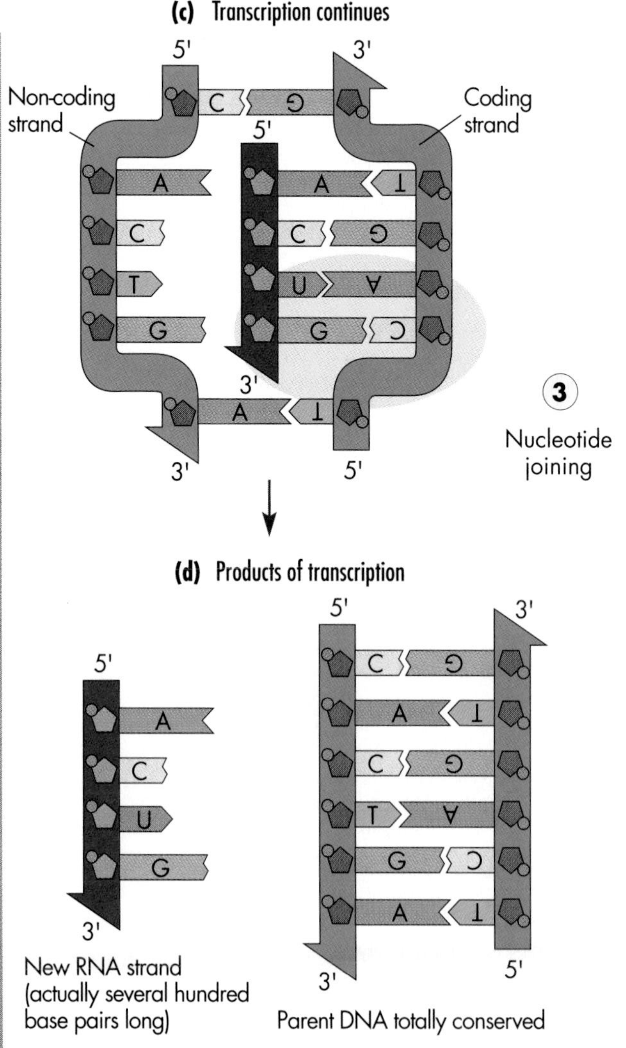

Comparison of Transcription and DNA Replication

If you compare Figure 6.6 to Figure 7.3, you can see that the transcription of DNA into RNA is somewhat similar to the DNA replication that takes place before a cell divides. In both cases, the two DNA strands unwind and separate, nucleotides diffuse to the site and line up by base pairing, and a polymerase enzyme joins the new nucleotides into a nucleic acid strand. Furthermore, the polymerase enzymes in both cases copy in the same direction (5′ to 3′) along a single DNA strand.

Transcription does differ from DNA replication, though, in several ways. Here are the differences in brief:

During DNA replication...	During transcription...
... the entire DNA molecule—hundreds of genes—is copied	... only a portion of the DNA—often a single gene—is transcribed at any one time
... both strands of a DNA molecule are copied	... only one of the two separated DNA strands is transcribed
... only one copy of each gene is made	... a single gene may be copied thousands of times. The cell needs to make large quantities of certain proteins such as AT3, for example, so that they can move out of the cell to be carried throughout the body in the bloodstream.
DNA replication occurs only in the S phase of the eukaryotic cell cycle	Transcription occurs throughout interphase, in G_1, S, and G_2

Introns and Exons

Harry Meade wasn't the first geneticist to isolate and study the human *AT3* gene, although he later worked with it extensively. Early workers found that the gene is located at a specific position on human chromosome 1 (Fig. 7.4a). The gene is initially transcribed into an RNA molecule about 19,000 base pairs long. This long RNA is called the primary transcript (Fig. 7.4b). Next, an amazing thing happens to this initially transcribed RNA—enzymes cut the primary transcript and discard parts of it while splicing other portions together, forming a spliced RNA (Fig. 7.4c). Biologists chose the term **introns** for the portions of the gene that are initially transcribed but then are spliced out of the primary transcript. *AT3* has five introns. Biologists chose the separate term **exons** for the portions of the gene that are transcribed and then spliced together and appear in the final spliced RNA. *AT3* has six exons. Introns *intrude* into the gene but don't appear in the final RNA and so they aren't translated into protein. In contrast, exons are *expressed*—they are usually translated into protein. Because introns are removed, the spliced RNA from the *AT3* gene is only 1,479 base pairs long—just 11 percent of the primary transcript's original length. Most genes in eukaryotic cells have introns. In contrast, prokaryotic cells generally don't have these extra DNA segments in their genes. Introns are important because, by facilitating the swapping of exons, they may help proteins evolve new functions. In the *AT3* gene, the spliced RNA undergoes further processing on each end and becomes the mature RNA. RNA then leaves the nucleus and heads for the cytoplasm, where it is translated into the *AT3* protein.

introns
the portions of the gene that are initially transcribed, but then are spliced out of the primary transcript

exons
the portions of the gene that are transcribed and then spliced together and appear in the final RNA

Harry Meade had to keep all these details about introns, exons, and RNA splicing in mind while designing a new method to induce goats to make human AT3 protein. We'll see how he did all of this a bit later. First, though, let's explore the details of the translation process itself, starting with the types of RNA .

transcription *translation*
DNA → RNA → Protein

LO³ Translation: Constructing a Polypeptide

Transcription actually makes several types of RNA molecules and all three different types are involved in translation, the synthesis of polypeptide chains that become active proteins or parts of proteins. Here are the three types and what they do.

Types of RNA

Protein synthesis requires three kinds of RNA: messenger RNA, transfer RNA, and ribosomal RNA. All three are made in the nucleus by gene transcription and then move to the cytoplasm.

messenger RNA (mRNA) an RNA molecule that carries the information to make a specific protein; mRNA is transcribed from structural genes and is translated into protein by the ribosomes

transfer RNA (tRNA) a small RNA molecule that translates a codon in mRNA into an amino acid during protein synthesis

Messenger RNA

Messenger RNA, or mRNA, carries genetic information from DNA in the nucleus to the ribosomes in the cytoplasm, where it directs protein synthesis. We just saw, for example, how the AT3 gene is transcribed into RNA. This RNA becomes mRNA when the introns are edited out (see Fig. 7.4). Mature mRNA then leaves the nucleus for synthesis of the AT3 protein in the cell cytoplasm. Because mRNA has a sequence of nucleotide bases that is complementary to a gene on the DNA molecule, it carries the same information as the gene but in complementary form. Messenger RNA molecules are generally from about 1000 to 10,000 nucleotides long, and this type of RNA makes up only 5 percent of the total RNA in the cell.

Transfer RNA

Transfer RNA, or tRNA, first picks up amino acids in the cytoplasm and then aligns them on a ribosome in the exact order specified by the information in the mRNA. These functions (picking up amino acids then ordering them) occur at opposite ends of the tRNA molecule. At one end of the tRNA, a specific enzyme catalyzes a reaction that attaches the tRNA to one of the 20 amino acids (Fig. 7.5a). At the other end of the tRNA, a group of nucleotides joins the tRNA to a specific position in the RNA (Fig. 7.5b).

Each tRNA molecule carries only one specific kind of amino acid. Because there are 20 different amino acids, there must be at least 20 different tRNAs (some amino acids are transported by more than one kind of tRNA). Molecules of tRNA are only about 75 nucleotides long, much shorter than mRNAs.

Figure 7.4
Messenger RNA: Transcribed from the *AT3* Gene

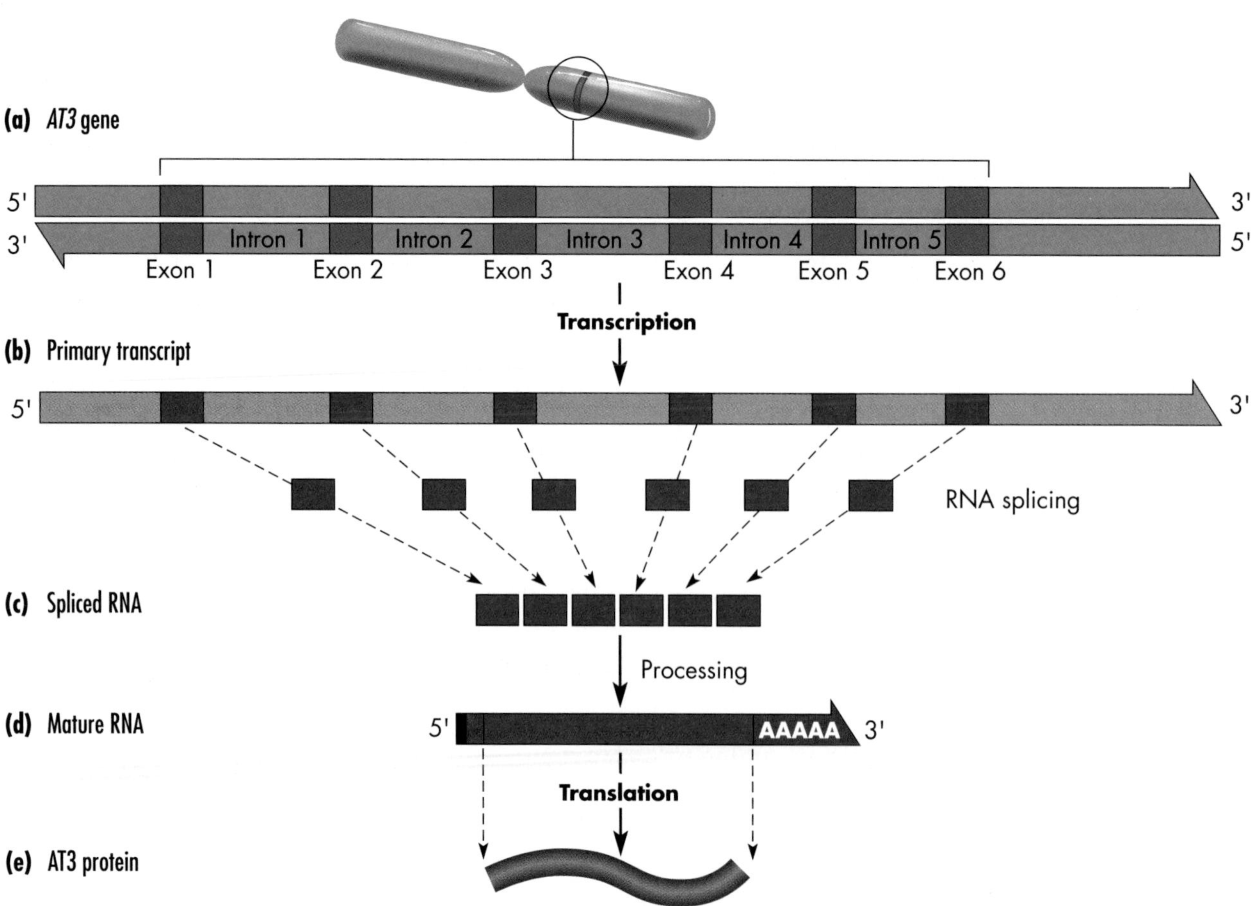

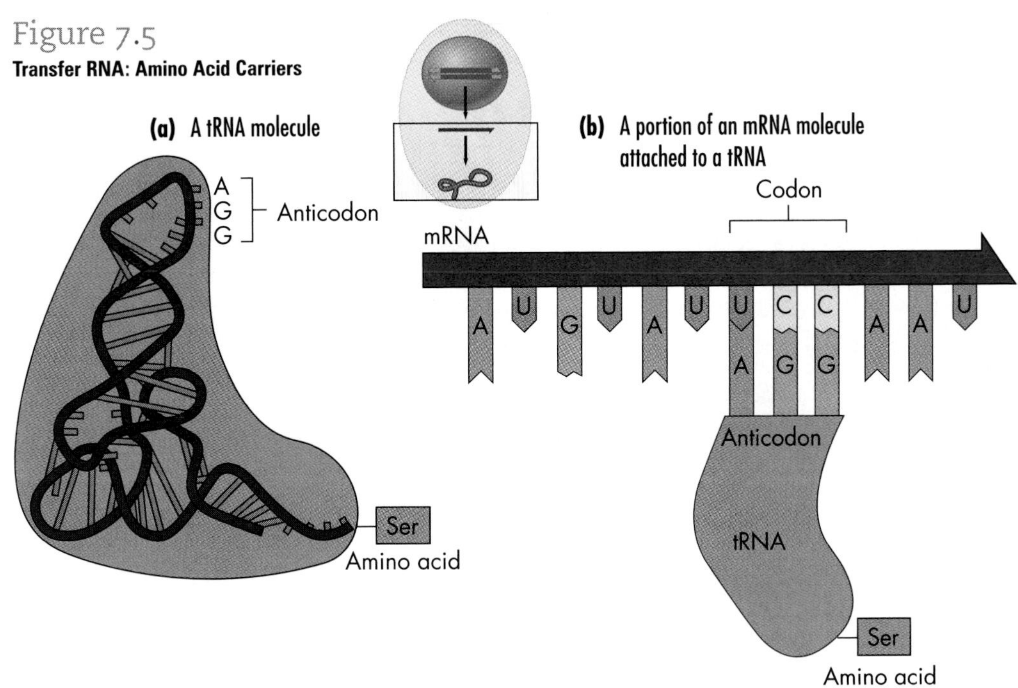

Figure 7.5
Transfer RNA: Amino Acid Carriers

codon
three adjacent nucleotide bases in DNA or mRNA that code for a single amino acid

anticodon
the three bases in a tRNA molecule that are complementary to three bases of a specific codon in mRNA

ribosomal RNA (rRNA)
an RNA molecule that is a structural component of ribosomes and is involved in protein synthesis

Codons and Anticodons During protein synthesis, sequences of three adjacent bases on mRNA specify the insertion of a particular amino acid into the growing polypeptide chain (Fig. 7.5b). A set of three nucleotides in mRNA that specifies the position of an amino acid in a protein is called a **codon**. For example, the first codon in the mRNA for antithrombin III protein is AUG, the second is UAU, and the third is UCC (Fig. 7.5b).

Since tRNA recognizes a codon in mRNA, it has to have a sequence complementary to the codon. This portion is called an **anticodon**, and is three adjacent nucleotides in tRNA that bind to a codon in mRNA (Fig. 7.5b). For example, a tRNA that carries the amino acid serine has, at the opposite end, the anticodon AGG. The AGG anticodon pairs up with the UCC codon on an mRNA.

The pairing of codon with anticodon is the key to how amino acids line up in the right sequence during protein synthesis. The order of the codons in mRNA specifies the order in which tRNAs will bring over and line up specific amino acids. The lined-up amino acids are then joined to each other by the ribosome, which contains the third type of RNA.

Ribosomal RNA and Ribosomes

Ribosomal RNA, or **rRNA**, is a key component of ribosomes, the microscopic "workbenches" on which proteins are forged. Ribosomal RNA works together with messenger RNA and transfer RNAs to translate genetic information into proteins. Eukaryotic ribosomes are about one-half protein and one-half rRNA. Ribosomal RNA is the most abundant type of RNA (80 percent) in the eukaryotic cell.

As the cell's workbenches, the ribosomes support mRNAs and tRNAs all the while genetic information is being translated. They also help to link amino acids to each other in a growing polypeptide chain. Approximately 75 proteins are wrapped around the

Figure 7.6

Ribosomes: Sites of Protein Synthesis

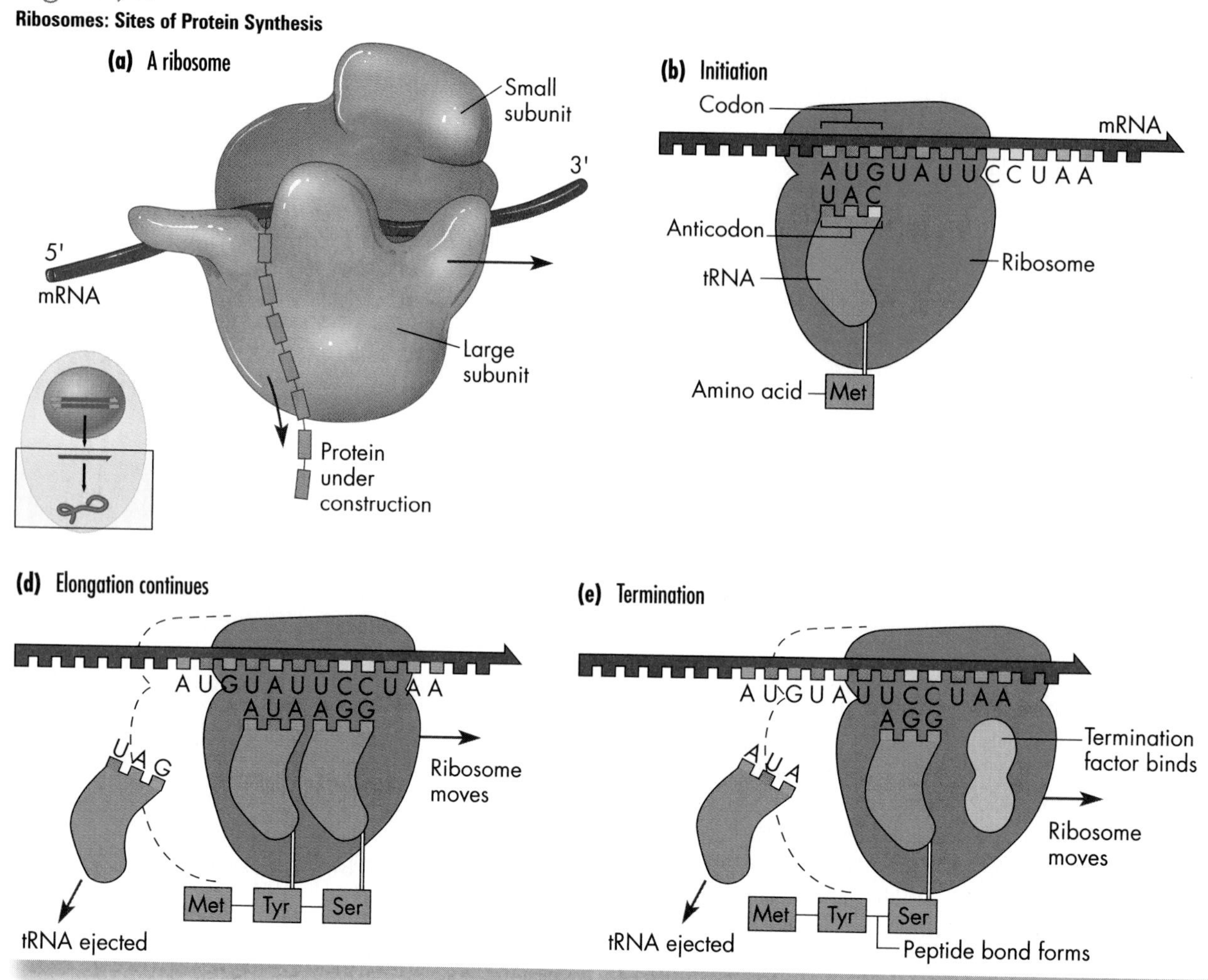

rRNAs, forming the beadlike structures of ribosomes (Fig. 7.6).

Each ribosome consists of one large and one small subunit separated by a groove (Fig. 7.6a). An mRNA strand "threads" through the groove, and the ribosome moves along the mRNA like a pulley on a rope. As it moves, the ribosome joins amino acids to the growing polypeptide chain one by one in the order specified by codons in the mRNA. Recent evidence suggests that rRNA itself helps to catalyze the addition of each amino acid to the next. Usually, several ribosomes at a time move down a single mRNA molecule like cars in a caravan, each ribosome producing a single polypeptide chain.

Cells contain tens of thousands of ribosomes, located wherever proteins are being synthesized; this can be in the cytoplasm of eukaryotic cells or in the DNA-containing region of prokaryotic cells. (In prokaryotes, translation of an mRNA begins at one end even before the other end of the mRNA has been completely transcribed.)

Translation: Protein Synthesis

Once Watson and Crick discovered the elegant double helix structure of DNA (see Chapter 6), the secret of the molecule's information storage capacity could be unlocked: it lies in the order of the nucleotide bases. Later work then showed that the sequence of nucleotides transcribed from DNA into mRNA determines the sequence of amino acids in proteins. Protein synthesis is one of the most important topics in biology, so let's explore it with the AT3 protein as our example.

As you read this, cells in your liver are probably cranking out messenger RNAs bearing the genetic information from the AT3 gene in your DNA that are ready to join with ribosomes and begin making the anti-blood-clotting protein. Here are the stages.

Stages of Protein Synthesis

The building of a polypeptide chain for a protein involves four stages: initiation, elongation, termina-

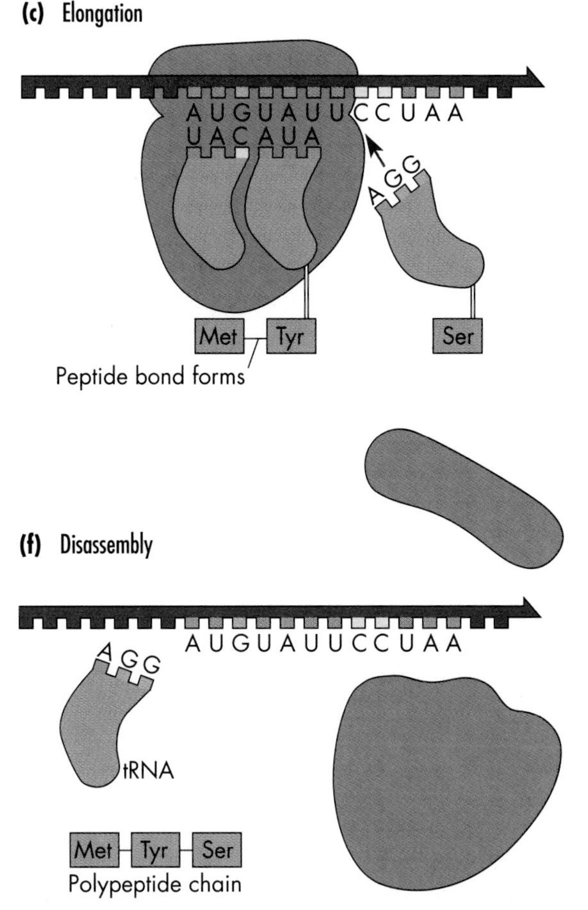

tion, and disassembly. Figure 7.6b–f illustrates each of these stages.

During initiation, the machinery needed for protein synthesis assembles itself to form the workbench, the ribosome (Fig. 7.6b). Specifically, the two ribosomal subunits come together on the mRNA, and the first tRNA floats over and binds to the mRNA. This first tRNA totes along the amino acid methionine, which is the first amino acid of most polypeptides. It binds to the start codon, AUG, on the mRNA.

During elongation, the next tRNA diffuses in, lining up its amino acid with the previous one (Fig. 7.6c), and the ribosome joins the amino acids together by peptide bonds (Fig. 7.6d). Simultaneously, the first tRNA is ejected from the complex as the ribosome moves one codon along the mRNA (Fig. 7.6d). Elongation continues, as the ribosome joins more and more amino acids together. Figure 7.6 shows just 3 amino acids being joined, but for the AT3 protein your liver produces, the elongation steps are repeated over and over until the polypeptide chain is 464 amino acids long. Amazingly, it takes only about 20 seconds for a protein of this size to be forged on the ribosomes in your liver cells.

Termination occurs when the ribosome reaches a stop codon in mRNA (Fig. 7.6e). In the case of AT3, the stop codon is the three nucleotides UAA. At that point, a termination factor protein binds to the ribosome–mRNA complex rather than a tRNA, and this brings the growth of the polypeptide chain to a halt.

In the final step, disassembly, the newly formed polypeptide falls away from the ribosome and the workbench components disassemble (Fig. 7.6f).

You can see the central dogma at work in Figure 7.6: the sequences of nucleotide bases in mRNA specify the order of tRNAs, each one toting a particular kind of amino acid. Just as DNA carries a genetic code, so too does each mRNA, with bases complementary to those in DNA.

The Genetic Code

We've seen how a human liver cell can assemble an AT3 protein. But if Harry Meade splices the human gene for AT3 into a goat, will the goat cells translate the human mRNA into an identical protein? Do different species, in other words, speak the same genetic language? The answer is yes, and it's based on the nature of the genetic code.

The genetic code is like a dictionary cells use to define which amino acids will be translated from each sequence of three adjacent bases in mRNA—in other words, from each codon. We've seen that RNAs have an alphabet of only 4 letters (the bases A, C, G, and U), while proteins have 20 amino acids. How can these 4 nucleotides code for all 20 different types of amino acids found in proteins?

In trying to understand the genetic code, biologists

initiation
the first main stage of protein synthesis, during which tRNA associates with a ribosome and an mRNA, forming a complex of many molecules

start codon
in DNA and mRNA, the codon that signals where the translation of a portion begins

elongation
the second main stage of protein synthesis, during which the sequential enzymatic addition of amino acids builds the growing polypeptide chain

termination
the third main stage of translation (protein synthesis), during which the growth of the polypeptide chain comes to a halt when the ribosome reaches the stop codon

disassembly
the end stage of protein synthesis during which the newly formed polypeptide falls away from the ribosome and the ribosome's "workbench" components disassemble

genetic code
the specific sequence of three nucleotides in mRNA that encodes an individual amino acid in protein; for example, the code for methionine is AUG

Figure 7.7

The Genetic Code

Codon			Amino acid
U	U	U	Phe
U	U	C	Phe
U	U	A	Leu
U	U	G	Leu
U	C	U	Ser
U	C	C	Ser
U	C	A	Ser
U	C	G	Ser
U	A	U	Tyr
U	A	C	Tyr
U	A	A	STOP
U	A	G	STOP
U	G	U	Cys
U	G	C	Cys
U	G	A	STOP
U	G	G	Trp

Codon			Amino acid
A	U	U	Ile
A	U	C	Ile
A	U	A	Ile
A	U	G	Met (START)
A	C	U	Thr
A	C	C	Thr
A	C	A	Thr
A	C	G	Thr
A	A	U	Asn
A	A	C	Asn
A	A	A	Lys
A	A	G	Lys
A	G	U	Ser
A	G	C	Ser
A	G	A	Arg
A	G	G	Arg

Codon			Amino acid
C	U	U	Leu
C	U	C	Leu
C	U	A	Leu
C	U	G	Leu
C	C	U	Pro
C	C	C	Pro
C	C	A	Pro
C	C	G	Pro
C	A	U	His
C	A	C	His
C	A	A	Gln
C	A	G	Gln
C	G	U	Arg
C	G	C	Arg
C	G	A	Arg
C	G	G	Arg

Codon			Amino acid
G	U	U	Val
G	U	C	Val
G	U	A	Val
G	U	G	Val
G	C	U	Ala
G	C	C	Ala
G	C	A	Ala
G	C	G	Ala
G	A	U	Asp
G	A	C	Asp
G	A	A	Glu
G	A	G	Glu
G	G	U	Gly
G	G	C	Gly
G	G	A	Gly
G	G	G	Gly

reasoned that if each individual nucleotide coded for only one amino acid, then there would only be four possible amino acids in proteins. We know, though, that proteins have 20 different amino acids. If a series of two bases encoded each amino acid (for example, AU for one, CA for another, and so on), the system would encompass 16 amino acids (4 × 4), still not enough. Biologists realized that RNA "words" made up of combinations of 3 nucleotides would allow for 64 possible combinations (4 × 4 × 4), and this would be more than enough to code for the 20 amino acids observed in proteins. In other words, even though the RNA alphabet contains only 4 letters, it could spell out 64 different three-letter words (codons) just as the 26 letters of our alphabet can be combined to make hundreds of thousands of words.

Since there are 20 amino acids in nature, this assumption seemed correct. But was it? And how could biologists figure out which codon translates into which amino acid? Experimenters in the laboratory created artificial genetic messages with RNA and mixed them in test tubes with ribosomes and other translational machinery that they harvested from bacterial cells. They made the artificial message UUUUUUUUUUUU, for example, and found that the "machinery" translated it into the polypeptide Phe–Phe–Phe–Phe. From this they learned that the codon UUU specifies the amino acid phenylalanine. By working through the entire set of 64 codons, they were able to work out the specific amino acid that each codon encodes. This is the genetic code shown in Figure 7.7.

Look closely at that figure: Is each amino acid coded for by a single codon? Or are there amino acids that are encoded by more than one codon? Are there any codons that encode more than one amino acid? Do all codons encode amino acids? Through this short exercise, you've probably already discovered that serine (Ser), taking one example, is encoded by six different codons UCU, UCC, UCA, UCG, AGU, and AGC. Obviously, the genetic code is redundant, and this has an impact on certain genetic mutations. The mutation of a UCU codon to UCC, for example, would not change the amino acid encoded because both UCU and UCC encode serine. This mutation, then, would be unlikely to affect the organism.

Perhaps your inspection of the table also uncovered this fact: Although an amino acid can have more than one codon, there are no codons that encode more than one amino acid. Each codon "spells out" just one single amino acid. In addition, three codons—UAA, UAG, and UGA—spell "stop" instead of

© ISTOCKPHOTO.COM

specifying amino acids; they are the stop codons we saw earlier, and they terminate a growing polypeptide chain (review Fig. 7.6e). As we also saw earlier, AUG is the start codon and it usually encodes methionine wherever it occurs.

Reading the Genetic Message

Geneticists studying the genetic code got a surprise when they tested a genetic message with a repeating triplet motif, such as AGCAGCAGCAGCAGC. They expected this message to translate into one polypeptide, but in fact, it encoded three different ones, each made up of just one kind of amino acid: either all serine (codon AGC), all alanine (codon GCA), or all glutamine (codon CAG). This test tube experiment makes sense only if translation can start at any place along the message:

...AGC AGC AGC AGC AGC...

...A GCA GCA GCA GCA GC...

...AG CAG CAG CAG CAG C...

Experiments like this proved that once translation starts at one point, it does indeed read off bases in groups of three and without overlapping or skipping bases. The experiments also showed that the starting place for translation determines the meaning of the message. The genetic message AGCAGCAGCAGCAGC can be divided into codons in the three ways you just saw, and each different way is called a reading frame. How does a reading frame determine the meaning of a message? Just consider for a moment howc hangesi nreadingf ramec ana lterthism essages pelledo utinEnglis hletters.

A cell needs to "know" where to start reading, and cells do have a special codon that signifies where translation should start; thus it establishes the reading frame. This start codon, AUG, also codes for the amino acid methionine (see Fig. 7.6a). In fact, methionine is the first amino acid in nearly every polypeptide chain.

The Genetic Code Is Almost Universal

Returning to Dr. Meade's research problem, do human cells and goat cells have the same genetic code? Yes. Meade observed that the human *AT3* gene, transcribed and translated in goat cells, produces the exact sequence of human AT3 protein. Humans and goats clearly must use the same genetic code. Furthermore, earlier experiments in bacteria, plants, and in many other life forms revealed that nearly all organisms use that same genetic code. (A few codons do differ from the norm in certain mitochondria and some protists.) It is an amazing testament to the unity of all living things that our human cells "speak" exactly the same genetic language as bacterial cells. Each codon is translated into the very same amino acid, whether the codon resides in a human liver cell, a goat mammary gland cell, or even an *E. coli* bacterial cell. The nearly universal nature of the genetic code is expected if all living things hark back to the same ancestral cells that originated very early in Earth's history. The genetic code was apparently developed at the dawn of life and has been inherited intact for billions of years.

reading frame
three ways a sequence of three nucleotides in a gene can be divided into contiguous, nonoverlapping sets of three nucleotides in codons

chromosomal mutation
a genetic change that affects large regions of chromosomes or entire chromosomes

single-gene mutation
changes in the base sequence of a single gene

Gene Mutation

Dr. Meade had been working for years on an anticlotting drug based on large quantities of the protein AT3. One use for the drug would be heart bypass patients who tend to form blood clots during surgery. But there are other potential uses, too, including the treatment of people with a mutation of the *AT3* gene. With our background of discussions to this point, we can now define a *mutation* (see Chapter 5) as a change in the base sequence of an organism's DNA. People with a mutation in the *AT3* gene can't form the AT3 protein, and thus tend to develop blood clots at inappropriate locations. These clots can block the blood supply to various tissues and cause severe, life-threatening diseases.

Geneticists recognize two general categories of mutations. Chromosomal mutations affect large regions of chromosomes, or even entire chromosomes, and so alter the locations of many genes. Chromosomal mutations include changes in chromosome structure or number (as in Down syndrome). Single-gene mutations, or changes in the base sequence of a single gene, alter individual genes, sometimes disrupting gene function, but sometimes having no effect at all.

Kinds of Single-Gene Mutations

We covered several types of chromosome mutations in Chapters 4 and 5, so this section focuses on single-gene mutations.

Figure 7.8

Mutations in Genes

(a)–(c)

(a) Original DNA

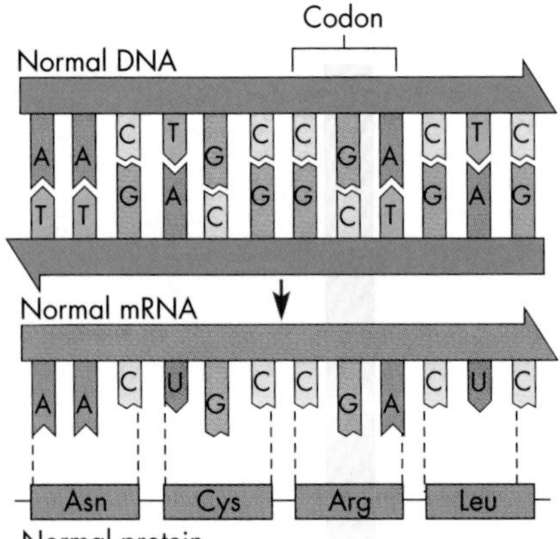

(d)–(e)

(d) Original DNA

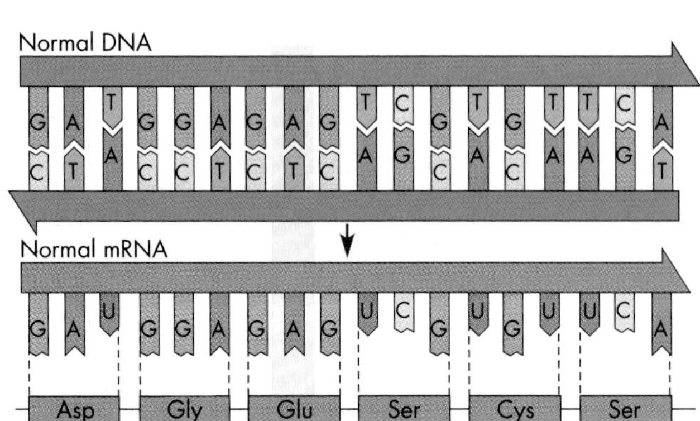

(b) Base substitution leading to amino acid replacement

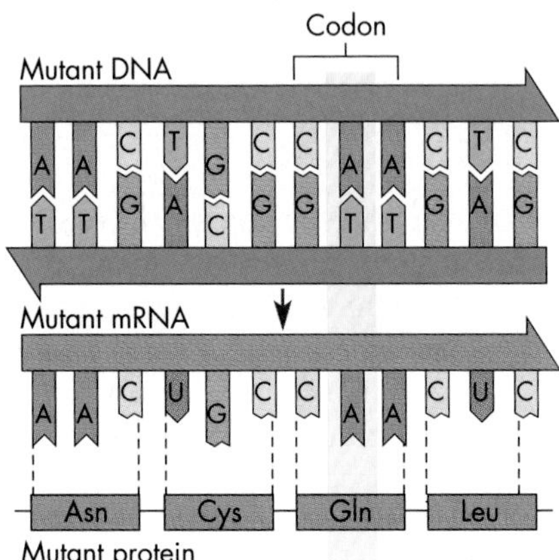

(e) Base deletion leading to a frame shift

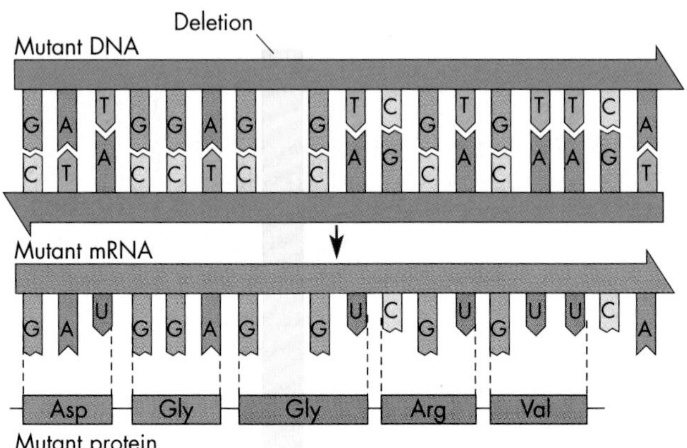

(c) Base substitution leading to premature termination

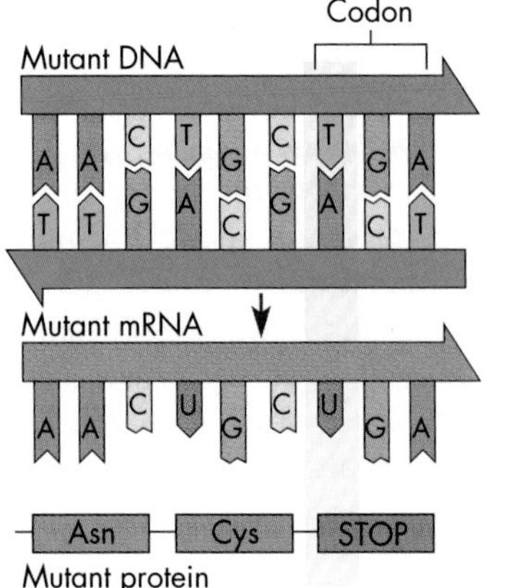

base substitution mutation
a type of mutation that occurs when one base pair replaces another

Base Substitution Mutation A common sort of mutation called a **base substitution mutation** occurs when one base pair replaces another. For example, in constructing the AT3 protein, the 129th codon is normally CGA for the amino acid arginine (Fig. 7.8a). In certain human families, however, this codon is changed to CAA, which encodes glutamine (Fig. 7.8b). This amino acid alteration changes the shape of the AT3 protein so that it does not inhibit clotting anymore, and the family members often inappropriately form clots in veins. In certain other families, a base pair substitution in the same DNA codon changes the original CGA to TGA. The complementary codon in the mRNA then changes to UGA, which means STOP (Fig. 7.8c). The resulting AT3 protein is too short to do its job.

Base Deletions and Insertions We just saw how base substitutions change one base pair into another. In contrast, *base deletions* and *base insertions* remove or add base pairs to the gene. For example, some people have a single base pair deleted from their AT3 gene, making the gene one base pair shorter than normal

(Fig. 7.8d,e). Changes like this can alter the reading frame of an mRNA, garbling the amino acid sequence from the point of the mutation all the way to the end of the protein. Shifts in reading frame can be caused by the deletion or insertion of one or two base pairs, but not *three*. Can you see why?

The Origin of Mutations

Mutations can obviously be destructive. So where do they come from? Mutations arise either through spontaneous errors that occur as DNA is replicating, or later, through damage to the DNA by physical or chemical agents. As DNA is replicating, the wrong base can be inserted into the growing helix. Usually, a repair by the DNA polymerase enzyme edits out this kind of error. Sometimes, though, physical and chemical agents called mutagens change DNA structure. Mutagens include ultraviolet rays from the sun, chemicals in cigarette smoke, and even many natural substances from plants or fungi. For example, aflatoxin, a natural compound found in moldy peanuts, is an extremely potent mutagen.

Despite the mutagens in our environment, certain enzymes usually repair the type of DNA damage they cause. When these enzymes themselves are defective, however, the consequences can be dire. Some people inherit the skin disease xeroderma pigmentosum and have a mutation that blocks the body's ability to form a critical DNA repair enzyme. As a result, their cells quickly accumulate mutations. One young Navajo girl, for example, suffered from this condition and had dozens of skin cancers where UV light from the sun has mutated the DNA in her skin cells, and the cells weren't able to repair these mutations.

Cancer-causing substances are known as carcinogens and they often act by generating mutations as well. A mutagen has the potential to cause cancer if it affects a gene that controls cell growth. Because people and bacteria have such similar DNA metabolism, nearly anything that can cause mutations in bacteria can also cause mutations in people. Because of this link, researchers have invented a test for the carcinogenicity (cancer-causing ability) of chemical and physical agents that exposes certain bacteria to suspected carcinogens and then measures the bacterial mutation rates. Researchers have used this test, called the Ames test, to catalog thousands of substances that are mutagenic and also potentially carcinogenic. The list includes asbestos, ultraviolet light, nitrates in foods that become carcinogenic during cooking, aflatoxin from moldy peanuts, and other food contaminants. People who expose themselves to known mutagens, such as too much sunlight or the toxins in cigarette smoke, place a heavy burden on their DNA repair enzymes and magnify their risk of developing cancer.

If Dr. Meade's company succeeds in producing enough human AT3 protein for a new drug, it may benefit people who form too many blood clots due to mutations in their AT3 genes. Meade must also be sure that in manipulating the human AT3 gene, he and his colleagues don't cause new mutations that could lead to a defective protein. Mutation is an important element in both gene action and in genetic manipulation. And as we saw in Chapter 1, and will see again and again in the book, mutations are the source of genetic variation and of evolutionary change. But mutation is not the only difficult issue in genetic engineering. There is another one involving the basic control of gene action: how can you get a protein made in the human liver (AT3) to be made in a goat's mammary gland and then excreted in its milk?

mutagen
physical and chemical agents, including ultraviolet rays from the sun, chemicals in cigarette smoke, and even many natural substances from plants or fungi, that can change DNA structure

carcinogen
a cancer-causing substance

Ames test
a test for carcinogenicity (cancer-causing ability) of chemical and physical agents that exposes certain bacteria to suspected carcinogens and then measures the bacterial mutation rates

gene regulation
the process that controls how and when each gene is turned on and off in each living cell

LO4 How Are Genes Controlled?

Normally, the liver is the only tissue that makes large amounts of the anticlotting protein AT3. Before Meade could splice the gene into a goat and get the animal to make AT3, he needed a very thorough understanding of how genes are regulated in cells. Luckily, researchers have extensive knowledge of gene regulation, the process that controls how and when each gene is turned on and off in each living cell.

Gene Regulation in Prokaryotes

Biologists first studied gene regulation in bacteria and, as a result, understand gene control in those organisms in the greatest detail. They've learned that bacteria have streamlined and highly sophisticated mechanisms of controlling gene expression, and this allows them to change the kinds of proteins they make quickly and frequently. Let's consider, for example, the *E. coli* bacteria growing in your intestine, and let's say you just finished eating a bowl of yogurt or ice cream. The bacteria in your intestine are now

operon
the collective term for a regulatory protein and a group of genes whose transcription it controls in bacterial cells

repressor
a protein that binds to an operator and blocks transcription of the adjacent gene

operator
a series of nucleotides that bind to a repressor protein, thereby preventing the initiation of gene and transcription from the adjacent promoter

floating in a sea of lactose, the main sugar in milk. In order to use that sugar, the bacteria quickly make enzymes that help break down lactose once it is taken into the bacterial cell.

A bacterial cell that uses resources efficiently can grow and divide quickly. But an *E. coli* cell that makes unnecessary proteins—say, the enzymes for breaking down lactose when that sugar is not present—expends energy and materials that could go toward cell reproduction; thus, it grows and divides more slowly. As simple as bacterial cells are in structure and function, they have **operons**, elegant gene regulation systems consisting of regulated clusters of genes acting in a coordinated function. Operons allow the cells to make only the types and amounts of proteins they need when and where they need them.

The best-studied operon regulates the way cells break down lactose. Around 1950, biologists knew that an enzyme called beta-galactosidase in *E. coli* cells breaks down lactose into the two simple sugars glucose and galactose. *E. coli* cells growing in the absence of lactose (for instance, in your intestines *before* you eat yogurt) contain none of this enzyme. However, in the presence of lactose (right after the yogurt), each bacterial cell produces thousands of beta-galactosidase molecules within about 20 minutes. Somehow, lactose *induces* the formation of the sugar-digesting enzyme as well as two other functionally related proteins.

In 1960, through elegant experiments, French geneticists François Jacob and Jacques Monod determined how *E. coli* bacteria regulate the synthesis of beta-galactosidase. They found that a certain protein in the *E. coli* cell called the **repressor** can sense lactose in its surroundings. If lactose is absent, the repressor protein prevents the transcription of mRNA for the beta-galactosidase enzyme (Fig. 7.9a). If lactose is present, however, the repressor protein allows the DNA to start churning out copies of mRNA for the enzyme (Fig. 7.9b). Careful experiments showed that the repressor acts by binding to DNA. Specifically, the repressor, in the absence of lactose, binds to the **operator**, a series of base pairs in DNA near the gene that encodes the lactose-digesting enzyme (Fig. 7.9a). When the repressor binds to the operator DNA, it covers the promoter DNA, the base pairs to which RNA polymerase binds. (Recall that RNA polymerase is the enzyme that copies DNA into RNA.) Because RNA polymerase can't bind the promoter, there is no transcription. Without transcription, there can be no mRNA. And without mRNA, there can't be an extra lactose-digesting enzyme. But when the bacteria are growing on lactose, that sugar binds to the repressor (Fig. 7.9b), and the binding alters the shape of the repressor so that it can no longer bind to the operator. This then unblocks the promoter. Now RNA polymerase can bind to the promoter and transcribe the gene, and this transcription, in turn, is translated into the lactose-digesting enzyme. The bacterium can now use the energy from lactose for growth.

Figure 7.9
Gene Regulation in a Bacterial Cell

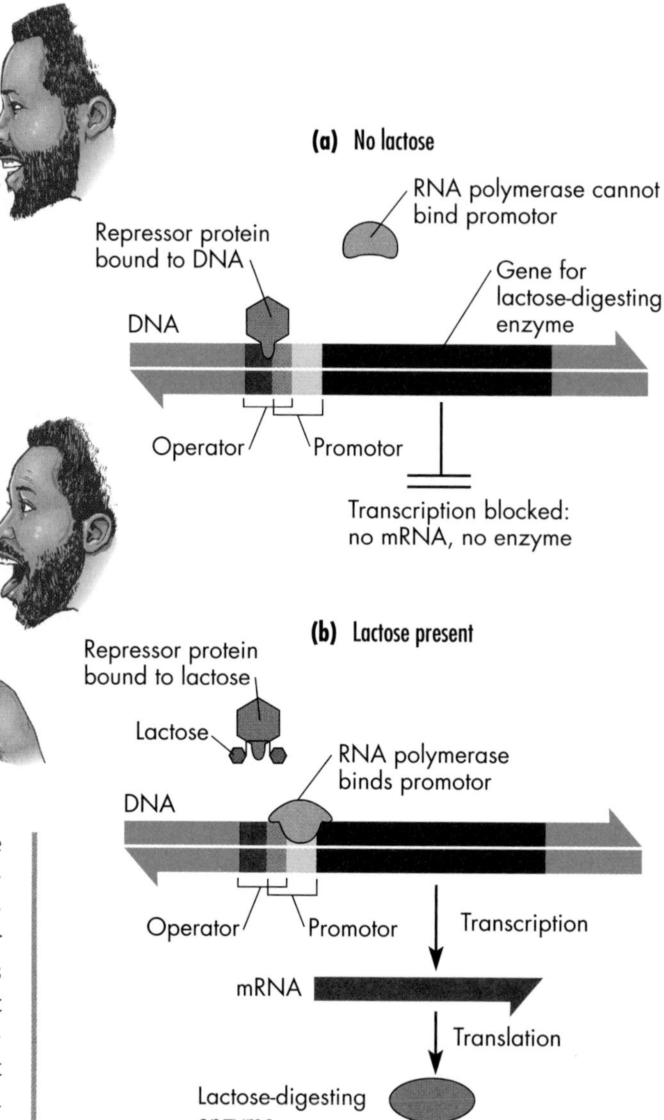

Figure 7.10
Regulating Eukaryotic Genes: *AT3* and Goat's Milk

recombinant DNA technology
the techniques involved in excising DNA from one genome and inserting it into a foreign genome

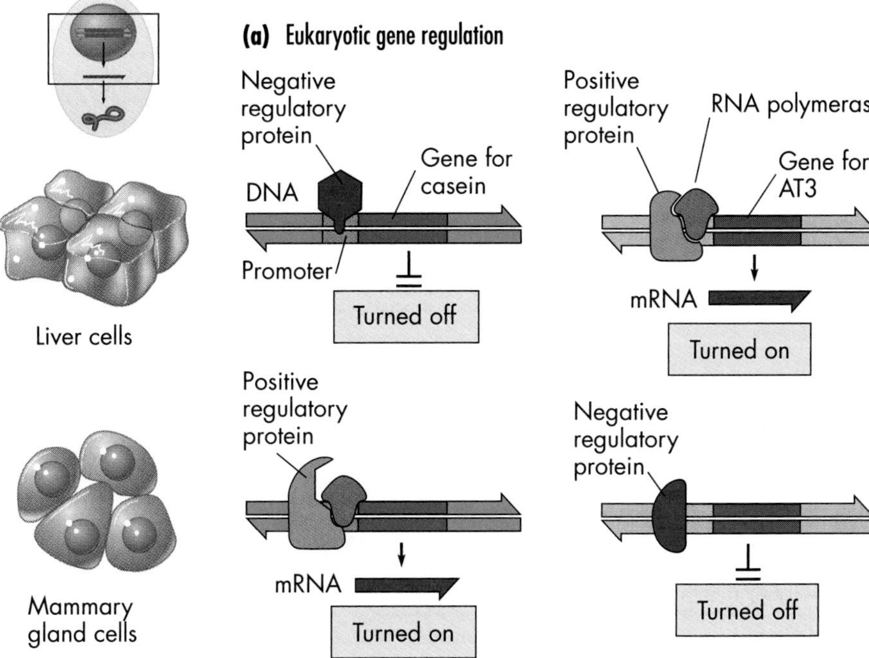

Jacob and Monod gave the name "operon" to the protein signaler, the operator DNA sequence to which it binds, the promoter to which RNA polymerase binds, and the group of genes transcribed together in a single RNA in that part of the genome. This work was so fundamental to understanding how genes are controlled that Jacob and Monod won a Nobel Prize in 1965.

Gene Regulation in Eukaryotes

Back now to the question of how multicellular organisms turn on certain genes in certain tissues so that a person, let's say, makes AT3 in her liver but not in her mammary glands, and milk protein (casein) in her mammary glands but not her liver. All your body cells, in other words, have the same genes in the cell nucleus but only certain genes are available for transcription in each cell. Why? It turns out that eukaryotic cells like your own have regulatory mechanisms similar to the ones we just saw in bacteria: regulatory proteins bind special regions of DNA called promoters, regulatory elements, and enhancers (Fig. 7.10a). This binding then alters the action and the rate of gene transcription in a way that is specific to each cell type. So by applying these principles of eukaryotic gene regulation, we can see that Harry Meade had to isolate the regulatory portion of the casein gene and somehow hook it up to the coding portion of the AT3 gene (Fig. 7.10b).

We've finally arrived, now, at the place where Dr. Meade was able to use all his background understanding on gene function, gene regulation, and protein synthesis along with the modern genetic tools of recombinant DNA technology, methods for combining DNA segments (usually from different species) in the laboratory. Let's see how he did it.

LO5 Recombinant DNA Technology

DNA molecules recombine with each other in nature (Chapter 4). Bottom line, that's what sex is about. But viable sexual reproduction exchanges

Figure 7.11

How Geneticists Create Recombinant DNA Molecules

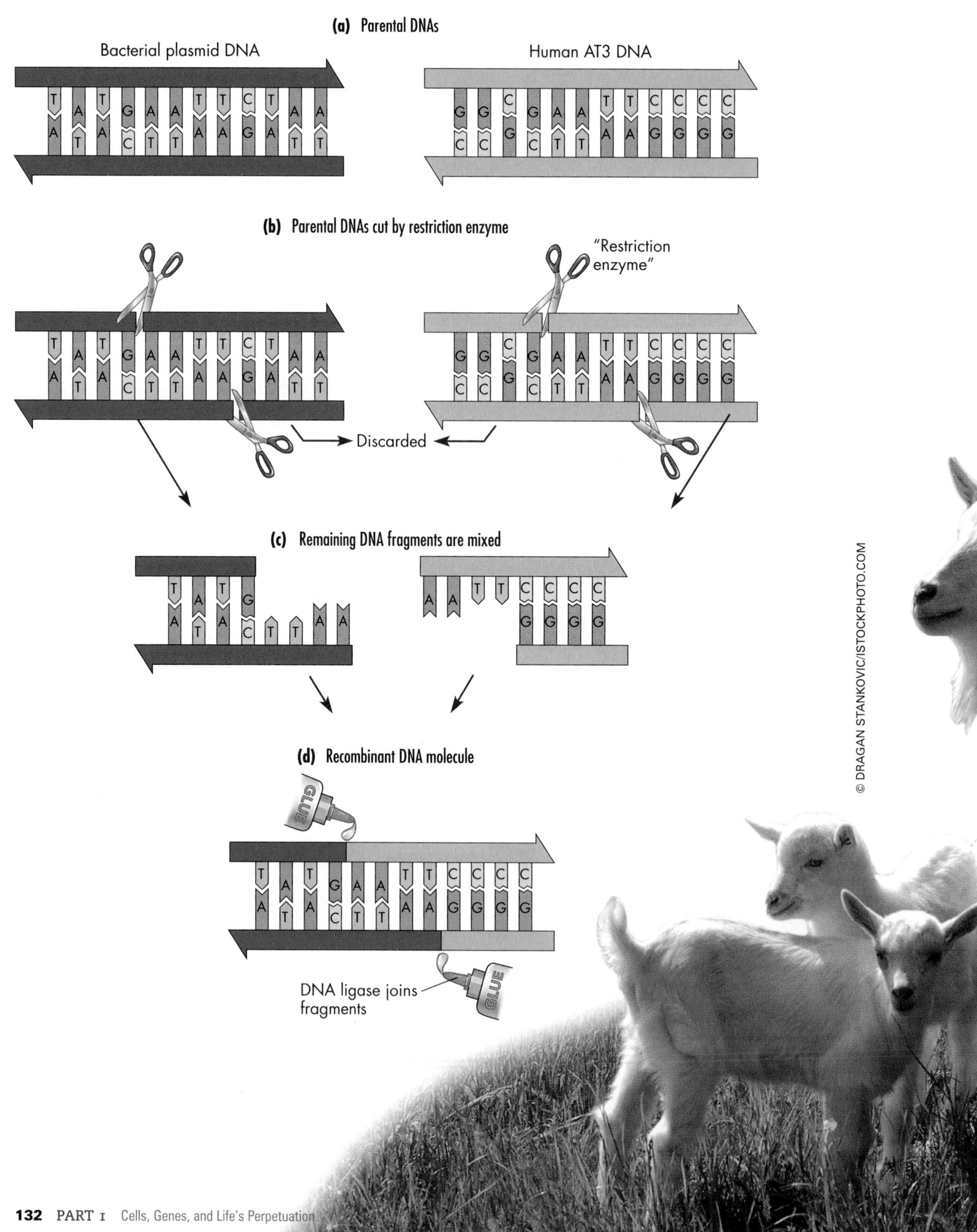

corresponding DNA fragments *within a single species.* In the early 1970s, biologists realized that they could deliberately recombine DNA molecules from different species in the lab. Here's how.

Constructing Recombinant DNAs

The first step in any attempt to recombine DNA from different species is to produce millions of copies of a single desirable piece of DNA, that is, to **clone** the DNA. Because Harry Meade wanted to combine a human *AT3* gene with the goat casein promoter, he started with molecular cloning.

Molecular cloning is accomplished by splicing the desired DNA fragment (a gene, a promoter region, or what have you) into a **vector**, which is usually a virus or plasmid that occurs in nature. A **plasmid** is a simple circle of DNA, about 3,500 base pairs long, that replicates independently inside a bacterial cell. That independent replication makes them handy for genetic engineers and so do several other traits: plasmids contain genes that make a bacterial cell resistant to certain antibiotics, such as tetracycline and ampicillin. These can serve as a marker, as we'll see. Plasmids can also move readily from one bacterial cell to another. And they can accept pieces of foreign DNA, carry them along, and replicate them wherever they go. So a geneticist could insert a gene like *AT3* into a plasmid for cloning.

It's important to keep in mind that the same properties that make plasmids so useful in the genetics lab make them a disaster in hospitals, because they can carry antibiotic-resistant genes from one bacterial species to another and one patient to another. As resistance spreads, antibiotics can very quickly become ineffective at preventing disease.

Plasmids and cloning are just two tools of recombinant engineers. They also have special enzymes to cut DNA molecules in specific places and paste the fragments together, as well as ways to insert new DNA into foreign cells.

Cutting the DNA

To cut DNA molecules, researchers use special molecular "scissors," proteins called **restriction enzymes** that they isolate from bacteria (Fig. 7.11b). Bacteria naturally produce restriction enzymes and use them to cut certain viral DNAs into pieces; this restricts the types of viruses that can infect them. A restriction enzyme recognizes a few specific base pairs in a row wherever they occur in a DNA molecule. The enzyme then cleaves the DNA at a consistent place in or near the recognized sequence. Some restriction enzymes cut the DNA in a staggered fashion, so that one strand of the double helix sticks out beyond the other (Fig. 7.11b). When the weak hydrogen bonds between the strands break, the two DNA fragments can separate, and single-stranded ends protrude (Fig. 7.11c).

clone
all cells derived from a single parent cell; a replica of a DNA sequence produced by recombinant DNA technologies

vector
in molecular genetics, a virus or plasmid used to carry pieces of DNA into cells; in medicine, an organism such as a mosquito that transmits a disease-causing microbe such as the malarial protist

plasmid
a circular piece of DNA that can exist either inside or outside bacterial cells but can reproduce only inside a bacterial cell; plasmids are used extensively in genetic engineering as carriers of foreign genes

restriction enzyme
a naturally occurring enzyme that cuts DNA at precise points

DNA ligase
a protein that joins DNA strands together end to end during DNA recombination and replication

Joining the DNA

Biologists call the little tails protruding from the two ends of a staggered cut "sticky ends." These ends act "sticky" in that they can easily re-form hydrogen bonds with complementary base pair sequences that stick out on the ends of other DNA molecules. In fact, any two DNA molecules with complementary sticky ends can join together, no matter how unrelated are the rest of the two DNA molecules. By making these sticky ends on pieces of DNA from organisms as different as humans, goats, and bacteria, biologists can recombine the DNA molecules from the different species. Figure 7.11c shows in pale blue a small portion of the human *AT3* gene cut by a restriction enzyme called EcoRI. The figure shows in dark blue a DNA fragment from a bacterial plasmid similarly cut by the same restriction enzyme.

Hydrogen bonds can quickly form between the AATT base sequence of the human DNA and the complementary TTAA sequence of the plasmid DNA, and can hold the two fragments together (Fig. 7.11d). The biologist then adds the enzyme **DNA ligase**, which acts like a glue dispenser to cause strong covalent bonds to form between the opposing ends of the two molecules. (In nature, DNA ligase repairs damaged DNA.)

probe
a radioactive nucleic acid made in the laboratory that is complementary to a gene; a molecular genetic tool used during cloning

Transgenic Goats that Make Human AT3 Protein in Their Milk

Harry Meade and his collaborators had cloning, plasmids, molecular "scissors," and a "glue dispenser" to use in constructing a recombinant DNA molecule, and their work required three phases. In the first phase, they cloned the *AT3* gene. In the second phase, they spliced the human *AT3* to goat casein promoter. In the third phase, they introduced the recombinant gene into goat cells and produced the desired human protein.

Phase 1: Cloning AT3

To clone *AT3*, Harry Meade and colleagues first isolated RNA from human liver cells (Fig. 7.12, Step ①). This included mRNA for the *AT3* gene, but also mRNAs for albumin and for thousands of other proteins made by liver cells. In Step ②, they mixed this mRNA with a solution of free base pairs and a special enzyme, reverse transcriptase. (Recall from Chapter 2 that this enzyme makes a DNA copy from an RNA template. This resulted in double-stranded complementary DNA copies, or cDNA. Next, they cut plasmids and the cDNA with a restriction enzyme to open the circle and then inserted the human DNA fragments into plasmid molecules (Step ③). In Step ④, they transferred the plasmids, along with the hitchhiking human DNA, into bacterial cells, a process called *transformation* (review Fig. 6.2). Finally, they cultured the bacterial cells, which now contain the plasmid combined with human DNA, on a petri dish (Step ⑤). Each cell divided repeatedly and produced a colony, a clone of cells with millions of identical copies, all descended from the same founder cell with the same DNA. Because there are thousands of different genes expressed in liver cells, each bacterial colony can have a cDNA from a different human gene. We said earlier that plasmids can carry antibiotic resistance. In this case, the researchers added an antibiotic to the bacterial growth medium so that any bacterial cell without the plasmid inside would die and not form a colony.

Eventually, thousands of bacterial colonies grew on the plate, and Meade and colleagues had to find the one (or ones) containing the *AT3* gene. In Step ⑥, they made a **probe** for the *AT3* gene, a radioactive nucleic acid that is complementary to the gene. They didn't know the gene's precise DNA sequence, but because they had the AT3 protein, they knew its amino acid sequence. By using the genetic code in reverse, they could write down DNA sequences that could encode a specific portion of the AT3 amino acid sequence. They could then artificially synthesize a radioactive DNA with this sequence. In Step ⑦, they could let this labeled DNA bind (or hybridize) to its

Figure 7.12
How to Clone a Human Gene

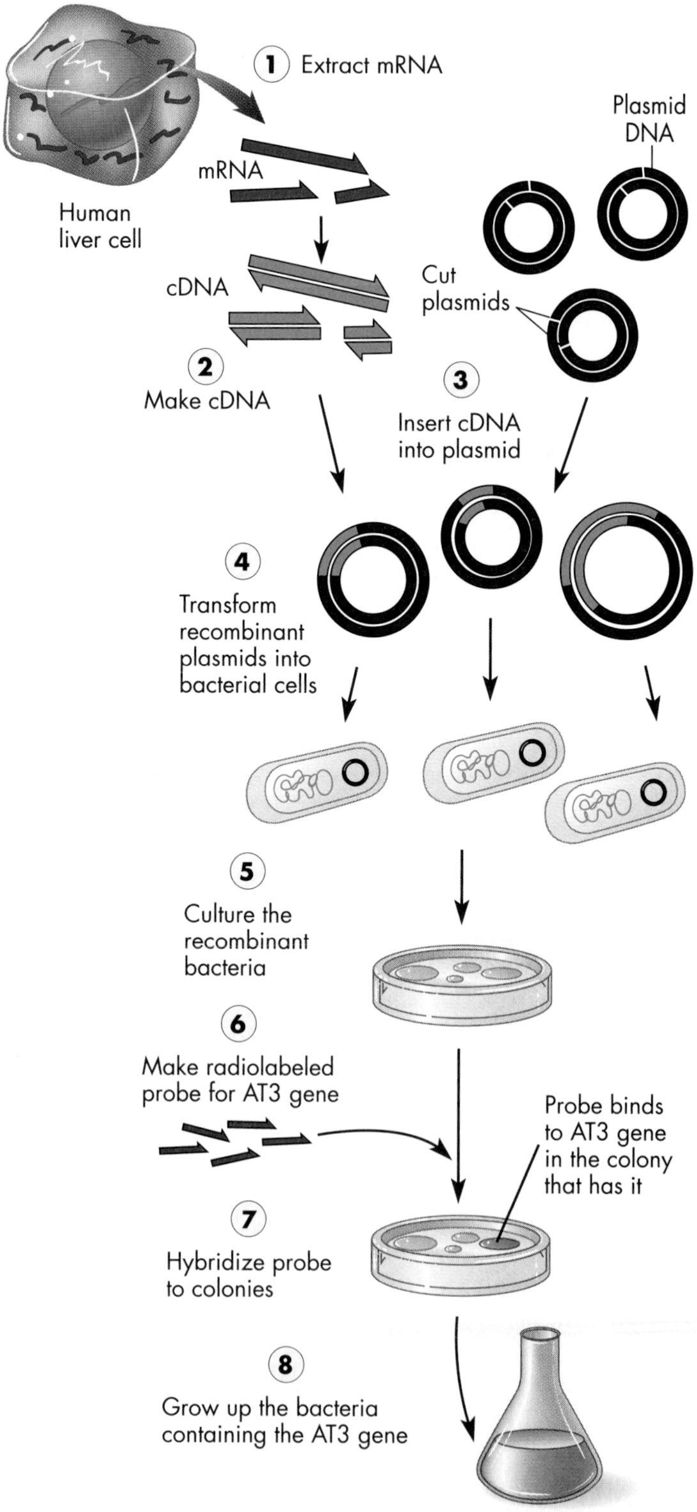

complementary DNA in the colonies, specifically that colony (and only that one) producing the *AT3* gene. Once they identified and isolated this colony (Step ⑧), they had millions of copies of the human *AT3* gene in the plasmids of millions of bacterial cells. They had cloned the gene.

Phase 2: Getting Proper Expression of the Cloned Gene

Using similar methods, Meade's group isolated and cloned the goat casein gene, including its promoter sequence. With molecular scissors and glue, they cut the genes to leave sticky ends on both, and then joined them together (see Fig. 7.11b–d). If they were lucky, the casein regulatory sequences would drive expression of the *AT3* gene during protein synthesis, so that when the goat made milk (including the most abundant protein in milk, casein), it would make human anticlotting protein as well.

Phase 3: Introducing Recombinant DNA into Goats

Now it was time to introduce the recombinant molecule into living goats. The researchers first isolated plasmids containing the casein–*AT3* recombinant gene from the bacteria, sucked up the recombinant DNA into a very small needle, and injected squirts of it into a number of goat embryos (each one a single cell and not yet divided into the first two, four, or eight cells of the embryonic animal). In preparation for the injection, they had flushed the embryos from the reproductive tracts of donor nanny goats. Next they transferred the DNA-injected embryos into the uteruses of surrogate nannies (recipient females) (Step ②) and allowed the embryos to develop into full-term baby goats. In Step ③, they tested the newborn kids for the presence of the injected DNA. Because injected DNA is rarely and randomly incorporated into goat chromosomes, most kids with the recombinant DNA will have it in one place in a single chromosome and hence be heterozygous for the desired DNA. About 5 to 10 percent of the offspring contained the recombinant DNA and could be called **transgenic animals** (animals bearing foreign genes inserted by recombinant DNA methodologies). They raised these transgenic goats and then mated pairs of them. Billy goats, obviously, can't produce milk like their sisters, but they're useful for breeding.

When the transgenic females became pregnant and gave birth, they started producing milk and many of them produced the human AT3 protein in their milk. Some of their newborn kids also had the recombinant gene in double doses—they were homozygous and were also transgenic—a second generation (Step ④). Finally, the research team started collecting milk from the lactating nannies and purifying the protein from the milk (Step ⑤). With enough of the AT3 protein, it's now possible to produce a prescription drug that can block inappropriate clotting during heart surgery or in other patients.

Who knew?

Carcinogens cause mutations in genes that regulate cell growth, causing growth to be uncontrolled. Thus, carcinogens are mutagens, although not all mutagens are carcinogens; other mutagens can kill a cell rather than making it cancerous.

transgenic animals
animals bearing foreign genes inserted by researchers using recombinant DNA methodologies

CHAPTER

8

Reproduction *and* Development

Learning Outcomes

LO[1] Explain the role of animal mating behavior in sexual reproduction

LO[2] Describe physical characteristics involved in sexual reproduction

LO[3] Compare and contrast the differences in male and female gametes

LO[4] Name and describe the developmental stages of an animal embryo

LO[5] Describe the events in human embryological development

LO[6] Describe the types of developmental changes that continue after birth

“Human sexuality shapes our bodies, our behavior, and our life-long experiences.”

Medical Help with Conception

Marcy and Stephen were a couple with a plan. They planned to travel, buy a house, and build their careers. Later, when Marcy turned 30, they would start their family. Everything went according to their careful plan except that six years into the marriage, they were still childless. Eventually they sought out Dr. Frances Batzer, an infertility specialist.

What do you know?

Why do you think more U.S. businesses are involved in infertility research than contraception research?

Nearly one in five American couples has difficulty conceiving a baby—that's more than 5.3 million couples in all. In 40 percent of infertile couples, the woman has reproductive problems; in another 40 percent of cases, the man has them; in the rest, the infertility lies with both partners or is of unknown cause. Reproductive medicine offers the chance to help them conceive—*in vitro* fertilization: literally, *fertilization,* or the fusion of egg and sperm, but in a glass dish. The principles of *in vitro* fertilization are fairly simple: Retrieve eggs from the would-be mother or a donor. Collect sperm from the prospective father or a donor, and then unite sperm and eggs in a laboratory dish to bring about fertilization. Allow some cell divisions to take place in the fertilized egg (zygote; see Chapter 4). Finally, transfer the very early embryo into the mother's womb or that of another woman (a surrogate). When the pre-embryo implants itself in the wall of the womb, pregnancy begins. The result is the marvelous transformation from individual egg and sperm to zygote, embryo, fetus, and finally, after birth, a human baby. This transformation depends on **development**—the process by which offspring become bigger and more complex—from the single-celled stage to adulthood and sexual maturity. Development is not only a fascinating subject in itself, but it also forms an important bridge between the genes we have been discussing and the whole organs and organisms we'll talk about in the rest of the book. Through the unfolding process of development, genes direct the building of organs as complex and important as the human brain, hand, and eye.

Let's face it—whether in our day-to-day lives, or in books, movies, the media, even billboards, we humans are keenly interested in sexual attraction, sexual activity, and reproduction. In fact, human sexuality shapes our bodies, our behavior, and our life-long experiences. Since sexuality reflects our evolution within the animal kingdom, we'll set the stage by discussing animal reproduction. Next we'll see where human anatomy and behavior fit in. Then we'll follow the normal sequence of development, from eggs and sperm all the way to a new baby, child, and adult. Along the way, you'll discover the answers to these questions:

- ☑ How do animal mating behaviors maximize the chances that egg and sperm will unite?
- ☑ What are the main parts and processes of the male and female reproductive systems and which hormones control them?

development
the process by which an offspring increases in size and complexity from a zygote to an adult

ovum
(pl. ova) an unfertilized egg cell

pheromone
a compound produced by one individual that affects another individual at a distance; for example, female moths secrete pheromones that attract males

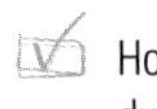 How do sperm and egg unite during human fertilization?

 How do fertilized eggs become embryos?

 How do an embryo's organs take shape in the proper places within the organism?

 What are the events of human pregnancy and birth?

 How do our growth, maturation, and aging complete the human life cycle?

LO1 Sexual Reproduction: Mating and Fertilization

In animals as different as zebras and zebra fish, the reproduction game is essentially the same: a tiny, mobile male gamete or *sperm* cell (see Chapter 4) fuses with a relatively huge ovum (pl. *ova*) or egg cell and triggers the development of a new individual. The meeting of egg and sperm requires that individual organisms mate at an appropriate time and in a way that is typical to and successful for their species. Individual egg and sperm cells must make contact and fuse through a series of precise events based on the way the reproductive cells (gametes; see Chapter 4) are built and behave. So how does animal behavior ensure this mating and fusion?

Some animals, such as hydras, flat worms, and sponges sidestep the problem most of the time by reproducing asexually, with new individuals arising from old through regeneration or budding. In others like earthworms and sea slugs, the animals are *hermaphrodites:* each individual makes both eggs and sperm. As self-sufficient as this sounds, however, pairs of hermaphrodites usually get together and reciprocally exchange sperm and fertilize each other's eggs.

Cooperation and timing are obviously important for earthworms and sea slugs, and the same is doubly true for most other kinds of animals that make either egg or sperm, but not both. First, potential partners must recognize each other through sight, smell, or sound. The male peacock's tail, the lion's mane, the elephant seal's proboscis, and the stickleback fish's red stripe are all examples of visual cues that evolved and act as attractants to the opposite sex of their own species. Cricket chirps and bird songs are familiar auditory cues. And mammals and insects have evolved hundreds of odorous chemical compounds called pheromones (FEAR-o-moanz) that communicate sexual receptivity and attractiveness.

In many water-dwelling species, such as salmon and sea urchins, fertilization is external; that is, the animals deposit eggs and sperm directly into the surrounding environment (usually water), where some gametes meet by chance and fuse. In frogs and salamanders, the male clasps the female firmly for prolonged periods, and this touching stimulates the couple to release their gametes simultaneously. This synchronized release makes it much more likely that healthy eggs and sperm will meet because unfertilized eggs survive only a short time. In land-dwellers, such as most mammals, birds, reptiles, insects, and snails, fertilization is internal: the male deposits sperm directly into the female's genital opening, and the gametes meet in a tube or a chamber. Internal fertilization helps ensure that sperm will be concentrated and protected within the female's body until viable eggs are available for fertilization.

LO2 Male and Female Sexual Characteristics

Human males and females have different sets of physical characteristics involved in sexuality and reproduction. Our primary sexual characteristics are our reproductive organs, which are capable of passing along part of an individual's set of genes to the next generation. In contrast, our secondary sexual characteristics are our external features, such as enlarged muscles in males and milk-producing breasts in females. These secondary characteristics are not directly involved in sexual intercourse, but they can play significant roles in attraction, nursing, and other reproductive behaviors. As a society, we already devote huge amounts of attention to these outward (secondary) traits and how to enhance them. So here we're going to concentrate on the reproductive organs themselves.

As you read in Chapter 4, each sex has a pair of *gonads*—reproductive organs that produce sex cells and sex hormones. Through meiotic divisions (see Chapter 4), the male gonad—the *testis* (pl. *testes*)—produces sperm, and the female gonad—the *ovary*—produces eggs. Both sexes also have various ducts and glands that transport and store sex cells and, in the female, that nurture the developing embryo. All of these structures are involved in human fertilization.

Male Reproductive System

A male's reproductive organs have a straightforward job: produce and transfer sperm into the female reproductive tract. The gonads that produce the sperm cells, the testes, also secrete the primary male hormone, **testosterone**. Testosterone not only aids in sperm production but also helps bring about and control male secondary sexual characteristics, from facial and body hair growth to muscle enlargement, voice deepening, and (in some males) aggressive behavior.

Testes: Sperm-Producing Organs

Most males have a pair of testes that develop inside the body cavity. Shortly before birth, they descend into the *scrotum,* an external sac between the thighs (Fig. 8.1). The lower temperature in the scrotum, outside the body, is necessary for active sperm production, and an elevated temperature of this sac, and the testes inside, can cause sperm development to stop temporarily. A fever, for example, can kill hundreds of thousands of sperm cells.

Each testis is an oval structure about 4cm (1.5 in.) long. Packed inside are about 400 highly coiled, hollow tubes called *seminiferous tubules,* each around 70 cm (28 in.) long. The walls of the seminiferous tubules contain sperm-generating cells, or *spermatogenic cells.* These meiotic cells undergo the special cell divisions of meiosis that reduce the number of chromosomes from two sets (diploid) to one set (haploid) (see Chapter 4). The resulting haploid cells develop into male gametes (sperm). While sperm cells are developing within the tubule walls, other larger cells embedded nearby sustain, surround, and nourish the sperm. These support cells are called *Sertoli cells* (Fig. 8.1c). The tissue that encases the seminiferous tubules also houses *interstitial cells,* which produce the male steroid hormone testosterone, a chemical relative of cholesterol that stimulates male sexual characteristics and behavior.

testosterone
hormone produced by the testes in vertebrate males; stimulates embryonic development of male sex organs, sperm production, male secondary sex characteristics, and male behaviors

urethra
(Gr. *ouerin,* to urinate) the tube that carries urine and releases it to the outside; in males this tube also carries sperm

penis
in mammals and reptiles, the male organ of copulation and urination

Accessory Ducts and Glands: Sperm Delivery Route

Sperm travel from the seminiferous tubules in each testis into a coiled tube, the *epididymis,* attached directly to the top of the testis (Fig. 8.1). Here the sperm mature and develop the ability to swim (motility) (Fig. 8.1a and b). When a male is sexually stimulated, sperm are washed rapidly from the epididymis down a system of ducts—like logs in a flume—and are forcefully spewed from the body. Contractions in the walls of the epididymis push sperm into a connecting tube, the *vas deferens* (45cm, or 18in., long), a sperm duct that also contracts and continues propelling the sperm (Fig. 8.1a and b). A physician (usually a urologist) severs this sperm duct when performing a *vasectomy;* this procedure permanently prevents sperm from exiting the body and so acts as a form of sterilization, even though it doesn't prevent the production of sperm or male hormones. (The hormones continue to circulate normally, and the sperm, blocked from exit, are broken down and reabsorbed by the reproductive tissues.) The sperm ducts from the two testes merge into the ejaculatory duct. Glands secrete buffering fluids that combine with sperm to become the semen that is ejaculated. Secretions from one set of glands, the seminal vesicles, regulate the pH of semen and stimulate muscular contractions in the female reproductive tract. Secretions from the chestnut-shaped *prostate gland* help neutralize the acidity of the female reproductive tract. This natural acidity protects a woman's delicate tissues from microorganisms but also tends to inhibit sperm swimming.

A final set of glands, the bulbourethral glands, adds yet another alkaline fluid to the sperm, and when a man becomes sufficiently aroused, semen finally exits the **urethra**, a tube that runs through the penis.

The Penis

The **penis** has a dual role: transporting urine and semen to the outside and becoming erect for semen delivery. During erotic stimulation, cylindrical columns of erectile tissue in the penis fill with blood, and the penis becomes stiff like a balloon filled with water (Fig. 8.1a and b). This happens because cells in the penis release the gas nitric oxide (NO), which causes a second messenger, a nucleotide (see Chapter 2), to prompt muscles in the spongy erectile tissues to relax. This relaxation allows blood to "pool up" in the penis, enlarging and stiffening it. To sustain an erection, a man must produce the nucleotide faster than a naturally occurring enzyme can break it down. In many impotent men, however, the enzyme wins out. That's where the much-publicized drug Viagra comes in: Viagra blocks the enzyme's action for a while, tipping the balance in favor of the nucleotide and allowing erection.

An average ejaculation from the penis produces about 3 or 4mL of semen (about a teaspoonful) and usually contains 120 to 400 million sperm. Men with

Figure 8.1
Male Reproductive System

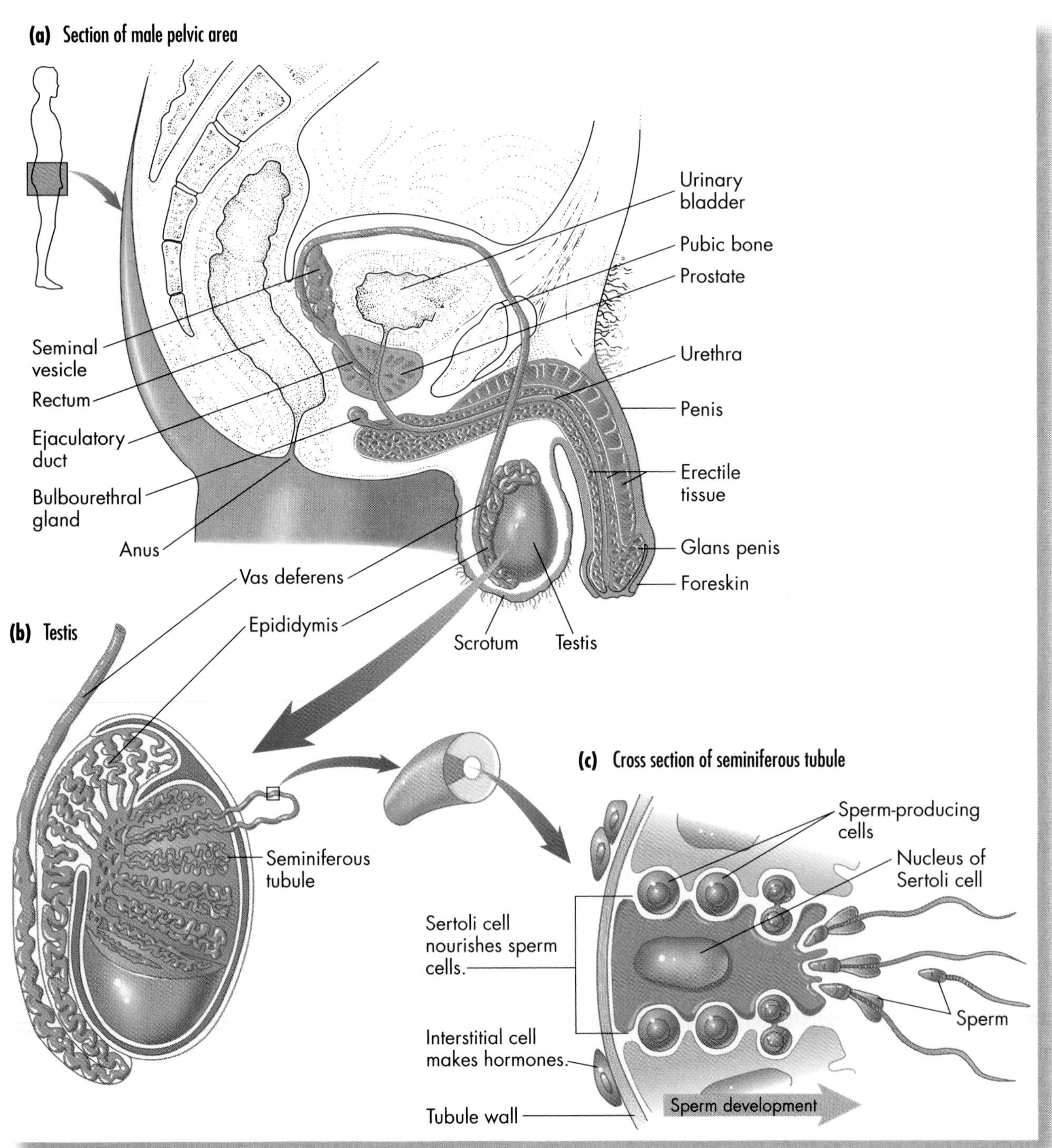

fertility problems often have a sperm count below 100 million sperm per ejaculation.

How the Body Controls Sperm Production

Hormones from the brain and testes work together to regulate the timing of sperm production. If sperm are ejaculated or if the testes fall behind in their production of the wriggling gametes:

(Step 1) The brain secretes luteinizing hormone (LH) and follicle-stimulating hormone (FSH) into the bloodstream.

(Step 2) Blood vessels carry these brain hormones to the testes.

(Steps 3 and 4) The brain hormones cause testicular tissue to step up production of both sperm and testosterone.

(Step 5) When the testosterone concentration (also carried in the bloodstream) rises above a certain level, it helps block further release of the brain hormones.

The lower concentrations of LH and FSH then trigger a decline in the testes' production of sperm and testosterone. When ejaculation releases sperm, the hormone levels fall below the set point once again, and the cycle begins anew. This type of regulatory cycle is called a *negative feedback loop* because it maintains a fairly constant level of a substance (here, testosterone) through an opposing mechanism—when levels rise, it brings them back down; when they fall, it brings them back up.

Female Reproductive System

A female's reproductive system is more complicated than a male's because it has many more jobs to do: Produce eggs and release them at appropriate intervals. Receive sperm. Nourish, carry, and protect the embryo as it develops. Eventually deliver a new organism into the world. The keys to all this feminine reproductive work are anatomy and hormones.

Production and Pathway of Eggs

The day a female is born, her body already contains about 2 million egg cell precursors. The eggs reside in two solid almond-shaped organs called *ovaries,* which lie inside the body cavity just below the waistline (see Chapter 4; also Fig. 8.2a and b). Each ovary contains immature egg cells called *oocytes,* which develop into mature eggs. Helper cells called *follicular cells* surround and supply materials to the oocytes (Fig. 8.2c). An oocyte surrounded by these helper follicular cells constitutes a unit called a follicle (Fig. 8.2c). At birth, each ovary houses about 1 million follicles whose oocytes are arrested in an early division stage of meiosis.

When a girl reaches puberty at about the age of 12 or 13, a process called ovulation begins to take place in her body every 28 days or so and will continue until she is about 50. During a woman's monthly cycle, one or more follicles in one of a woman's ovaries enlarge, the oocyte(s) matures, and at the time of ovulation (usually around day 14) one follicle ruptures, and a mature egg—now called an *ovum*—bursts into the body cavity (Fig. 8.3). (An ovum is the largest rounded human cell; it is about the size of the dot over this i.) Ovulation is usually accompanied by an increase in *basal body temperature* or the body temperature at rest (such as first thing in the morning). Women seeking to avoid or assist conception often take their temperatures at midcycle to look for this indicator of ovulation.

The follicle cells left behind in the ovary after ovulation enlarge and form a new gland, the corpus luteum (Latin for "yellow body"). Most birth control pills work by preventing ovulation. Because a baby girl has all of the egg cell precursors she'll ever have already in place, those cells age throughout her life. Genetic mistakes and malfunctioning organelles can accumulate as the years go by, however, and this explains why "old eggs" are one of the most important factors in infertility.

Every month, after a woman's ovum is released from the ovary, it floats toward the fringed opening of one of two tubes called oviducts (also known as fallopian tubes in humans) (Fig. 8.2b). The tubes have a small range of motion and actually assist conception by sweeping across the ovary surface and picking up an ovulated egg in a frilly-looking tube opening.

luteinizing hormone (LH)
hormone produced by the anterior pituitary that acts on the ovaries to stimulate ovulation and the synthesis of estrogen and progesterone and on the testes to stimulate testosterone production

follicle-stimulating hormone (FSH)
hormone produced by the anterior pituitary that acts on the ovaries to simulate growth of the ovarian follicle and in testes, stimulating sperm production

follicle
an oocyte (immature ovum) and its surrounding follicular cells

ovulation
the release of a mature ovum from the ovary

corpus luteum
the group of follicle cells left behind in the ovary after an egg has been released; if the egg is fertilized, the corpus luteum remains active in the ovary, producing hormones that help to maintain the pregnancy; if the egg is not fertilized, the corpus luteum degenerates

oviducts (fallopian tubes)
the tubes along which ova travel from the ovary to the uterus; usually the site of fertilization

embryo
an organism in the earliest stages of development; in humans this phase lasts from conception to about two months and ends when all the structures of a human have been formed; then the embryo becomes a fetus

fetus
an unborn offspring after the embryonic period, possessing all organs in rudimentary form; in humans, an embryo that has reached its eighth week

Oviducts are lined with millions of cilia that assist conception in another way. They act like paddles, sweeping the fertilized ovum toward the *uterus,* the thick-walled chamber where the embryo develops. Fertilization usually takes place in the oviduct as a phalanx of swimming sperm meets, and one penetrates a slowly moving ovum. Most barrier-type contraceptives, such as condoms and diaphragms, prevent this encounter between egg and sperm. Whether or not an ovum has been fertilized, cilia sweep it into the muscular, pear-shaped uterus. If the egg has been fertilized, it begins dividing and the very early embryo implants or burrows into the uterine wall, where it develops into an embryo and fetus. In many women who have difficulty conceiving a baby, the tubes are blocked by scar tissue from a prior infection.

During nearly all of a woman's monthly cycles, the ovum remains unfertilized: instead of implanting into the uterine wall, it degenerates. The remains of the egg may be discharged from the uterus through the cervix, the neck of the uterus. Just beneath the *cervix* lies the

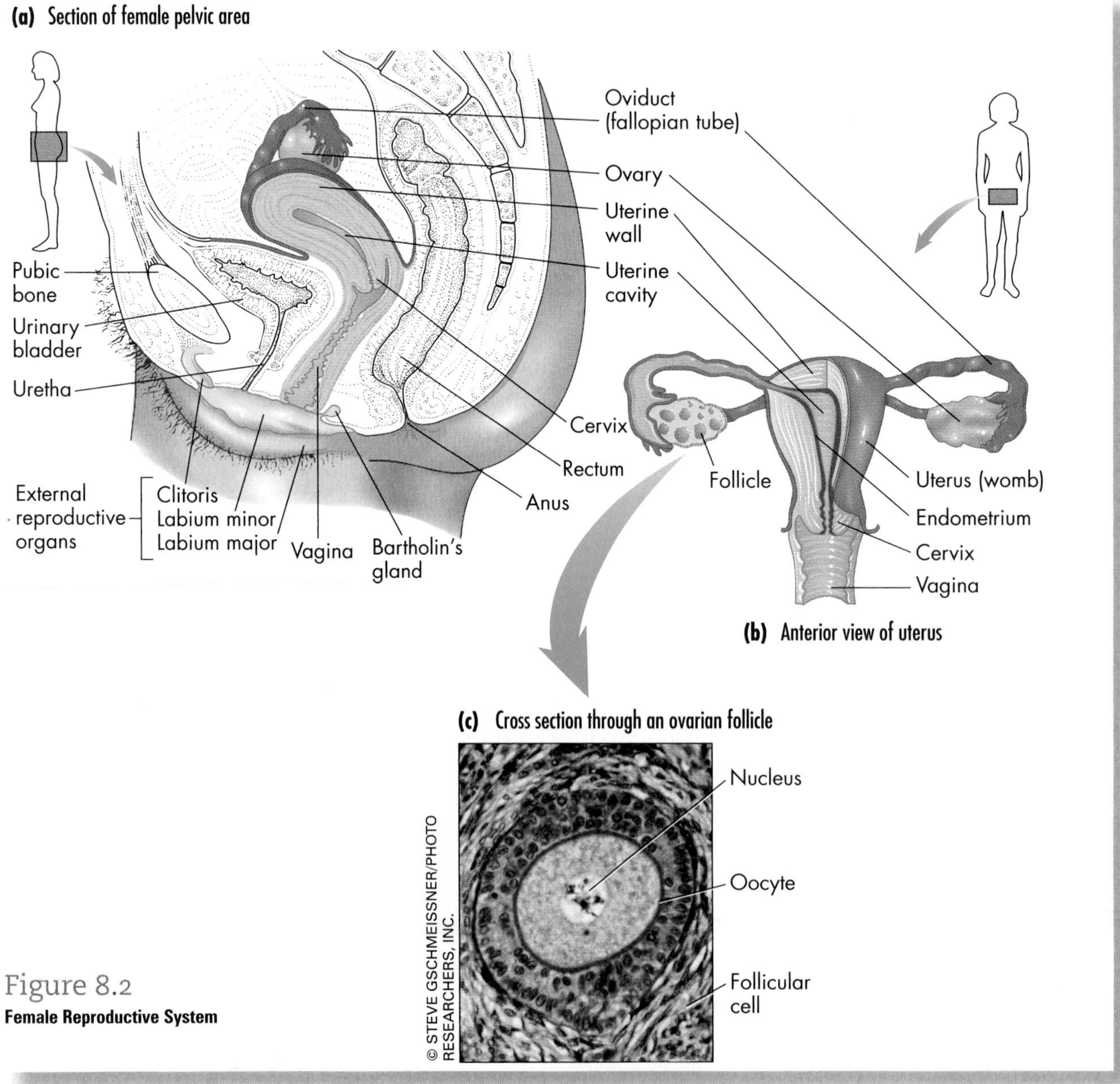

Figure 8.2
Female Reproductive System

vagina—a hollow, muscular tube that receives the penis during intercourse, conveys uterine secretions (including monthly menstrual flow), and serves as the stretchable birth canal through which the fetus passes during childbirth.

The vaginal opening is surrounded by external reproductive structures that include protective tissues called the *labia minor* and *labia major,* as well as tissues sensitive to sexual stimulation, such as the clitoris. Also included are the lubricating Bartholin's glands (see Fig. 8.2a).

The Menstrual Cycle: Hormonal Control of the Ovaries and Uterus

Just as the moon and tides have regular, natural cycles, women's bodies have a regular monthly cycle called the *menstrual cycle* (Fig. 8.3). During each menstrual cycle, the inner lining of the uterus, the *endometrium,* thickens in preparation for pregnancy just before the ovary releases an egg (ovum). If the ovulated ovum is not fertilized, the uterine lining sloughs off, and menstrual bleeding occurs, beginning a new 28-day cycle.

As part of the menstrual cycle, the ovaries produce rising and falling levels of the female steroid hormones estrogen and progesterone.

vagina
muscular tube that receives the penis during copulation, forms part of birth canal, and channels uterine secretions

estrogen
a female sex hormone produced by the ovary; prepares the uterus to receive an embryo, and causes secondary sex characteristics to develop

progesterone
a female sex hormone secreted by the corpus luteum in the ovary that stimulates uterine wall thickening and mammary duct growth

Figure 8.3
How Hormones Control the Menstrual Cycle

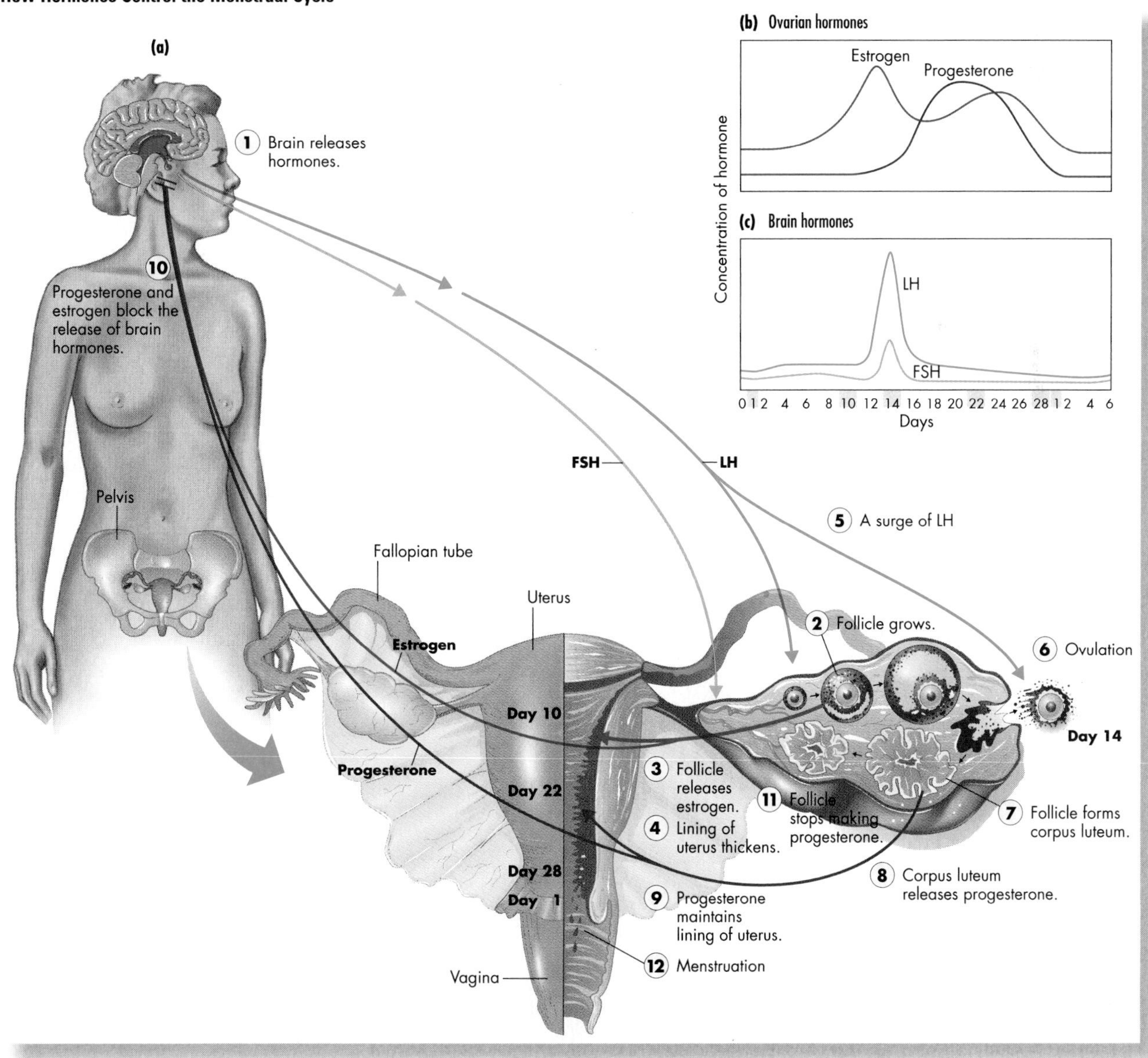

The female brain makes LH and FSH chemically identical to those produced in the male brain. Together, these hormones drive a feedback loop that produces the fairly regular monthly cycle.

Whole books have been written on the intricacies of the female hormonal cycle, but here we'll just give a quick overview. Refer to the numbered steps in Figure 8.3 as you read along: Hormones from the brain (LH and FSH; Step ①) cause a follicle to grow in the ovary (Step ②) and to release estrogen (Step ③); the estrogen, in turn, travels through the bloodstream and causes the lining of the uterus to thicken (Step ④). A surge of LH (Step ⑤) then induces ovulation (Step ⑥). The ovary's follicular cells mature into the corpus luteum (Step ⑦) and then make and release progesterone (Step ⑧), which maintains the uterine lining (Step ⑨). Progesterone and estrogen also block the release of the brain hormones in a negative feedback loop (Step ⑩). Soon, if there is no pregnancy, the corpus luteum stops making progesterone (Step ⑪), the uterine lining sloughs off, and menstruation begins (Step ⑫). When the progesterone level drops, the brain is once again free to release its hormones, and the cycle starts over again.

Despite "la difference" between males and females, both share certain hormonal similarities: They both produce the same brain hormones that control reproduction. They both produce reproductive hormones that help make eggs and sperm available and bring about sexual development in teenagers. And in both sexes, hormones functioning in a negative feedback loop control the production of gametes.

LO3 Human Fertilization: The Odyssey of Eggs and Sperm

A human egg is the largest rounded cell in a female's body. An egg cell's large size helps it perform its many jobs:

- The egg donates a haploid nucleus containing one set of chromosomes to the new embryo. In some animals, the egg protects the developing embryo inside jellylike protein coatings, strong fertilization membranes, sacs of fluid, or hard or leathery shells.
- *Yolk,* which contains rich stores of lipids, carbohydrates, and special proteins, nourishes the embryo and usually provides all the embryo's mitochondria.
- The egg supplies the cytoplasm for the fertilized egg (or zygote; see Chapter 4) and in it the machinery for making new proteins—a machinery that allows rapid cell division.

In many animals, the egg's cytoplasm contains special substances that control development. An ovum is like a computer loaded with a software program, ready to play out the actions of development as soon as the sperm's entry starts the program running.

Compared to the egg, a sperm is one of the smallest cells in a male's body, stripped down to just the essential elements needed to perform superb penetration (Fig. 8.4). These include (1) a compact haploid nucleus; (2) several mitochondria, which provide energy; (3) a long, lashing flagellum (tail) that propels the sperm; and (4) a sac of enzymes, called the *acrosome,* that digests a path through the egg's protective outer coatings. Some men are nearly sterile because their sperm swim slowly and weakly; others are completely sterile because their sperm have abnormal tails or oversized heads or can't penetrate the egg's protective coat. A sperm's streamlined size and shape reflect its narrow function: to reach the egg, penetrate its coating, and deliver a haploid nucleus (containing one set of chromosomes) into the egg's cytoplasm. This penetration and delivery of the nucleus, together with the activation of the resting egg, make up the events of fertilization.

Fertilization

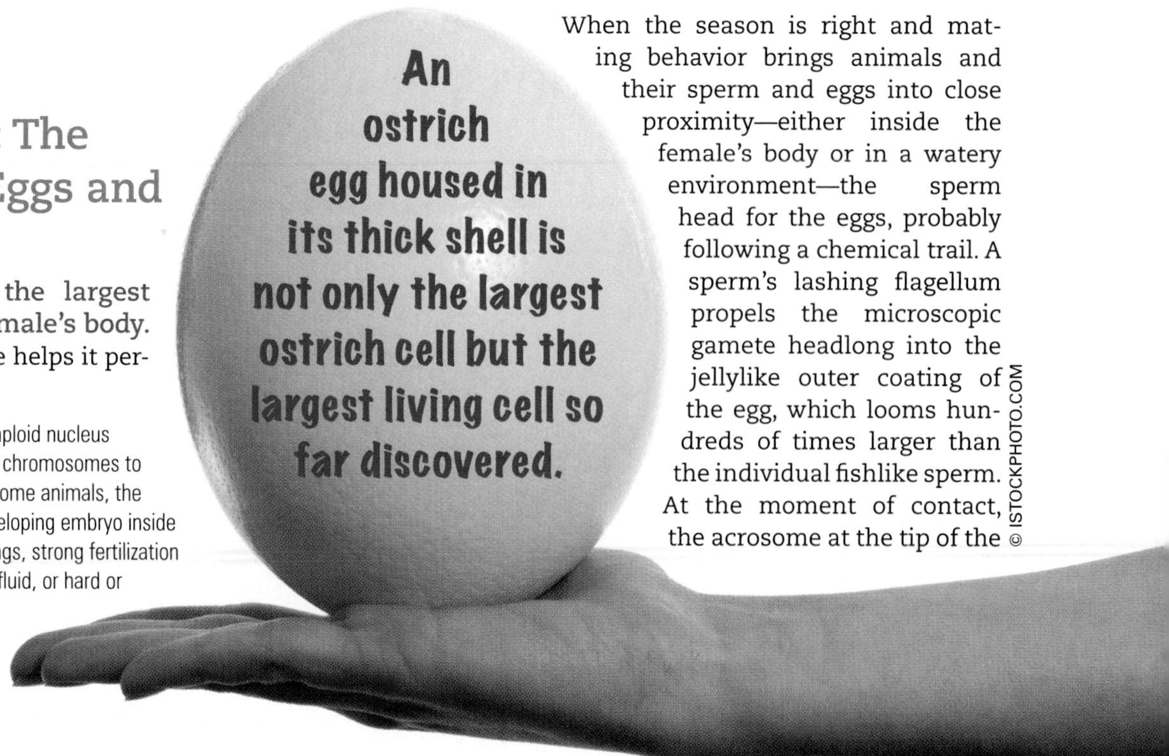

When the season is right and mating behavior brings animals and their sperm and eggs into close proximity—either inside the female's body or in a watery environment—the sperm head for the eggs, probably following a chemical trail. A sperm's lashing flagellum propels the microscopic gamete headlong into the jellylike outer coating of the egg, which looms hundreds of times larger than the individual fishlike sperm. At the moment of contact, the acrosome at the tip of the

An ostrich egg housed in its thick shell is not only the largest ostrich cell but the largest living cell so far discovered.

Figure 8.4
The Sperm

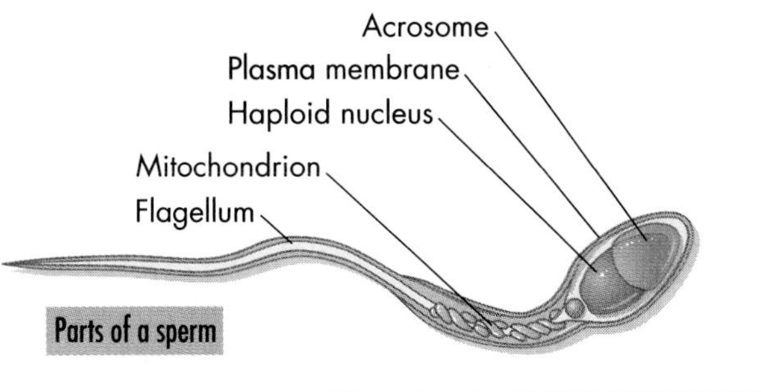

> A sperm's streamlined size and shape reflect its narrow function: to reach the egg, penetrate its coating, and deliver a haploid nucleus (containing one set of chromosomes) into the egg's cytoplasm.

sperm head releases enzymes that help the sperm to penetrate the coatings on the egg's surface. The plasma membranes of the egg and sperm fuse. Then the sperm head—including the nucleus and the centriole (see Chapter 4; the organelle that helps organize the cell's inner scaffolding)—plunges into the egg's cytoplasm. The sperm's tail, including its energy-producing mitochondria, remains outside the egg. (This is why the egg donates all of an embryo's mitochondria.) Beginning precisely at this site of sperm penetration, a wave of molecular signals sweeps over the egg and triggers the rapid construction of a barrier, the *fertilization membrane* (see Fig. 8.5). This essentially marks the victory of one sperm because the barrier prevents additional sperm from entering the egg. The sperm and egg nuclei fuse, establishing a novel genetic combination that is diploid (contains two sets of chromosomes)—and a new zygote.

In humans, fertilization normally occurs as the egg moves along the oviduct and swimming sperm encounter the huge target. After fertilization, the newly formed zygote continues its journey through the oviduct until it reaches the uterus four or five days later. The process of egg release (ovulation) has already caused the ripe follicle (the corpus luteum) to make the hormone progesterone and in turn, through this chemical messenger, to maintain the monthly buildup of the uterine walls (see Fig. 8.3, Steps (6) to (9)). As a result, the fertilized egg can implant in the well-prepared uterus and begin the remarkable transformation from a single fertilized egg cell into a

Figure 8.5
Fertilization: Egg and Sperm Fuse

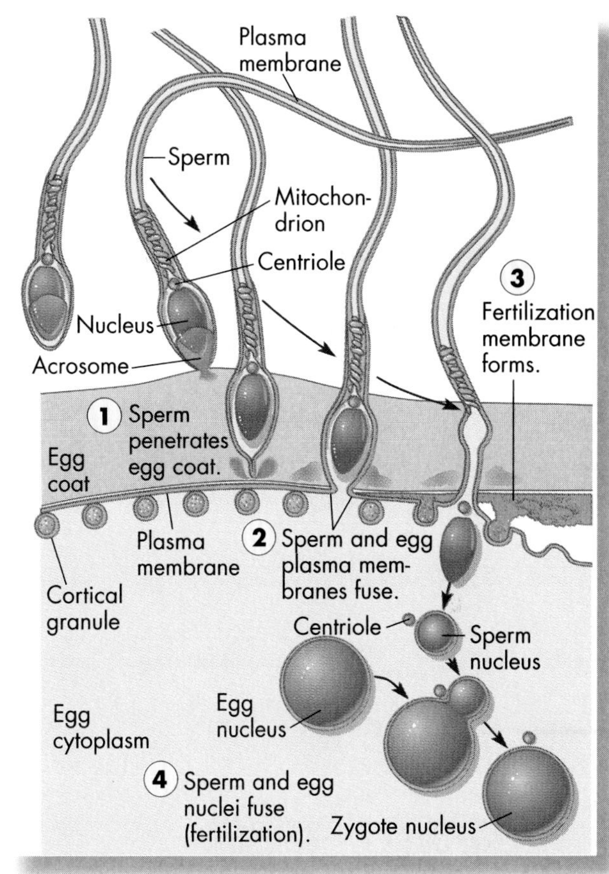

complex animal—first an embryo and then a fetus—through the mechanisms of development.

LO4 Embryonic Development

We've seen how sperm and egg form, and we've explored the marvelous instant of fertilization. But fertilization is over in seconds—and then what? What ingenious biological mechanisms start to unfold to turn a single fertilized egg into a baby with its intricate little ears, wrinkled hands, tiny nose, and sophisticated heart and brain? If you were watching the process unfold under a microscope, you'd see embryonic *development*—the process by which an offspring increases in size and complexity from a fertilized egg to a complex organism—and its four stages: cleavage, gastrulation, organogenesis, and growth. So let's look at each stage now, then at the underlying mechanisms that bring them about and transform a fertilized egg into an embryo, fetus, and baby.

Figure 8.6 depicts the main events of development in a frog. Each frog embryo develops into a tiny,

Figure 8.6

An Overview of Developmental Stages and Processes in a Frog

Sperm

(a) Formation of egg and sperm

Egg

Eggs are produced in ovaries, and sperm are produced in testes of mature females and males.

(b) Fertilization

Sperm and egg fuse.

(c) Cleavage

Mitotic cell divisions in the fertilized egg create a hollow ball of cells, the blastula.

Cross section of blastula

Ectoderm
Mesoderm
Endoderm

Tube-within-a-tube stylized body plan

(d) Gastrulation

Future gut

Blastula cavity

Anterior

Dorsal

Future mouth

Ventral

Posterior

Ectoderm
Mesoderm
Endoderm

Yolk plug, site of future anus

Future head

Neural folds

Future belly

Future tail

The ectoderm folds into a tube and forms the spinal cord.

Cell rearrangements in the blastula cause three tissue layers to form by cells moving from outside to inside of embryo.

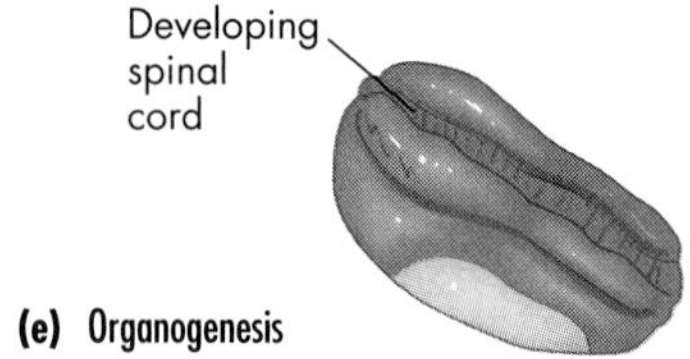

(e) Organogenesis

Cells continue to divide, migrate, change shape, and rearrange themselves, generating organ shape.

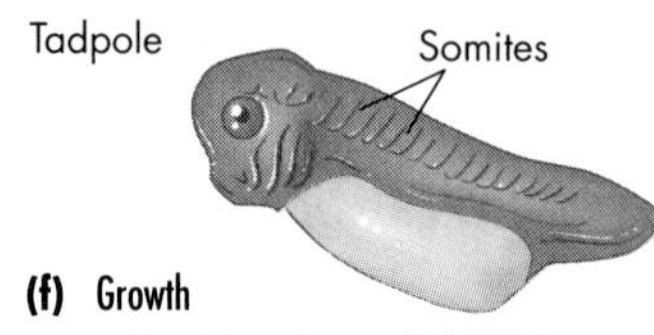

(f) Growth

Different cells accumulate different sets of proteins and assume their specific functions. The embryo grows.

wriggling tadpole inside a transparent jelly coat, and the stages are easily seen. But the processes are the same, in principle, for the embryos of worms, flies, frogs, humans, and most other animals. In fact, much of what biologists know about human development they learned first from animals in the lab.

Four Stages of Development

Cleavage: Cell Division Begins

If you were watching the marvel of development unfold step by step, the most obvious thing you'd see is a change in cell number: one fertilized egg must become millions of cells in an embryo, billions in a fetus, and trillions in a child and adult. This enormous increase takes place through a series of cell divisions. In the first few division cycles, the fertilized egg cleaves or divides in two; these so-called *daughter cells* (see Chapter 4) cleave again to form four cells; those form eight, and so on. Based on this cleaving, the first few division cycles are called **cleavage** (Fig. 8.6c). In many animals, cleavage produces a ball, called a *blastula,* which is no bigger than the original fertilized egg but is made up of many cells that are smaller than the original egg (Fig. 8.6c). As the original egg cell contents become separated into different cells of the blastula, molecular information that will direct the shaping of the embryo sometimes ends up in different regions of the "ball." Because of this regional information and its subsequent control of embryonic shape, cleavage helps organize development at a very fundamental level.

Gastrulation: Three Layers Form and Migrate

If you were watching the steps of development, you'd see another obvious change, as well: cell position. Embryos and babies are not just random blobs of cells but are, instead, precisely shaped, and their cells are exquisitely organized in space. The most basic feature of this organization is that an early embryo has three main body layers that give rise to all the organs and tissues of the developing, enlarging individual. The process that converts a hollow ball, the blastula, into a three-layered embryo is **gastrulation**, which means, literally, "the formation of the stomach." Indeed, the three layers become arranged as a tube within a tube within a tube. The stomach and digestive tube lie on the inside; a tube of blood vessels, muscles, bones, kidneys, and other organs come to surround the digestive tube; and an outer tube, the skin, covers both of the others. Cell movements in the hollow blastula give rise to these three nested tubes (Fig. 8.6d).

These cell movements of gastrulation are a living sculptural process. Cells at a particular spot on the blastula begin to dive inside the ball, pulling their neighboring cells in with them (Fig. 8.6d). The result is three primary tissue layers (Fig. 8.6d):

- The inner layer, or *endoderm* ("inner skin"), will produce the digestive tract and parts of the liver and lungs that are derived from it.
- The middle layer, or **mesoderm** ("middle skin"), will form the blood vessels, kidneys, and reproductive organs, as well as the body's muscles and most of the bones.
- The outer layer, or *ectoderm* ("outer skin"), will become the outer parts of the skin and the nervous system.

At the embryo's front end, the cells eventually punch through to form the mouth. At the opposite end, where the cells first in-pocketed, the anus forms. Now the embryo has an orientation—a head-tail axis—as well as its basic layers. Hard as it is to imagine, each of us went through this same sculpturing process in the second week of life.

Organogenesis: Organs Unfold

As gastrulation ends, the three embryonic layers are poised to generate the next phase of development, **organogenesis**, or the formation of the body's organs and tissues with proper shapes, positions, and functions (Fig. 8.6e). This stage can last several weeks and bring into existence all the embryo's systems for moving, digesting, exchanging air, expelling wastes, protecting itself from disease, and so on. Refinements and maturation of these organs and systems will continue throughout development, well into youth and adolescence. A cell-to-cell communication process (induction) helps determine where organs like the spinal cord and brain will form. A set of cell-sculpturing processes (morphogenesis) brings about the proper shape of the organs once positioned. This, for example, allows the human brain to take on its hallmark contours. It also causes the *somites,* segmentally repeating blocks of mesoderm, to become your vertebrae (Fig. 8.6f). Pattern formation helps shape a whole region of the embryo, so several parts—such

cleavage
the early, rapid series of mitotic divisions of a fertilized egg resulting in a hollow sphere of cells known as the blastula

gastrulation
the movement of cells in the embryo that generates three cell layers—the ectoderm, mesoderm, and endoderm—each layer in turn giving rise to specific body organs and tissues

mesoderm
the middle cell layer of an embryo; gives rise to muscles, bones, connective tissue, and reproductive and excretory organs

organogenesis
the formation of organs during embryonic development

growth
an increase in size

as the features of the face and regions of the brain—are in proper relationship to each other. And the cells of the organs take on their specialized functions through the differentiation process so that, for example, cells in the eye detect light and cells in the stomach wall secrete acid and not vice versa.

Let's look at the very earliest organ formation in the animal embryo, the origination of the nervous system, with its spinal cord, senses, nerves, and brain. That set of organs will one day control all our actions, reactions, thoughts, and feelings, but it begins just after gastrulation as nothing more than a sheet of nondescript ectoderm cells on the surface of the embryo (Fig. 8.7a). Suddenly, cells in this sheet begin to thicken and then fold up into a minute, hollow cylinder, the *neural tube*. First, the layer of cells folds, as if it were forced in from the sides like two wrinkles in a bed sheet (Fig. 8.7b, c). The two folds fuse at the top, and then bud off from the ectoderm to form the neural tube, an embryonic structure that will, in turn, form the spinal cord and brain (Fig. 8.7d). Cells forming the neural tube assume a wedge-like shape, helping the sheet roll up into a tube.

In the human embryo, the neural tube first closes in the mid trunk and then zips up toward the front and down toward the rear. This process is not complete until about 29 days after fertilization in a human embryo. If the human neural tube does not completely close at the back end (posterior) of the embryo, spinal bones (vertebrae) do not grow to encircle the unclosed portion of the tube, and the spinal cord can squeeze out of the gap. The result is *spina bifida*, or open spine, the most common severe major birth defect among live-born infants, affecting one in every 2,000 live births.

Growth: The Organism Enlarges to Adult Size

Along with the emergence of embryonic organs comes growth or expansion in size. The cleavage that divides a fertilized egg into a ball of cells is only the beginning of cell division in the embryo. A major size increase is needed, and cell division—usually rapid—continues until hatching or birth and then throughout the development of the young organism. A blue whale, for example, increases in weight 200 millionfold from fertilized egg to adult. In the whale, as well as virtually all other organisms, growth is due to an increase in cell number, not the size of individual cells (see Chapter 2). Much of the growth goes on after birth or hatching when the limits to food and expansion room are removed.

Figure 8.7
Neural Tube Formation in a Frog Embryo

(a) The sheet begins to fold

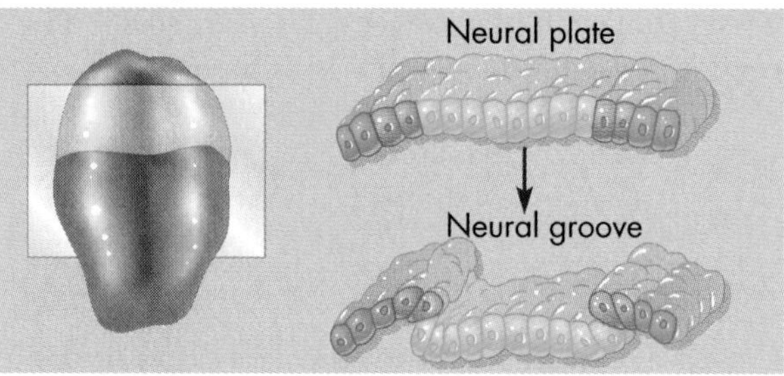

(b) The neural folds move together

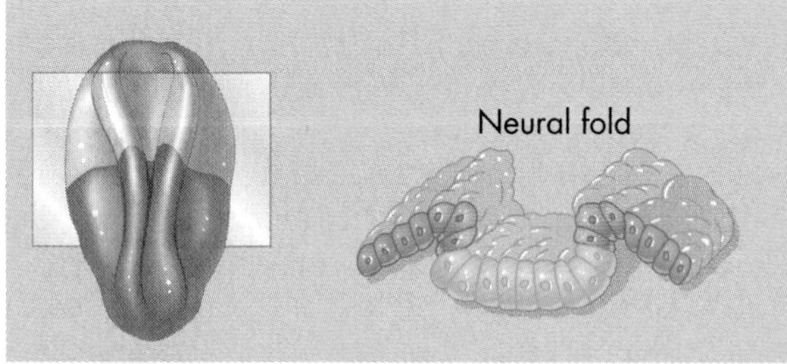

(c) The folds touch

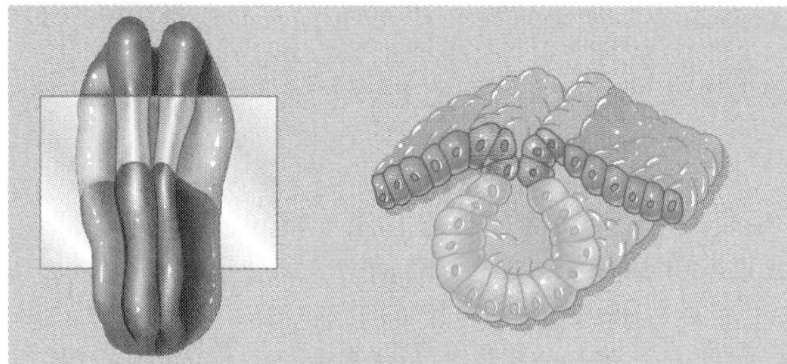

(d) The tube forms and crest cells disperse

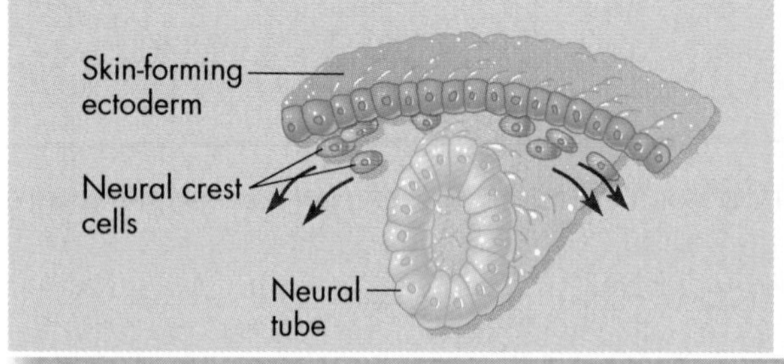

LO5 Human Development

After natural conception, the fertilized human egg divides while it moves along the oviduct toward the uterus (Fig. 8.8). In an assisted conception, the early divisions take place in a lab dish. But in both cases, by day 4 the embryo consists of about 30 cells, and a cavity has formed in its middle (Steps ④ and ⑤). The embryo, now called a *blastocyst* (Step ⑤), has two groups of cells: (1) the inner cell

Figure 8.8
The Marvel of Human Development

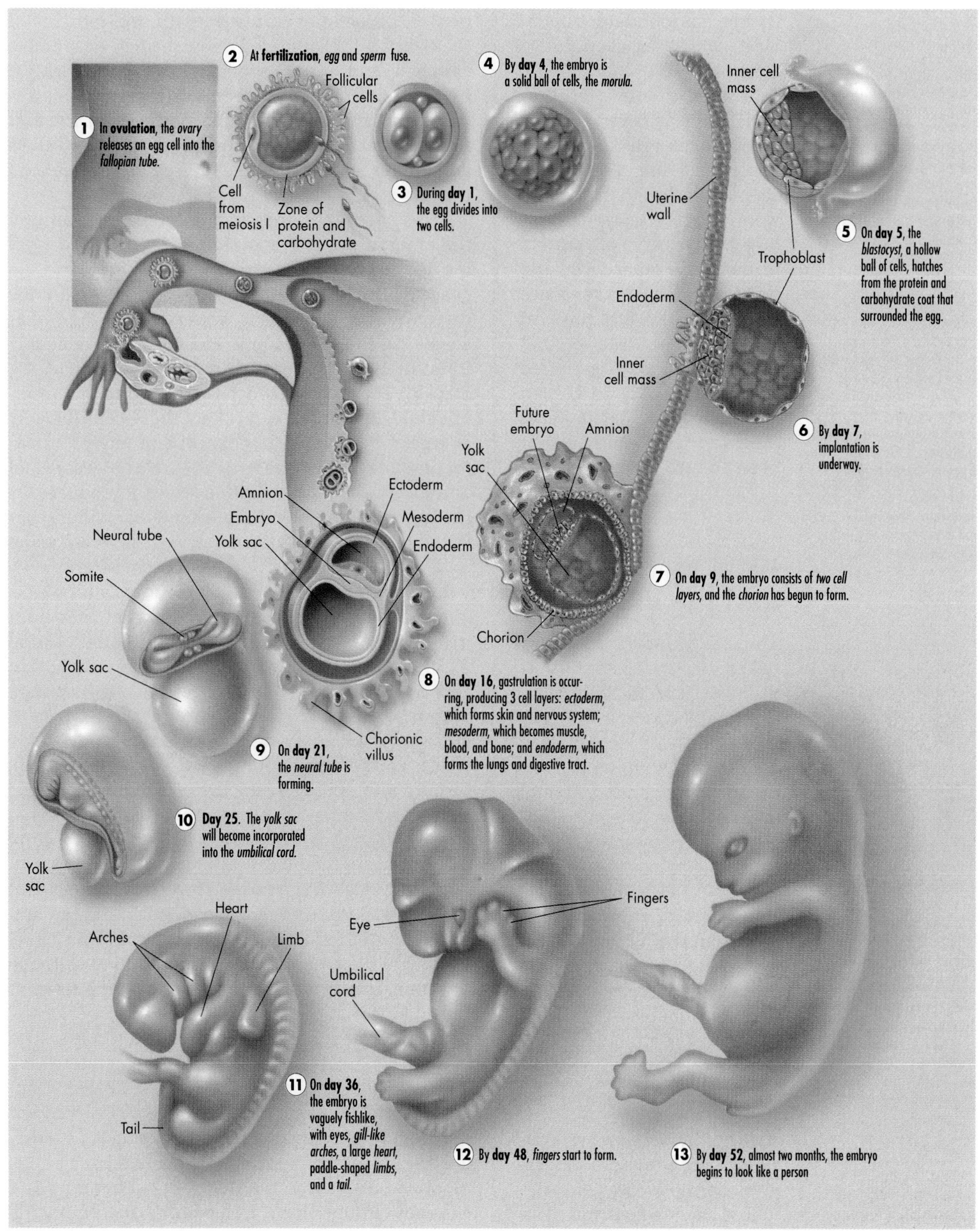

implantation
attachment of a blastocyst to the uterine wall, where it is nourished and develops into an embryo and fetus

pregnancy
begun by the implantation of the developing embryo in the uterine wall, a series of developmental events involving close cooperation between mother and embryo that transforms a fertilized egg (zygote) into a baby

chorion
a fluid-filled sac that surrounds the embryos of reptiles, birds, and mammals; in placental mammals it gives rise to part of the placenta and produces hCG, a hormone that maintains pregnancy by preventing menstruation

amnion
an important fetal membrane, resembling a hollow ball; the ball encloses a cavity (amniotic cavity) of fluid (amniotic fluid) where the embryo grows during pregnancy

human chorionic gonadotropin (hCG)
a hormone produced by the chorion of a developing embryo that maintains pregnancy by maintaining the secretion of estrogen from the ovary and thus preventing menstruation

amniocentesis
a procedure for obtaining fetal cells from amniotic fluid for the diagnosis of genetic and other abnormalities

chorionic villus sampling
a procedure for removing fetal cells from the chorion in the developing placenta for the diagnosis of genetic defects

mass, which develops as the embryo proper, such as the frog embryo in Figure 8.6; and (2) the surrounding trophoblast, which forms nutritive tissue.

Six days after fertilization, the blastocyst consists of about 100 cells. By now, it has moved into the uterus; soon it attaches to the uterine wall, secretes enzymes that break down a small portion of the wall's outer layer, and burrows into it. This sequence establishes the first physical connection between mother and young and is called **implantation** (Fig. 8.8, Step ⑥). It marks the beginning of **pregnancy**: the development of an embryo in the uterus.

Fetal Membranes and the Placenta

The embryo has a three-way ticket to survival in the form of a double-layered, fluid-filled sac that surrounds and nurtures it with an outer layer or **chorion** and an inner layer or **amnion** (Fig. 8.8, Steps ⑦ and ⑧). The embryo's surrounding layer of trophoblast cells contributes to the chorion, and this layer does three things: (1) It absorbs nutrients from the mother's blood and passes them on to the rapidly dividing embryo; (2) it contributes to the dark, spongy placenta, which sustains the embryo throughout the nine months of pregnancy; and (3) it produces the hormone **human chorionic gonadotropin** (hCG). This hormone prevents the onset of a new menstrual cycle, which would flush the implanted embryo from the uterus. Home pregnancy tests use a simple but ultrasensitive system for detecting hCG, and 99 percent of the time these tests accurately reveal a pregnancy that is just two or three weeks old. As hCG enters the mother's bloodstream (starting at about day 10 after fertilization), it causes the corpus luteum, the remains of the ovarian follicular cells, to continue to produce estrogen and progesterone. These hormones, in turn, maintain the uterine lining, prevent menstruation, and allow pregnancy to continue for the first two months. "Morning after pills" such as RU486 and newer drugs stop the uterus from receiving progesterone and thereby block pregnancy.

The chorion is the first layer to form around the embryo, but it is soon joined by a second layer inside the first, the amnion (Fig. 8.8, Step ⑧). The space within the amnion becomes the amniotic cavity, which encloses a watery, salty fluid that suspends the developing embryo in a relatively injury-free environment during pregnancy (Fig. 8.8, Step ⑨). The embryo sloughs off cells into this amniotic fluid. Physicians can collect these cells (usually at weeks 15 to 16 of pregnancy) and analyze them for possible genetic abnormalities in a process called **amniocentesis**. As the chorion grows, it produces little fingers of tissue called *villi* (singular, villus) that become enmeshed with maternal tissue. By using an alternate diagnostic procedure called **chorionic villus sampling** earlier during pregnancy (at weeks 9 to 12), a physician can collect chorion cells from these fingers and look for genetic or chromosomal problems in the fetus. The enmeshment of chorion and maternal tissue forms the dark-red, spongy **placenta**, an exchange site where a thick tangle of embryonic blood vessels encounters blood-filled spaces in the uterine lining. The placenta, in turn, forms a vital link allowing the diffusion of CO_2, O_2, nutrients, and wastes between mother and embryo (Fig. 8.9). The umbilical cord is a lifeline connecting the embryo to the placenta .

In the placenta, the embryo's blood does not mingle with the mother's blood, but nutrients and oxygen pass from her blood across embryonic vessel walls into the embryo's blood supply. In the reverse direction, carbon dioxide and other wastes pass from the embryo into the placenta. After a few weeks, the placenta begins to produce estrogen and progesterone. These hormones maintain the uterine lining and block the production of LH and FSH. This blockage, in turn, prevents menstruation and the end of pregnancy. The corpus luteum then slowly degenerates and stops releasing estrogen and progesterone. At the 8-week stage, the major organ systems and external features have formed, and the embryo is now called the *fetus*. Fetal development continues until about 9 months. The three 3-month periods of pregnancy are called *trimesters*.

Developmental Stages in a Human Embryo

During the next weeks and months, the nervous system and other organs form, and the tiny masses are transformed into organisms with a characteristic human shape.

The First Three Months

The formation of body organs begins as the neural tube rolls up midway through the third week of pregnancy. Primitive blood cells and blood vessels have already formed. During the fourth week (see Fig. 8.8, Step 10), the heart begins to pump blood. By the fifth week, the primitive brain looks like a miniature, lumpy inchworm, and its cells are dividing at such an inconceivably fast rate that 50,000 to 100,000 new neurons are generated each second. The limb buds are now visible, and the intestine is a simple tube. The liver, gallbladder, pancreas, lungs, eyes, nose, and brain begin to form. Although the embryo is only about 5mm (less than 0.25in.) in length from crown to rump, it is already about 10,000 times larger than the fertilized egg.

When the embryo (now officially called the fetus) reaches its eighth week (see Fig. 8.8, Step 13), most of its organs are present, and during the rest of gestation they simply enlarge and mature. The fingers and toes are well formed by the mechanisms of limb development we saw earlier, and the head begins to lift away from the chest. The first indications of skeletal bone formation can be seen during the ninth week, and by then the fetus can also bend its body, hiccup, and respond to loud sounds with increased movements. At ten weeks, it can move its arms, open its jaws, and stretch. At this point, the fetus is about 23mm (1in.) in length from crown to rump. The first trimester ends at 12 weeks; by this time, the fetal pulse is detectable, the fetus is about 40mm (1.5in.) long, and it can yawn, suck, swallow, and respond to touch with movements and quickened heartbeat. The primary sex organs develop during the first trimester, but you can't tell a boy from a girl at that stage because the gonads, sex ducts, and external genitalia look the same in both sexes. The chromosomal tests we mentioned earlier (amniocentesis and chorionic villus sampling), however, can determine gender at this stage.

placenta
in mammals, the spongy organ rich in blood vessels by which the developing embryo receives nourishment from the mother

The Second Three Months

The second and third trimesters of pregnancy are devoted to the increase in size and maturation of the organs developed during the first trimester. During

Figure 8.9
Fetal Membranes and the Placenta

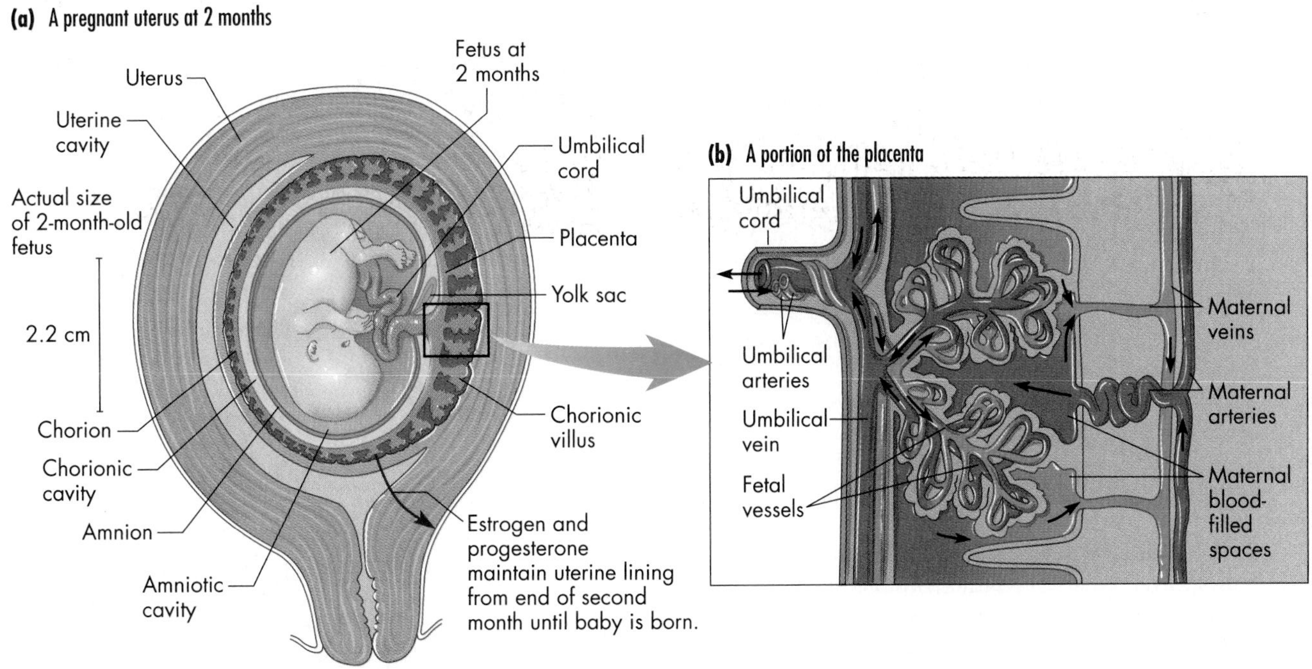

the 12th week, the mother feels her uterus enlarging. By the end of the 15th week, the fetus measures about 56mm (2.25in.) from crown to rump. By the 16th week, its face looks "human." From the 20th to the 24th weeks, the body becomes covered with downy hair, a stethoscope detects fetal heart sounds, and the fetus responds (through increased motion and heart rate) to its mother's voice. The lungs have formed but do not yet function, and the gripping reflex begins, along with tactile explorations of the umbilical cord, the amniotic sac, and the fetus's own face, feet, and toes.

Getting Ready for Birth

At the beginning of the third trimester, the fetus's eyelids open, its eyebrows and eyelashes form, and it can detect light. At this stage, with almost two months to go, a baby born prematurely would have at least a 10-percent chance of survival. The fetal brain grows rapidly throughout pregnancy, and eventually the outermost layer—the cerebral cortex—fills 80 percent of the skull, with two thirds of its surface area tucked into elaborate grooves and creases. During the final couple of weeks, fetal activity decreases because of its size and the lack of space in the mother's uterus. The fetus is said to have come to term, about 280 days after the mother's last menstruation before conception. When the mother is carrying twins or other multiples, the term of pregnancy is often shorter, and the babies are smaller.

Birth and Lactation

As labor and delivery approaches, the fetus prepares itself and its mother's body for the coming events. All normal fetuses store special brown fat around the neck and down the back to help generate heat after expulsion from the warm uterus. Special reserves of carbohydrates accumulate in the heart and liver as a source of nourishment until the baby can suckle milk. The placenta secretes a hormone that prepares the mother's mammary glands to produce milk. The muscles of the mother's uterus become more excitable, contracting periodically in "false labor," building strength as the time of birth approaches.

If development stopped at birth, we'd all be perpetual babies.

During the last trimester, the mother's uterus develops a 100-fold increase in sensitivity to the hormone oxytocin (Fig. 8.10). Oxytocin is produced in the hypothalamus in the brain and is transported to a pregnant woman's pituitary gland, which releases it. When sensitivity to oxytocin reaches a threshold, labor—the physical effort and uterine contractions of childbirth—begins. Nerves

Figure 8.10

Birth: Hormones Trigger Contractions and Delivery

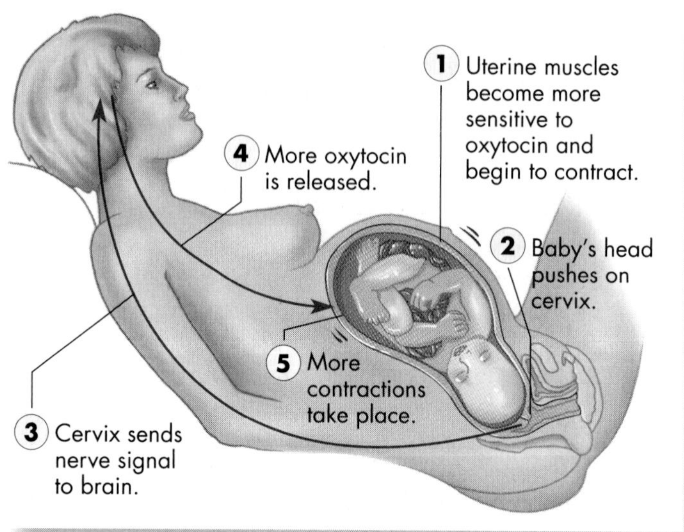

in the cervix signal the mother's brain to release more oxytocin, which causes the uterine muscles to contract even more. Oxytocin also stimulates the uterine wall to release prostaglandins—hormones that stimulate uterine contractions still further. This causes the release of even more oxytocin in a positive feedback loop that causes more and more oxytocin to be released until the final event, birth, ends the loop.

As contractions grow longer, stronger, and more regular, the cervix widens. Each contraction starts at the upper end of the uterus and moves downward toward the cervix, which the baby's head pushes farther and farther open as it moves toward the vagina, or birth canal. This period can span 12 hours or longer, or it can speed by in minutes. Uterine squeezing eventually causes the amniotic sac to burst (the "water breaks"), and when the cervical opening reaches a width of about 10cm (4in.), delivery is usually only minutes away. With considerable pushing of her abdominal muscles, the mother is able to force the baby's head past the pelvic opening, allowing the baby's shoulders and hips to emerge.

LO6 Growth, Maturation, and Aging

If development stopped at birth, we'd all be perpetual babies. Instead, development and growth continue throughout childhood and adolescence, ushering in significant change and maturation, especially in the brain and reproductive organs. At birth, the brain is the most developed organ and is 25 percent of its final adult weight. By age 3 it has reached 75 percent of its adult size and weight, and by age 10, 90 percent. Body weight lags far behind brain weight, advancing from 5 percent of adult weight at birth to only 50 percent at age 10. In girls between the ages of 11 and 13, and in boys between 13 and 15, the dramatic changes of sexual maturation or puberty take place; these include maturation of the reproductive system and development of the secondary sexual characteristics we talked about earlier.

With modern nutrition, sanitation, and health care, the physical maturity that starts at the end of adolescence begins an adult life that continues for 50 to 75 years or longer. The peak performance of all the body's organ systems usually occurs between the early twenties and the early thirties. Sometime after age 30, the process of aging—a progressive decline in the maximum functional level of individual cells and entire organs—begins to accelerate. At about age 50, females experience menopause, a cessation of the menstrual cycle, while males may experience a decrease in their ability to maintain an erection. Sometime after age 60 (or as late as 70 or 75 in active adults with healthy lifestyles), people reach senescence, or old age. The decline in cell and organ function is then less gradual and more profound.

What Causes Aging?

We know from watching our parents and grandparents—and even our cherished pets—that aging is natural and inevitable. But what makes it happen? Biologists have two main sets of ideas about aging: genetic clock hypotheses and wear-and-tear hypotheses. Many experts support the idea of a genetic clock, arguing that there is a genetically specified timetable for aging and death. Just as genes regulate the timing of organ formation, they also influence the rate at which organ function slows. To other biologists, the evidence suggests that the most likely cause of aging is wear and tear: the accumulation of random errors in the replication and use of DNA and in the synthesis of proteins, or the accumulation of metabolic byproducts that disable enzymes, other proteins, and lipids. Informational errors may be due to environmental insults such as exposure to sunlight, radiation, or chemicals. Unfortunately, researchers are still a long way from understanding the basic mechanisms of aging.

puberty
the maturation of the sex organs and the development of secondary sex characteristics, such as breasts in females, and facial hair and a deep voice in males

aging
a progressive decline in the maximum functional level of individual cells and whole organs that occurs over time

senescence
a condition characterized by a profound decline in cell and organ function that occurs with aging

Who knew?

Worldwide, overpopulation is a greater concern than infertility. However, in developed countries such as the United States, there are more couples who have the financial ability to pursue medical help for problems with fertility. Also, it is more common in the United States for women to put off pregnancy until they are older; it is in this age group that infertility is most common.

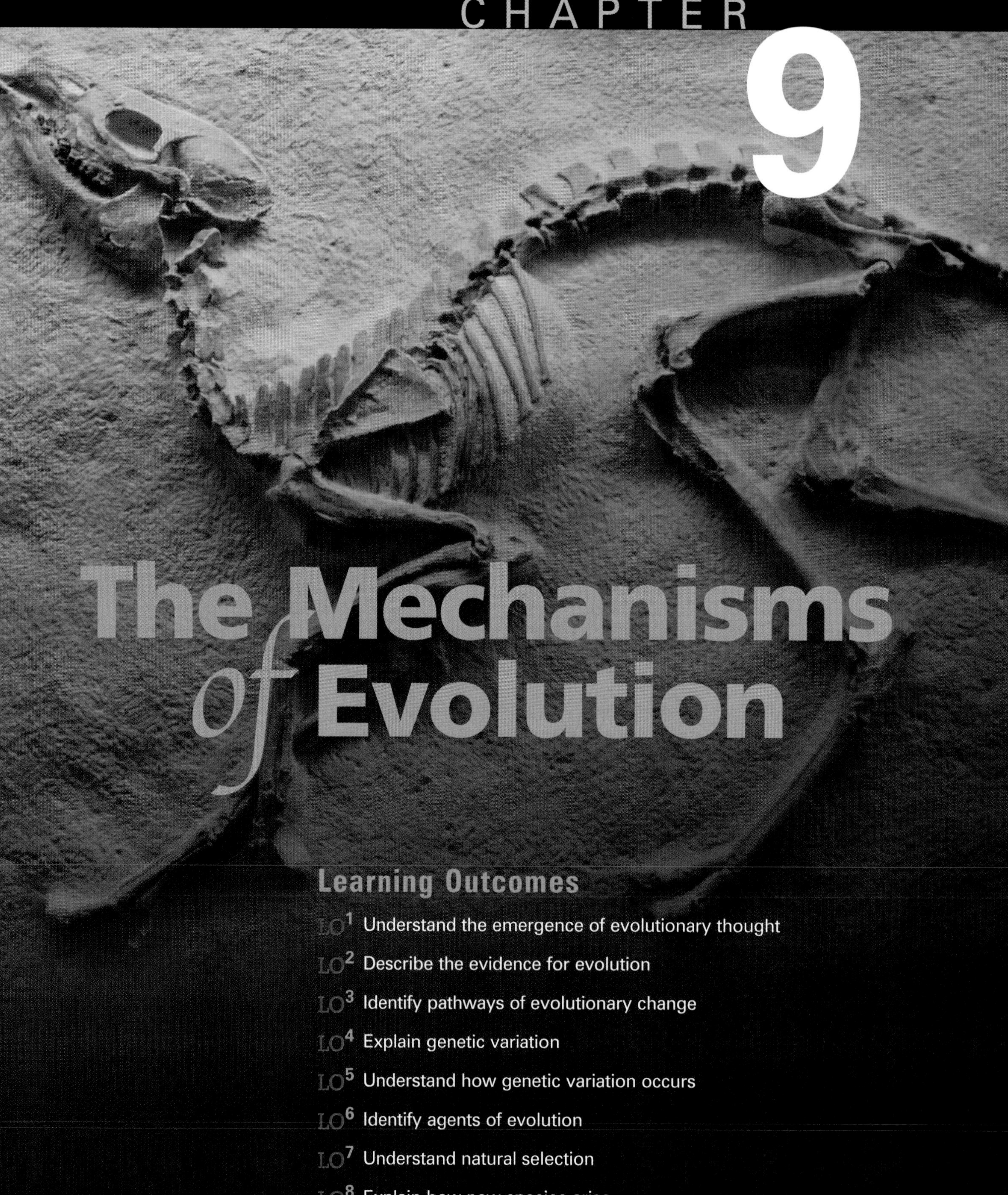

CHAPTER 9

The Mechanisms of Evolution

Learning Outcomes

LO[1] Understand the emergence of evolutionary thought

LO[2] Describe the evidence for evolution

LO[3] Identify pathways of evolutionary change

LO[4] Explain genetic variation

LO[5] Understand how genetic variation occurs

LO[6] Identify agents of evolution

LO[7] Understand natural selection

LO[8] Explain how new species arise

“Species that exist today are not ancestors to other currently living species; the ancestral species are long extinct.”

Resistance on the Rise

To Wayne Chedwick *Staphylococcus aureus*—a common bacterium—is an unseen enemy that robbed him of his livelihood, his ability to walk, and parts of his feet. In any college lecture hall, movie theater, or ballpark, up to 10 percent of the crowd will have “staph” populations living harmlessly in their noses and throats. But *S. aureus* can quickly turn against people—regular carriers and others—and cause boils, ulcerated wounds, bone infections, even blood poisoning. Mr. Chedwick has had them all.

His doctors prescribed various antibiotics, of course. But the *Staphylococcus* strain colonizing Wayne’s foot bones can’t be killed by most of the usual antibiotics. The reason? The strain he contracted has, over time, acquired genes for drug resistance: it has evolved new and dangerous genetic traits.

At the Federal Centers for Disease Control and Prevention in Atlanta, Dr. Fred Tenover has been carefully tracking these new cases of antibiotic resistance and considers them “worrisome” and potentially “a major threat.” Each year, says Tenover, 2 to 3 million people pick up an infection while in the hospital. Over 70 percent of these “are resistant to at least one antibiotic,” he explains, and the infections claim 90,000 lives annually.

These cases of near and total antibiotic resistance have profound medical implications, conjuring a world where simple infections take lives as they often did before penicillin became widely available in the 1940s. But the cases have obvious biological significance, as well. They are modern examples of evolution by natural selection, our subject in this chapter.

By exploring this chapter, you’ll see how natural selection brings about antibiotic resistance, but also how it engenders Earth’s stunning diversity of plant and animal life. You’ll encounter Charles Darwin and Alfred Russell Wallace once again, as we did in Chapter 1. You’ll see the patterns of evolution that established the dinosaurs and Ice Age mammals. And you’ll see how natural selection brings about new species, eliminates poorly adapted ones, and helps outfit organisms with successful survival tools such as genes for resisting antibiotics. Along the way, you’ll find the answers to these questions:

What do you know?

Why is extinction the likely fate of any species?

- How did naturalists develop the concept of evolution?
- What is the evidence for evolution?
- What are the various patterns of evolutionary change?
- How does genetic variation arise?
- What principles govern the inheritance of genetic variation in populations?
- What are the mechanisms of evolution and how do they work?
- How does natural selection fit populations to their environments?
- What are species and how do they originate?

inheritance of acquired characteristics the long-disproved belief that changes in an organism's appearance or function during its lifetime could be inherited

LO1 Emergence of Evolution as a Concept

An explanation for how such antibiotic resistance evolves in bacteria helps us understand evolution, or changes in gene frequencies in a population over time. But does evolution operate the same way in other organisms? And how does evolution account for the tremendous diversity of life? There are millions of kinds of insects, hundreds of thousands of kinds of plants, and so on. Did evolution produce them all and if so, how?

Until the end of the 18th and beginning of the 19th centuries, most naturalists believed that each species had been created separately and had remained unchanged from their creation to the current day. They thought, for example, that striped bass and sparrows were created at the beginning of the world and have remained exactly the same ever since.

About this same time, however, scientific exploration of the natural world was already uncovering facts that contradicted the notion of a single creation event and unchanging species. If all types of organisms were created at a single place and at a single point in time, then *why were there different groups of organisms in Earth's different regions?* If all organisms were created at one time, then *why would ancient extinct organisms be different from modern living species?* If each species had been created individually and never changed, then *how could the same basic bone plan be the best design for swimming* and *flying* and *fine manipulation?* Faced with such puzzles, a few naturalists began to suggest that populations of organisms might have changed, or evolved, over time.

Lamarck: An Early Proponent of Evolutionary Change

French naturalist Jean Baptiste Lamarck (1744–1829) was the first writer to attract much attention to the idea that species (including humans) are descended from other species. Lamarck also attributed this change to natural laws, not to miracles, and proposed a hypothetical mechanism for how this change could come about.

Lamarck suggested two main points. First, he proposed that the physical needs of an animal determine how its body will develop. Second, Lamarck proposed that changes in organ size caused by use or disuse would be inherited. For example, the longer neck a giraffe acquired through stretching to reach leaves to eat would be passed on to its offspring. Later biologists called Lamarck's hypothesis the inheritance of acquired characteristics.

Biologists in the mid-1800s did not yet understand cellular mechanisms of heredity. But two English naturalists, Charles Darwin and Alfred Russell Wallace, were able to make detailed observations in various parts of the globe and devise a remarkable and elegant explanation for how living things evolve.

Voyages of Discovery

Darwin and Wallace lived in a time when many areas of science were advancing quickly. One area of excitement was the growing conviction that the natural world is in a constant state of change. Fossils discovered in the late 1700s and early 1800s convinced many scientists that the forms of plants and animals had changed, or evolved, over time. One active fossil hunter from the era, Mary Anning, discovered the first plesiosaur, an extinct marine reptile, as well as many other fossils that contributed to the debates over evolution. Geologists of that same period also found evidence of physical evolution that contradicted Biblical timetables established for creation. They documented mountain building, erosion, and volcanic eruptions to prove that our planet must be millions, not thousands, of years old and must be undergoing slow but continuous change.

It was in this diverse social and scientific context that Darwin and Wallace independently observed unfamiliar organisms on their travels, and—unaware of each other's ideas—drew nearly identical conclusions about how the mechanisms of evolution might occur.

Darwin was passionately interested in natural history. He became the ship's naturalist for Captain Fitzroy of the HMS *Beagle* on a five-year voyage (1831–1836) to map the coastline of South America. At stops during the voyage, Darwin explored South American jungles, plains, and mountains. A seminal stop on the trip was the Galápagos, a cluster

of black volcanic islands west of Ecuador inhabited by a small but unique set of plants and animals. Darwin was impressed by the variable shapes and colors the members of a single species could show, and he wondered how such variety could have arisen. He returned to England with notebooks full of sketches and observations, convinced that the living plants and animals he observed could illuminate the origin of species. Alfred Russell Wallace had also sailed to South America to collect plants, insects, and other natural specimens. Like Darwin, he had noticed on his travels the striking variations among populations of individual species. Also like Darwin, Wallace had devised a theory of evolution by natural selection to explain what he saw.

By 1858, Wallace had published several papers containing portions of a theory of evolution that included natural selection, the notion of changes over time from a common ancestor, and the idea of survival of the fittest. Then the following year, Darwin released his monumental and well-documented book, *On the Origin of Species by Means of Natural Selection*.

Darwin and Wallace left the world a priceless legacy with their two key ideas: (1) descent with modification—the notion that all organisms are descended with changes from common ancestors; and (2) natural selection, the increased survival and reproduction of individuals better adapted to the environment.

Descent with Modification

Darwin thought of evolution as a branching tree with the more recently evolved organisms at the tips of the outer branches and the more ancient or extinct species at the base and on lower branches. This contrasts sharply with the form predicted by Lamarck or by advocates of creation. It also contradicts the still widespread but wrong idea that evolution is a ladder from "lower forms," like worms and fish, to "higher forms," like birds, dogs, and humans. Species that exist today are *not* ancestors to other currently living species; the ancestral species are long extinct. The biggest conceptual advance Darwin and Wallace put forward was the *mechanism* for how descent with modification could occur.

Evolution by Natural Selection

Darwin's and Wallace's experiences in then-remote regions such as South America and the Galápagos Islands convinced them that evolution had occurred. But Darwin returned to England still puzzling over some mechanism that could cause evolution. He eventually put together two indisputable facts based on observations, and drew a far-reaching conclusion:

Fact 1. Individuals in a population vary in many ways, and some of these variations can be inherited.

Fact 2. Populations can produce many more offspring than the environment's food, space, and other assets can possibly support. This causes individuals within the same population to compete with each other for limited resources.

Darwin's Conclusion. Individuals whose hereditary traits allow them to cope more efficiently with the local environment are more likely to survive and produce offspring than individuals without those traits. As a result, certain inherited traits become more common in a population over many generations.

descent with modification the notion that all organisms are descended with changes from common ancestors

natural selection the increased survival and reproduction of individuals better adapted to the environment

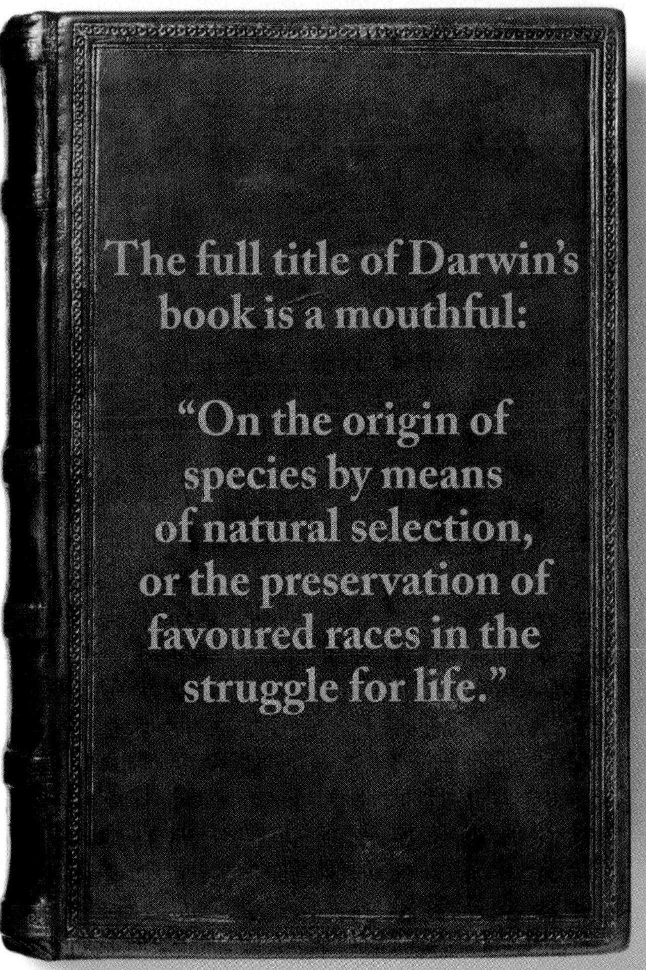

fossils
traces or remains of a living thing from a previous geologic time

sedimentary rock
layer upon layer of sand and dirt accumulated over thousands or millions of years and eventually forming rock layers that sometimes entomb fossilized organisms from different eras

Darwin (and later Wallace) used the term "natural selection" to describe the pressure exerted by environmental factors. Their term also applied to the greater reproductive success displayed by better-adapted individuals. Both chose this term because nature "selects" the favorable traits that are passed on to the next generation. The favorable traits, or *adaptations,* include body parts, behaviors, and physiological processes that enable an organism to survive in its current environment. Favorable traits are modified or maintained as a result of natural selection because they improve an organism's chance of surviving and reproducing successfully. Biologists now accept the principle of natural selection as a main mechanism behind evolution in nature.

People sometimes get confused, thinking that natural selection *causes* the variations within a population. However, short necks, long necks, antibiotic resistance, and other traits already exist in populations as a result of mutations or, sometimes in bacteria, of picking up foreign genes (transformation). Natural selection simply results in the best-competing individuals: those with the traits best adapted to current environmental conditions survive and produce more offspring in the next generation.

Later in the chapter, we'll return to a more detailed explanation of how natural selection favors certain genes. First, though, let's examine some of the evidence for evolution by natural selection amassed by Darwin, Wallace, and biologists after them.

> Natural selection simply results in the best-competing individuals: those with the traits best-adapted to current environmental conditions survive and produce more offspring in the next generation.

LO2 Evidence for Evolution

Evolution has such tremendous power to explain so many different aspects of biology that most biologists accept it as a fact. Researchers have found supporting evidence for evolution in the fossil record, in the anatomy of plants and animals, in molecular genetics, and in the geographical distribution of organisms.

The Fossil Record

Fossils are traces or remains of living things from a previous geologic time. Fossils give scientists tangible evidence of what past organisms looked like, and when and where they lived. A fossil can be nothing more than the trail, preserved in rock, of an animal that once slithered across the muddy bottom of some ancient lake or sea. Other fossils preserve leaf prints, footprints, casts, outlines, or sometimes even soft body parts in rock after the perishable organic matter is removed and replaced with minerals. The most familiar fossils, however, form from hard, decay-resistant structures such as shells, bones, and teeth.

Fossils and Sedimentary Rock

Although some fossils turn up in ice, peat bogs, or tar pits, fossil hunters find most fossils in sedimentary rock, which forms as layer upon layer of sand and dirt accumulated over thousands or millions of years. Newer layers press down on older ones, until pressure and heat cement the dirt, clay, and/or sand particles together, gradually changing them into rock.

Paleontologists discovered one of the richest fossil beds in North America in the Badlands region near the border of North and South Dakota, and learned from it priceless lessons about the evolutionary process (Fig. 9.1). Today, the Badlands region contains grasslands, high plateaus, and rock outcroppings. About 70 million years ago, however, the area lay beneath a shallow sea. The warm, expansive waters were densely populated by ammonites, extinct shelled relatives of squids and octopuses. By about 30 million years ago, the seas had retreated, and huge rhinoceros-like animals, the titanotheres, roamed the land. A few million years after that, a three-toed, knee-high ancestor of the horse, named *Mesohippus,* grazed where the larger titanotheres had roamed. Finally, fossils reveal that sheeplike oreodonts became abundant by about 25 million years ago. Paleontologists found each type of fossil at a particular level, suggesting that each of the different kinds of animals lived during a well-defined period.

If organisms evolved over immense time spans, then paleontologists ought to be able to find fossils intermediate in form between major groups of today's organisms (Fig. 9.2). And indeed they have. A famous example involves the 150-million-year-old crow-sized *Archaeopteryx.* This animal looked like a small dinosaur in nearly all respects, but it had feathers like a bird (Fig. 9.2c). A recently discovered series of fossils from China shows animals very similar to the velociraptors from the movie *Jurassic Park.*

Figure 9.1
Layers in Time

THERE IS NO EVOLUTIONARY RULE THAT SAYS THAT LIVING THINGS PROCEED FROM THE SIMPLE TO THE COMPLEX.

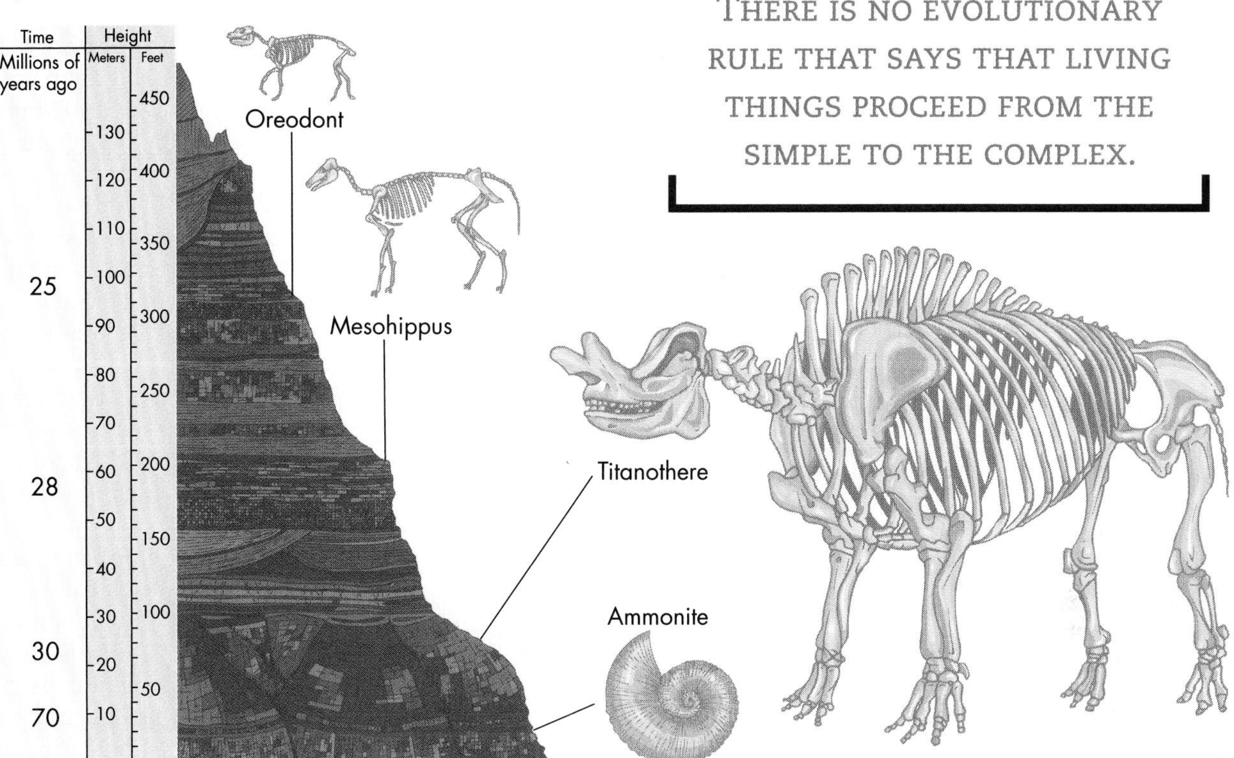

The tip of the animal's tail, however, was shaped like the feather-supporting tail stump in today's birds. Based on the similarities between birds and dinosaurs, some paleontologists now classify birds as a type of living dinosaur!

Fossil evidence supports evolution in several ways:

- Different organisms lived at different times.
- Past organisms were different from today's living organisms.
- Fossils in adjacent rock layers are more similar to each other than to fossils in distant layers.
- Many intermediate forms like *Archaeopteryx* have surfaced.
- In general, older rocks contain simpler forms, and younger rocks contain more complex ones.

As we saw before, though, there is no evolutionary rule that says that living things proceed from the simple to the complex. Instead, species have proceeded from being well-suited to their environments to being better suited—or else disappearing.

Evidence from Comparative Anatomy

If evolution occurs by descent with modification, that is, inheritance with changes, then one would predict the anatomy of living species to resemble that of extinct relatives, but with changes. And that is just what biologists see by studying two types of anatomical evidence: homologous structures and vestigial organs.

Homology: Organs with Similar Origins

Similar structures on different organisms help support the argument for descent with modification. Think, for a moment, about the forelimbs of various mammals and the functions they allow: human hands can deftly manipulate keyboards, surgical tools, paintbrushes; a cheetah's front legs help it run 116 kilometers (72 miles) per hour for short sprints; a whale's flippers allow it to dive powerfully and gracefully; a bat's wings enable it to fly. As different as they are, each type of limb is made up of the same skeletal elements (Fig. 9.3).

Fossil evidence supports the hypothesis that the varied forelimbs of mammals arose from the forelegs of ancestral five-fingered amphibians and became modified by natural selection in ways that facilitated different tasks. The idea that this same set of bones—one in the upper part of the limb, two in the lower part, and five digits—are the very best arrangement for manipulation *and* running *and* swimming *and* flying seems unreasonable to most anatomists. Elements (such as legs, flippers, and wings) in different species that derive from a single

homologous elements elements (such as a leg, flipper, or wing) in different species that derive from a single element in a common ancestor

vestigial organ a rudimentary structure with no apparent utility but bearing a strong resemblance to structures in probable ancestors

element in the last common ancestor of those species are called **homologous** (*homo* = same) **elements**. Homologous organs can have different functions but similar genetic blueprints, and can be constructed in much the same way in the embryos of different species.

Figure 9.2
Fossil Intermediates

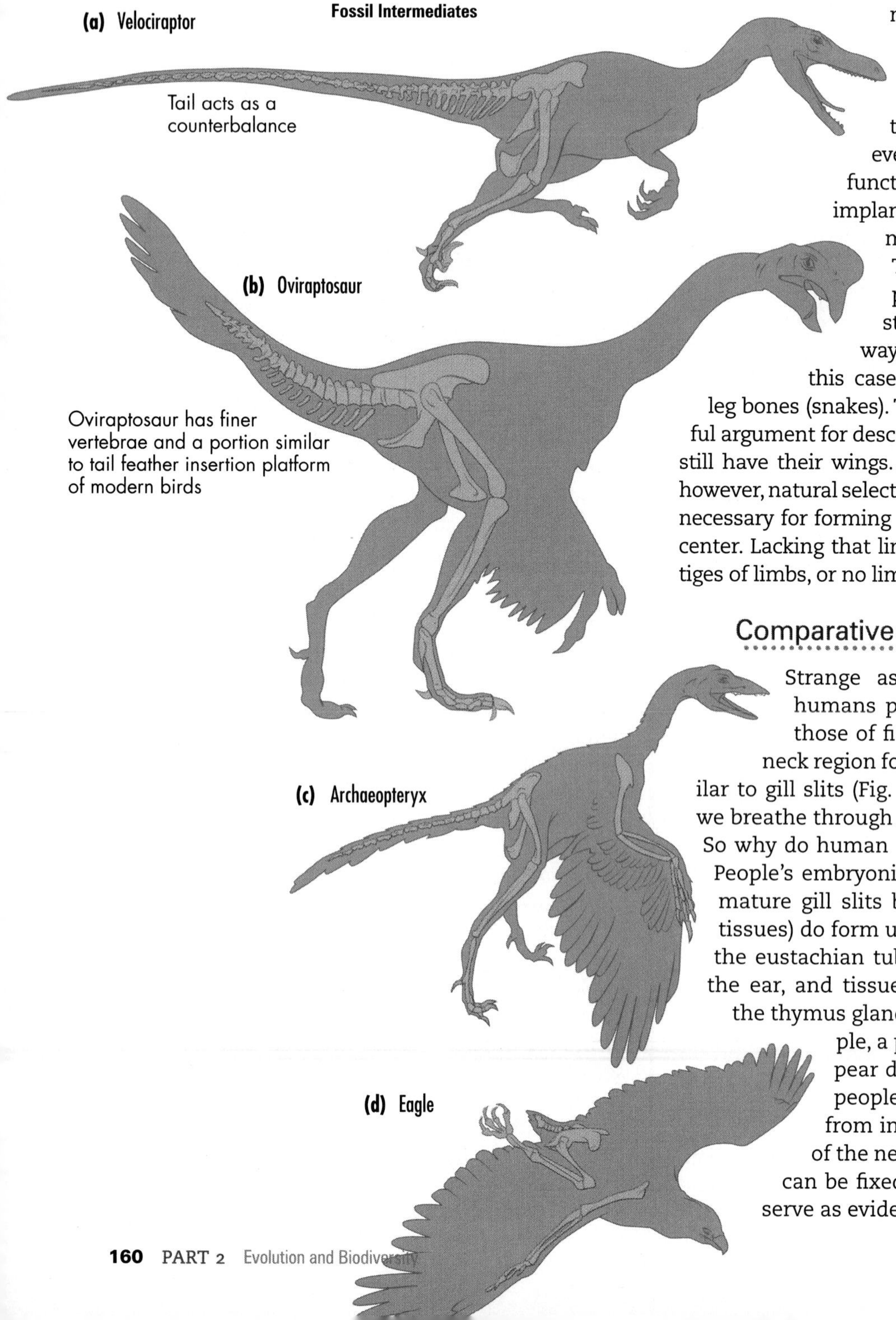

(a) Velociraptor

Tail acts as a counterbalance

(b) Oviraptosaur

Oviraptosaur has finer vertebrae and a portion similar to tail feather insertion platform of modern birds

(c) Archaeopteryx

(d) Eagle

Vestigial Organs

What does your appendix have in common with your wisdom teeth and with a snake's hipbones? The answer is they're all **vestigial organs** (veh-STIHJ-uhl; Latin *vestigium* = footprint, trace): or rudimentary structures with no apparent use in the organism, but strongly resembling useful structures in probable ancestors. You may have lost both your appendix and wisdom teeth because these structures can be worse than useless if they become infected or impacted. We humans also have the vertebrae (spinal bones) for a tail and for muscles that can wiggle our ears, despite their uselessness today for our survival. In the same way, python snakes have vestigial hipbones and thighbones, even though the bones no longer function. Curiously, a snake embryo implanted with a chick limb bud starts making signals to build legs. This shows that as different as pythons are from chickens, they still share developmental pathways for homologous organs, in this case, wings (chickens) and useless leg bones (snakes). This type of sharing is a powerful argument for descent with modification. Chickens still have their wings. In snakes and their ancestors, however, natural selection has removed alleles of genes necessary for forming the limb bud, with its signaling center. Lacking that limb bud, snakes make only vestiges of limbs, or no limbs at all.

Comparative Embryology

Strange as it seems, as embryos, we humans pass through stages similar to those of fish embryos, during which our neck region forms grooves and pouches similar to gill slits (Fig. 9.4a,b). Like most vertebrates, we breathe through lungs, not gill slits (as fish do). So why do human embryos form these pouches? People's embryonic neck pouches don't become mature gill slits but they (along with adjacent tissues) do form useful organs: one pouch forms the eustachian tube leading from the mouth to the ear, and tissue between the pouches forms the thymus gland and the tonsils. In some people, a pouch fails to change or disappear during development, and these people are born with a tube leading from inside the mouth to the outside of the neck like a fish. This abnormality can be fixed surgically and does, at least, serve as evidence of evolution.

Figure 9.3
Comparative Anatomy Supports Descent with Modification

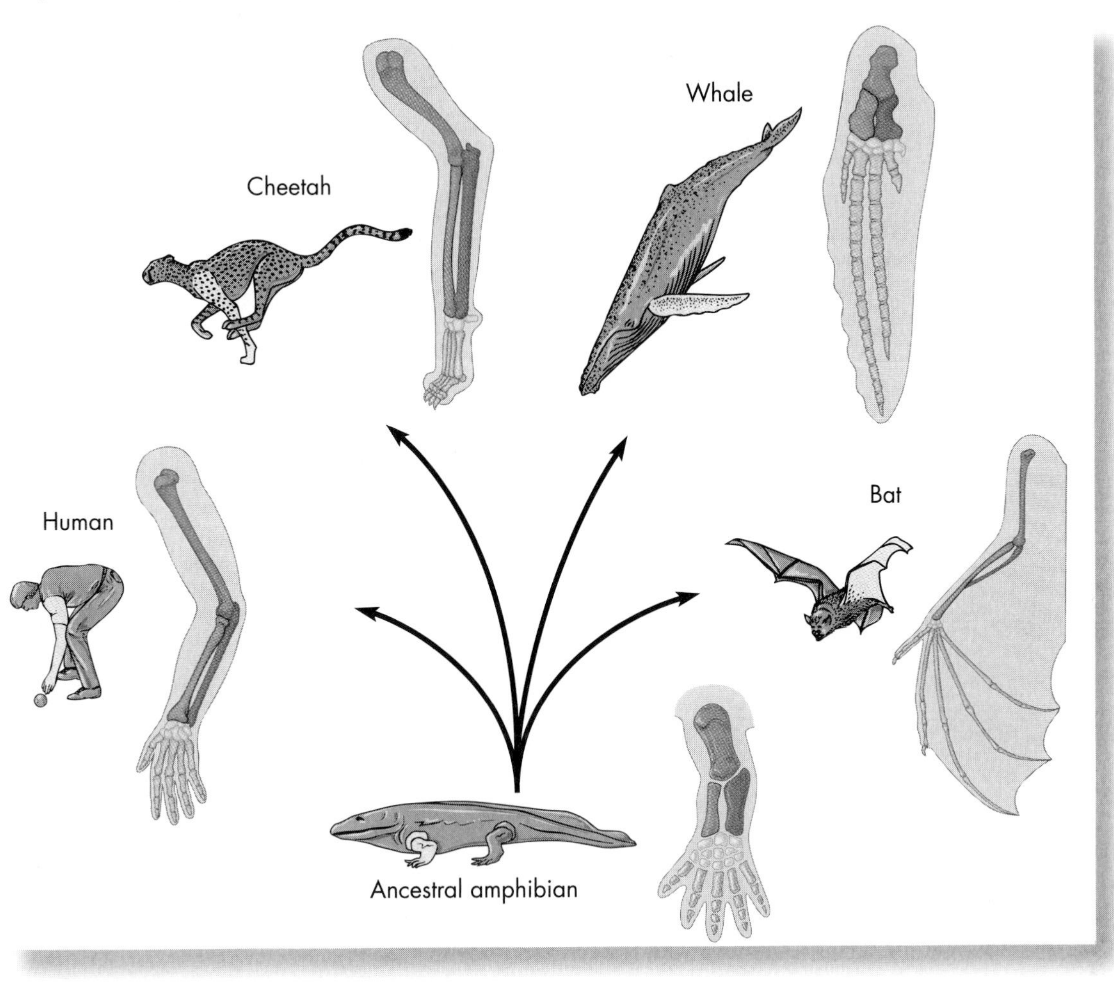

Figure 9.4
Comparative Embryology, Gills Slits, and Descent

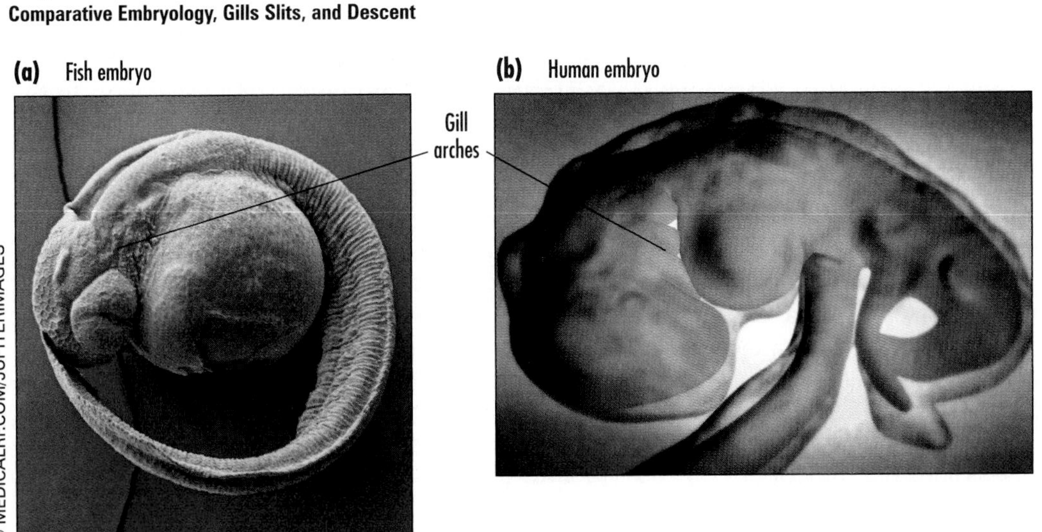

biogeography the study of the geographical location of living organisms

Evolutionary theory can easily explain why all vertebrate embryos look similar to fish embryos for a while, right down to the pouches. The last common ancestor of humans and ray-finned fish (such as bluegill or bass) lived 450 million years ago, and the embryos of this ancestor developed gill pouches like today's human and fish embryos. In the lineage giving rise to humans, changes in gene frequencies driven by natural selection caused the pouches and nearby tissues to develop into the eustachian tube, thymus, and tonsils. In fish, natural selection modified the same embryological structures into efficient oxygen-gathering gills. Embryology therefore supplies evidence that organisms often inherit features that originated in ancient ancestors but became genetically modified over time.

Evidence from Molecules

If all life evolved from a common ancestor that underwent gradual genetic changes, then all living organisms today should share basic features at the molecular level. In fact, all organisms have the same four bases in DNA, the same 20 amino acids in proteins, virtually the same genetic code, and so on. By comparing the same protein or the same gene in dozens of different species, molecular geneticists can construct molecular family trees. On such a molecular tree, they would place two very similar proteins or genes on branches diverging from each other more recently, and two less similar proteins or genes on branches that diverged longer ago. Biologists then assume that the organisms from which the genes or proteins came share the same evolutionary relationships as the molecules on the tree.

Molecular family trees powerfully support evolution because totally independent molecular and fossil data show common patterns of relationships between living and extinct species. Biologists have even been able to isolate DNA from some fossils. For example, researchers have extracted DNA from 30-million-year-old termites trapped in amber, the fossilized sap of ancient conifer trees. The DNA from the fossil termites was very similar to DNA from living termites that share the same physical features, and very different from DNA in living termites with radically different physical features. These results are exactly what evolutionary biologists expected to find—that ancient species are more similar to some of today's species and less similar to others.

Evidence from Biogeography

When Darwin and Wallace sailed separately to foreign lands, their first clue about evolution was **biogeography**, or the geographic distribution of organisms. In the Galápagos Islands, for example, Darwin was astonished by the unusual plants and animals he saw: finches using cactus spines to pry insect larvae out of dead wood; gigantic tortoises that varied in form from island to island; iguanas that swam in the sea and fed on algae instead of living exclusively on dry land. He saw no frogs or salamanders, however, and the only native mammals he observed were a few bats and one type of mouse.

Darwin compared these Galápagos organisms to the plants and animals he had seen while visiting the Cape Verde Islands off the coast of Africa. Even though the two island groups had similar climates, sizes, and the same dark volcanic soil, their organisms were of totally different species, genera, and taxonomic families. Rather than being related to *each other's* fauna and flora, each island's organisms were more closely related to those inhabiting the nearest *mainland*. Ironically, though, those mainland areas were very different from the neighboring islands in climate, soil, and other ecological details. So why were the island animals more closely related to those of the mainland than those on other similar islands?

Most 19th-century naturalists accepted the idea of special creation, and would explain the distribution of organisms by suggesting that each individual island had its own individual creation event related somehow to the creation event on the nearest mainland. But why would that creation event have discriminated against frogs, salamanders, and most mammals in a place like the Galápagos? Darwin thought it unlikely that creation overlooked amphibians. Instead, it seemed much more likely to him that the Galápagos harbored only organisms that had flown, swam, or rafted over from the nearest continental shore on floating vegetation to the newly forming volcanic islands. Encountering no competition on the young islands, these "colonists" could then diverge into new species as they adapted to different island environments through natural selection. Like Wallace, Darwin had begun his travels believing that every species was created in its special place, but saw evidence in biogeography that argued against it and the notion of unchanging species.

Throughout his travels in the Galápagos and Cape Verde Islands, Darwin also noted the powerful influence of the environment. For example, in both places, species of plants that grew as low, green herbs on the mainland had close relatives on the islands that were woody and treelike. He saw examples of unrelated organisms, as well, displaying similar characteristics in response to common environmental conditions. Australia, South America, and Africa, for instance, all have large, flightless birds with heavy bodies and long necks: the emu, the rhea, and the ostrich, respectively (Fig. 9.5). Why wouldn't the ostrich ever be created in South America, Darwin wondered? Why would an open grassland environment in Africa produce an ostrich but in South American a rhea and in Australia an emu? Why would three types of flightless birds have been created when surely one would have sufficed for the different continental grasslands? Darwin suggested that descent with modification led to the different species on the different continents, with similar physical and behavioral adaptations resulting from similar selection pressures.

LO3 Pathways of Descent

We've seen all sorts of evidence for evolution from fields as different as paleontology, anatomy, embryology, genetics, and biogeography. Now we need to focus on *how* it supports the principles of evolution. This section focuses on how evolutionary change takes place over time. What are the patterns of evolutionary change? How fast do organisms evolve? And how do extinctions affect evolution? The answers will help us understand the many pathways of descent over time.

Figure 9.5
Big Birds and Biogeography

Patterns of Evolution

divergent evolution the splitting of a population into two reproductively isolated populations with different alleles accumulating in each one; over geologic time, divergent evolution may lead to speciation

The *Staphylococcus* bacteria infecting Wayne Chedwick's feet could have become resistant to antibiotics in several small steps as mutations accumulated or in one jump if the cells acquired genes for antibiotic resistance from other bacteria. In multicellular organisms, evolution can also take place at different rates and can follow a number of converging, diverging, or radiating patterns.

Gradual Evolutionary Change

One species can gradually change into a new species as genetic differences accumulate in a population slowly over many generations (Fig. 9.6a). A certain genus *(Globoratalia)* of marine protists called *foraminifera* is a good example. From about 10 million to 5.6 million years ago, the chalky shells secreted by these saltwater species changed very little in shape. Gradually, over the next 0.6 million years, they underwent rather rapid change in size and shape into what experts consider a new species, which remains alive and nearly unchanged in today's Indian Ocean.

In a bacterium like *Staphylococcus*, gradual change can occur as a species is exposed to a low level of antibiotic, and a random mutation results in an altered gene. This mutation could change a gene that, for example, encodes an enzyme that helps digest a compound related in structure to penicillin. Now altered slightly, the enzyme can slowly digest the antibiotic, too, allowing the cell to resist low (but not high) concentrations of the antibiotic. This mutated cell will divide and leave more daughter cells. In time, another mutation in the same gene might occur that enables the enzyme to digest the antibiotic even more rapidly, and thus allow the cell to resist higher and higher doses of the antibiotic. Eventually, the accumulating population of daughter cells could be classified as a different strain. If similar change happens in many different proteins, the cells might be classified as a new species.

Divergent Evolution

Sometimes, a single population splits into two or more populations and the isolated populations start to accumulate genetic differences gradually. This is **divergent evolution**. Millions of years ago, before the Colorado River cut a mile-deep gash through what is now the high

adaptive radiation evolutionary divergence of a single ancestral group into a variety of forms adapted to different resources or habitats

convergent evolution evolution of similar characteristics in two or more unrelated species; often found in organisms living in similar environments

plateau of northern Arizona, a single species of squirrel occupied the area. As the gash deepened into the Grand Canyon, it became an uncrossable barrier and the squirrels were divided into two separate populations. Genetic changes accumulated and the two squirrel populations diverged into two separate species that have inhabited opposite rims of the Grand Canyon ever since (Fig. 9.6b).

Adaptive Radiation

Sometimes several populations of a single species diverge simultaneously into a variety of species (Fig. 9.6c). This process, adaptive radiation, may take place when one ancestral species invades new territories that allow it to exploit a variety of environments and different ways of life. Hawaiian honeycreepers are a good example. One ancestral species of the colorful birds colonized the Hawaiian Islands and radiated into about 20 different species, each adapted to survive on different foods.

Convergent Evolution

Two or more dissimilar and distantly related lineages can evolve in ways that make the organisms resemble each other superficially the way the streamlined bodies of sharks and porpoises do (Fig. 9.6d). A good example of this process, convergent evolution, involves the squirrel-like sugar gliders of Australian forests and the flying squirrels of Europe, Asia, and North America. Both kinds of animals soar through forests on thin folds of skin that extend along both sides of the body from wrist to ankle. Like all of Australia's original mammals, sugar gliders are marsupials that grow in the mother's pouch (marsupium), while flying squirrels are placental mammals. These two lineages diverged more than 150 million years ago.

Geologists agree that the Australian continent separated from the rest of Earth's continents more than 50 million years ago, and had few, if any, placental mammals at the time of the separation. Without competition from placental mammals, Australian marsupials radiated into a great number of species, many of which look and act like placental mammals on other continents that inhabit similar environments. In this case, sugar gliders resemble flying squirrels closely in ways associated with the gliding life style, but their most recent common ancestor lived over 150 million years ago. Convergent evolution can also happen when different bacterial species inhabit "environments" (including hospital patients) with high concentrations of antibiotics. The different bacteria—say, different staphylococcal and streptococcal species—can independently undergo mutations that allow them to survive one or more of the drugs. Thus they can converge, independently, on an antibiotic-resistant phenotype such as the ability to survive methicillin or vancomycin.

The Tempo of Evolution

How fast do old species converge, diverge, or radiate into new ones? One idea suggests that all lines change at about the same constant rate over time. More recent analyses, however, imply that structural changes often occur in fits and starts.

Phyletic Gradualism

Traditionally, evolutionary biologists thought that after a population splits into two, natural selection

Figure 9.6

Patterns of Descent in Evolution

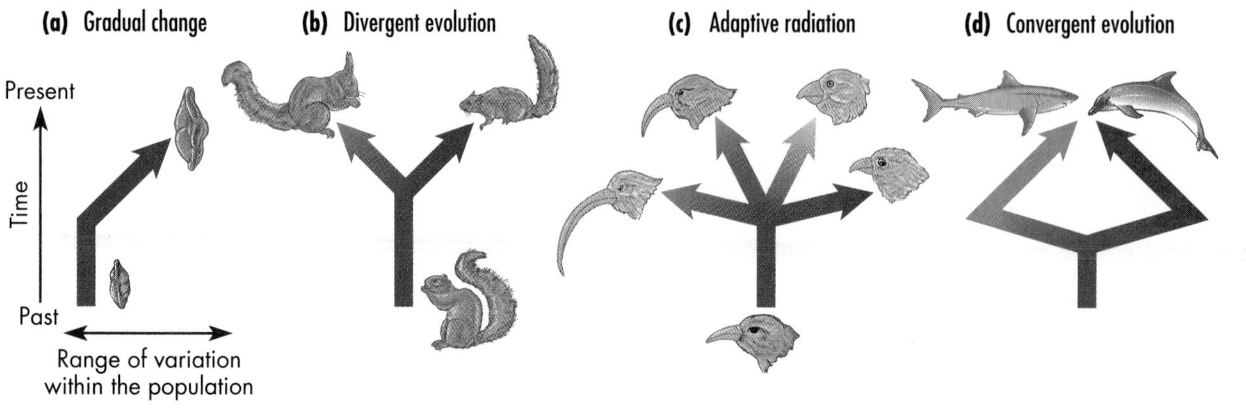

and random genetic changes would cause the new subgroups to diverge from each other at about equal and constant rates. In this model, termed **phyletic gradualism**, the giraffe's short-necked ancestor would have slowly acquired giraffelike qualities as genetic changes gradually accumulated.

Punctuated Equilibrium

Starting in the 1970s, some evolutionary biologists pointed out that if new species usually form gradually, then the fossil record should show numerous intermediate species. This certainly does occur, but paleontologists often find surprisingly little evidence of continually, gradually transforming lineages with a complete series of intermediate forms. Instead, fossils often reveal that species exist for millions of years with little change and that very rarely these long periods of equilibrium are interrupted, or "punctuated," by great phenotypic changes, resulting in new species.

The **punctuated equilibrium** model has three main tenets:

1. Changes in body form evolve rapidly in geologic time.
2. During **speciation** (the formation of a new species), changes in body form occur almost exclusively in small populations, and new species that result are quite different from their ancestral species.
3. After the burst of change that brings about speciation, species keep basically the same form until they become extinct, perhaps millions of years later.

phyletic gradualism
the concept that morphological changes occur gradually during evolution and are not always associated with speciation; distinct from punctuated equilibrium

punctuated equilibrium
the theory that morphological changes evolve rapidly in geologic time; in small populations, the resulting new species are distinct from the ancestral form; after speciation, species retain much the same form until extinction; distinct from phyletic gradualism

speciation
the emergence of a new species; speciation is thought to occur mainly as a result of populations becoming geographically isolated from each other and evolving in different directions

Models Compared

The family trees for the okapi and giraffe can help you compare the two hypotheses. Phyletic gradualism predicts that the lines eventually becoming okapis and giraffes began to diverge slowly (Fig. 9.7a). Gradually, changes accumulated that eventually prevented the two groups from interbreeding, thus making two separate species. Changes continued to accumulate gradually in both species until the okapi and giraffe emerged as we know them. In contrast, punctuated equilibrium predicts that offshoots of the initial evolutionary line split off repeatedly (Fig. 9.7b). These split-offs took place as small, isolated populations splintered from the main group, underwent bursts of evolutionary change, and then maintained new features for long periods (usually) before the line died out.

In fact, it seems quite likely that evolution proceeds in *both* ways. Evolution may often take place when a small

Figure 9.7
Okapis and Giraffes: Two Models of Descent

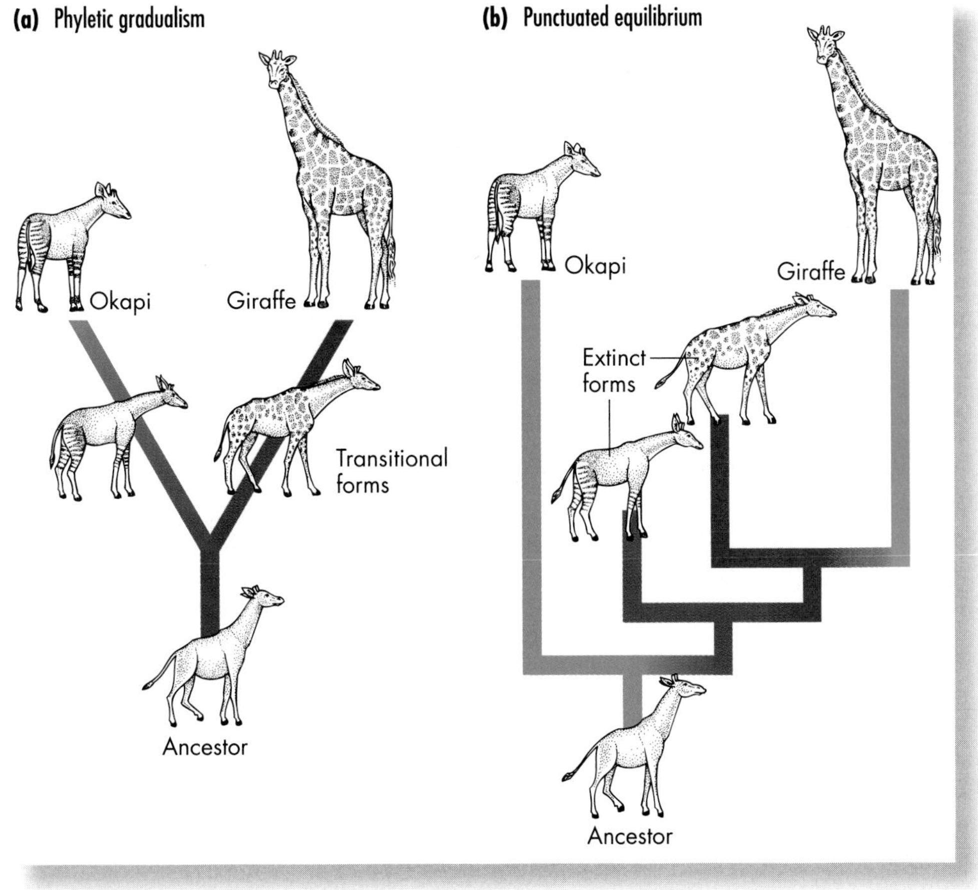

synthetic theory of evolution
in the 1930s and 1940s, biologists began to combine evolutionary theory with genetics; the synthetic theory suggests that (1) gene mutations occur in reproductive cells at high enough frequencies to impact evolution; (2) gene mutations occur in random directions unrelated to the organism's survival needs in its environment; (3) natural selection acts on the genetic diversity brought about by such random mutations

neutral mutations
single-gene mutations that leave the gene's function basically intact and neither harm nor help the organism

gene duplication
one method that may lead to the evolution of new genes with new functions; this can occur when a chance error in DNA replication or recombination creates two identical copies of a gene

gene pool
the sum of all alleles carried by the members of a population; the total genetic variability present in any population

population becomes isolated geographically or reproductively from other members of its species. Gradually, over many generations (but nonetheless a very short stretch of geologic time), the isolated population accumulates genetic changes. Once enough change has occurred and the new species is fitted to its environment, the species may stay essentially the same for millions of years.

Whether evolution is slow and gradual or fast and punctuated in any particular case, the fundamental raw material giving rise to the change is the same: genetic variation.

LO4 Genetic Variation: The Raw Material of Evolution

In the 1930s and 1940s, biologists began to combine evolutionary theory with genetics in the so-called **synthetic theory of evolution**. It suggests that:

(1) Gene mutations occur in reproductive cells at high enough frequencies to impact evolution.

(2) Gene mutations occur in random directions unrelated to the organism's survival needs in its environment.

(3) Natural selection acts on the genetic diversity brought about by such random mutations.

To understand how modern biologists see evolution, we need to understand how genetic variation occurs in the first place and how it is maintained in a population.

Sources of Genetic Variation

Mutations, new base pair sequences, and new combinations of genes (all of which we encountered in previous chapters) can lead to the varied genetic raw material that evolution acts upon.

Single-Gene Mutation

Mistakes or alterations in the DNA sequence of a single gene can (but don't always) alter how that gene functions (Fig. 9.8a). Some of these so-called *single-gene mutations* are **neutral mutations**: they leave the gene's function basically intact and neither harm nor help the organism. Other mutations produce favorable effects that benefit the organism. Many mutations, however, are either harmful or lethal; they reduce, modify, or destroy the function of a gene necessary for survival.

New Genes with New Functions

Mutation can alter old genes. But how do new genes arise with new functions? One way is through **gene duplication**. Sometimes, a chance error in DNA replication or recombination creates two identical copies of a gene, and mutation in one or both of these copies can change their nucleotide sequences such that one gene maintains the original gene function while the other acquires a related, new function (Fig. 9.8c).

Recombination

Mutation can create new versions (new alleles) of old genes. The process of *genetic recombination*, the shuffling of existing genetic material, can rearrange those alleles further (Fig. 9.8b). Recombination might bring together advantageous alleles of different genes into the same cell.

It is important to keep in mind that recombination merely shuffles existing alleles into new combinations but keeps the frequency the same (in the absence of selection), while mutation can change the frequency (commonness) of alleles in a population's **gene pool** (all the genes in a population at any given time) by changing one allele into another.

How Much Genetic Variation Exists?

At the end of the 20th century, geneticists for the first time compared corresponding large portions of a chimpanzee and a human chromosome to see how different and variable our DNA sequences are. Using as subjects 30 chimpanzees and 69 humans who represented all the world's major language groups, geneticists found that, on average, any 2 chimpanzees differed at about 13 places in the stretch of 10,000 base pairs, while any 2 humans differed only at about 3.7 places. These results show two things. First, there is a high rate of genetic variation within a species. For example, you differ from an unrelated person by about a million base pairs over your entire

Figure 9.8

The Sources of Genetic Variation

(a) Single-gene mutation

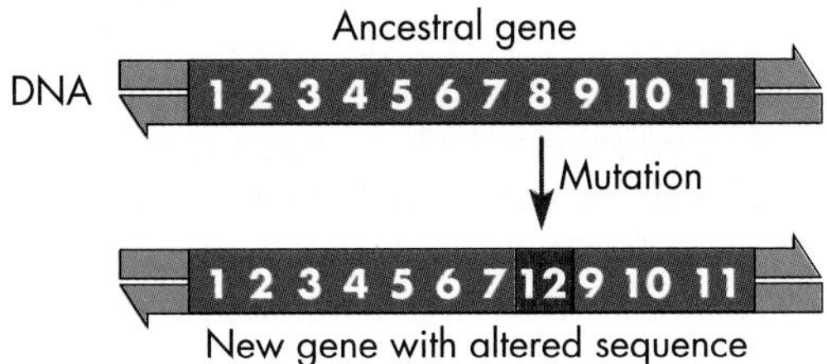

(b) Recombination

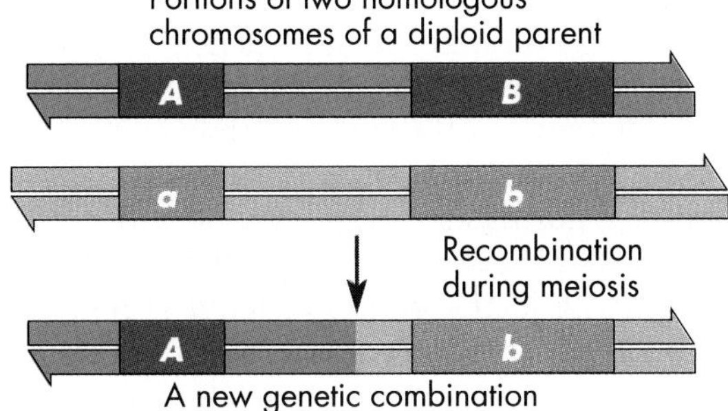

(c) Duplication and divergence

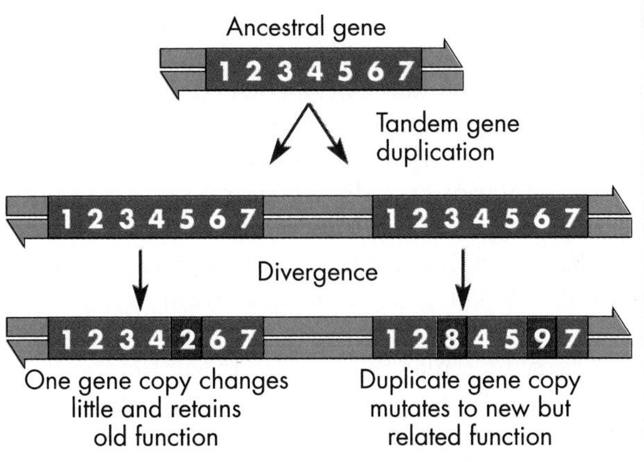

genome. This amount of genetic variation is probably enough to provide the material for natural selection. Second, different species can harbor different amounts of genetic variation. Chimpanzees are about four times as variable as humans. Assuming equal mutation rates, this might reflect a more ancient origin of the chimpanzee populations than of the human populations.

Some animals have very little genetic variation. For example, cheetahs, the handsomely spotted cats of the African savannah that are the world's fastest runners, have less genetic variation than most other mammals, about 100 times less than humans do. This genetic uniformity may someday propel cheetahs toward extinction. Because each individual is so similar genetically, a single new virus, for example, to which all cheetahs were vulnerable could wipe out their entire population. Furthermore, because genetic variation is required for evolution to occur, cheetahs have much less chance for evolutionary change than organisms with abundant genetic diversity.

Hardy-Weinberg principle in population genetics, the idea that in the absence of any outside forces, the frequency of each allele and the frequency of genotypes in a population will not change over generations

Hardy-Weinberg equilibrium a proposed state wherein a population, in the absence of external pressure, has both stable allele and stable genotype frequencies over many generations

LO5 How Is Genetic Variation Inherited in Populations?

For a century, biologists have called the mathematical model describing how genes behave in populations the **Hardy-Weinberg principle**, in honor of G.H. Hardy and of E. Weinberg, scientists who discovered the principle independently.

Biologists say that a population is in **Hardy-Weinberg equilibrium** when, in the absence of external pressure, it has both stable allele frequencies and stable genotype frequencies over many generations. (Appendix C presents simple equations that allow the analysis of any ratios in the starting population.)

Hardy-Weinberg Equilibrium

In the absence of outside influence, the frequencies of alleles do not change over the generations, and the frequencies of different genotypes do not change after the first generation.

How Do Biologists Use the Hardy-Weinberg Principle?

Using the Hardy-Weinberg principle, a biologist can predict allele frequencies in populations, and, in turn, can determine whether or not a population is evolving. The allele frequencies stay the same, and hence the population won't evolve, as long as it is free of outside influences. These "outside influences" are obviously important and a key to evolution. So what are they?

Five conditions must hold in order for allele and genotype frequencies to remain constant over many generations:

(1) no mutation,

(2) no migration into or out of the population,

gene flow the incorporation into a population's gene pool of genes from one or more other populations through migration of individuals

genetic drift unpredictable changes in allele frequency occurring in a population due to the small size of that population

(3) large population size,

(4) random mating, and

(5) no natural selection at work.

How likely is it that all five conditions will be met and that allele frequencies in a population will remain unchanged (nonevolving) generation after generation? Zero. Populations in nature *do* evolve, and allele frequencies do change within populations over time. The Hardy-Weinberg principle is useful only as a theoretical standard—an unchanging baseline—to compare against real populations in real environments that have outside influences.

LO6 The Agents of Evolution

Although the Hardy-Weinberg equations only apply to sexually reproducing diploid eukaryotes, the basic logic is similar for a population of *Staphylococcus* bacteria lurking on your skin, in your throat, or deep in your nasal cavities. These cells can have plenty of genetic variation and the potential to show many different traits, but that variation doesn't work alone to bring about change in a population of cells evolving toward antibiotic resistance. First something has to *act* on the genetic raw materials, the heritable variation. That something can be any of the outside factors we just saw—*mutation, migration, chance in small population size, nonrandom mating,* or *selection*. Working alone or together, these can alter allele frequencies and upset the Hardy-Weinberg equilibrium. Let's see how.

Mutation as an Agent of Evolution

By changing an original allele into a new one, mutation can alter allele frequencies in a population. The main importance of mutation to evolution, however, is not tiny changes in allele frequencies. Instead, it is the new phenotype that the new mutation may cause, and how natural selection acts on it.

Gene Flow: Migration and Alleles

When organisms migrate from one population to another one nearby, they may take alleles away from the first group and introduce them into the second. Biologists call this change in allele frequencies due to migration in or out **gene flow**.

Archaeologists have applied the gene flow concept to a mystery of human history. Several decades ago, archaeologists began to dig up a certain type of decorated beaker at widely separated sites in Europe and the Middle East where humans lived about 9,000 years ago. It seems that ancient peoples originally made the vessels in the Middle East. Over several thousand years, however, their production and use spread across Europe. Archaeologists wondered what exactly spread: Was it simply the knowledge of how to make and decorate this new pottery? Or did the Middle Eastern potters themselves migrate slowly across the continent? To answer this question, human geneticists sampled allele frequencies for 95 genes in various populations now living across Europe. They found gradients of allele frequencies as one travels from southeast to northwest. These gradients suggest that the potters themselves migrated towards the northwest, intermarrying with the local populations as they went. A single holdout tribe, the Basques living at the border of what is now Spain and France, resisted intermarrying with the migrating Middle Easterners. Today they alone display the language and genetic makeup of the prehistoric, premigration Europeans!

Gene flow is especially important in the evolution of bacterial antibiotic resistance. Many antibiotic resistance genes are present on the small circles of DNA called plasmids. These DNA circles can escape from one species of bacterium infecting a hospital patient, for example, then encounter another bacterial population in a dirty hospital drain and enter a second species by the process of transformation (Chapter 6). If another patient or staff member touches a towel, glass, or bedpan from that sink, he or she could pick up the second type of bacterium now carrying a plasmid bearing the antibiotic resistance genes. These genes, in effect, will have entered a new population and changed allele frequencies. As we've seen, they could also endanger people's lives.

Genetic Changes Due to Chance

Genetic drift is a term that refers to changes in allele frequencies that happen by chance and can't be predicted. Drift occurs most dramatically in small populations. Two types of genetic drift are the bottleneck effect and the founder effect. Let's look at each one.

Bottleneck Effect

Genetic drift can affect real-world organisms through a mechanism called a *population bottleneck*. Before we define it formally, let's consider the cheetah again.

Until about 10,000 years ago, cheetahs had a large population inhabiting all of Africa and the Middle East, and stretching into Asia. About ten millennia ago, however, the population apparently crashed due to disease, drought, or overhunting by humans, leaving only a few thousand of the beautiful felines. Since then, cheetah populations have rebounded but genetic diversity remains low (despite the few mutations that have accumulated since the crash) because all of the living cheetahs derive from the same few survivors. Biologists call a situation like this, in which a large population is slashed and then recovers from a few survivors, a **population bottleneck**. They call the reduced genetic diversity based on the few surviving original alleles the **bottleneck effect**. If the chance survivors have allele combinations that leave them susceptible to certain diseases, then the population's long-term future can be in doubt. Elephant seals are another example of the bottleneck effect. Hunted to near extinction in the late 1800s, a remaining population of just 20 animals has now recovered to tens of thousands, but their genetic diversity is extremely limited.

Founder Effect

Another type of genetic drift stems from the long-term isolation of a population founded by a few individuals. This kind of genetic drift is called a **founder effect**, because a few individuals split off from a large population and founded a new, isolated one. Nevertheless, both populations continue to exist. Because the small group of "founders" bears such a small fraction of the larger population's alleles, it may be a skewed sample, genetically speaking.

Nonrandom Mating

A population's mating tendencies—who mates with whom—can alter the ratio of heterozygotes and homozygotes. One type of nonrandom mating occurs when relatives are more likely to mate with each other than with unrelated individuals, a situation we call **inbreeding**. In many species, one member tends to mate with another that was born nearby, for example, frogs born in the same pond or people living in the same valley. Chances are good, therefore, that the two are related to some degree and have more identical alleles than nonrelatives. As a result, homozygotes become more common and heterozygotes become less common than expected by chance. If these individuals are homozygous for deleterious recessive alleles, then the population will have many more individuals that are less fit than in a normally breeding population, a situation called **inbreeding depression**. (In heterozygotes, these alleles would remain hidden by the dominant alleles.)

© ERIC ISSELÉE/ISTOCKPHOTO.COM

LO7 Natural Selection Revisited

How do populations become better adapted to their environments? Natural selection can alter a species' evolutionary trajectory in at least three different ways. In **directional selection**, a population shifts toward one extreme form of a trait (Fig. 9.9a). In the last 4 million years, for example, cheetahs have become half their former size, presumably because natural selection favored alleles that constantly pushed cheetah weight downward.

A second mode of natural selection is called **stabilizing selection**, because it results in individuals with intermediate phenotypes, as extreme forms are less successful at surviving and reproducing (Fig. 9.9b).

population bottleneck a situation arising when only a small number of individuals of a population survive and reproduce; therefore only a small percentage of the original gene pool remains

bottleneck effect the reduced genetic diversity that results from a drastic drop in a population's size

founder effect in evolutionary biology, the principle that individuals founding a new colony carry only a fraction of the total gene pool present in the parent population

inbreeding nonrandom mating that occurs when relatives mate with each other rather than with unrelated individuals

inbreeding depression a situation of weakened genetic viability that occurs when a population has many more individuals that are less fit than in a normally breeding population

directional selection a type of natural selection in which an extreme form of a character is favored over all other forms

stabilizing selection a mode of natural selection that results in individuals with intermediate phenotypes; under these selection pressures, extreme forms are less successful at surviving and reproducing

disruptive selection
a type of natural selection in which two extreme and often very different phenotypes become *more* frequent in a population

reproductively isolated
every true species in nature fails to generate fertile progeny with other species; the result is reproductive isolation

reproductive isolating mechanisms
any structural, behavioral, or biochemical feature that prevents individuals of a species from successfully breeding with individuals of another species

The third mode of natural selection acts opposite to stabilizing selection. In **disruptive selection**, two extreme phenotypes become *more* frequent in a population (Fig. 9.9c). For example, a butterfly species that tastes good to birds might mimic in coloration other butterfly species that are distasteful to birds. Some of the good-tasting mimics occur in a blue form, and others in an orange form. Mimics of either foul-tasting model are likely to escape being eaten. In contrast, intermediate forms mimic neither of the bad-tasting species and are more likely to be eaten by predators. This has caused alleles for the intermediate phenotype to decrease in frequency over time.

All five agents of evolution we've been exploring—mutation, migration, genetic drift, nonrandom mating, and selection—can nudge a population away from Hardy-Weinberg equilibrium.

LO8 How Do New Species Originate?

According to modern evolutionary theory, a *species* is a group of populations that interbreed with each other in nature and produce healthy and fertile offspring. New species arise when one group of organisms becomes **reproductively isolated** from another—it fails to generate fertile progeny with other true species. But what isolates species in the first place?

Reproductive isolating mechanisms are biological features that keep the members of species A from successfully breeding with the members of species B. Some reproductive isolating mechanisms act before fertilization and others act after fertilization.

Looking at isolating mechanisms that act before fertilization, some are behavioral. For example, dozens of frog species can live in the same area, but one species will mate only in ponds, one species only in running water, and a third species only in shallow puddles.

Other mechanisms that act before fertilization can be mechanical or chemical. For example, sometimes one species' sperm can't fertilize another species' eggs because a specific chemical surrounds the egg and allows binding only by sperm from the same species.

Figure 9.9
Three Modes of Natural Selection

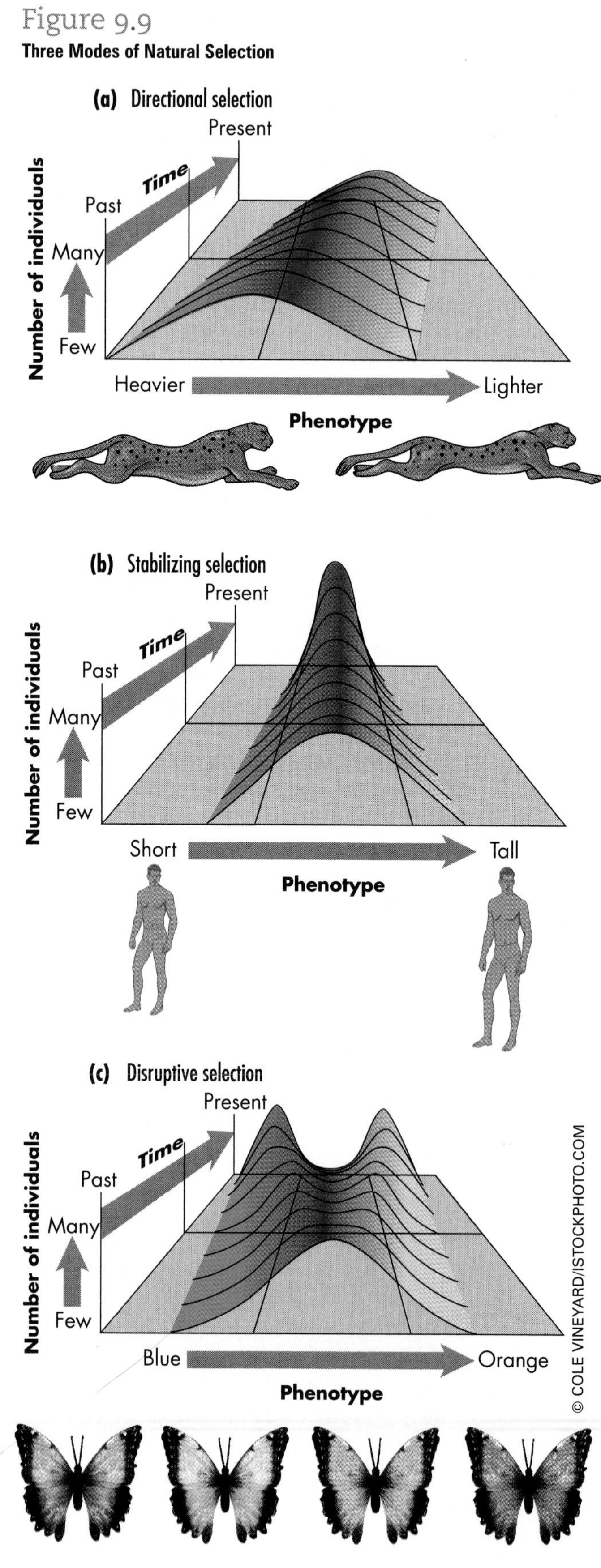

Looking at reproductive isolating mechanisms that operate after fertilization, sometimes different species mate and offspring result, but they die or are themselves infertile. Many American frogs, for instance, can interbreed, but genetic differences between the species lead to abnormal zygotes and tadpoles that do not grow to adulthood. In contrast, the mules that result from the mating of a male donkey with a female horse are very hardy hybrids, noted for strength and endurance. They usually don't produce offspring with other mules, horses, or donkeys, however, because chromosomes can't pair normally during meiosis. This prevents a flow of genes between donkeys and horses, and keeps the species separate.

How, then, do reproductive isolating mechanisms arise and lead to new species?

The Origin of New Species

Most biologists believe that many species arise after populations are split apart geographically and then evolve based on the reproductive isolating mechanisms we just discussed.

A physical barrier, such as a river, a desert, or different zones of vegetation can separate populations of a single species and prevent gene flow between them. The split populations grow more and more distinct from each other as mutation, genetic drift, and adaptation cause different sets of characteristics to accumulate. Eventually, the differences are great enough to prevent matings even if the two populations come into contact again later. This mechanism is called *geographical speciation*, or **allopatric speciation** (*allo* = different + *patra* = native land) (Fig. 9.10). If the isolated populations are small, the founder effect can come into play and so can punctuated equilibrium, because in the small, separated populations, the pace of evolution can speed up for a while before stable species emerge.

It's easy to see how species can evolve where populations are totally separate (allopatric). But species can evolve even without any geographical barriers separating the populations. This is called **sympatric speciation** (*sym* = same + *patra* = native land). In plants living in the same geographic region, new species can sometimes arise in a single generation based on **polyploidy**, an increase in the number of chromosome sets in a cell. A genus of wildflowers called *Clarkia* provides a good example of polyploidy. Geneticists discovered that one species, *Clarkia concinna*, has seven pairs of chromosomes in a diploid set, while *Clarkia virgata* has five pairs in a diploid set. If these two species mate, the hybrid progeny gets seven haploid chromosomes from one parent and five from the other. The hybrid is sterile because the two different sets of chromosomes don't pair and move correctly during meiosis. At some time in the past, however, a **tetraploid** hybrid plant

allopatric speciation the divergence of new species as a result of geographical separation of populations of the same original species

sympatric speciation a situation in which a population diverges into two species after a genetic, behavioral, or ecological barrier to gene flow arises between subgroups of the population inhabiting the same region

polyploidy an increase in the number of chromosome sets in a cell

tetraploid the condition of having four sets of chromosomes

Figure 9.10

Geographical Barriers and Allopatric Speciation

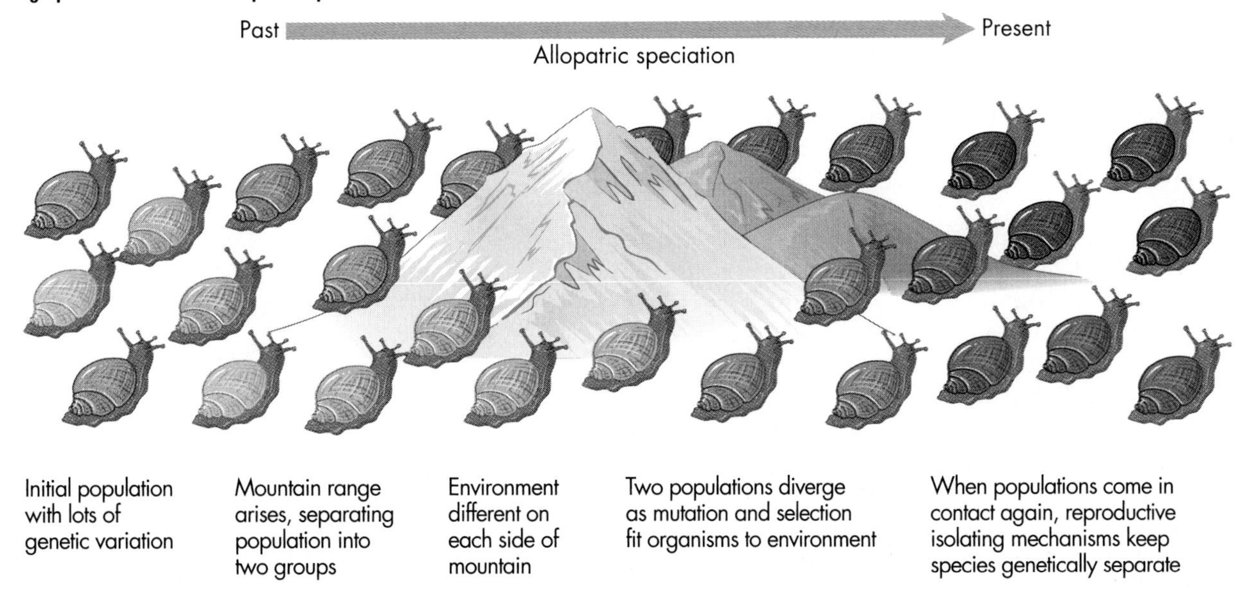

arose in which the chromosome count had doubled from 12 (7 + 5) to 24. Every chromosome in the new tetraploid hybrid, *Clarkia pulchella*, had a partner, so meiosis could occur normally. The species can't produce fertile hybrids with either diploid species *C. concinna* or *C. virgata*, but it can fertilize itself and produce new seeds. In this case, a new species was formed in one step within one original population without a geographical barrier. Sympatric speciation is relatively common in plants, and has contributed to the development of wheat, cotton, and some of our other important crops.

Evolutionary Pathways of Antibiotic Resistance

As we've explored evolution in this chapter, we've seen how natural selection can bring about new species and even higher orders of organisms. We've seen that the heavy use of antibiotics in hospitals, for example, can act as an agent of natural selection, "choosing" as parents of the next generation potentially dangerous cells that can't be killed. What, then, does our exploration of evolution suggest for dealing with the life-threatening problem of antibiotic-resistant bacteria?

Researchers have shown that under normal conditions (an environment free of antibiotics), antibiotic-resistant bacteria are poor competitors. They're actually at a disadvantage when they're *not* challenged with antibiotics because normal bacterial cells can outcompete them. In many cases, they'll die out, just like cavefish with fully developed eyes. What's the evolutionary implication here? It's this: one logical way to fight resistance is to cut the load of antibiotics in our everyday environments. By some estimates, half of all prescribed antibiotics are unnecessary, and medical consumers need to cut back on their demands. Farmers also use large amounts of antibiotics to stimulate the growth of farm animals and to protect fruit trees from bacterial diseases.

Who knew?

Species have only three possible fates: stay the same, change, or die. If species stay the same, it usually indicates an unchanging environment; but over thousands or millions of years, environments inevitably change with geological processes and climatic variations. If species change, over time they eventually become different enough to qualify as different species; the original species, having disappeared, becomes extinct. If species are wiped out in sudden environmental changes, then extinction has occurred. In any case, eventually extinction occurs.

Mutations occur in any given gene about 1 in every 100,000 to 1,000,000 times the gene replicates. In a bacterial population of 1,000,000 cells, therefore, between 1 and 10 of them might have a mutation in an antibiotic resistance gene. If bacterial populations are kept smaller through vigilant hygiene, it is that much less likely that a mutation will arise.

For patients like Wayne Chedwick, with serious bone infections that threaten to destroy toes and feet, effective antibiotics are a lifeline. Researchers continue to develop new drugs for which resistance has not yet evolved. Unfortunately, bacteria will eventually evolve resistance to any new antibiotics; we can only slow that process by using the principles of natural selection to our own advantage.

"Half of all prescribed antibiotics are unnecessary, and medical consumers need to cut back on their demands."

CHAPTER 10

Learning Outcomes

LO 1 Review evidence for how Earth formed and life emerged

LO 2 Explain how biologists think the first cells arose

LO 3 Explain how eukaryotic cells and multicellular life arose

LO 4 Understand how scientists organize life's diversity

LO 5 Review the current biodiversity crisis in light of historical mass extinctions

Life's Origins *and Biodiversity*

"Earth had a favorable composition, geological activity, size, and distance from the sun for life to arise and flourish."

Hawaii's Disappearing Crown Jewels

If you visit the Hawaiian islands and hike the mountainsides, you might see one of the world's strangest plants, the Haleakala silversword—one of the country's most famous endangered plants. The story of the silversword has important answers for the mysteries surrounding life's origins and diversity.

What do you know?

What would the physical condition of today's Earth be if life had never evolved?

The story starts almost 6 million years ago when molten rock from Earth's core shot upward through a massive vent and piled lava higher and higher into the sea. By 5.6 million years ago, it built up once more to become the future island of Kauai.

After its formation, the island of Kauai was a prime piece of uninhabited real estate. The land began filling in with plants and animals that flew, swam, or rafted over from older islands and continents. At some point, a bird in what is now California accidentally picked up a sticky seed on its feathers. The seed was from a nondescript herb called a *tarweed*. The bird flew west, eventually landing on the nearly new island, and the tarweed seed plopped off into a lava crack. There it survived, grew, and founded the silversword lineage, which includes the Haleakala silversword.

These strangely spectacular rosettes once dotted Hawaii's volcanic craters by the thousands. But voyagers and settlers released sheep, goats, and pigs onto the islands, and the animals eventually ate all but a few dozen plants clinging to the highest, steepest crater walls. One quarter of the silversword species are now in danger of extinction.

Botanist Robert Robichaux has devoted more than a decade to saving Hawaiian silverswords by raising seedlings and replanting them in the Haleakala crater and elsewhere. They are "the crown jewels of the Hawaiian flora," he says, and they "offer us a unique window into the evolution of life's diversity. If we lose them, we lose that window."

The silverswords' story helps illuminate our current subjects: life's origins and splendid variety, and the imminent threats to that diversity. Along the way, you'll find the current answers to some of biology's most fascinating puzzles and to our questions for this chapter:

- How did Earth form and how did life emerge here?
- How did the first cells evolve?
- How did multicellular life arise from early microbes?
- How did life's grand diversity emerge from those early ancestors?
- How do biologists classify the millions of living and extinct species?
- What is today's biodiversity crisis?

LO1 Earth's Formation and the Emergence of Life

In many ways, the birth of an island mimics the origin of dry land on our planet billions of years ago. That primal land formation was just one step, however, in a vastly longer process that included the generation of the planet itself, and even before that the inception of the universe.

A space probe launched in 1989, the Cosmic Background Explorer (COBE), provided supporting evidence for a widely accepted cosmological theory about how the universe began, the so-called Big Bang. This is the idea that a cataclysmic event took place about 13 to 15 billion years ago, during which all matter in the universe—concentrated into a single mass by unknown processes—underwent an enormous explosion and began to expand rapidly. The COBE probe confirmed that the expansion continues today throughout the universe.

Between 5 and 7 billion years after the Big Bang, a giant disk made up primarily of hydrogen and helium, the two lightest elements, spun in our corner of the universe. Gravitational attraction among atoms led to an accumulation of matter at the disk's center. The resulting compaction created a new star, our sun. At various distances from the center of the spinning disk, planets condensed from the cold gas and interstellar dust orbiting far from the pale young sun. Earth, the third planet from the sun, formed about 4.6 billion years ago.

The Birth of Earth

The newly formed Earth was nothing but a huge, frigid ball of ice and rock, devoid of oceans and atmosphere. The planet began to warm as radioactive elements in the rock decayed, releasing heat deep within Earth. The enormous, crushing pressure of gravity and the fiery impact of meteors bombarding the planet also warmed Earth into a molten mass that did not cool for several hundred million years.

As Earth cooled, a thin crust of rock formed at the surface like the rubbery skin on a bowl of pudding. Molten rock, or lava, frequently erupted through the crust, and clouds of gases issued from the planet's interior. Vapor clouds from volcanoes formed Earth's first atmosphere, and atmospheric chemists think it probably consisted of these compounds: carbon dioxide (CO_2), nitrogen (N_2), water vapor (H_2O), hydrogen sulfide (H_2S), and traces of ammonia (NH_3), and methane (CH_4). (Recall from Chapter 2 that carbon, hydrogen, oxygen, nitrogen, sulfur, and phosphorus are the major elements in living things.)

As Earth's surface and surrounding atmosphere cooled, water vapor in the atmosphere condensed and fell to the surface as rain, eventually cooling the surface further and creating a huge shallow ocean dotted with sharp volcanic cones that spewed out more lava and gases. The sun was about one quarter its current intensity, and looked pale in the sky. It is in this watery environment, with its air lacking molecular oxygen (O_2), that biologists believe life arose.

Life as we know it—based on carbon compounds and liquid water—could have arisen only on a planet with enough of both. Of all the planets in our solar system, we can only be certain at this point that Earth had a favorable composition, geological activity, size, and distance from the sun for life to arise and flourish.

Life's Emergence

With its liquid water and its atmosphere containing life's basic elements, early Earth was primed for living things to emerge. But how did these elements become organized into biological molecules and how did those, in turn, organize into living systems? Biologists have many competing theories and no definite answers, but they do agree on one thing: life originated in a series of small steps, each following the laws of chemistry and physics. The origin of life is a unique branch of biology because biologists can't directly observe how life formed nearly 5 billion years ago on our planet, which has changed so dramatically. They can conduct experiments to recreate the assumed conditions of early Earth in the lab. But they can still show only how life *might* have arisen, not how it *did* arise for sure, even with the laws of physics and chemistry as a guide.

> "If we represent the time span from Earth's formation to the present as a single 24-hour day, then Earth formed just after midnight, the heavy meteor bombardment stopped about 3:55 a.m., and cells had already evolved well before 5:30 a.m."

Why not look for direct physical evidence of the earliest life-forms and how they emerged in very ancient fossils? A record of the telltale chemical events may well have been laid down over 4 billion years ago in sediments

Figure 10.1
Stromatolites: Primitive Colonies

that then became fossilized. Unfortunately, meteors left over from the formation of our solar system frequently flamed into Earth's atmosphere until about 3.8 billion years ago, and these no doubt melted most or all of the earliest fossils and the chemical evidence locked inside them. One of the meteors was so huge—about the size of Mars—that it knocked loose the material that formed our moon. The meteor showers pelting Earth also pounded the newly formed moon, leaving thousands of craters you can see on a clear night through binoculars.

Biologists think that the ancestors of today's living organisms evolved shortly after the intense meteor bombardment stopped. What's their evidence? Paleontologists have found fossils about 3.6 billion years old at a place called North Pole, Australia, and the preserved remains closely resemble living, stumplike stromatolites: mounds of minerals deposited layer upon layer by colorful prokaryotic cells called *cyanobacteria,* or blue-green algae (Fig. 10.1). These fossils show that living things must have evolved quickly from nonliving building blocks, when our planet was extremely young. If we represent the time span from Earth's formation to the present as a single 24-hour day, then Earth formed just after midnight, the heavy meteor bombardment stopped about 3:55 a.m., and cells had already evolved well before 5:30 a.m.

Biologists, then, can be pretty sure about the *when*. So the next question concerns the *what*: what happened during those first few hundred million years after the streams of meteors stopped falling? We mentioned nonliving building blocks; the first steps would have been the forming of biological molecules from inorganic compounds (Table 10.1).

carbonaceous chondrite
a class of meteorites containing various kinds of organic molecules

How Simple Organic Molecules May Have Formed

The story of prebiotic ("before life") chemistry has a first act: the origin of small molecular building blocks such as simple sugars and amino acids and their assemblage into the large molecules (macromolecules) that characterize living organisms. No one was sure this type of spontaneous formation was even possible until the 1950s when a 23-year-old chemistry student named Stanley Miller came up with an idea. He tried to simulate in the lab the process of organic molecule formation from inorganic raw materials. Miller filled a large glass flask with several compounds that the chemists of his day thought were the elements making up Earth's first primitive atmosphere: hydrogen, water (H_2O), traces of ammonia (NH_3), and methane (CH_4) (Fig. 10.2). For about a week, Miller passed electrical discharges through this system to simulate lightning.

Amazingly, a pinkish sludge collected in the apparatus and contained large quantities of several amino acids; some lactic acid, formic acid, and urea; and traces of other biologically significant compounds. Atmospheric chemists now think that the early atmosphere was different from the one Miller recreated and might have included carbon dioxide (CO_2). Nevertheless, this classic experiment and later variations on it still proved that some organic molecules can be produced from nonbiological raw materials and physical processes like simulated lightning.

In the decades since Miller's historic test, space scientists have collected a different kind of evidence from outer space that proves the same point: organic molecules can form in the absence of Earthly life. Astronomers have discovered enormous interstellar clouds containing organic molecules. They've also collected hundreds of **carbonaceous chondrites**, a class of meteorites containing various kinds of organic molecules. In 1969, one of these dense, charred meteorites fell in Australia. Chemists immediately examined it using careful test methods that protected the specimen from contamination by Earthly substances. Intriguingly, this object from deep space contained organic compounds that closely matched the ones in the pinkish sludge in Stanley Miller's flask. This evidence from space shows that organic molecules can form from inorganic raw materials acted upon according to the universal laws of physics and chemistry. In fact, scientists think that carbonaceous chondrites and other meteorites may even have carried huge amounts of organic material to Earth's surface during the long period of intense meteor bombardment.

How Macromolecules May Have Formed

In the scientific story of life's origins, the next step after small organic molecules would have been macromolecules. Biologists think that small organic subunits would have joined, forming larger polymers such as proteins and nucleic acids. Long polymers like these tend to be fragile, especially in water. But chemists have shown that when they freeze solutions of amino acids dissolved in water or heat them to drive off water molecules, small polypeptides (chains of amino acids) form spontaneously. Picture the shore of a black volcanic island (like those in the Hawaiian Islands) but billions of years ago when Earth was newly formed. Perhaps a "primordial soup" made up of seawater and dissolved organic molecules collected in the ancient tide pools along that lava rock shore. The sun's heat could have concentrated those precursors via evaporation, or they could have frozen at night. Drying or freezing could have forced out water and may have provided the energy to form amino acid chains (polypeptides) and other macromolecules. Experimenters have also shown that under certain conditions, organic molecules can become trapped on the surfaces of clay particles. With the input of light or heat energy, these molecules can then react and become concentrated and organized. Organic molecules trapped on iron pyrites ("fool's gold") might also act this way, leading to the same end: the formation of large organic molecules such as polypeptides, polynucleotides, polysaccharides, and perhaps lipids—the complex raw materials of life.

Table 10.1

Steps in the Evolution of Life

Event	Billions of Years Ago	Consequences
Full diversity of life forms present	0.5	Complete ozone screen; atmosphere like today's; 20% oxygen; large, active fishes, land plants
Shelled animals and early land plants appear	0.55	Diversity evident in fossil record; 10% oxygen in atmosphere
Multicellular organisms appear	0.67	Fossils and tracks made; oxygen and ozone accumulate in atmosphere to about 7%
First eukaryotic cells appear	2.1	Mitosis, meiosis, genetic recombination, and aerobic respiration occur; 2% oxygen in atmosphere
Oxygen-tolerating blue-green algae appear	2.0	Ozone screen begins to form; iron deposits appear on earth's surface
Strong evidence of photosynthetic organisms	2.8	Oxygen is given off into atmosphere, but still is less than 1%
Autotrophs, methane-generating bacteria, and sulfur bacteria appear; suggestive evidence of photosynthesis	3.6	Little change in atmosphere
The origin of life	3.8	Primordial atmosphere lacks oxygen

Figure 10.2
Spark Chamber Experiments: Attempts to Simulate Earth's Primordial Conditions

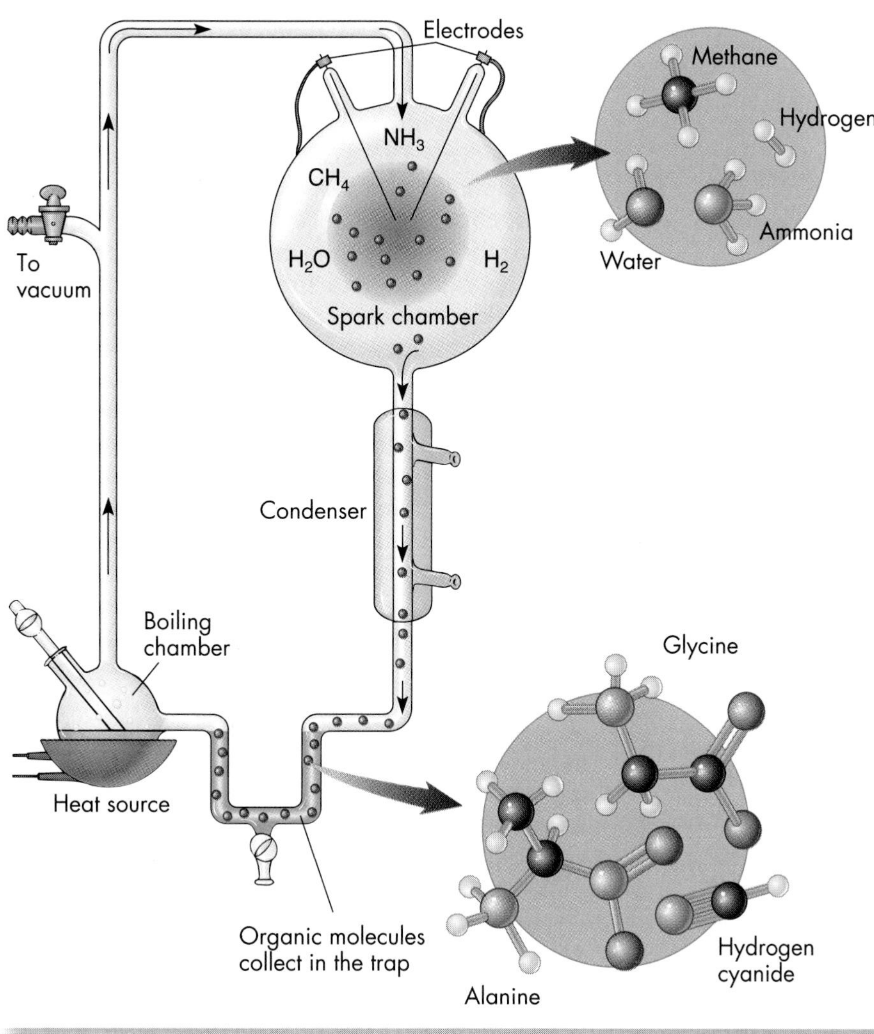

the evolution of self-copying molecules a circular problem. So which did come first, proteins or nucleic acids, and how did they start copying themselves and/or each other?

Many biologists are convinced that the first systems of self-replicating molecules were based on the nucleic acid RNA (Fig. 10.3a). Experiments conducted in the 1980s showed that certain types of RNA can act as catalysts. These RNA catalysts, or ribozymes, can snip themselves in two and join with nucleotides or other RNA molecules (Fig. 10.3b). In more recent experiments, researchers have even shown that the ribosomal RNA in today's cells can catalyze the assembly of amino acids into protein (Fig. 10.3c). These studies suggest a stage in the evolution of life some biologists call the RNA world, during which RNA molecules would have catalyzed their own replication as well as the joining of amino acids into proteins.

Self-Replicating Systems

It's not much of a stretch to explain how small organic molecules might have accumulated and linked up into proteins and nucleic acids (although their stability was probably poor and their existence fleeting). As we saw in Chapter 1, however, self-replication is another fundamental feature of life's chemistry. So researchers studying the origins of life also need to explain how primitive organic chains could have begun making copies of themselves. Here is the most popular explanation.

RNA Catalysts

In today's cells, nucleic acids can't replicate without the help of protein enzymes. Proteins, however, require nucleic acids for their synthesis. This makes

Self-Copying Molecules and Natural Selection

Biologists think that over eons of this natural selection, RNA molecules would have become tuned by natural selection for stability and rapid replication. Eventually, the more stable form of nucleic acid, DNA, took over the information storage task of RNA (Fig. 10.3d), and proteins became the common catalysts within cells (Fig. 10.3e).

LO² The First Cells

In our story of evolving life, we're not yet at the point of living cells but only of self-replicating organic molecules. As we've seen again and again, true cells have genes that encode proteins, they are surrounded by a plasma membrane, and they have complex metabolic pathways. So how did all of these features emerge from systems of self-replicating molecules in an RNA world or some other scenario?

Membranes Evolve

Every living cell is enclosed within a plasma membrane—a flexible, semipermeable lipid wrapper. Membranes can keep harmful substances out and useful chemicals in. A membrane, for example, could keep a protein that speeds RNA replication close to that particular RNA molecule (Fig. 10.3e). Without that boundary, the protein could diffuse away, and not be concentrated enough to replicate the RNA.

In living cells, membranes arise from the preexisting membrane of a parent cell. So how could membranes have formed before life started? Researchers have heated together polypeptides and phospholipids, and shown that membranelike sheets can form spontaneously. These can fold back on themselves to form membrane-enclosed compartments and under certain conditions, can even divide. Substances in the surrounding medium can become trapped inside these membranes and then apportioned when the compartments divide. Suppose that self-replicating RNAs were trapped in such a membrane: those compartments might then have the advantages of producing more RNAs, dividing faster, and leaving more compartments behind with protective membranes. If some RNA molecules kept membranes around themselves better than others, then a primitive natural selection mechanism would promote the improvement of membranes.

Figure 10.3

The Origins of Life: The "RNA World" Hypothesis

(a) RNA forms

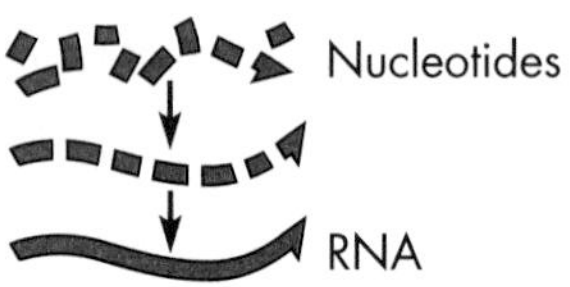

(b) Ribozymes catalyze RNA replication

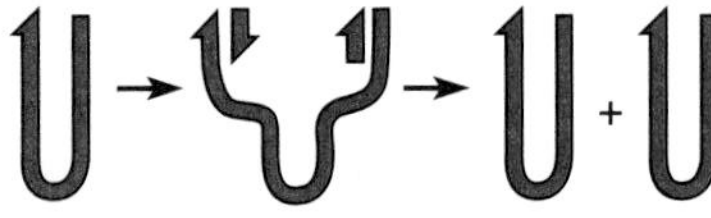

(c) RNA catalyzes protein synthesis

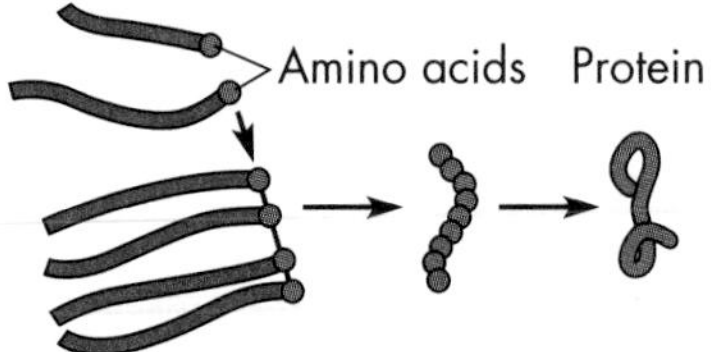

(d) RNA encodes both DNA and proteins

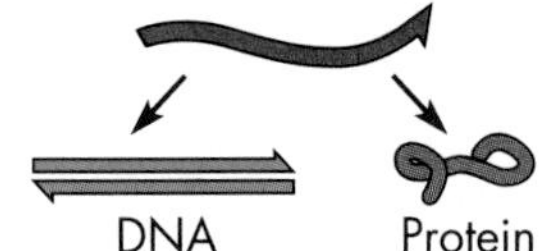

(e) Proteins catalyze cell activities

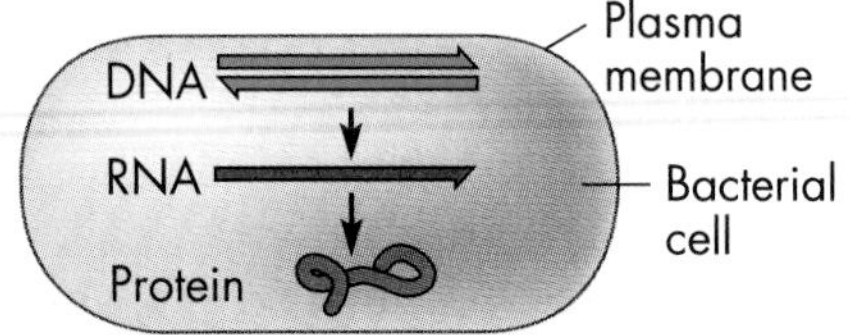

Origin of Metabolic Pathways

The same kinds of selective processes that favored membranes could have led to metabolic pathways. Biologists suspect that initially, primitive self-replicating compartments could have gotten all needed materials from the primordial soup: nucleotides, amino acids, lipids, and so on. But eventually, some substances may have started running out. Let's say, then, that a membrane-enclosed RNA by chance encoded a protein that could act upon the abundant material Y and change it into the needed but scarcer material X. This membrane-enclosed unit would have an obvious selective advantage. When substance Y became depleted, another round of genetic change and natural selection could favor a protein that could change substance Z into substance Y. Through this kind of selective pressure, metabolic pathways could have evolved: $Z \rightarrow Y \rightarrow X$, and so on. Somewhere, at some point, replication, protein synthesis, and metabolism became so closely integrated inside membrane-bound compartments that the resulting units were indistinguishable from living cells. At that point, life had begun on Earth.

The pathway from the formation of organic molecules to the probable first cells in their organic soup was a long and complicated one—a pathway that surely involved millions of small evolutionary steps. Modern experiments can never demonstrate exactly how life did evolve, but they can show that life's origin adhered to the laws of chemistry and physics and the environmental conditions existing on Earth 4 billion years ago.

Life Alters Earth

Early living organisms appear to have dramatically altered the physical conditions on Earth itself and paved the way for later life-forms.

Geologists have discovered ancient rocks containing organic carbon deposits that were probably substances from the earliest fossilized cells. These rocks formed in ocean sediments about 3.8 billion years ago, shortly after meteorites stopped obliterating all traces of earlier events. The metabolism in these early cells would have had to be anaerobic because, as we've seen, the primordial atmosphere lacked free oxygen. Also, these earliest living systems were probably **heterotrophs**. That is, instead of generating their own food molecules, they obtained nourishment from external sources, in this case from carbon-containing organic molecules that fell to Earth in meteors or appeared when lightning shot through Earth's first atmosphere.

Eventually, there would have been so many primitive heterotrophs removing organic molecules from their watery environment that the supply would have run out: all the free organic compounds would have been incorporated into biological molecules. For the first time—but far from the last—organisms would have changed their environment.

Evolution of Autotrophs

As heterotrophs depleted the primordial oceans of organic materials, they would have run short of carbon and energy supplies. Natural selection would then have favored **autotrophs**, organisms that could produce their own organic molecules from simpler inorganic precursors, such as methane and carbon dioxide.

A subset of autotrophs called **chemoautotrophs** can generate organic molecules and also trap energy using inorganic compounds such as hydrogen sulfide and iron. By studying organisms from Pele's Vent and other deep-sea hydrothermal settings, biologists have determined that all currently living organisms share a common ancestor, a primitive chemoautotroph that probably lived at temperatures near the boiling point of water. These ancient progenitors probably used carbon dioxide as a carbon source and used sulfur and hydrogen gas as energy sources (rather than light and water as do photosynthetic cells). Biologists conclude that they could survive only in oxygen-free (anaerobic) environments and they apparently lived deep enough below the surface of oceans, lakes, and hot springs so that the water layer above them screened out the highly energetic ultraviolet light that can destroy complex biological molecules.

Although chemoautotrophs do well in some environments today, the sunlight at Earth's surface provides an abundant and available source of energy. Very early in the course of life's history, autotrophic organisms began to tap this rich resource. Evidence from fossilized remains shows that photosynthetic cells had already evolved by 3.6 billion years ago—amazingly, just a few hundred thousand

heterotroph
(Gr. *heteros*, different + *trophos*, feeder) an organism, such as an animal, fungus, and most prokaryotes and protists, that takes in preformed nutrients from external sources

autotroph
(Gr. *auto*, self + *trophos*, feeder) an organism, such as a plant, that can manufacture its own food

chemoautotroph
an organism that derives energy from a simple inorganic reaction

Hot Spring in Yellowstone National Park, Wyoming

endosymbiont hypothesis the idea that mitochondria and chloroplasts originated from prokaryotes that fused with a nucleated cell

years after the end of the meteor bombardment period. Photosynthetic cyanobacteria are among today's simplest photosynthetic cells, and scientists have uncovered the earliest remains of distinct cells in 3.5-billion-year-old sedimentary rocks.

Photosynthesizers Change the Atmosphere

The early photosynthesizers were still very, very simple cells that probably used hydrogen sulfide and hydrogen gas as electron donors in their energy metabolism (review Chapter 3). These raw materials would have run out at some point, too, but water was essentially unlimited and any organism that could extract electrons from water could thrive in many more environments. During photosynthesis, today's cyanobacteria pick up electrons from water and release oxygen as a waste product. Based on fossils of cells closely resembling modern cyanobacteria, biologists think that cells able to split water molecules during photosynthesis and release oxygen gas into the environment had evolved by 2.8 billion years ago. This metabolic development had dramatic implications: it would allow living things to modify the planet's atmosphere in a major way.

Oxygen Tolerance and Aerobic Cells

For over a billion years after oxygen-producing, photosynthetic cells evolved, the atmosphere would have accumulated little or no oxygen. There is proof, instead, that the oceans literally rusted: iron compounds suspended in ancient oceans reacted with the free oxygen given off by the photosynthesizing cells and this formed iron oxides, or rust, which fell to the ocean floor by the ton. Only after most of the iron in seawater was bound to oxygen in this way would O_2 have built up in the atmosphere. Ironically, since oxygen is poisonous to anaerobic cells, this initial buildup would have harmed both the early heterotrophs and the photosynthesizers themselves.

About 2 billion years ago, as oxygen levels continued to build, thick-walled cells that could probably resist damage from oxygen lived and died in the Lake Superior area. Fossils of these cells tell us that mechanisms had begun to evolve (in this case, thick walls) that could ward off the toxic effects of oxygen given off by cyanobacteria and accumulating in the oceans and atmosphere.

This oxygen buildup would have had two more consequences. First, it would have acted as an agent of natural selection favoring the energy pathway of aerobic respiration, with oxygen as the final electron acceptor (review Chapter 3). Second, the accumulating oxygen in the upper atmosphere formed an ozone screen that had a major impact on the global environment. Sunlight striking gaseous oxygen (O_2) creates ozone (O_3), which in turn absorbs ultraviolet light. This screen protects living things from some of the damage that ultraviolet light inflicts on DNA and allowed cells to start living closer to the planet's surface rather than hidden around deep ocean vents.

LO3 Evolution of Eukaryotic Cells and Multicelled Life

Living things cropped up 3.6 billion years ago and have inhabited Earth for 80 percent of its history to the present. For more than half that time, the most complex life-forms were prokaryotic cells. Not until about 2.1 billion years ago did eukaryotic cells appear. Multicellular organisms—animals, plants, and most fungi—evolved only about 900 million years ago.

Biologists have carefully examined the stepwise bursts of evolution that led to the first eukaryotic cells, to their diversification into single-celled groups, and then to their descendants, the multicellular eukaryotic organisms. Some biologists hypothesize that the earliest eukaryotic cells lost the ability to make tough cell walls but got toughness and rigidity from different sources, namely the nuclear envelopes and cytoskeletons. These new features also would have allowed them to engulf food particles in pockets of the plasma membrane (review Chapter 2). Nuclear envelopes may have originated as invaginations of the plasma membrane—in-pockets that came to surround and protect the host cell's naked DNA. Fossils show these envelopes starting about 2.6 billion years ago (Fig. 10.4a,b).

Eukaryotes probably engulfed something else at least 2.5 billion years ago: the precursors to mitochondria and chloroplasts (Fig. 10.4c–e). To explain how this happened, biologists have proposed the **endosymbiont hypothesis** (*endo* = within + *symbiont* = live together). This is the idea that mitochondria and chloroplasts originated as free-living bacterial cells. A generalized ancestral cell type then "swallowed" them. Together, the host and guest organisms formed a single organism with each member adapted to the group arrangement and benefiting from it. Nonphotosynthetic prokaryotes, once engulfed, could have evolved into mitochondria (see Figure 10.4d), while the photosynthetic ones could have given rise to chloroplasts (Fig. 10.4e).

Figure 10.4
The Endosymbiont Hypothesis

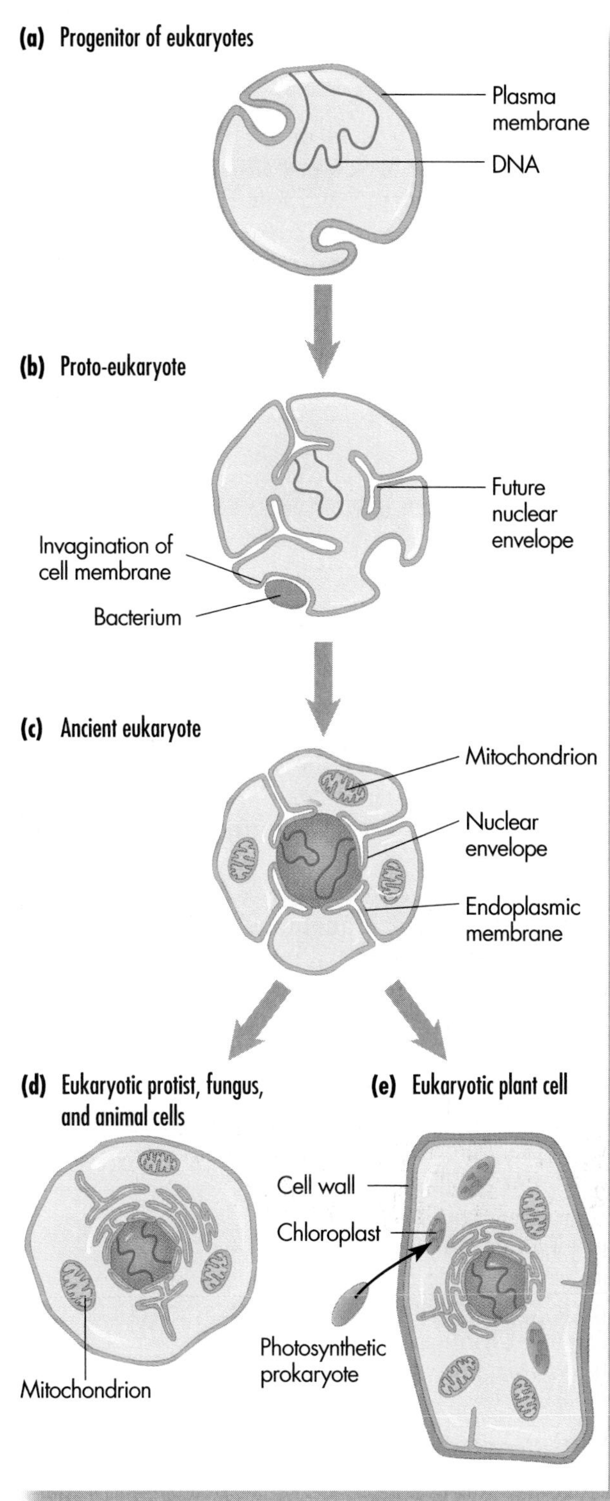

So, about 1.9 billion years ago cells contained mitochondria and chloroplasts. Their radiation into numerous groups of single-celled organisms, however, happened far later, about 1 billion years ago. Some biologists suggest that the evolution of sex led to the burst in diversity. The recombining of genetic material during meiosis and fertilization permitted much more genetic variability. Hence, sexual reproduction could have sped the pace of evolution considerably.

Cambrian Explosion
in just a few million years at the beginning of the Cambrian period, over 500 million years ago, all of the major animal phyla we see today began to leave preserved remains in the fossil record; in the hundreds of millions of years since, no new body plans have appeared

Simple Animals Emerge

About 580 million years ago simple animals first appeared in the fossil record. Paleontologists suspect that this surge in diversity coincided with higher oxygen levels in the atmosphere. The first fossilized animals were soft-bodied. It's somewhat mysterious that soft-bodied animals became fossilized 580 million years ago, considering that the soft body parts of later organisms are seldom fossilized. Some paleontologists suspect that there were no scavenging animals during these times, and so dead animals lay scattered on the ocean floor much longer than they would today—and thus were more likely to become fossilized. About 540 million years ago, this all changed: in a process dubbed the Cambrian Explosion, animals with hard body parts emerged and left many more fossils. The armoring of Cambrian animals suggests that predators had begun to roam the seas. Those with tougher skins or shells would have been devoured less often by predators than soft-bodied neighbors. Armored animals could therefore have reproduced more effectively and left more offspring bearing their genes.

Large Plants and Animals Arrive

By 400 million years ago the atmosphere was essentially like today's, with about 20 percent oxygen content. The ozone screen was in place, and large, complex life-forms were appearing in profusion. Large fishes swam in the ancient seas, and primitive land plants grew along moist shores. Within another 200 million years, arthropods, amphibians, reptiles, birds, and mammals would move about the continents, and large stands of conifer trees and flowering plants would grow abundantly, including the California tarweeds, the ancestors of Hawaiian silverswords.

Geologic Change Affects Life

The evolution of living things changed the atmosphere, rusted the iron in ancient oceans, and produced the ozone shield. Simultaneously, though,

Archean Era
the geological era spanning the time between Earth's origins and 2.5 billion years ago

Proterozoic Era
the geologic era from 2.5 billion years ago to about 580 million years ago

Paleozoic Era
the geologic era from about 580 million years ago to 245 million years ago

plate tectonics
the geologic building and moving of crustal plates on Earth's surface

Pangaea
a single supercontinent that existed on Earth about 245 million years ago

end-Permian extinction
the mass extinction that took place at the end of the Permian period, about 213 million years ago

Mesozoic Era
the geologic era from about 245 million years ago to about 65 million years ago

Cenozoic Era
the current geologic era, starting about 65 million years ago

Earth was changing from its own geologic processes and these influenced the evolution and diversification of living things. Geologists describing these events have divided Earth's history into eons, eras, periods, and epochs, based on sudden changes in the fossil record. For our purposes, however, the names of just five time periods are useful to remember:

1. Archean ("ancient") reaches back to the formation of the oldest known rocks.
2. Proterozoic ("earlier life")
3. Paleozoic ("old life")
4. Mesozoic ("middle life")
5. Cenozoic ("recent life")

The next few sections describe each period, as does Table 10.2.

Archean and Proterozoic Times

In the 500 million years or so between the origin of Earth and the beginning of the Archean Era, meteor impacts and volcanoes probably prevented life from emerging. Some scientists, in fact, refer to that hellish early environment as the "Hadean" (from "Hades"). But compared to that, the Archean was a peaceful time when the first continents rose from the sea. During the Proterozoic Era, beginning about 2.5 billion years ago, a phase of mountain building began and lasted for nearly two billion years. The organisms that had evolved earlier began to diversify.

Paleozoic Era

About 580 million years ago the Paleozoic Era began suddenly with the Cambrian Explosion. This was the nearly simultaneous appearance in the fossil record of all the major animal phyla that are alive today. Sometime before this sudden appearance of life-forms, plate tectonics, the process that moves about the landmasses of the Earth's crust by seafloor spreading and subduction, began (Fig. 10.5).

About 245 million years ago Earth's landmasses converged into a single supercontinent called Pangaea ("all Earth") (see Table 10.2). As the plates collided, landmasses buckled and tilted, lifting up huge mountain ranges; draining warm, shallow inland seas; and altering the size and position of ocean basins. The resulting climate changes were a calamity for both sea and land animals. About 75 percent of the families of amphibians (salamanders and relatives), 80 percent of the reptiles (lizards, snakes, and relatives), and 96 percent of all marine species became extinct. Paleontologists call this cataclysm the end-Permian extinction, because it marked the end of the Permian period in the Paleozoic Era, and they consider it history's greatest mass extinction.

Mesozoic Era

Pangaea broke apart in the Mesozoic Era between 245 and 65 million years ago separating populations of organisms and providing great opportunities for evolutionary change. When Pangaea existed, for example, mammals had not yet evolved the placenta, the organ that nourishes the young in the uterus. Instead, offspring were born at a very early developmental stage, found their way to their mother's pouch, or marsupium (Latin for "purse"), where they were nourished by milk. Small primitive pouched mammals or marsupials thrived all across Pangaea, as well as on the giant landmass of Australia after it broke apart and drifted away. Thousands of species of mammals with placentas evolved on other continents, but not on Australia. Without competition from placental mammals, the marsupials of Australia radiated into a rich collection of unique animals, including kangaroos, koalas, and wombats. In most parts of the world, marsupials lost the competition with placental mammals and disappeared. In North America, for example, opossums are the only surviving descendant marsupials.

Along with the breakup of Pangaea came the emergence and diversification of the largest land animals ever to live, the dinosaurs. A warming climate around the globe engendered great swampy forests containing tropical plants and trees, flowering plants, birds, and mammals.

Cenozoic Era

A catastrophe ended the Mesozoic and ushered in the current era, the Cenozoic Era, about 65 million years ago. In history's second-greatest mass-extinction event, one half of the world's species, including the dinosaurs, died out. Current research suggests that at this time, a meteor crashed into the gulf coast of Mexico. The resulting airborne debris darkened the

Table 10.2

History of the Earth

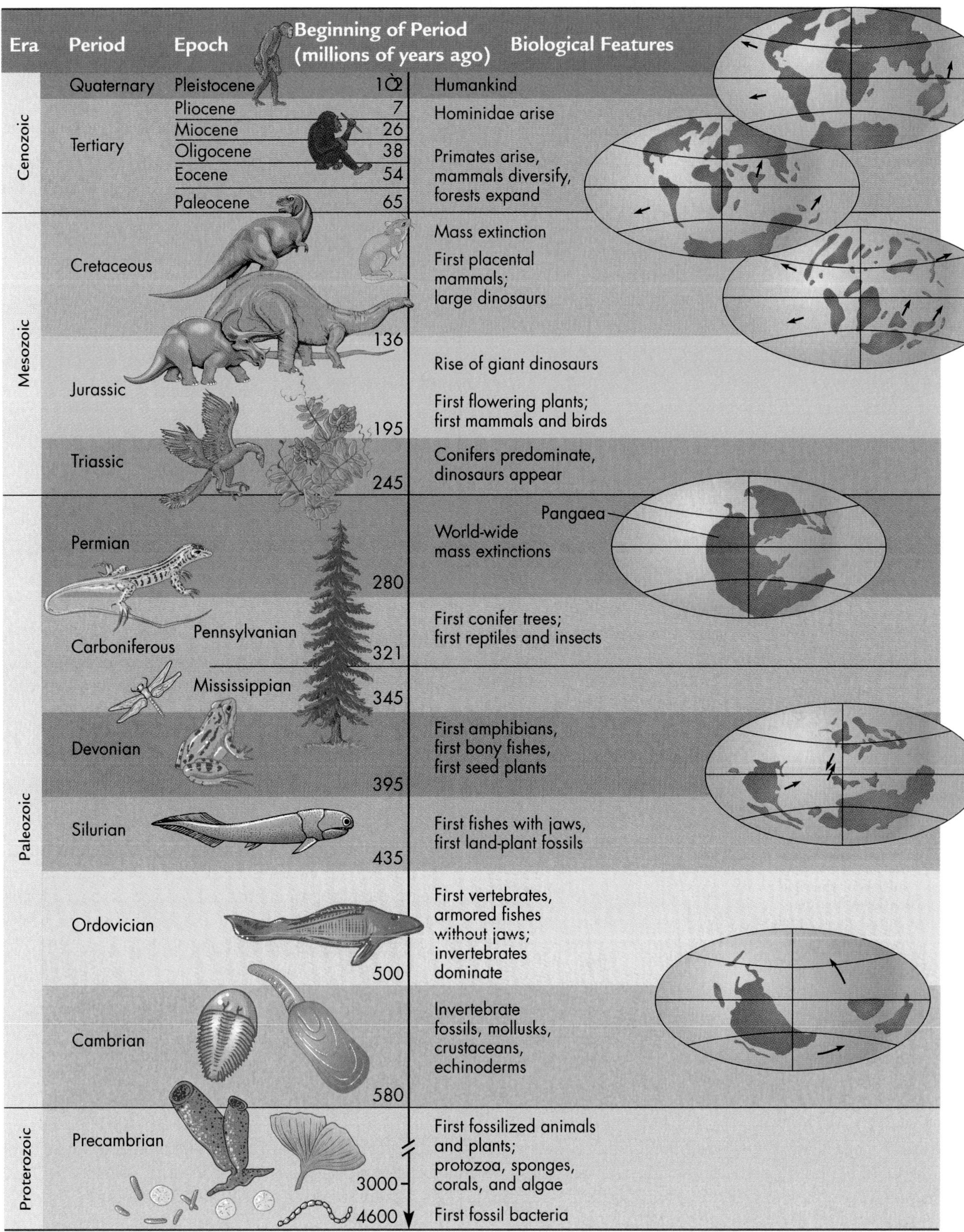

Era	Period	Epoch	Beginning of Period (millions of years ago)	Biological Features
Cenozoic	Quaternary	Pleistocene	1Ò2	Humankind
	Tertiary	Pliocene	7	Hominidae arise
		Miocene	26	
		Oligocene	38	Primates arise, mammals diversify, forests expand
		Eocene	54	
		Paleocene	65	
Mesozoic	Cretaceous		136	Mass extinction First placental mammals; large dinosaurs
	Jurassic		195	Rise of giant dinosaurs First flowering plants; first mammals and birds
	Triassic		245	Conifers predominate, dinosaurs appear
Paleozoic	Permian		280	World-wide mass extinctions
	Carboniferous	Pennsylvanian	321	First conifer trees; first reptiles and insects
		Mississippian	345	
	Devonian		395	First amphibians, first bony fishes, first seed plants
	Silurian		435	First fishes with jaws, first land-plant fossils
	Ordovician		500	First vertebrates, armored fishes without jaws; invertebrates dominate
	Cambrian		580	Invertebrate fossils, mollusks, crustaceans, echinoderms
Proterozoic	Precambrian		3000	First fossilized animals and plants; protozoa, sponges, corals, and algae
			4600	First fossil bacteria

binomial nomenclature a system of naming organisms that assigns each species a two-word name consisting of a genus name followed by a species name

genus (pl. genera) a taxonomic group of very similar species of common descent

species a taxonomic group of organisms whose members have very similar structural traits and who can interbreed with each other in nature

family a taxonomic group comprising members of similar genera

order a precise arrangement of structural units and activities; also, in taxonomy, a taxonomic group comprising members of similar families

class a taxonomic group comprising members of similar orders

phylum (pl. phyla) a major taxonomic group just below the kingdom level, comprising members of similar classes, all with the same general body plan; equivalent to the division in plants

division a taxonomic group of similar classes belonging to the same phylum, which is often called a division in the kingdoms of plants or fungi

kingdom a taxonomic group composed of members of similar phyla, i.e., Animalia, Plantae, Fungi, and Protista

domain a taxonomic group composed of members of similar kingdoms

sky around the globe, and could have drastically cooled Earth's climates, leading to the extinctions. When the skies cleared, birds, mammals, and flowering plants continued diverging into large and varied groups of species. Plate tectonics continued in the Cenozoic Era, bringing the continents to their present positions (Table 10.2). These crustal shifts continued to separate groups of organisms, rafting them to entirely new locations and conditions, and further encouraging the emergence of new plant and animal groups.

LO4 Organizing Life's Diversity

More than 200 years ago, a Swedish botanist named Carolus Linnaeus developed a classification system that assigned every organism then known to science to a series of increasingly larger, more general and all-inclusive groups. He and his contemporaries did the assigning by observing similarities in organisms' sizes, shapes, colors, and other tangible traits.

Linnaeus also began consistently to use a system for naming each type of organism—the system of **binomial nomenclature** (*binomial* = two names). This system assigns each species a two-word name, such as *Homo sapiens*, our own species' name, or *Argyroxiphium sandwicense*, the name of the Haleakala silversword (Fig. 10.6). The first part of an organism's name is the genus (pl., *genera*), a term that may include several similar species. The second part of the name, the species, works along with the genus term to designate only one type of organism. A species is also a unique group within a genus whose members share the same set of structural traits and can successfully reproduce in nature only with members of the same species. The genus refers to a group of very similar organisms related by common descent from a recent ancestor and sharing similar physical traits.

Figure 10.5
Plate Tectonics: Continents Colliding

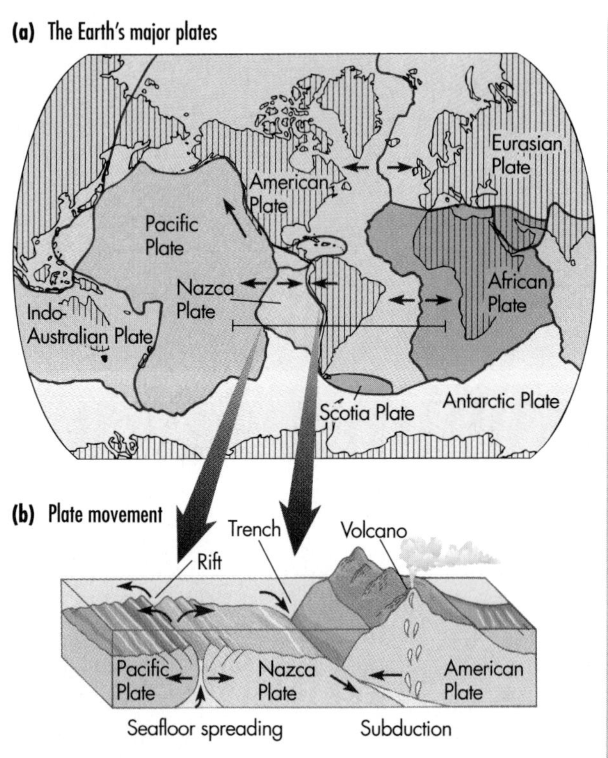

All genus names begin with a capital letter, and both genus and species names are written in italics or underlined. Biologists usually write or say the full scientific name the first time they refer to it in a discussion. After that, though, they often abbreviate the genus and simply use *A. sandwicense*, for example, or *H. sapiens*.

The Hierarchy of Classification

Each organism fits into a series of categories, or taxonomic groups, based on shared characteristics. These are arranged in a hierarchy of broader and broader shared traits, and assigned to the **species**, **genus**, **family**, **order**, **class**, **phylum** (or **division**), **kingdom**, and **domain**. To see how this works, check out all the taxonomic groups for *A. sandwicense* and *H. sapiens* in Figure 10.6.

Did Ken pour coffee on Fran's good shirt?

[domain, kingdom, phylum, class, order, family, genus, species]

Criteria for Classification

A taxonomist's job might seem easy—just group organisms according to their shared characteristics. But which characteristics should they use? Is flower color a good measure? Should they group together all plants with yellow flowers—yellow silversword with dandelions and yellow roses, for example? Or how about flower size: should they group silverswords and sunflowers because both have large flower heads made of small, individual flowers? The species is an objective category, because biologists can often test whether two different species can interbreed successfully in nature. But the higher levels—genus, family, and so on—are a matter of opinion, albeit the well-educated opinions of taxonomists. So how do they decide what to include in different genera or families?

The answer is that taxonomists usually want to lump together organisms in a way that reflects their evolutionary history—their position on a family tree of organisms, or a **phylogeny**. This method, in turn, is called the *phylogenetic approach*.

phylogeny
the study of the evolutionary history of different groups of organisms

Figure 10.6
The Hierarchy of Classification

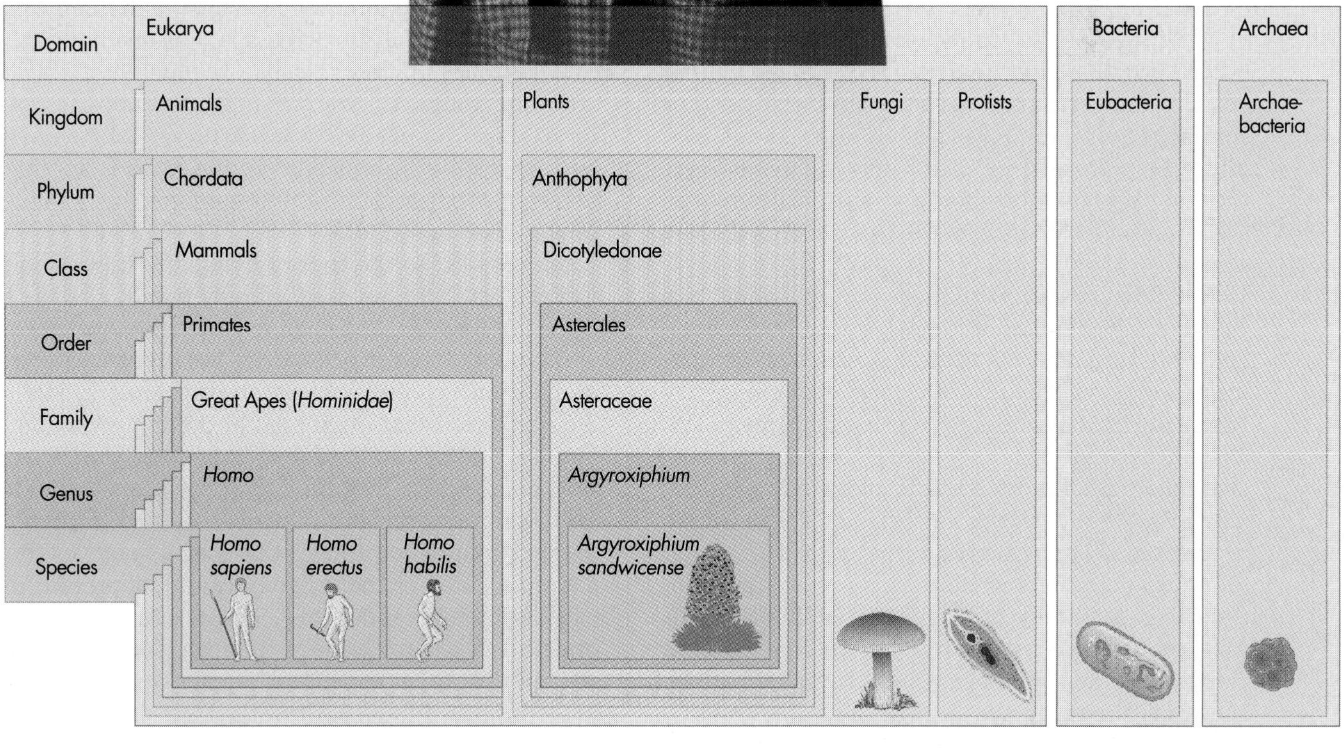

Domain	Eukarya				Bacteria	Archaea
Kingdom	Animals	Plants	Fungi	Protists	Eubacteria	Archaebacteria
Phylum	Chordata	Anthophyta				
Class	Mammals	Dicotyledonae				
Order	Primates	Asterales				
Family	Great Apes (*Hominidae*)	Asteraceae				
Genus	*Homo*	*Argyroxiphium*				
Species	*Homo sapiens*, *Homo erectus*, *Homo habilis*	*Argyroxiphium sandwicense*				

cladistics
a method of reconstructing patterns of descent using traits or characteristics thought to have evolved only within the group under consideration; for example, hair evolved only in mammals

clade
a taxonomic group made up of an ancestral organism and all the descendents derived from it

ancestral trait
a trait derived from an ancestral group

outgroup
a closely related species whose lineage diverged before all of the members of the group in question; used to infer ancestral traits

derived trait
a newly originated inherited change

analogous trait
a similar trait possessed by two different groups but not by ancestors of the groups

homologous trait
a similarity shared by two species due to inheritance from the last common ancestor

Figure 10.7
Using Cladistics: Identifying Shared Derived Traits

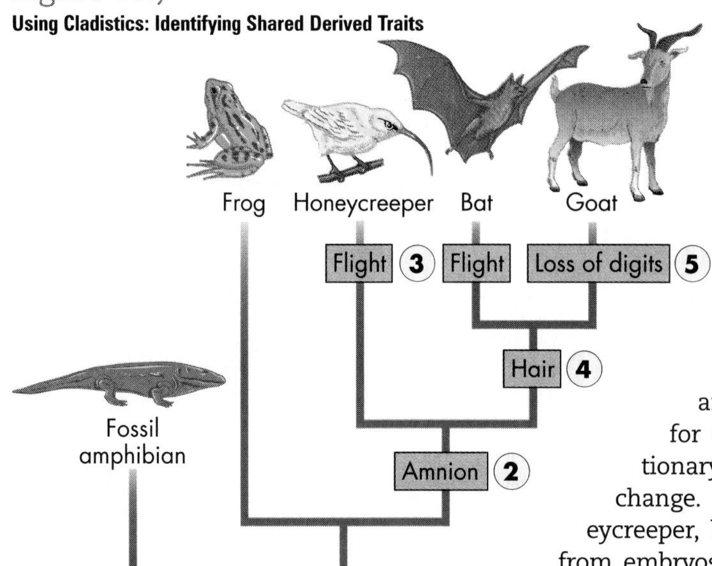

Cladistics: Classification by Shared Derived Traits

To make a family tree or phylogeny, it makes sense to put the most closely related species on the tips of nearby branches and more distantly related organisms at the tips of branches further down on the tree. So biologists need to identify groups sharing recent common ancestry, and to do this, they often use a procedure called **cladistics**. Cladistics is the study of **clades**, taxonomic groups made up of an ancestral organism and all the descendents derived from it.

Let's look at several animals—a frog, a small bird called a Hawaiian honeycreeper, a goat, and a bat—and work through a cladistic approach to constructing a family tree (Fig. 10.7). A cladist looks among the traits exhibited by such a group of organisms to find shared, derived traits. All four of these organisms share the trait of limbs, including legs and forelimbs (Fig. 10.7, Step 1). Thus, to a cladist, the presence or absence of limbs would not be a good character for determining relationships among the frog, honeycreeper, bat, and goat. Possession of limbs, in this case, represents an **ancestral trait**, a characteristic displayed by an ancient ancestor common to all the species in the group. We can infer ancestral traits by looking at an **outgroup**, a closely related species whose lineage diverged before all the members of the group in question. In this example, extinct fossil amphibians could serve as an outgroup; because they had limbs, we can conclude that limbs are ancestral.

In contrast to an ancestral trait, cladists look for a **derived trait**, an evolutionary novelty, a newly inherited change. For example, the honeycreeper, bat, and goat all develop from embryos enclosed in an amnion, a protective membrane (see Chapter 8). Frog embryos, however, lack an amnion. The amnion, therefore, is a trait that is shared by birds, bats, and goats and was derived (evolved) after they diverged from the frog lineage. The shared derived trait of the amnion groups the honeycreeper, bat, and goat in the same clade (Fig. 10.7, Step 2).

Next, the cladist looks among the species for additional shared, derived traits, but questions arise. Should we group bats with birds because they both fly? Or should we group bats with goats because they both have hair? The concept of shared, derived traits helps us out. Fossil evidence shows that flying evolved independently in ancestors of the bird and the bat. Flight in birds and bats is thus an **analogous trait**, a similar trait possessed by the members of two different groups but not by their last common ancestor (Fig. 10.7, Step 3). The yellow flower color shared by certain silverswords and certain roses is another analogous trait, and so is the streamlined body shape of porpoises and sharks.

In contrast to the trait of flying, the trait of hair is a good trait for building a family tree by cladistics (Fig. 10.7, Step 4). Body hair is a **homologous trait**, a similarity that two species share because they inherited it from a common ancestor. Furthermore, hair is a derived trait that originated in the lineage shared by bats and goats. A cladist, therefore, would use hair, a shared, derived trait, to group goats and bats in the same classification category. Finally, additional derived traits appeared in the goat lineage, as, for example, three of the original five digits present in each limb were lost (Fig. 10.7, Step 5). Usually taxonomists use as many traits as they can identify among the species under consideration, always identifying shared, derived traits to draw phylogenetic trees.

Figure 10.8

Molecular Phylogenies: A Powerful New Approach

(a)

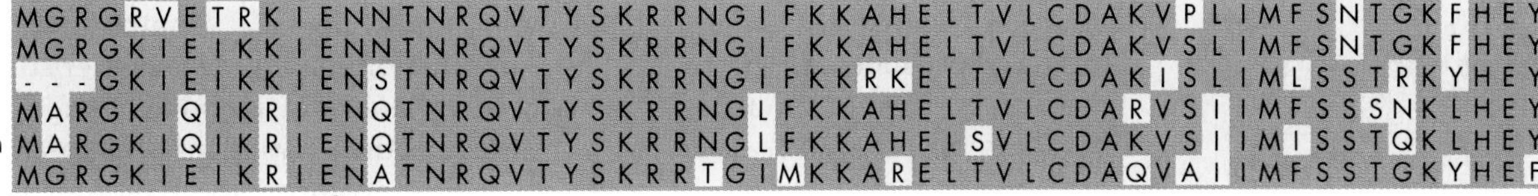

*Each of the 20 amino acids is represented by a different letter.

(b)

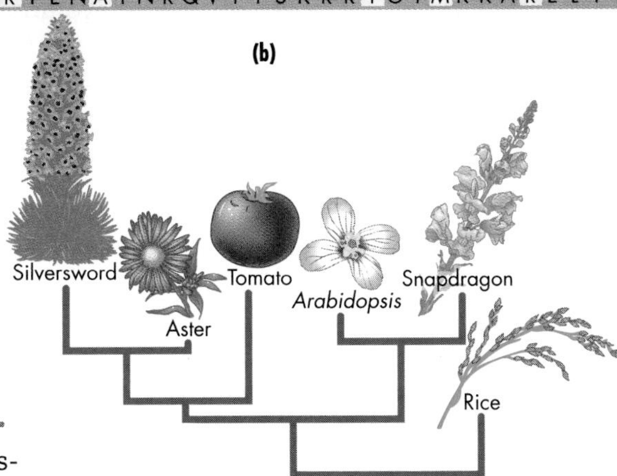

Molecular Phylogenies

Toes, hair, and other distinctive physical traits are very handy for classifying organisms, but most living things don't have limbs, let alone toes. All, however, have DNA and/or proteins, so biologists have recently been creating family trees by comparing DNA or protein structures. And instead of counting toes, they look at molecular traits such as the orders of nucleotides or amino acids. Biologists still look for shared, derived orders of subunits. For example, biologists isolated a silversword gene called *apetala-3* that helps control petal formation in the flowers. Once they isolated the gene, they looked at the protein it encodes and compared part of its amino acid sequence with those of petal-producing proteins in other plants (Fig. 10.8a). They found mostly identical sequences (shaded in the figure), but a few amino acids did differ from plant to plant. The silversword's petal protein was closer in amino acid sequence to the aster protein, so they grouped these together on the tree. They placed snapdragons on a lower branch, and placed rice, with the least similar sequence, diverging near the base of the tree. Gene phylogenies like these reveal that the closest relative to members of the Hawaiian silversword alliance is the tarweed, a flowering plant in the aster family living in California and the Pacific Northwest.

The Tree of Life

Two centuries ago, the great classifier Carolus Linnaeus considered all organisms to be either plants or animals. The more biologists observed living organisms, however, and the more they learned about their cellular and genetic structures, the more species they discovered that fit outside the plant and animal kingdoms. One of the biggest breakthroughs in understanding the relationships of all living organisms to each other came with the sequence comparisons of ribosomal RNA (rRNA). The sequences of rRNA base pairs (see Chapter 7) are more useful for constructing family trees than the amino acid sequences of proteins. This is because rRNA is found in all organisms and evolves rather slowly—so slowly that researchers can track the sequence changes and use them to determine the branchings that occurred in the distant past. This, in turn, allows them to construct more accurate family trees. Studies of rRNA confirm that all organisms alive today occupy the tips of branches stemming from three main trunks on the tree of life (Fig. 10.9). These three great lineages—or *domains*—consist of the Bacteria (or *Eubacteria*), the Archaea (or *Archaebacteria*), and the Eukarya (or eukaryotes).

The Bacteria and Archaea are both single-celled prokaryotes—cells lacking membrane-enclosed nuclei. Protists, plants, fungi, and animals all fit into the domain Eukarya because they all contain eukaryotic cells.

Bacteria
one domain including single-celled prokaryotes; the other such domain is Archaea

Archaea
a domain including single-celled prokaryotes that have cell membranes and other biochemical and genetic traits different than those of the domain Bacteria

Eukarya
the domain including all organisms whose cells possess a true nucleus and other membrane-bound organelles

Protista
generally single-celled organisms with nuclei, such as paramecia and amoebas

Fungi
the kingdom comprising multicellular heterotrophs such as mushrooms or molds, as well as unicellular yeasts, that decompose other biological tissues

Plantae
the kingdom comprising multicellular photoautotrophs such as mosses, ferns, and flowering plants

Animalia
the kingdom of animals

The base of the Eukarya contains a diverse group called the Protista. These are generally single-celled species with nuclei, such as *Euglena,* paramecia, and amoebas. The Fungi kingdom includes multicellular heterotrophs, such as mushrooms and molds, as well as unicellular yeasts. Fungi decompose other biological materials, develop from spores, and lack flagella. The *Plant* kingdom, or Plantae, is comprised of multicellular photoautotrophs, including some algae, and mosses, ferns, and flowering plants. Finally, the *Animal* kingdom, or Animalia, includes the multicellular heterotrophs, which lack cellulose, usually exhibit movement, and develop from embryos. Animals include the groups encompassing sponges, insects, clams, birds, and mammals.

Notice in Figure 10.9 the somewhat surprising fact that Archaea and Eukarya are more closely related to each other than either is to Bacteria. This reflects the belief that some ancient Archaean cell probably gave rise to the eukaryotic cell nucleus, or at least donated the nuclear genes that control genetic information processing (review Fig. 10.4). Another curious fact is that the familiar organisms we see every day—birds, trees, grass, flowers, dogs, mushrooms, flies—are just tiny branches at the periphery of the tree of life. The vast expanse of evolutionary diversity, determined through molecular analyses, lies in the single-celled organisms. Those simple living things are far more diverse genetically from each other and from larger organisms than are all the familiar plants, animals, and fungi.

In the next three chapters of the book we explore the kingdoms and domains in greater detail. Here, we look at the very serious problem of species loss.

LO5 The Biodiversity Crisis

Over the past 230 years, biologists have discovered, named, and recorded 2 million species of living organisms. This has been basically a hit-or-miss prospect, based on where naturalists

Figure 10.9
Domains, Kingdoms, and Life's Family Tree

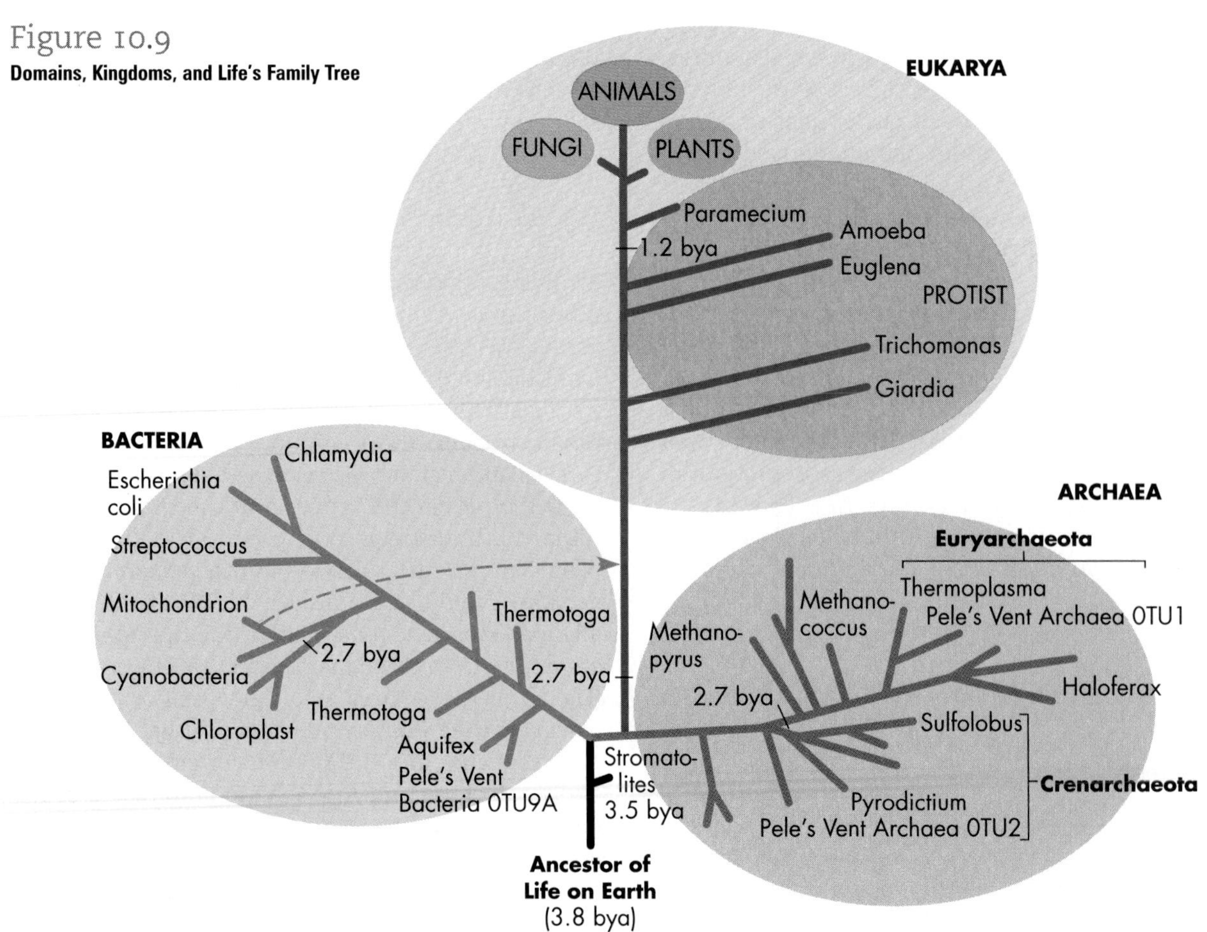

have gone, what they've been interested in collecting, and how much time (and funding) they've had for studying new and unknown organisms. Of the 2 million known species, about 5 percent are single-celled prokaryotes and eukaryotes. About 22 percent of the known species are fungi and plants. And about 70 percent are animals (Fig. 10.10). Most of the animals are invertebrates (they lack backbones). The sea harbors the most varied groups of invertebrates. By far the largest number of species are land-dwelling animals, and of these, over a million named so far are insects. Ours is a planet of terrestrial insects, and there are probably millions of species still to be discovered and classified.

For decades, biologists estimated that an accurate number of total species would be about 3 to 5 million, based on models of the likely diversity in temperate zones plus a multiplier of 2 for the number of tropical regions. This estimate lost favor, though, when biologists went out to study the deep ocean floor and the treetops of tropical rain forests. There could easily be 30 *million* species of insects and other invertebrates to describe in the rain forests alone—the vast majority undescribed and unnamed at this point. Adding in tropical plants, seabed organisms, and all the other undiscovered species in nontropical climatic zones, the figure of 50 million new species begins to look possible—even conservative.

Why should we care if Earth has up to 50 million species but only 2 million have been named? A minor reason is we're going to need a lot more biologists to discover and catalog these species before they become extinct as people rapidly disrupt and destroy large sections of tropical rain forests. Second, we could lose an irreplaceable genetic reservoir, including sources of new foods, drugs, and fibers for paper and cloth before we even find them. Third, we could get a tremendous ecological shock in the future, because we don't know how important this great degree of diversity is to the stability of all interdependent life. Is a huge number of species important for ecosystems to function? Are ecosystems with large numbers of species more stable and long-lived than those with fewer species? We think the answers are yes, but we don't know how many species are needed and how many can be lost before an ecosystem fails.

Figure 10.10
Life's Diversity: Planet of the Insects

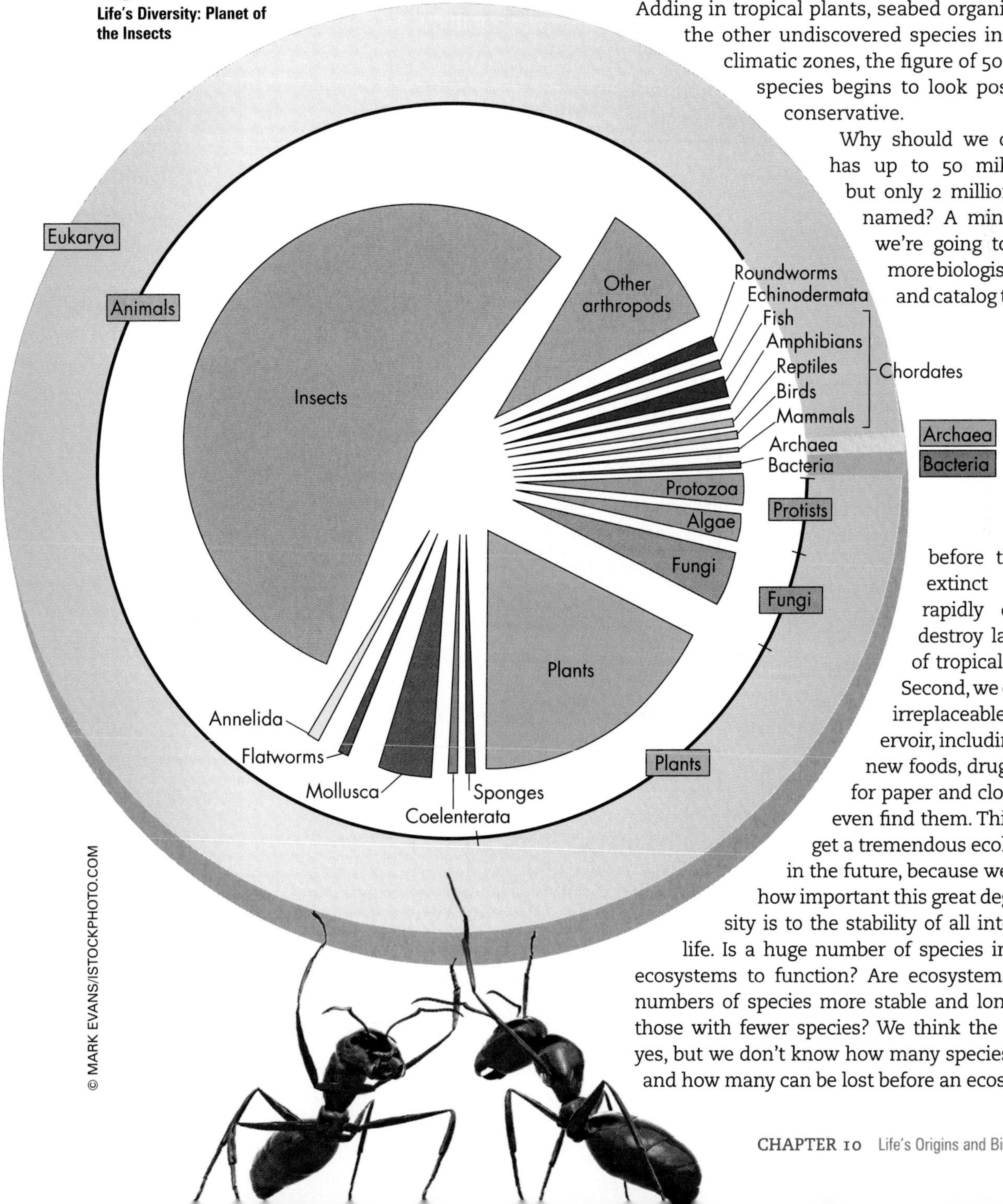

mass extinction the disappearance of large numbers of species

The Threat of Extinction

Biologists consider Hawaii and the other Pacific Islands to be among the world's 25 biodiversity "hot spots," small regions together containing the greatest percentage of Earth's biodiversity. Some other hot spots are the tropical Andes, the Mediterranean basin, and Indonesia and surrounding islands.

Just as many silversword species are threatened by people's farming and land development and by introduced grazers like pigs and goats, human activities are threatening biodiversity in the other 24 hot spots. In all these areas—tropical rain forests, coral reefs, drylands, and other life zones—people are destroying habitats faster than scientists can survey and describe them. Habitat destruction takes away interrelated plant, fungal, and bacterial systems as well as animals' feeding and nesting sites. When the last plant or the last breeding pair of a species disappears, the species becomes extinct. Since large areas contain more species than small areas, we can directly translate the loss of massive tracts of habitat into staggering species extinction. Some biologists estimate that up to 20 percent of the world's species could disappear over the next 30 years due to the doubling of human populations in tropical and semitropical zones. Species are disappearing at an alarming rate (more details in Chapter 14). The extent of the loss, in fact, may begin to approach some of prehistory's great extinction episodes. The main difference is that we are causing this one, and we're killing things off at a much faster rate of loss than in past extinction events. But the past events were unstoppable, and this time, we are making drastic changes to land, sea, and atmosphere that could be avoided.

We can, of course, learn from historic mass extinctions, the extinction of many groups of organisms over relatively short spans of geologic time. They can serve as models for what we can expect during and after the current biodiversity crisis. Figure 10.11 shows a series of mass extinction events that have punctuated the history of life. Following each episode, a burst of evolution brought about many new forms, replacing the extinct ones. A good example is the radiation of the mammals following the disappearance of the dinosaurs. Past extinctions also teach us that during a massive die-off, species don't disappear entirely at random. In the extinctions that took place about 10,000 years ago after the last ice age, land animals with large bodies—for example, mammoths,

Figure 10.11
Extinction Events and Radiation

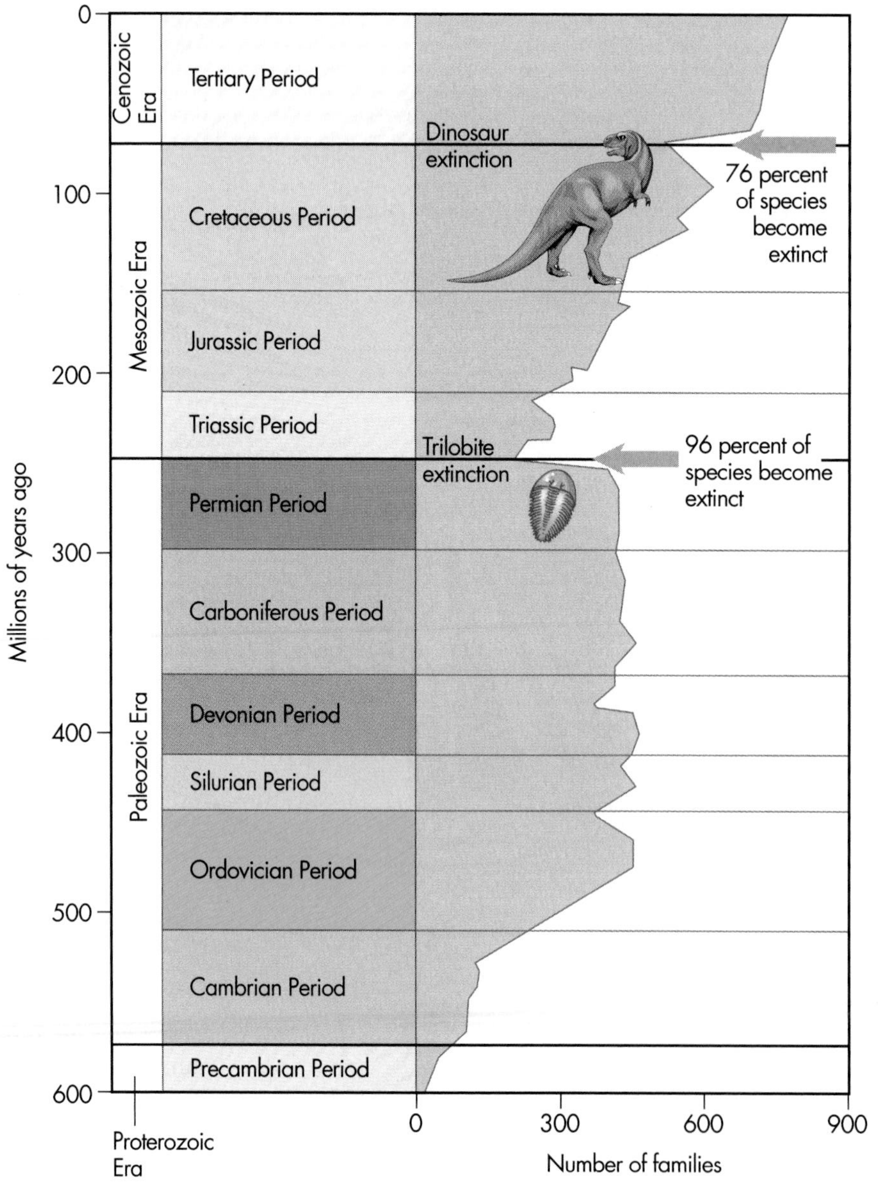

mastodons, and saber-toothed tigers—tended to die out faster than smaller land animals. This suggests that today's extinction crisis will claim many of our favorite animals first, including rhinoceroses, giraffes, pandas, and a number of whale species.

Another lesson from the past is that genera living in many large regions tend to survive mass extinctions, while genera localized to particular small regions are especially vulnerable. If a genus is confined to one small corner of the world, like the silverswords native to Haleakala volcano, then altering this small region will eliminate critical habitat and could easily endanger or wipe out the organisms in the genus. Rare, locally adapted species have tended to be our most useful sources for new medicines and foods.

A final lesson we can take from the fossil record of extinctions is that the permanent losses we cause in the next 50 years or so will not be recouped for millions of years.

The permanent losses we cause in the next 50 years or so will not be recouped for millions of years.

The Importance of Taxonomy

Along with the changes we've been wreaking upon ecosystems through pollution, habitat destruction, and the killing of organisms for food, hides, trophies, and wood, we've given new importance to the science of taxonomy. Our survival, and that of the millions of species we are currently endangering, depends on our understanding how ecosystems change as species die out and are replaced. As we just saw, though, we're missing even the roughest estimate of how many species exist, let alone a complete list of what they are and how they live, die, and interact in the environment.

As we've also mentioned, discovering, categorizing, and investigating new species has practical significance beyond pure biological understanding. Researchers have developed many cancer-controlling drugs and other modern medicines from chemical compounds in rare plants. Nutritious fruits, roots, and seeds are no doubt growing in obscurity around the globe right now, and some of those could well become staple foods of the future and an important adjunct to the three or four dozen staple crops our entire species depends for survival.

The Nature Conservancy, World Wildlife Fund, National Science Foundation, and many other organizations are directing and funding efforts in biodiversity hot spots. But governments and private groups must commit more resources to combating overpopulation, pollution, and habitat destruction. And as citizens we must approve the funding, inform ourselves, and participate in direct and indirect ways. Maybe some of the readers of *LIFE* will even become motivated to study taxonomy. It is truly a science with a deadline.

Who knew?

The changes that have occurred due to the activities of living organisms can be linked to the evolution of oxygen-generating photosynthesis; these changes include the precipitation of iron from seawater, the creation of the outer atmospheric blanket of ultraviolet-absorbing ozone, and the relatively high percentage of molecular oxygen in the modern atmosphere. Without life, iron would still be abundant in ocean water, there would be no ozone layer to protect Earth's surface from ultraviolet radiation, and the atmosphere would have practically no free oxygen.

CHAPTER 11

Single-Celled Life

Learning Outcomes

LO[1] Describe the biological properties and diversity of prokaryotes

LO[2] Identify the relationship between living cells and noncellular biological entities

LO[3] Identify the relationships of protists to each other and to plants, animals, and fungi

LO[4] Explain how a knowledge of single-celled life forms contributes to the fight against disease

"Many prokaryotic organisms can cause diseases, but far more prokaryotes benefit the environment and our bodies."

The Wily Scourge: Malaria

"In the middle of the night," Mark, a volunteer teacher in Nigeria, recalls, "I woke up with a pounding headache. I felt like someone was jackhammering inside my skull. It was painful to move—in fact, to exist! My forehead was burning up, but I felt chilled off and on, too. I started looking back at my life," he says, convinced that he was dying.

What do you know?

Why are antibiotics effective against bacteria but not against viruses?

Mark dragged himself to a nearby hut and awakened his sleeping friend, a nurse named Linda. After hearing Mark's symptoms, she drowsily pronounced them "just malaria." A blood test confirmed the presence of malaria parasites in his blood. For a week he took the classic malaria treatment, quinine, which he recalls as "nasty, bitter stuff that makes you urinate and makes your ears ring." Eventually, the infection ended. But another began just before Mark finished his two-year commitment.

Mark knew the risks of malaria he faced in rural Nigeria—the same, in fact, as throughout much of Africa and parts of South America and Asia. "Every week or two, some student or teacher came down with malaria," he recalls, and one student's infant daughter died from complications. What Mark found surprising was that he had caught malaria twice, even though he was taking preventative medicine every week. The drug he was taking "was supposed to be 90 percent effective," Mark said, "plus I slept under a mosquito net every night and my house had good screens—all items that most Africans can't afford." Yet still he caught two cases of malaria.

"The disease is a major catastrophe for Africa," says Dr. Daniel Goldberg, a well-respected malaria researcher. "It afflicts nearly 500 million people a year there and on other continents and kills about 2 million a year, mostly children." Researchers like Goldberg believe that the elusive malaria parasite, *Plasmodium falciparum,* has caused half of all human deaths throughout history, making it our species' master scourge.

How can a single-celled organism slay and debilitate with such impunity? You'll find out as you explore this chapter's topic: the evolution and diversity of single-celled life forms. You'll see that *P. falciparum* is but one fascinating species among millions of microbes, and you'll learn about its interrelationships with and differences from other protists, prokaryotes, and viruses. By touring this largely microscopic realm, you'll find answers to these questions:

- ☑ Why are prokaryotes so important to life on Earth?
- ☑ Where do viruses fit into the evolutionary tree of single-celled organisms?
- ☑ How is the malaria parasite related to prokaryotes, to other protists, and to larger organisms?
- ☑ How can a knowledge of prokaryotes and protists help researchers and doctors fight humankind's greatest killer?

bioremediation bacterial breakdown of environmental pollutants

LO[1] Prokaryotes and Their Importance to Life on Earth

Malarial parasites are protists, eukaryotic single-celled organisms that are members of a stunningly diverse kingdom. To fully appreciate the protists—especially complex ones like *Plasmodium falciparum*—we need to set the stage first with the simpler single-celled organisms called *prokaryotes*. This group is even more diverse than the protists in terms of their genetics, biochemistry, and habitats—so much so that it takes two domains, Bacteria and Archaea, as well as several kingdoms to encompass them (Fig. 11.1). Many prokaryotic organisms can cause diseases, but far more prokaryotes benefit the environment and our bodies. One type may even help save people from malaria!

Prokaryotes are organisms whose genetic material is not contained within a nuclear envelope (see Fig. 2.10). Prokaryotic cells are usually much smaller than eukaryotic cells. For example, an intestinal prokaryote such as *Escherichia coli* is only about one one-thousandth the size and volume of a malarial parasite. Despite their small size, though, prokaryotes are profoundly important to the environment, medicine, and industry. Let's start by exploring their significance, then look at their general biological properties and diversity. As you read along, note that in the past, biologists used the word "bacteria" (little "b") to refer to all prokaryotes. Today, however, they use "prokaryote" for members of both domains Bacteria and Archaea, and use the capital "B" to designate the domain name, Bacteria (or as some biologists say, Eubacteria).

Importance of Prokaryotes

Prokaryotes have an immense ecological impact. Billions of prokaryotic cells in nearly every square meter of soil, water, and air release oxygen into the atmosphere and recycle carbon, nitrogen, and other elements. Prokaryotes consume enormous quantities of dead animals, fungi, plant matter, and human and animal wastes. They can also consume pesticides and pollutants that would otherwise poison our environment.

For decades, biologists have documented the prokaryote's enormous natural recycling role. More recently, field researchers have developed ways to encourage more bacterial breakdown of environmental pollutants (a process they call **bioremediation**). After oil spills, cleanup crews can seed the contaminated shoreline with a "fertilizer" containing nitrogen and phosphorus. These nutrients, in combination with the organic compounds in the shoreline sludge, promote the growth of local prokaryotes with a natural appetite for greasy hydrocarbons, and within a couple of weeks, the sprayed areas dramatically improve.

Prokaryotes have great medical importance, both negative and positive, in addition to their immense environmental role. They cause hundreds of human diseases, including staph and strep infections, blood

Figure 11.1
Family Tree of the Prokaryotes

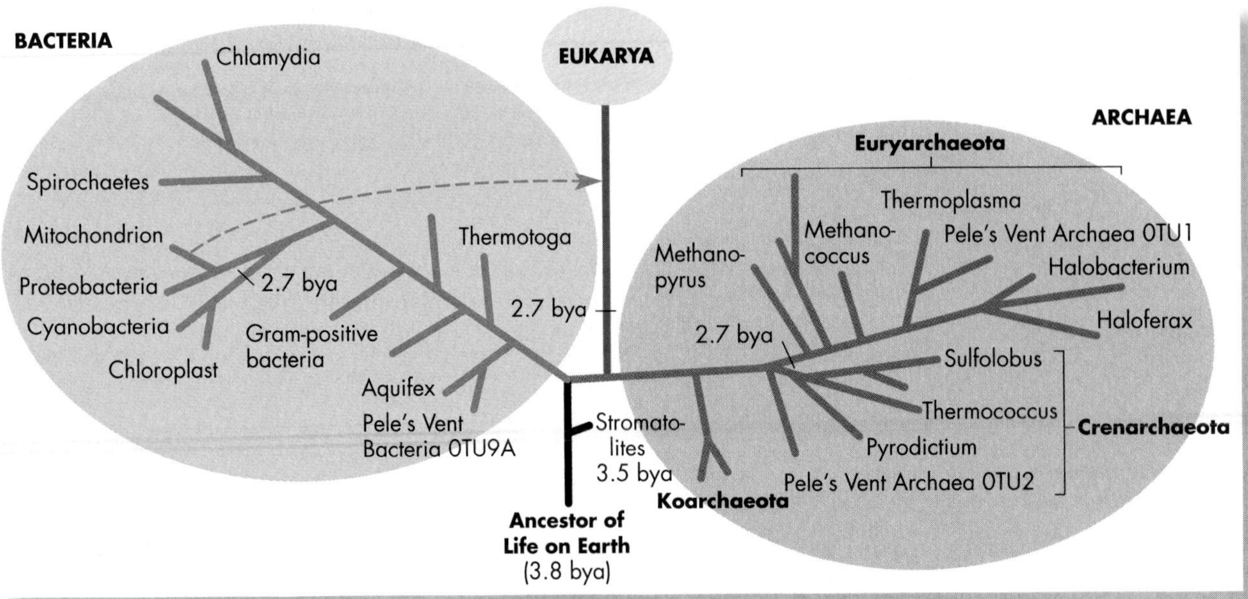

poisoning, tetanus, venereal diseases, and dental caries, as well as thousands of diseases in other animals and plants.

Despite this tremendous negative medical impact, harmful or pathogenic species are a small fraction of all prokaryotes, and many of the single-celled organisms supply materials crucial to our health and survival. For example, *Escherichia coli* bacteria, along with other inhabitants of the human gut, produce vitamins K and B_{12}, riboflavin, biotin, and various cofactors that we probably absorb and use. Residents like *E. coli* may also blanket the intestinal walls so heavily that they block other dangerous microbes from passing into the bloodstream. Most plant-eating animals—cattle, sheep, rabbits, koalas, termites, and others—couldn't digest grasses, leaves, or wood without prokaryotes in their intestines actually breaking down the cellulose. You've probably demonstrated the importance of your own flora (as doctors call your internal microbes) if you've ever taken large amounts of antibiotics for an infection. The drugs probably killed off the beneficial organisms in your intestines and reproductive system and left you, temporarily, with digestive difficulties, diarrhea, and/or yeast infections (oral, vaginal, etc.) until the normal inhabitants returned and re-established a balance.

In the food and chemical industries, single-celled organisms help to produce cheese, yogurt, pickles, soy sauce, chocolate, and other foods, and to generate supplies of chemical compounds, such as butanol, fructose, and lysine. Commercial applications aside, much of what we know about how cells function comes from studying prokaryotes. They have small genomes and fast division rates, and are easy to grow in the laboratory. These traits have made the tiny cells favorites for investigating biochemistry, molecular biology, and genetic engineering—all fields that rely on enzymes, chromosomes, plasmids, and other components harvested from bacteria (see Chapter 7). Without their microscopic research subjects, biologists would be decades behind their current state of research.

Let's see, now, what all prokaryotes have in common. Let's also see what makes them so important to the environment and to other organisms.

Figure 11.2
Prokaryotic Cell

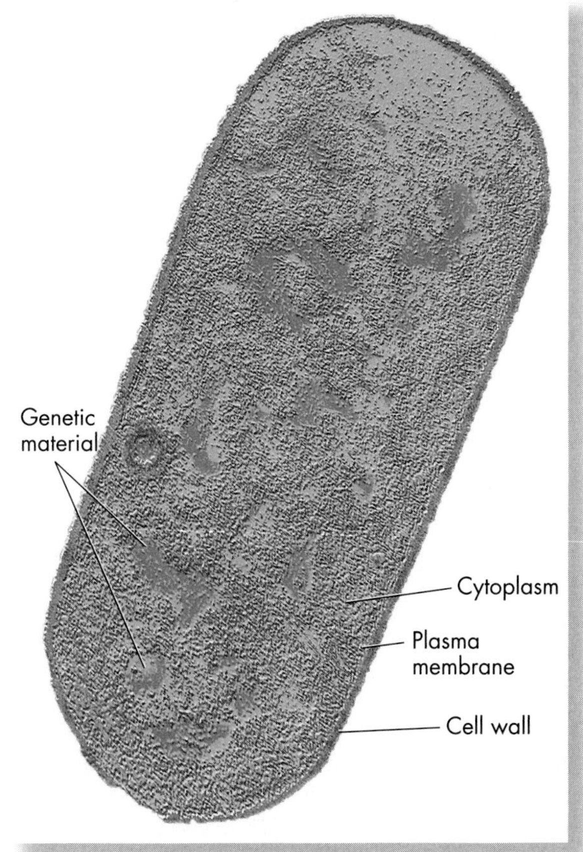

General Structure of Prokaryotic Cells

Biologists recognize thousands of prokaryotic species and assign them to the domains Bacteria and Archaea (Fig. 11.1). Bacteria and Archaea differ in basic ways, but the traits they share define the prokaryotic organism.

Cell Structure

Prokaryotic cells have an outer cell wall that surrounds the plasma membrane (Fig. 11.2). This membrane, in turn, surrounds a noncompartmentalized cytoplasm dotted with ribosomes. Prokaryotic cells lack a nuclear envelope and other membrane-enclosed organelles. A prokaryote's chromosome is a circular strand of DNA that is loose, not tightly complexed with spool-forming proteins like eukaryotic chromosomes (see Fig. 6.5).

Plasma Membrane

In prokaryotic cells, regions of the plasma membrane carry out roles accomplished by specific organelles in eukaryotic cells. Cellular respiration, for example, takes place on a part of the plasma membrane that folds inward toward the cell's interior, rather than in a mitochondrion. Likewise, photosynthesis takes place on infolded membranes instead of in separate chloroplasts. The membranes themselves are different, too; in Archaea, the membrane lipids link differently to the three-carbon head of the molecule (review Fig. 2.11).

peptidoglycans sugar-protein complexes occurring in prokaryotic cell walls

gram-positive cells prokaryotes containing peptidoglycans in a single broad layer; do pick up Gram's stain

gram-negative cells prokaryotes in which the peptidoglycan layer is covered by an outer sheet of proteins and lipopolysaccharides; don't pick up Gram's stain

Gram's stain a special stain that distinguishes gram-positive and gram-negative organisms

coccus (pl. cocci) a bacterial cell with a spherical shape

bacillus (pl. bacilli) (L. *bacilli,* a rod) a bacterial cell with a rod shape

spirillum (pl. spirilla) a bacterial cell with a spiral shape

vibrio (pl. vibrios) a bacterial cell with a curved rod shape

heterotroph (Gr. *heteros,* different + *trophos,* feeder) an organism, such as an animal, fungus, and most prokaryotes and protists, that takes in preformed nutrients from external sources

Cell Wall

In prokaryotes, the cell wall performs functions achieved by the eukaryote's cytoskeleton (see Fig. 2.10). In species within the domain Bacteria, the prokaryotic cell wall is a strong but flexible covering, made primarily of sugar-protein complexes called peptidoglycans. In so-called gram-positive organisms, these complexes occur in a single broad layer, while in gram-negative organisms, the peptidoglycan layer is covered by an outer sheet containing proteins and lipopolysaccharides (fat-sugar complexes). Species in the domain Archaea may have cell walls made of polysaccharide, protein, or glycoprotein, but they never contain true peptidoglycans.

Based on these cell wall variations and the differences they cause in permeability, gram-positive cells turn purple when exposed to a special stain called Gram's stain, while gram-negative cells turn reddish (Fig. 11.3). This staining technique is one of the first procedures a microbiologist will perform to help identify a newly found or unfamiliar prokaryote. Gram staining is important because it helps microbiologists classify prokaryotic organisms into groups of species. It's also important because gram-positive and gram-negative organisms are usually sensitive to different types of antibiotics (chemicals made by one microorganism that can slow the growth of or kill another microorganism).

Gram-positive cells

Gram-negative cells

© RICHARD J. GREEN/PHOTO RESEARCHERS, INC.

Figure 11.3
Gram Staining and Prokaryotic Cell Walls

BACTERIA
EUKARYA
ARCHAEA
Ancestor

Gram-positive and gram-negative organisms are usually sensitive to different types of antibiotics.

Forms of Bacterial Cells

Prokaryotes are usually spheres (cocci), rods (bacilli), spirals (spirilla), or curved, comma-shaped rods (vibrios) (Fig. 11.4). A pro-karyote's name will often give a clue as to its shape. For example, (a) *Enterococci* is spherical, (b) *Bacillus anthracis,* the cause of anthrax, is rod-shaped, (c) *Leptospira* is a slender spiral, and (d) *Vibrio cholerae,* the cause of cholera, is comma-shaped.

Nutrition in Prokaryotes

As a group, prokaryotes have great ecological versatility, able to consume an enormous range of energy sources, including biological materials and organic substances such as methane; inorganic substances, such as hydrogen sulfide; industrial chemicals such as hydraulic fluids and toxic herbicides; and cancer-causing wastes, including vinyl chloride and polychlorinated biphenyls (PCBs). They can also live in air, soil, water, and ice; inside rocks miles under the surface of the Earth; and in the bodies of virtually all other life forms.

Most prokaryotes are heterotrophs, obtaining organic nutrients from the environment. Many are

Figure 11.4
Diversity of Shapes among Bacteria and Archaea

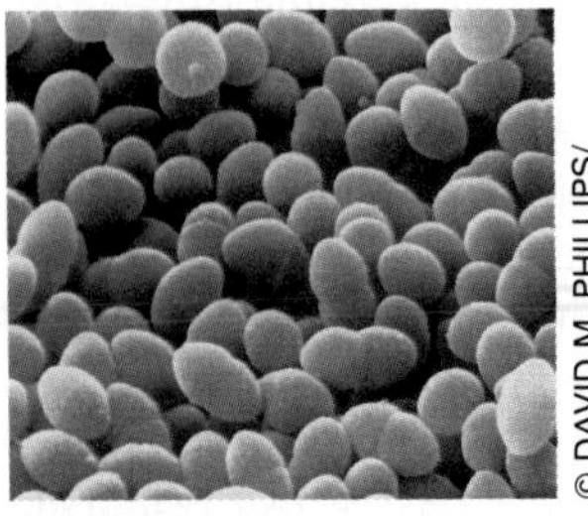

© DAVID M. PHILLIPS/PHOTO RESEARCHERS, INC.

(a) spheres (cocci)

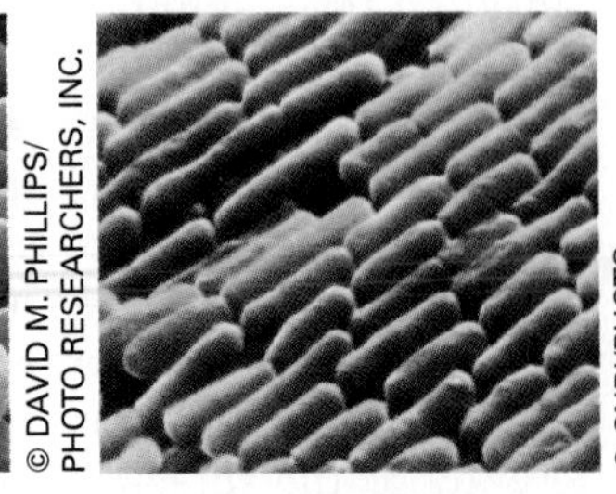

© CNRI/PHOTO RESEARCHERS, INC.

(b) rods (bacilli)

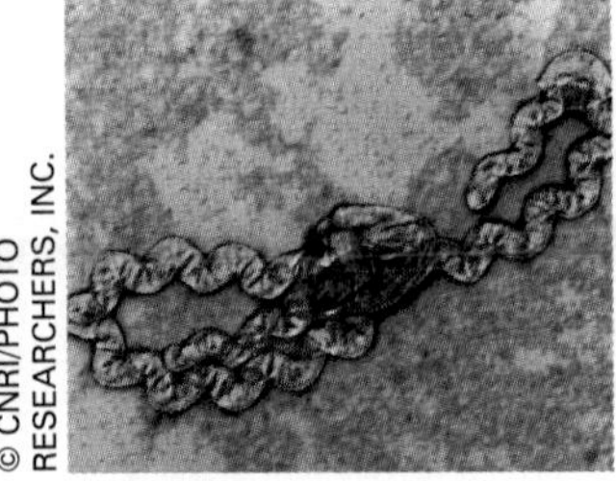

© EYE OF SCIENCE/PHOTO RESEARCHERS, INC.

(c) spirals (spirilla)

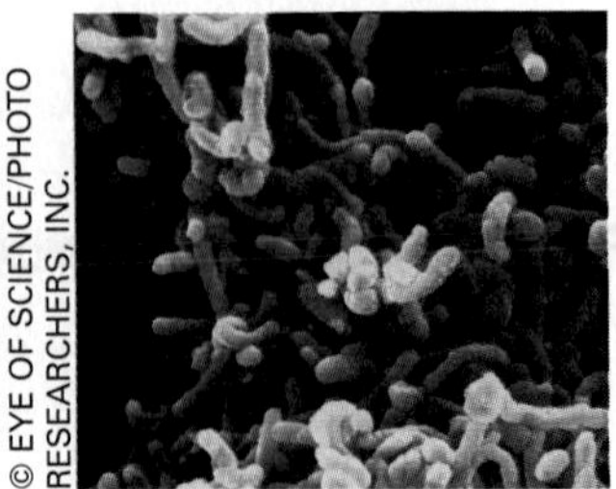

© STEM JEMS/PHOTO RESEARCHERS, INC.

(d) curved rods (vibrios)

also saprobes, that is, they live on dead or dying organisms and act as decomposers.

Thousands of prokaryotic species are symbionts (literally, "organisms that live together"), and reside on or inside other living organisms. The bacteria residing normally in your intestines are beneficial symbionts. Most pathogenic (disease-causing) bacteria are parasites that do harm to their hosts.

A few prokaryotes are autotrophs and make their own organic nutrients from inorganic materials in the environment. These include photoautotrophs, which capture energy from light. Green and purple photosynthetic bacteria and cyanobacteria are photoautotrophs. Others are chemoautotrophs, which capture energy from certain chemicals. Sulfur bacteria and methane bacteria are chemoautotrophs that can obtain their carbon from carbon dioxide and their energy from the bonds of inorganic compounds, such as hydrogen gas, hydrogen sulfide, sulfur, iron, and certain nitrogen compounds. As we saw in Chapter 10, chemoautotrophs living around deep ocean hydrothermal vents may resemble some of the earliest living cells.

Reproduction in Prokaryotes

Prokaryotes usually reproduce by binary fission, simply dividing in half into two identical offspring. Binary fission can be incredibly fast and efficient, with division occurring as often as every 20 minutes. For example, if a single *E. coli* bacterium and all its progeny continued dividing every 20 minutes, after just 48 hours there would be a mass of bacteria 4,000 times the weight of Earth. This doesn't happen, of course, because the availability of food, oxygen, and other resources limits their growth.

Although prokaryotes usually divide in half, they have several means of genetic recombination, including *conjugation,* the direct exchange of DNA through a strand of cytoplasm that bridges two cells, and indirect exchanges, such as transduction and transformation. *Transduction* is the transfer of genes from one bacterium to another via a virus, while *transformation* is the transfer of genes by the uptake of DNA directly from the surrounding medium. In Chapter 6, we saw how biologists use transformation to introduce recombinant DNA into foreign host cells (see Fig. 6.2).

Many prokaryotes have yet another reproductive trick that enables them to survive unfavorable conditions. Some species such as those that cause botulism, gas gangrene, and anthrax form small, tough-walled resting cells called endospores. Endospore formation is triggered by worsening environmental conditions. Endospores can survive extremes of heat, cold, drought, and even radiation for long periods; when conditions improve, they grow into new bacterial cells. Endospores are the reason surgical instruments must be sterilized with high heat and pressure, and also why canned foods must be processed exactly the right way.

The Domain Archaea

The organisms most similar to the earliest living cells are members of the domain Archaea. Most Archaea are anaerobes, cells that can't survive in the presence of oxygen. Many also inhabit Earth's most hostile environments and have pushed to the extreme the limits of a "livable" habitat. Biologists call such organisms extremophiles—extreme-loving organisms. Extremophiles can live in near-boiling water in hot springs, extremely salty lakes, and highly acidic or alkaline water and soils. Some Archaea also have weird biochemical properties. For instance, methanogens produce methane (CH_4, natural gas) as an essential component of their metabolism. Methanogens include the species *Methanobacterium ruminantium,* which inhabits a cow's digestive tract. Cows belch into the atmosphere the methane gas these prokaryotes produce, and it actually

saprobe
an organism that lives on decomposing organic matter

symbiont
an organism that lives in a close relationship with an organism of another species

parasite
a type of predator that obtains benefits at the expense of another organism, its host; a parasite is usually smaller than its host, lives in close physical association with it, and generally saps its host's strength rather than killing it outright

photoautotroph
an organism that captures energy from light

chemoautotroph
an organism that derives energy from a simple inorganic reaction

binary fission
asexual reproduction by division of a cell or body into two equivalent parts

endospores
heavily encapsulated resting cells formed within many types of bacterial cells during times of environmental stress

anaerobes
cells that don't require the presence of oxygen for every harvest

extremophiles
prokaryotes that survive in Earth's most extreme environments

methanogen
a type of archaebacterium that produces methane as a metabolic byproduct

© AMY STEBBINS/ISTOCKPHOTO.COM

thermophiles prokaryotes that thrive in very hot conditions

halophile a type of archaebacterium that can tolerate extremely high salt concentrations

actinomycetes single-celled organisms that produce antibiotic compounds

mycoplasma a type of the smallest free-living cells, these simplified members of the domain Bacteria lack cell walls, live inside animals, plants, and sometimes other single-celled organisms, and can cause a dangerous form of pneumonia as well as infections of the urinary tract and other organs

spirochetes bacteria with a distinctive spiral shape

chlamydia bacterial species that live inside animal cells and lack an ability to make their own ATP; can cause a sexually transmitted disease

cyanobacteria (sing. cyanobacterium) one of the blue-green algae; a photosynthetic, oxygen-generating, and nitrogen-fixing prokaryote

proteobacteria the largest and most diverse group in the domain Bacteria

rickettsias tiny, rod-shaped parasitic bacteria

capsid the protein coat that encases a virus

contributes a huge volume of a greenhouse gas, adding to our global climate change problem (which you'll read more about in Chapter 15).

Diversity among the Archaea

The domain Archaea consists of three main branches, which microbiologists call *kingdoms* (see Figure 11.1):

- Korarchaeota—only known by fragments of their DNA found in boiling hot springs; reside on some of the lowest branches on the tree of life and may be similar to the most recent ancestor of all surviving life forms on Earth
- Crenarchaeota—includes **thermophiles** (heat lovers) that use sulfur as an electron acceptor in their energy metabolism rather than oxygen (as most species do)
- Euryarchaeota—includes *methanogens,* methane-generating organisms, and extreme salt-loving organisms or **halophiles**

While the Archaea are a diverse and ancient domain of life, species in the domain Bacteria rival their variety and are much better known.

The Domain Bacteria

Biologists have determined that there are about 12 major evolutionary branches among the Bacteria (not all shown on Fig. 11.1). The most important groups of Bacteria of the 12 major branches based on RNA genes include:

- Primitive bacteria
  - *Aquifex*—extremophiles like the most primitive Archaeans; oxidize hydrogen or reduce sulfur in metabolism
  - Thermotoga—lives in nearly boiling or merely scalding conditions
  - *Deinococcus*—resistant to high doses of irradiation
- Gram-positive bacteria (all others listed are gram-negative)
  - *Streptococcus mutans*—causes tooth decay
  - *Streptococcus pyogenes*—causes strep throat
  - *Staphylococcus aureus*—causes staph infections
  - **Actinomycetes**—produce antibiotics, natural substances that fend off competing microorganisms
  - **Mycoplasmas**—smallest free-living cells, lack cell walls and live inside animals, plants, and other single-celled organisms; cause "walking pneumonia"
  - *Bacillus thuringiensis*—may provide a new means of controlling malaria; contains a gene for a protein called *Bt,* a toxin that accumulates in tiny crystals and kills insects that consume them
- **Spirochetes**—include the agents that cause Lyme disease and syphilis
- ***Chlamydia***—causes the most frequent sexually transmitted disease in North America
- **Cyanobacteria**—formerly called *blue-green algae;* contain chlorophyll *a,* which absorbs light in photosynthesis; contribute millions of tons of oxygen to the atmosphere and biologically usable nitrogen and carbon to the environments and organisms around them. Ancient cyanobacterium probably entered symbiotic relationship with eukaryotic cell and evolved into chloroplasts in green algae, which evolved into plants.
- **Proteobacteria**—largest and most diverse group in the domain Bacteria; includes
  - *Rhodospirillum*—purple in color, carry out photosynthesis with chlorophyll pigments very different from those in plant cells. Ancient purple bacterium probably gave rise to mitochondria in eukaryotes (review Fig. 10.4).
  - *E. coli*—common intestinal bacterium
  - *Rhizobium*—fix nitrogen in peas and beans
  - **Rickettsias**—parasitic bacteria transmitted by ticks, fleas, and lice; the cause of typhus fever, Rocky Mountain spotted fever, and other serious infections

LO2 Noncellular Biological Entities

Most of the organisms we've discussed so far consist of cells. This section explores biological entities, including viruses, virions, and prions, that aren't organized as cells and aren't alive by the characteristics of the living state we described in Chapter 1. Instead, they are nonliving parasites on and in living cells that probably evolved from them.

Viruses: Infectious Genes with a Coat

Viruses are geometric packages of genes that are 1,000 to 10,000 times smaller than the cells of *B. thuringiensis* or other prokaryotes. A virus particle is, in effect, a minute package of DNA or RNA surrounded by a protein coat, or **capsid**, and occasionally by other materials. Viruses infect cells by attaching part of the capsid to the cell's exterior wall or plasma membrane and then either entering the cell or simply injecting their DNA or RNA into the cell, leaving the capsid outside. Once inside, the viral genes commandeer the cell's protein-synthesizing machinery, sometimes stopping production of cellular proteins entirely and preempting the machinery for the production of new virus particles. Eventually, the cell may burst and die, or may survive and gradually release thousands of viruses that can then infect other cells. Most biologists consider viruses to be nonliving because these agents lack the machinery for self-reproduction or metabolism.

Treating Viral Disease

Because viruses aren't cells, they aren't killed by the antibiotics that control most disease-causing prokaryotes. Right now, in fact, there are really no medical cures for any viral diseases. One of the biggest challenges in modern medicine is to develop drugs that can fight viruses.

Figure 11.5
Protists and the Tree of Life

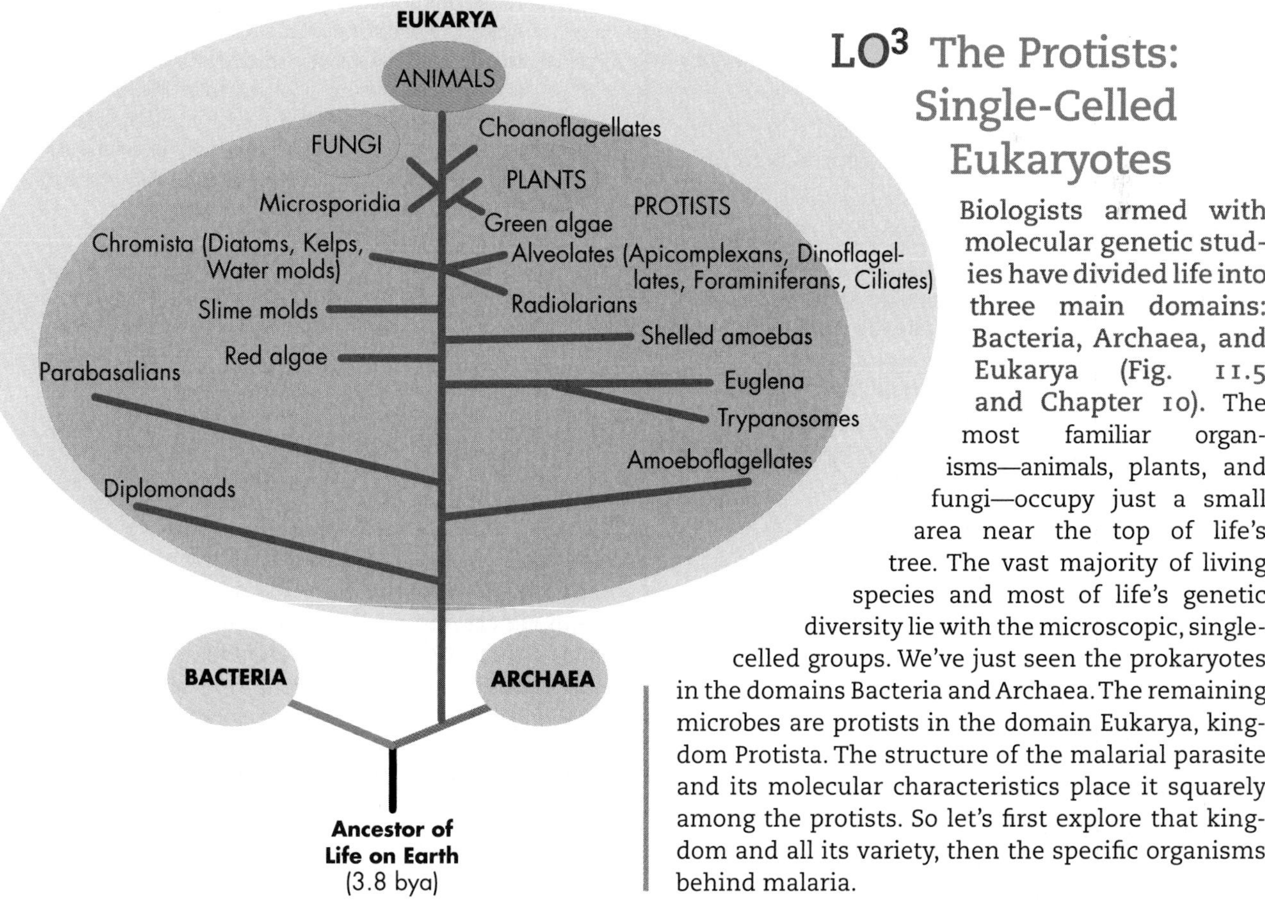

Viroids: Infectious RNAs

viroid
an intracellular parasite that affects plants and consists only of small RNA molecules without any protein coat

prion
an intracellular disease-causing entity apparently consisting only of protein and having no genetic material

The viroids are parasites that live inside plant cells and consist of a single molecule (an RNA). They are far smaller than the smallest virus. Viroid RNAs bind to RNAs in the host cell's ribosomes and disrupt the cell's protein synthesis. Some diseases of potatoes, cucumbers, citrus trees, and artichokes stem from viroid infection.

Prions: Infectious Proteins

Like viroids, prions also consist of a single molecule, but it is one of protein rather than RNA. Prions are the smallest disease-causing agents that can be transmitted from one animal to another. They are implicated in serious nerve and brain diseases, including mad cow disease in cattle. Biologists are not sure how prions reproduce or cause disease, and the main protection against them now is to avoid eating animal matter that might harbor prions.

LO3 The Protists: Single-Celled Eukaryotes

Biologists armed with molecular genetic studies have divided life into three main domains: Bacteria, Archaea, and Eukarya (Fig. 11.5 and Chapter 10). The most familiar organisms—animals, plants, and fungi—occupy just a small area near the top of life's tree. The vast majority of living species and most of life's genetic diversity lie with the microscopic, single-celled groups. We've just seen the prokaryotes in the domains Bacteria and Archaea. The remaining microbes are protists in the domain Eukarya, kingdom Protista. The structure of the malarial parasite and its molecular characteristics place it squarely among the protists. So let's first explore that kingdom and all its variety, then the specific organisms behind malaria.

diplomonads the group of protists that includes *Giardia,* a common human parasite

amoeboflagellates single-celled protists that live in water and soil, and usually display pseudopodia

pseudopodia limblike cellular extensions that help protists and certain blood cells move and feed

parabasalian ancient, primitive protists such as the organisms inside a termite's gut

kinetoplastids primitive protists with long whiplike flagella; includes the protist causing African sleeping sickness

euglenoids green, spindle-shaped protists with eyespots

stigma [1] the tiny, light-sensitive eyespot of a euglenoid; [2] The sticky top of a flower that serves as a pollen receptacle

true slime mold a type of funguslike protist characterized by a plasmodium, a mass of continuous cytoplasm surrounded by one plasma membrane and containing many diploid nuclei

plasmodium one form of a true slime mold that is a mass of continuous cytoplasm surrounded by one plasma membrane that moves slowly, like a giant amoeba; also, the genus of malarial parasites

cellular slime mold a type of funguslike protist, usually existing as free-living amoebalike cells, but aggregating into a multicellular fruiting body before producing reproductive spores

fruiting bodies spore-producing reproductive structures in many fungi

red algae small delicate aquatic protists that occur as thin filaments or flat sheets and produce red pigments

What Is a Protist?

Because of the many forms, biologists tend to define protists more by what the organisms are *not* than what they are and do:

1. Protists don't develop from embryos;
2. Protists don't have extensive cell differentiation among cell types of a species; and
3. Protists aren't animals, plants, or fungi.

Ancient Protistan Lineages

Recall the endosymbiont hypothesis from Chapter 10. This is the idea that the first eukaryotic cells may have arisen after the nucleus evolved and after a simple bacterium took up residence and led to mitochondria. Not long after that, ancestral eukaryotes began to live in the primordial seas and a number of lineages of protists diverged from them and diversified.

- **Diplomonads**—includes *Giardia,* a common human parasite found in untreated water
- **Amoeboflagellates**—live in water and soil, and usually display **pseudopodia** (literally, "false feet"), extensions that help the organisms move and feed
- **Parabasalians**—includes bacteria that produce and secrete cellulase, an enzyme that breaks down the cellulose in wood into glucose subunits (gives termites their ability to digest wood)
- **Kinetoplastids** (trypanosomes)—have long whiplike flagella that move the cell; a kinetoplastid called *Trypanosoma brucei gambiense* causes African sleeping sickness, spread through the bite of the tsetse fly; can evade the host's immune system by changing their surface coats—a trait we'll encounter again with the malarial parasite
- **Euglenoids**—like kinetoplastids, have long whiplike flagella that move the cell; named after *Euglena,* from the Greek words for "good eye" because each euglenoid has an eyespot, or **stigma**, with a light receptor that allows it to swim toward light to maximize photosynthesis

Multicellular Protistans Evolve

Most protists are single-celled organisms—but not all. Some protists stemming from the first primitive eukaryotes evolved a multicellular body form (or aspects of it) completely independently of the many-celled plants, animals, and fungi. The most common are the slime molds and the red algae.

- Slime molds—harvest food and energy by secreting digestive enzymes onto organic matter and then absorbing the digested material back into the cell
  - **True slime mold**—exists as a flat, fan-shaped mass of cells called a **plasmodium**; moves like a giant amoeba; has just one enormous plasma membrane surrounding thousands of diploid cell nuclei, which divide simultaneously during mitosis
  - **Cellular slime mold**—like true slime molds, move like amoeba; actively engulf and consume bacteria rather than secreting digestive enzymes onto organic matter. When food is scarce, the individual cells of a cellular slime mold join together and send up brightly colored, multicellular **fruiting bodies**, reproductive structures that can look like golf balls on bent tees (Fig. 11.6). Spores are produced in the balls, and these are scattered by wind or rain and germinate in new locations, dispersing the species. Ironically, recent molecular evidence suggests that cellular slime molds, including the much-studied *Dictyostelium discoidum,* are more closely related to the fungi than they are to many other protists
- Rhodophyta—also known as **red algae**; small, delicate organisms that occur as thin filaments or flat sheets of many cells with an ornate, fanlike appearance (Fig. 11.7); some rhodophyta are single-celled or grow in colonies of many cells. You may be surprised to find algae in a chapter about microbes rather than one on plants. But the term **alga** (pl., *algae*), comes from the Latin word for "seaweed," and it has been used to describe simple chlorophyll-containing organisms that live in water, whether the organisms are true plants or not

You can see from the branches of the eukaryotic tree of life that red algae are only distantly related to plants. Red algae generally live in shallow, tropical ocean waters, but a few species survive at depths of about 270 m (880 ft). These deep-sea denizens have reddish pigments in their cells that absorb light in the blue-green range—the only wavelengths that can penetrate to such great ocean depths. The red pigments absorb light energy and pass some of it to chlorophyll for use in photosynthesis (see Chapter 3). Red algae have several commercial uses. Nori, a seaweed, is a common ingredient in Asian cooking. Red algae produce a gel-like protein that forms the

Figure 11.6
Slime Molds

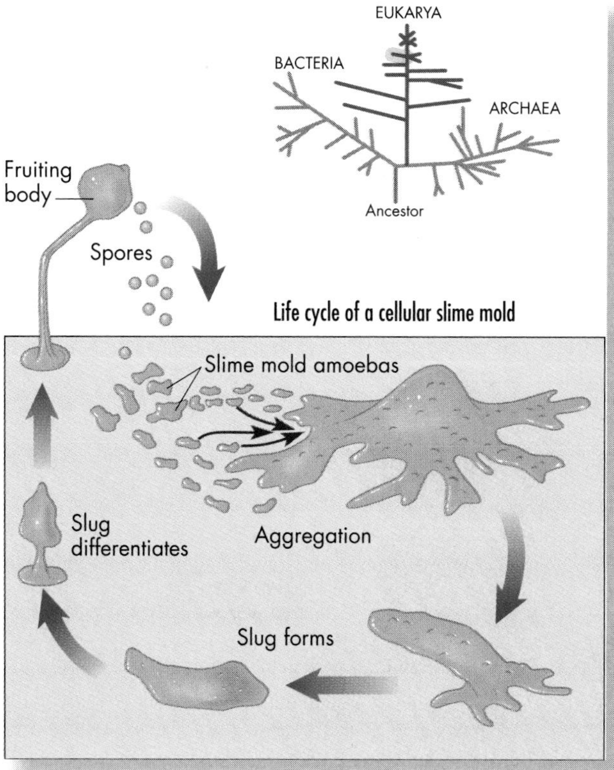

rubbery base of the laboratory growth medium called agar. Red algae also produce the starchy thickener called *carrageenan*, which chemists include in some cosmetics and paints, and which you might have noticed in the ingredient lists for some ice creams and puddings.

Figure 11.7
Red Algae: Deep Dwellers

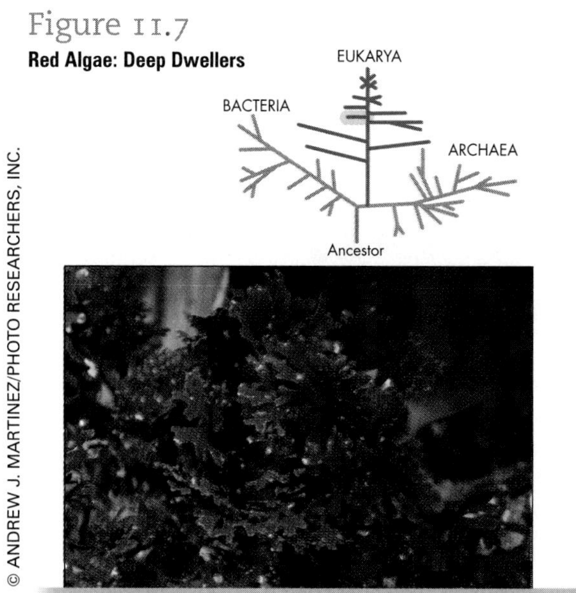

Complex Eukaryotes: Top Branches in the Eukaryotic Tree

We return, now, to the protists that cause malaria, as well as to a number of other complex, beautiful protists that branched off later on the tree of life than the groups we just explored.

alga (pl. algae) simple chlorophyll-containing organism, often single-celled; was probably the ancestor to the land plants

alveolates protists that include apicomplexans, dinoflagellates, foraminiferans, and ciliates

alveoli in protists, tiny membranous sacs under the plasma membrane

- **Alveolates**—a recently recognized group of protists that also includes armored marine cells, the dinoflagellates and foraminiferans, and miniature predators, the ciliates. Alveolates are named for a common cell structure: a system of membranous sacs (or **alveoli**) under their plasma membranes. The cause of malaria, *Plasmodium falciparum,* is part of a subgroup of alveolates, the Apicomplexa.
    - Apicomplexa—malarial parasites; has a complex of rings and tubules at the cell's apex. You can see these structures in the malarial parasite in Figure 11.8. Apicomplexans also have a sporelike stage in which the cells lack any means of locomotion. You might think this would reduce a protist's ability to get around or infect hosts, but in fact, this stage helps *P. falciparum.* You can follow this protist's complex life cycle in Figure 11.9 on the next page. It is truly remarkable how many different forms this single-celled organism takes as it moves from a mosquito's midgut to its salivary gland, then into a person's bloodstream. From there it enters liver cells, multiplies, then enters red blood cells and consumes the hemoglobin, usually destroying the red cell. Malaria is the most infamous apicomplexan, but other members of this group also cause serious human and animal diseases, including toxoplasmosis, the "cat box" infection that can

Figure 11.8
The Malarial Parasite: Which Kingdom?

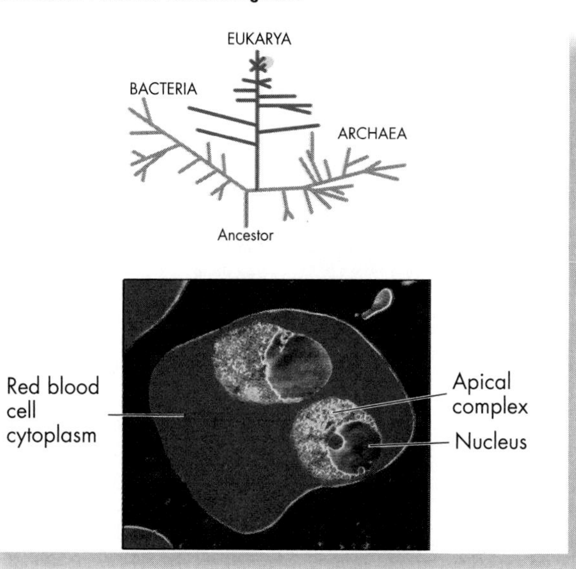

dinoflagellates protists with armor-like coverings and two flagella, one in a beltlike groove and the other trailing behind; often can cause red tides

red tides dense blooms of certain dinoflagellates that tint water red and produce deadly toxins

foraminiferans delicately shaped protists that live in the oceans and secrete usually whitish, calcium-based shells

cause a pregnant woman to deliver a malformed baby. The apicomplexans are the most significant alveolates because malaria claims more human victims than any other infectious disease.

- **Dinoflagellates**—marine protists that have two flagella that cause them to spin as they swim (*dinoflagellate* means "spinning cell with flagella"). One flagellum winds about the middle of the cell like a belt in a groove and causes the spinning motion, while another projects backward in a second groove and propels the cell forward (Fig. 11.10). Some dinoflagellates have internal plates of "armor," elaborately embossed vesicles filled with tough cellulose. Dinoflagellates can cause **red tides**, dense blooms that can tint the water blood-red and produce deadly toxins that act as nerve poisons. The poisons build up in fish and shellfish, killing them and poisoning people who gather and eat them. Government agencies in coastal areas of North America often ban collection and consumption of shellfish during May through August, when red tides can occur.
- **Foraminiferans**—The third group of alveolates are the *foraminiferans*, delicately shaped protists that live in the oceans and secrete whitish, calcium-based shells (or "tests") that can look like spiral seashells or chambered nautiluses. These pro-

Figure 11.9
The Complex Life Cycle of Malarial Protists

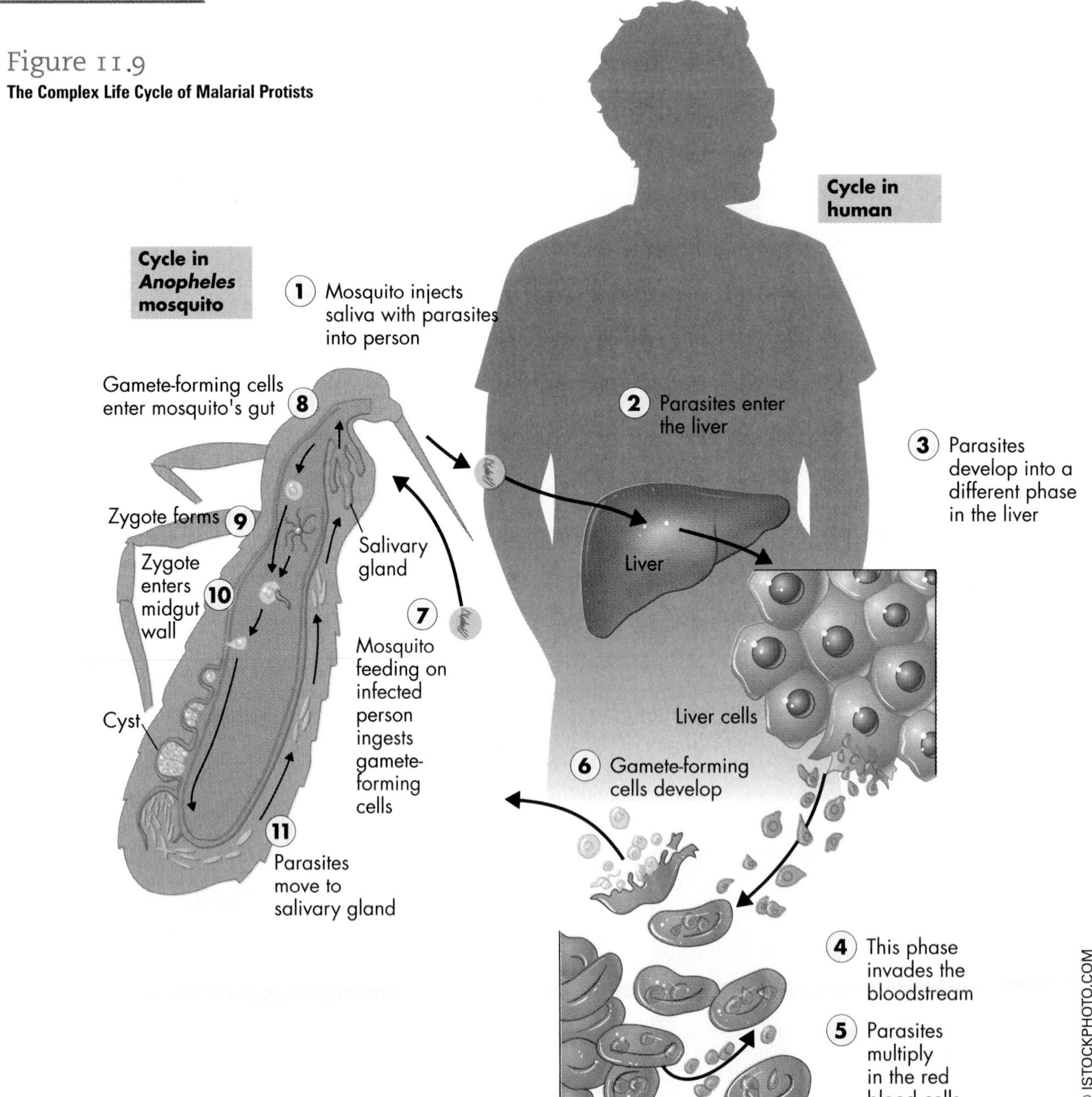

Figure 11.10
Alveolate Diversity

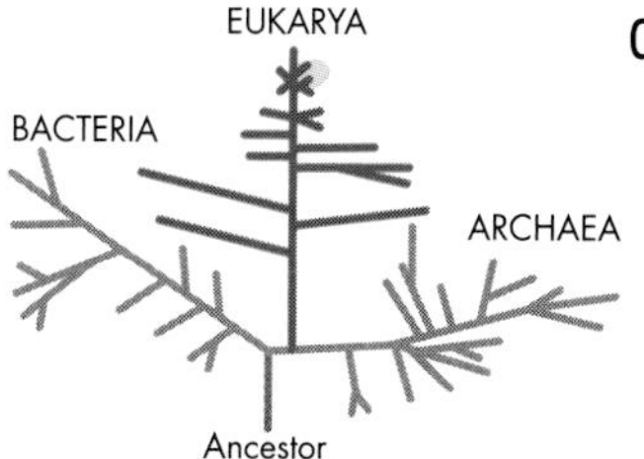

Ciliates: *Didinium* and *Paramecium*—the hunter and the hunted

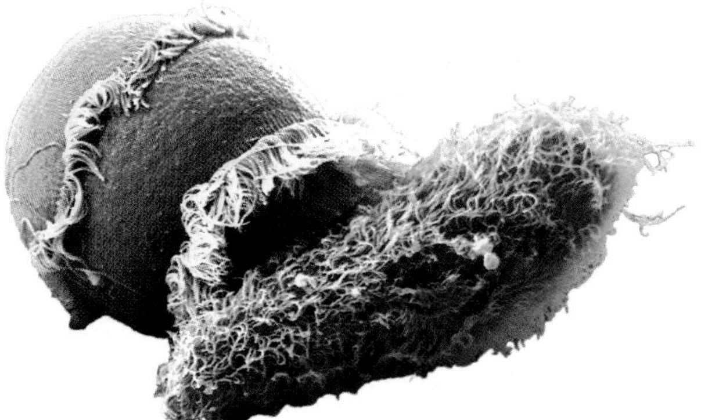
© EYE OF SCIENCE/PHOTO RESEARCHERS INC.

tozoa are so plentiful that when the organisms die, their shells pile up and contribute to forming thick limestone and chert deposits, including England's famed White Cliffs of Dover. The miniature shells often appear as fossils in rock layers and have contributed to our knowledge of Ice Ages and even of the current global climate change: Geologists have measured oxygen isotopes in foraminiferan shells and used them to reconstruct global temperature changes over the last 100,000 years.

- Ciliates—The fourth group of alveolates is the **ciliates**, some of Earth's most complex single-celled organisms. Ciliates are covered by hundreds of short, eyelash-like organs of locomotion. Cilia bend with coordinated, oarlike movements, propelling the cell or helping to sweep food particles into the ciliate's gaping "gullet" (Fig. 11.10). Ciliates have several organelles with functions analogous to an animal's organs: Anal pores discharge wastes. Microfilaments and microtubules support and contract like bones and muscles. Tiny toxic darts (**trichocysts**) spear and paralyze prey. Food vacuoles are filled with enzymes and function as digestive organs. Each ciliate also has a giant nucleus (biologists call it a polyploid **macronucleus**), containing many sets of chromosomes, that directs cell activities. One or more small diploid nuclei (micronuclei) undergo meiosis and are exchanged during sexual reproduction.

We've explored the alveolates, including the wily malarial parasites. However there are two more lineages of protists that branched off near the top of the eukaryotic family tree but did not lead to multicelled plants, animals, or fungi as did the other groups we'll explore shortly.

- **Radiolarians**—single-celled protists that produce beautiful silicon-based shells that look like miniature glass ornaments (see this chapter's opening photo). The many shapes of radiolarian shells provide clues that paleontologists have learned to read when judging rock ages.
- **Chromista** (also called *Stramenopila*)—mainly colored protists: most contain golden, brownish, and greenish pigment molecules that help them to photosynthesize. Diatoms, brown algae, and downy mildew are all Chromista.
    - Diatoms—**Diatoms** are the most common Chromista and perhaps the most exquisitely beautiful of the **phytoplankton** (literally, "floating plants"). Along with their chlorophyll pigments, most diatoms also have carotenoid pigment that gives them a golden color. Their cell walls contain silica (the main component of window glass) instead of cellulose (the main compound in plant cell walls). Diatoms also store oils rather than starchy compounds. Many people recognize diatoms because of their jewel-like, glassy shells. These shells fall to the ocean floor and accumulate in crumbly white sediments called **diatomaceous earth**,

A variety of diatoms.

© STEVE GSCHMEISSNER/PHOTO RESEARCHERS, INC.

ciliate
a protozoan whose cells have rows of cilia that are used in locomotion and in sweeping food particles into the mouth

trichocysts
tiny toxic darts in protists that spear and paralyze prey

macronucleus
in ciliates, a large nucleus containing many sets of chromosomes that control cell activities

micronucleus
in ciliates, one of several nuclei that undergo meiosis and are exchanged during sexual reproduction

radiolarians
single-celled protists that produce beautiful silicon-based shells

chromista
protists with golden, brownish, and greenish pigments

diatoms
phytoplankton that are members of Chromista and usually contain golden pigments

phytoplankton
photosynthetic microorganisms that live near the surface of marine and fresh water

diatomaceous earth
crumbly white sediments made up of diatom shells

brown algae
Chromista that inhabit cool, offshore waters and range from golden brown to dark brown to black

kelp
one of the largest members of the algal world; a brown alga

frond
the leaflike structure of an individual alga that collects sunlight and produces sugars; also refers to the large divided leaf on a fern

stipe
the stemlike structure that provides vertical support to an alga

holdfast
a rootlike anchor that attaches an alga to its substrate, such as a rock on the ocean floor

water mold
a type of funguslike protist containing several nuclei within a common cytoplasm and forming relatively large immobile egg cells; members of the Oomycota

hypha (pl. hyphae)
one of many long, thin filaments of cells that make up a multicellular fungus

green algae
protists with green pigments that are closely related to plants; also called Chlorophyta

microsporidia
among the simplest of eukaryotic cells, they live only inside animal cells

which manufacturers use in toothpaste, swimming pool filters, and other applications. Living diatoms are so abundant and ubiquitous that some biologists estimate they may contribute more oxygen to the atmosphere than all land plants combined.

- Brown algae—Another type of Chromista are the **brown algae**, multicellular inhabitants of cool, offshore waters. Most members of the group are small, however the **kelps** are brown algae that can grow 100 m (over 325 ft) long and float vertically like tall trees. Huge floating masses of the brown alga *Sargassum* thrive in the Sargasso Sea, a mass of still water in the Atlantic Ocean north of the Caribbean Sea. These kelp forests provide important breeding grounds and hiding places for many fish and shellfish.

  Brown algae range from golden brown to dark brown to black. They also contain green chlorophylls *a* and *c*, but the golden-brown carotenoid pigment colors the cells. Carotenoids also allow them to collect the blue and violet wavelengths of light that penetrate medium-deep water and thus to exploit an ocean environment too dim for many other organisms.

  Kelps and other brown algae often look surprisingly like plants because they have body parts analogous to parts of land plants: leaflike **fronds** collect sunlight and produce sugars; the stemlike **stipe** supports the organism vertically; and the rootlike **holdfast** anchors the organism to submerged rocks. Special tubelike conducting cells carry sugars produced in the fronds to the deeper parts of the kelp. The tubes function like the specialized tissues that transport materials inside land plants, but they're not related. In fact, molecular genetic analyses show, surprisingly, that plants are more closely related to animals than they are to brown algae!

- *Water molds*—Some chromista, such as the **water molds**, or members of the Oomycota (which means "egg mold"), lack colorful pigments and are not photosynthetic. Studies of their genetic material nevertheless show that they're closely related to other chromista. Water molds inhabit soil and water; some are single-celled, while others form fuzzy, branching filaments called **hyphae**. Like the true slime molds, water molds have several nuclei within a common cytoplasm and form large, immobile egglike cells, hence their Latin name. After fertilization, they form spores that disperse by swimming. One parasitic water mold struck Ireland in 1845–1847. This species causes *potato blight,* a disease that rapidly rots and kills growing potato vines. The Irish depended so heavily on potatoes for their daily diet that when potato blight struck, more than a million people starved, and 1.5 million more emigrated, mostly to North America.

Protists Closely Related to Plants, Animals, and Fungi

Plants, animals, and fungi, composing the three most familiar kingdoms, grace the top of the eukaryotic tree, are complex and multicellular and relatively recently evolved compared to many of the simpler forms we just looked at. But branching off from the same lineages that eventually gave rise to the plants, animals, and fungi are closely related protists with many similarities.

Chlorophyta

The chlorophyta, or green algae, are protists closely related to plants. We'll consider the green algae with the plants in Chapter 12, because they appear to form a single branch on the tree of life, and many botany courses cover them along with plants.

Microsporidia

Microsporidia are protists apparently closely related to fungi and are among the simplest of eukaryotic cells. These tiny organisms live only inside animal cells. Microsporidia have among the smallest of eukaryotic genomes (smaller than many bacterial genomes, in fact), and they lack several organelles including mitochondria, a stacked Golgi apparatus, and peroxisomes (small vesicles containing enzymes that break down peroxide, a byproduct of metabolism). Some biologists thought that microsporidia were eukaryotic relics, representing the primitive state before eukaryotes acquired mitochondria (review Fig. 10.4). However, by studying proteins from *Nosema locustae*, a microsporidium that lives in locusts, biologists have found the organisms to be closely related to fungi. Perhaps microsporidia lost

A giant kelp (brown alga).

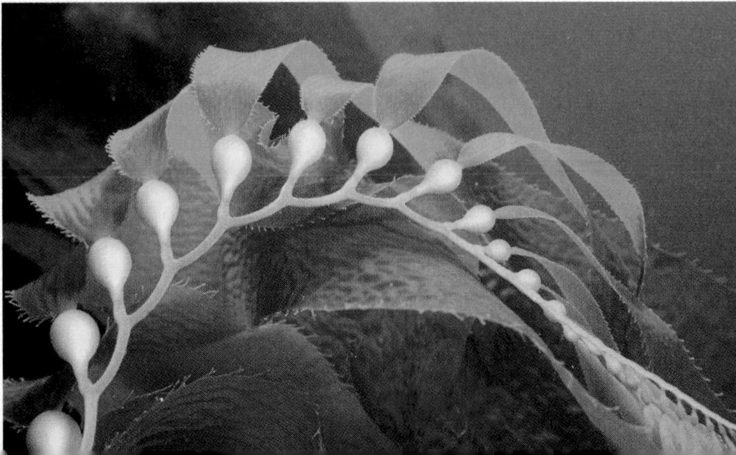

many genes and cell organelles as an adaptation to the parasitic life style.

Choanoflagellates

The **choanoflagellates**, or "collar flagellates," are protists closely related to animals. They are single-celled or colonial protists living in freshwater and in the oceans (see Chapter 13). The collar is formed by a ring of closely packed microvilli, small finger-like cell projections that encircle the flagellum. The wagging of the flagellum creates water currents that wisk prokaryotic cells into contact with the sticky collar; the cell then devours the ensnared bacteria by phagocytosis. A choanoflagellate looks very similar to the collar cells of sponges, the simplest animals (see Fig. 13.3). This, along with molecular genetic evidence, suggests strongly that choanoflagellatelike ancestors gave rise to all of the animals. Animals, of course, include humans subject to malaria—humankind's greatest medical scourge.

LO4 Fighting Humankind's Greatest Killer

Malaria is not only our species' most prolific killer, it is one of our oldest recorded maladies, appearing in Hippocrates's medical journals in the 5th century B.C. Efforts to fight malaria have a long history, too. In the mid-1600s, the natives of Lima, Peru, were already using bark from a tree in the Andean cloud forests to treat the disease. Spanish diplomats returned the bark to Europe, where it became known as quinine and remained the principle treatment until World War II, when researchers developed chloroquine and other synthetic substitutes. Unfortunately, resistance to multiple malaria drugs has evolved in many populations of the deadly protist *Plasmodium falciparum* (Fig. 11.11). Their major hosts, *Anopheles* mosquitoes, also evolved resistance to DDT and other insecticides. As a result, the incidence of malaria began rising dramatically in the late 20th century.

A detailed knowledge of the prokaryotes can help in controlling malaria, for example, how the gram-positive bacteria, *Bacillus thuringiensis*, can be used to kill mosquitoes. Do you suppose there's a virus somewhere that could specifically kill either the malarial parasite or malaria-transmitting mosquitoes? How could biologists go about looking for those viruses?

choanoflagellates
single-celled or colonial protists living in freshwater and in the oceans; each has a collar formed by a ring of microvilli

Malaria research is one of the most active areas of biology and a subject to watch now that you understand the players, both eukaryotic and prokaryotic.

Figure 11.11

Map Showing the Widespread Status of Drug-Resistant Malarial Parasites

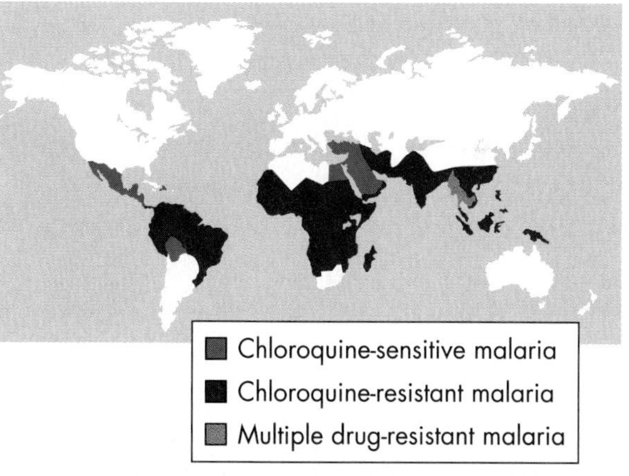

Who knew?

Antibiotics are chemicals that interfere with the metabolism of prokaryotes, often by disrupting cell walls or interfering with synthesis of the cell walls or other cell products. Since viruses don't have walls built like bacterial cell walls, and viruses don't synthesize their own products, antibiotics won't have any effect on viruses.

CHAPTER 12

Learning Outcomes

LO[1] Identify the form and function of fungi and how fungi interact with plants

LO[2] Describe the characteristics of plants and green algae

LO[3] Compare and contrast the life cycles of plants

LO[4] Recognize the successful adaptations of flowering plants

Fungi *and* Plants: *Decomposers and Producers*

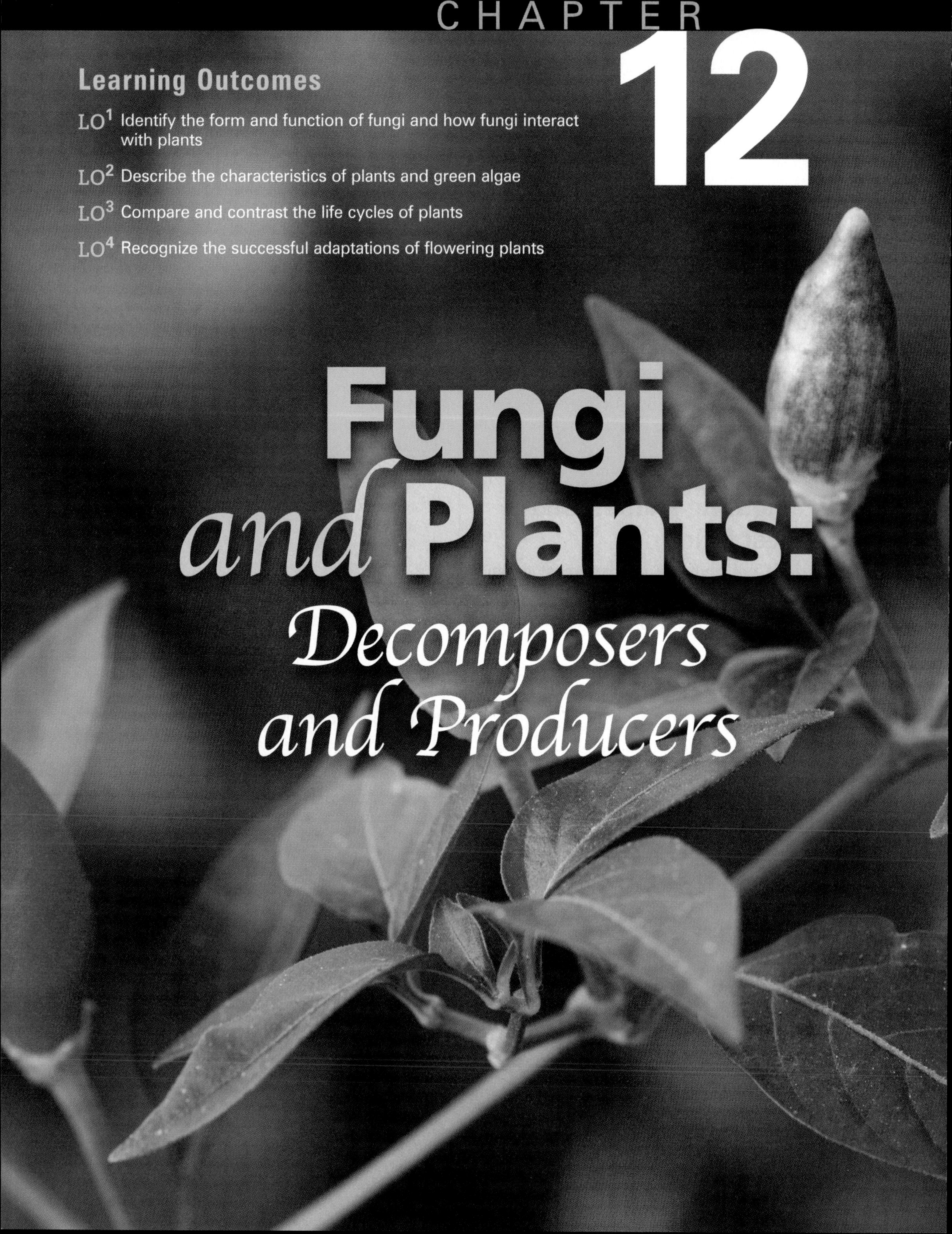

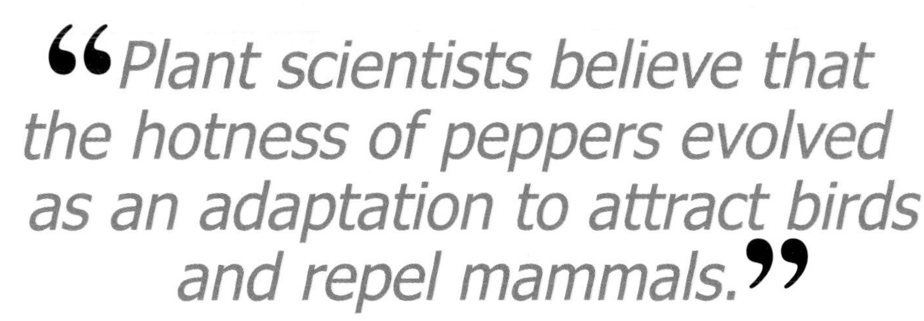

A Fiery Fascination

All chile peppers—from the mildest green and red bell peppers to the hottest habaneros (peppers one author refers to as "thermonuclear")—belong to the same genus, *Capsicum*. People in what is now South and Central America domesticated chile species starting 10,000 years ago. Today, plant breeders grow hundreds of varieties of chile plants, virtually all belonging to just one species, *Capsicum annuum*. Strange as it may sound, *Capsicum annuum* is the most widely spread plant species in the world, because chile peppers are the most widely used cooking spice.

What do you know?

Why are the flowering plants more diverse than any other plant group?

There's a great irony in the evolution and popularity of *Capsicum*. The genus is part of the "deadly nightshade" family, Solanaceae, whose members—including tobacco, potatoes, and tomatoes—are known for producing toxic compounds in their leaves and/or fruits. Chile peppers make none of these toxins. Instead, they generate spicy-hot alkaloids called *capsaicins* (cap-SAY-shins) in the pepper fruit's ovary tissue, which surrounds the seeds. Birds can't taste this pungency, and so they eat the fruits and seeds with gusto, then deposit the seeds unharmed in their droppings. In this way, birds act as beneficial seed-dispersal agents for pepper plants. Mammals, on the other hand, *can* taste the heat and virtually always avoid hot peppers after one nibble. This, too, benefits *Capsicum* plants, because unlike a bird's digestive tract, a mammal's stomach contains acids that will kill the seeds. Plant scientists believe that the hotness of peppers evolved as an adaptation to attract birds and repel mammals. How ironic, then, that we humans seek out the smoldering sensation of peppers—from jalapeño peppers, which are quite hot, all the way to mouth-blasting habanero peppers, which are over 40 times hotter! As you read on, you'll find out why so many humans love—sometimes even *crave*—a daily fix of hot chile flavor.

Chiles are a good case history for this chapter on plant diversity and evolution. Our goal here is to explore the major groups of fungi (sing., fungus) including the bread molds, yeasts, and mushrooms; green algae, such as the many types that undulate in tidepools; and plants, including mosses, ferns, pines, and flowering plants like chiles. We'll talk about plant evolution from ancient, single-celled progenitors, their relationships to each other, and their great ecological roles as decomposers and producers of organic matter in vast quantities. Even though chiles are just one small group in this varied sweep of fungal and plant species, they are an apt focus because: (1) They are associated with fungi in several ways, both beneficial and harmful.

fungi
the kingdom comprising multicellular heterotrophs such as mushrooms or molds that decompose other biological tissues

(2) Plants evolved from water dwellers to land dwellers, and chiles have most of the major adaptations to life on land. (3) Many plants coevolved with animals, and the reliance on birds as chile-seed dispersers is a good example.

As you explore chiles and their place in the spectrum of fungi, green algae, and plants, you'll find the answers to these questions:

- How do fungi make a living, and how do they reproduce?
- What are the characteristics of green plants that have allowed them to evolve and dominate most landscapes?
- What makes flowering plants—including the chiles—so successful?

LO1 The Lives of Fungi

If you uproot and examine a pepper plant, you will probably see small filaments or threads extending. Closer examination would reveal that these threads are not a living part of the plant but some other living entity that grows in the soil and attaches to the plant roots. These fine threads are actually the bodies of organisms belonging to the biological group called **fungi**, all of which occupy a single branch on the tree of life (Fig. 12.1). The group includes mushrooms, puffballs, yeasts, molds, and certain other organisms with a similar body plan, and the members are more closely related to animals than plants. So what are the filaments doing on and sometimes inside the pepper plant root cells? Do they help or hinder the plant's growth and its production of fiery fruits? Biologists have carried out experiments to answer these ques-

Figure 12.1
Fungi on the Tree of Life

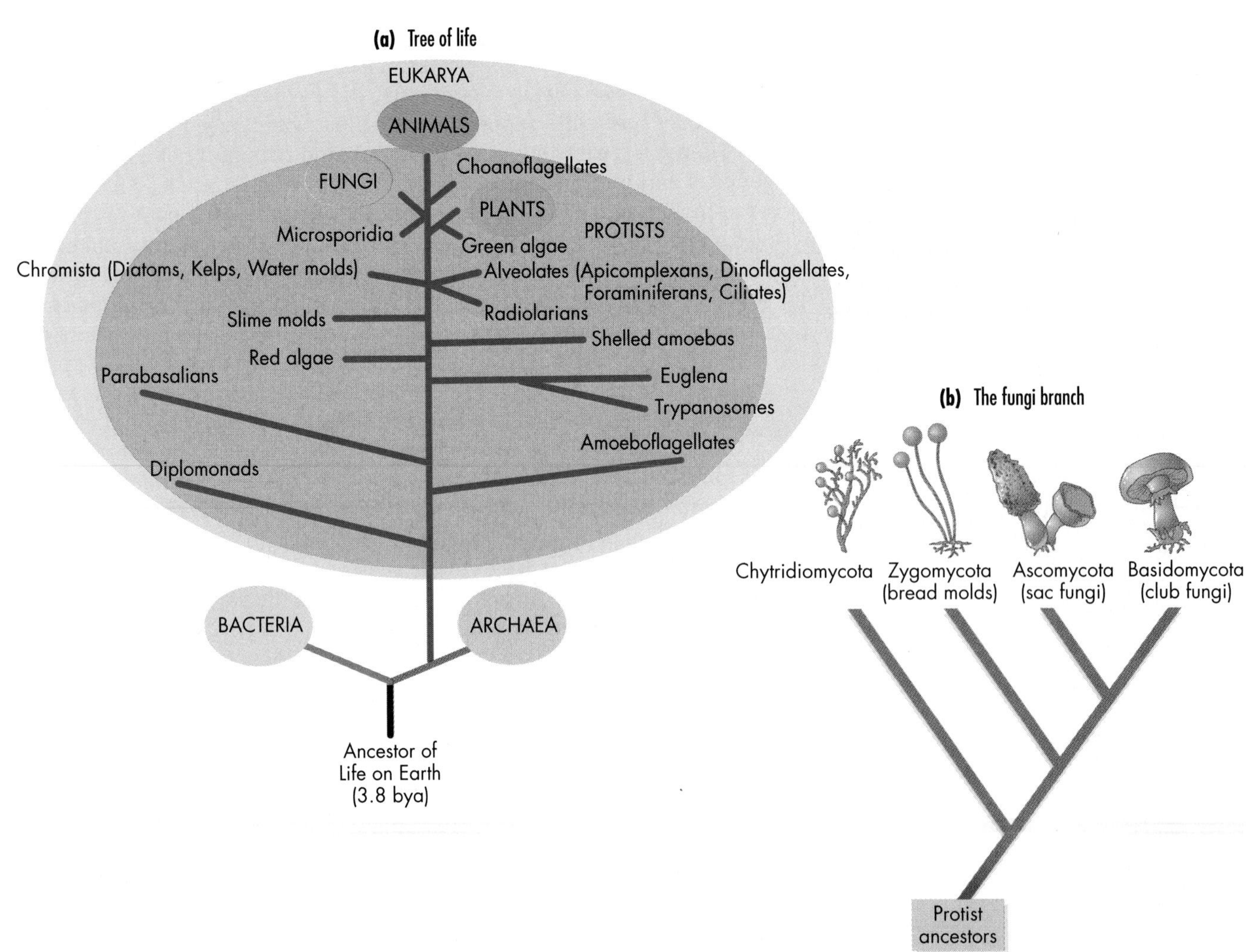

Figure 12.2
The Fungal Body

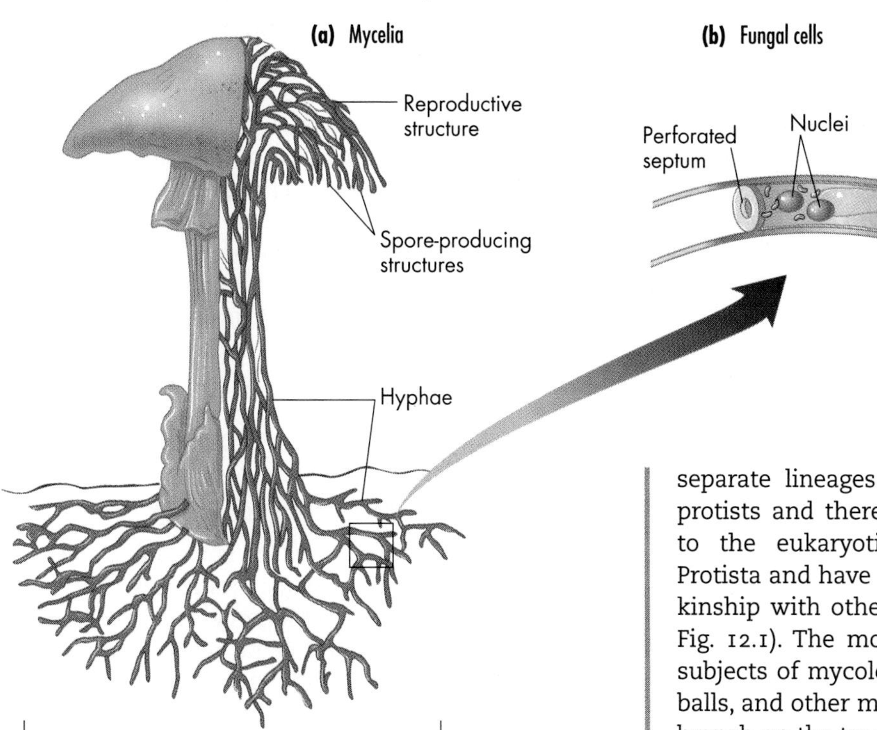

Eumycota
the "true fungi" branch on the tree of life

hypha (pl. hyphae)
one of many long, thin filaments of cells that make up a multicellular fungus

mycelium
the mass of filaments (hyphae) that makes up the body of a fungus

yeasts
generally single-celled fungi that reproduce by budding

chitin
a complex nitrogen-containing polysaccharide that forms the cell walls of certain fungi; the major component of the exoskeleton of insects and some other arthropods, and the cuticle of some other invertebrates

tions. But before we can fully understand the tests, we need a general background in the fungal body plan and lifestyle. In this section, then, we'll concentrate first on the body form of fungi, then see how they obtain energy, reproduce, and finally how they interact with chiles and other plants.

The Body Plan of Fungi

The long, slender fungal threads you unearthed on the roots of a chile plant are probably of the species *Glomus intraradices* and are typical examples of the fungal body plan. The filaments have these chief characteristics:

- They are eukaryotes (made up of eukaryotic cells).
- They are heterotrophs (they obtain their nourishment from other organisms).
- They develop a diffuse, branched, and tubular body.
- They reproduce by compact structures called spores.

Mycologists (fungal biologists) study organisms with some or all of these characteristics. As we saw in Chapter 11, mycologists also study water fungi and slime molds—organisms that apparently arose on separate lineages within the protists and therefore belong to the eukaryotic kingdom Protista and have only distant kinship with other fungi (see Fig. 12.1). The more frequent subjects of mycology—mushrooms, toadstools, puffballs, and other more familiar fungi—occupy a single branch on the tree of life called the **Eumycota** or true fungi. We'll consider only true fungi in this chapter.

In general, fungi consist of filaments called **hyphae** (HIGH-fee; sing., *hypha;* Fig. 12.2a,b). A fungal filament is made up of cells joined end to end, like the cars of a passenger train. Hyphae often interweave into a loose mat called a **mycelium** (pl., *mycelia*), rather like steel wool (Fig. 12.2a). You can see such a mat growing as a widening green or gray circle on the heel of a loaf of bread forgotten at the back of a refrigerator. Researchers have recently found that even the **yeasts**, which generally grow as single-celled fungi and reproduce by budding, will form hyphae under conditions of starvation.

Cell structure distinguishes fungi from prokaryotes (see Chapter 11) as well as from plants and animals. The cells of a fungus have membrane-enclosed nuclei, unlike the cells in the prokaryotic domains Bacteria and Archaea. Fungal cells also have cell walls, which distinguish them from animal cells. But the cell walls of fungi also distinguish them from plants, because fungal cell walls consist mostly of **chitin** (KI-tin), a nitrogen-containing polysaccharide, while plant cell walls are mainly cellulose. The tough chitin in fungal cell walls helps make fungi among the most resilient organisms of all the eukaryotes, able to withstand heat and drought. (The hard outer skeletons of insects are also made of tough chitin.) The cell walls that separate adjacent cells in a fungal filament are perforated by little holes (Fig. 12.3). These perforations

fruiting bodies spore-producing reproductive structures in many fungi

allow the cytoplasm of one cell to directly contact the cytoplasm of neighboring cells.

Even with their many interlinked cells, the bodies of most fungi are simple in structure and lack specialized tissues or organs such as the roots, stems, and leaves in plants. Fungi do, however, often form specialized reproductive structures. An example of such a reproductive structure, or fruiting body, is the familiar button mushroom cooks use in salads or on pizzas. The body of a button mushroom looks solid, but it actually consists of many fungal filaments packed tightly together like hair bound together into a pony tail. Below ground, an extensive, loose mycelium spreads beneath the mushroom, penetrating the soil for many square meters.

The *Largest?* We'll Just See About That!

A single individual fungus can be enormous. Mycologists near Mount Adams in Washington State have found a fungus that covers 1,500 acres of forest (about 600 hectares), weighs well over 100 tons (9,000 kilograms), and is between 400 and 1,000 years old. In the early 1990s, some fungal biologists suggested that this one individual fungus could be considered the largest living organism in the world. As we'll see later, plant biologists were quick to challenge this claim.

Fungal filaments or hyphae can grow very quickly: If you added together the growth at all the hyphal tips in a tangled fungal mat or mycelium, just one day's new growth could easily exceed 1 km (0.62 mi). This explains how mushrooms can literally spring up overnight on damp logs or soil. This meshwork of filaments gives a fungus a large surface-to-volume ratio, which enables even a very large mushroom to digest and absorb enough nutrients to grow rapidly.

Decomposing: How Fungi Make a Living

Although fungi often grow in the ground like plants, plants are autotrophs, while fungi are heterotrophs. Plants can extract energy and molecular building blocks directly from the physical environment and so can many kinds of prokaryotes. Fungi, however, must obtain their energy and materials by decomposing molecules first made by other organisms. Like animals, fungi obtain their nourishment by digesting the molecules of plants, or animals that eat plants. And like the cells of your digestive system, fungal cells secrete powerful enzymes into their surroundings that break down large organic molecules into smaller ones (Fig. 12.3). Again, like some of your intestinal cells, fungal cells absorb the smaller nutrient molecules through their cell membranes. This fungal feeding strategy seems so . . . animal. So how is it different? The answer is simple, and it involves *eating*. Most animals bring chunks of a plant or other organism into the mouth then *inside* the cavity of the gut before digesting it. A fungus, by contrast, digests nutrient matter *outside* of its own fungal body.

Most fungi decompose a wide range of nonliving

Figure 12.3
How a Fungus Makes a Living

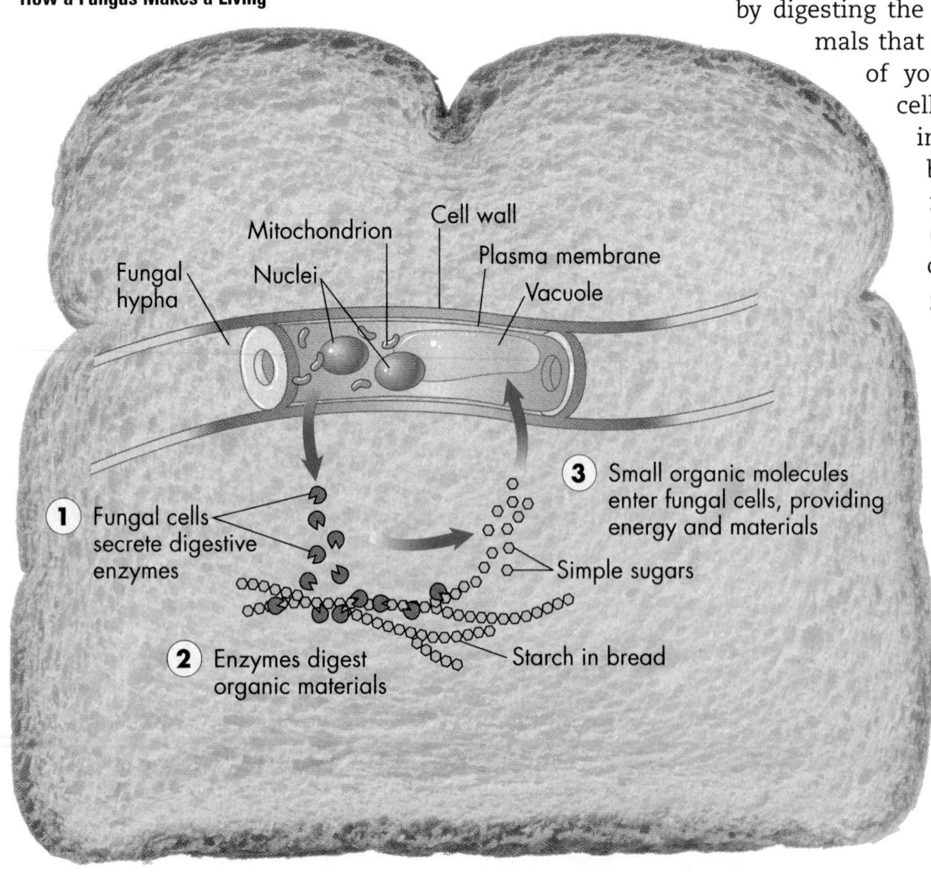

organic matter. Fungi consume everything from leather and cloth to paper, wood, paint, and other materials, slowly reducing old buildings, books, and shoes to crumbled ruin. In the process, they are inadvertent masters of recycling. During the long evolution of the fungi, however, some **saprobes**, or decomposers of nonliving tissue, evolved into **parasites**, or organisms that live on or in other living things, without contributing to the host's survival. Today, parasitic fungi are the main cause of plant diseases, with about 5,000 different fungal species inhabiting crops in fields, gardens, and orchards. As you'll see later, other types of parasitic fungi attack people and domestic animals, causing athlete's foot, ringworm, and vaginal yeast infections, to name a few.

Clearly, the dietary habits of fungi are far from fastidious. Their activities, however, are crucial for nature's balanced ecological cycles. Plants and photosynthetic bacteria, for example, remove carbon from the air and trap it in carbohydrates. If these were the Earth's only organisms, all carbon would soon be locked up in their cells and in their remains. This doesn't happen, however, because other organisms—largely fungi and bacteria—decompose plant matter and organic wastes as well as the remains of dead organisms and your toenail clippings. From these, they release into the environment small compounds containing carbon, nitrogen, phosphorus, and other elements required for life and they reenter the nonliving portion of the ecosystem, where they can eventually be recycled back into living cells.

The Sex Life of a Fungus

It's amazing how a colony of fungus can crop up on a slice of bread, on the underside of a board, or on an orange left too long in the fruit bowl. How did they get to these isolated places? Fungi reproduce and disperse by special microscopic reproductive structures called **spores**. Each tiny spore can germinate into a hypha and develop into a new colony of fungus (a mycelium). Spores are adaptations for survival; they are able to withstand extreme conditions of dryness, heat, or cold, they can disperse by wind or water to distant locations—like a board or a fruit bowl—and then produce a new fungus when conditions are favorable.

Some spores are produced without sex. These asexual spores are genetically identical to the parent and form by modifications of cells in the parent hypha. Fungi can also produce sexual spores (Fig. 12.4). Fungi are neither male nor female, but each haploid individual is one of two mating types, plus or minus (Fig. 12.4, Step (1)). The haploid cells of each mating type can grow into hyphae and reproduce by making asexual spores

saprobe
an organism that lives on decomposing organic matter

parasite
a type of predator that obtains benefits at the expense of another organism, its host; a parasite is usually smaller than its host, lives in close physical association with it, and generally saps its host's strength rather than killing it outright

spore
in eukaryotes, a reproductive cell that divides mitotically and produces a new individual; in prokaryotes, a resistant cell capable of surviving harsh conditions and germinating when conditions are once again favorable

plus/minus mating type
fungi are neither male nor female, but each haploid individual is one of two mating types, plus or minus

Figure 12.4
Generalized Life Cycle of Fungi

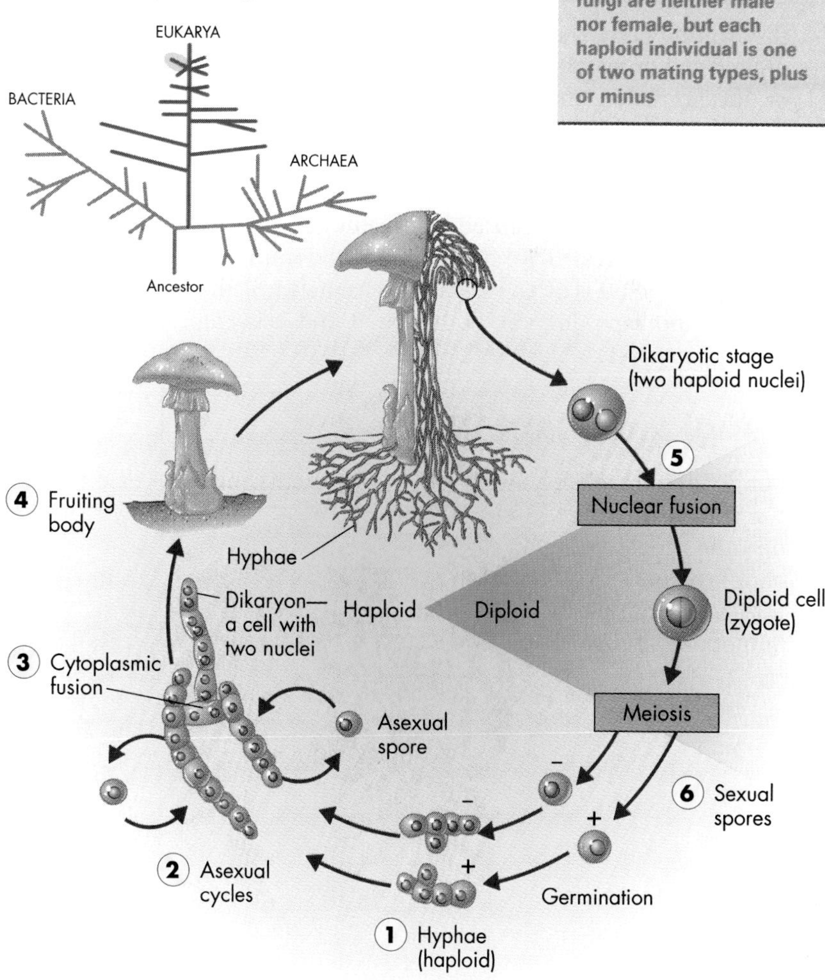

dikaryon
in fungi, a single cell containing two haploid nuclei in one cytoplasm resulting from the union of a haploid cell of the plus mating type with a haploid cell of the minus mating type

sexual fusion (nuclear fusion)
a process that produces a diploid cell or zygote through the fusing of two haploid nuclei; in fungi, this takes place within the fruiting body

mycorrhiza (pl. mycorrhizae)
an association between the root of a plant and a fungus that aids the plant in receiving water and nutrients and aids the fungus in receiving carbohydrates

symbiosis
a close interrelationship of two different species

(Step ②). Eventually, a haploid cell of the plus mating type can fuse with a haploid cell of the minus mating type, forming a single cell containing two haploid nuclei in one cytoplasm (a **dikaryon** (*di* = two + *karyon* = nucleus; Step ③). (Note that a dikaryon has two sets of haploid chromosomes in two separate nuclei, while a typical diploid cell has two sets of haploid chromosomes in a single nucleus.) Double-nucleated dikaryon cells can divide and form hyphae, and these threadlike structures can then join and give rise to a reproductive organ, the fruiting body (Step ④). The typical mushroom we recognize growing in a forest or field is an aboveground fruiting body. In special structures within the fruiting body, haploid nuclei can fuse, forming a diploid cell or zygote (a process called **sexual fusion** or **nuclear fusion**, Step ⑤). The diploid zygote usually undergoes meiosis, producing sexual spores, and the cycle is complete (Step ⑥). There are many variations of the fungal life cycle, and different types of fungi have different types of reproduction. But here are the essential features: The haploid stage dominates; the diploid phase is often relegated to a single cell, the zygote; and reproduction, both sexual and asexual, occurs by spores.

Fungi and Plant Roots

Fungi interact closely with other organisms, and the fungal threads on the roots of chile plants are just one example. Sometimes fungal interactions are beneficial; other times, they are harmful, causing diseases of plants and animals. What about the fungi you find on chile roots? Are they helpful or harmful? To find out, researchers sterilized one plot of soil to kill all the fungi present but left an adjacent plot untreated. Next, they planted chiles in the treated plot as well as in the adjacent control plot (which still contained fungi in the soil). At this point, they subdivided each of the two plots and spread phosphorus-containing fertilizer on some portions and left other portions unfertilized. When the chile plants from all four plots matured, the researchers dried and weighed the plants.

They found that when they provided phosphorus fertilizer, the chile plants grown in fungus-infected soil weighed approximately the same as those grown in sterilized soil. In the plots without added fertilizer, however, they saw a dramatic difference. The chile plants grown in fungus-infected soil were 91 times heavier than the plants grown in sterilized soil! What would you conclude from this experiment? The researchers deduced that the fungi helped the chile plants take up phosphorus, an element necessary for plant growth, when quantities of the element in the soil were limited to naturally occurring levels. When they added phosphorus fertilizer, however, this element was plentiful and no longer a limiting factor to growth, so the presence or absence of the fungi made no difference.

These experiments involved the fungal-root combination called **mycorrhizae** (myco-RYE-zee; sing., *mycorrhiza*, literally "fungus roots"), commonly associated with many plants. With this study, the researchers proved that mycorrhizae are highly beneficial for chile plant growth. The interaction between root fungi and the chile plants is a type of **symbiosis** in which two species have a mutually beneficial relationship. As we've seen, the plants benefit, but the fungus benefits, too, by

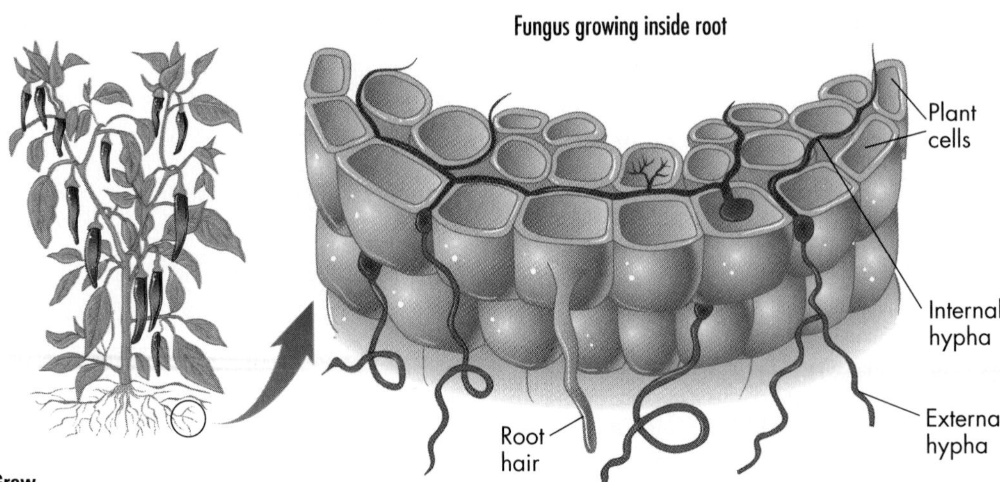

Figure 12.5
How Root Fungi Help Plants Grow

obtaining sugars from the plant roots for its own growth. In Figure 12.5 note the close physical association of fungus and root cell: a portion of the chile root mycorrhizal fungus actually invades the plant cells and obtains sugars and other carbon-containing molecules directly from the plant. In other plant roots, the fungi may remain outside the plant cells. Regardless, the plant helps the mycorrhizal fungus obtain organic nutrients, and the fungus helps the plant obtain phosphorus and probably water, as well. Biologists have discovered and named several hundred species of mycorrhizal fungi associated with the roots of perhaps 90 percent of all land plants. Because fossils of the oldest land plants have fossilized mycorrhizae among the roots, botanists think that the fungus–plant interactions may have helped the transition of plants from water to land.

Lichens: More Mutual Benefit

Have you ever looked carefully at the bark of an old tree or the sides of old gravestones? The tree bark or stones are often encrusted with gray, orange, or greenish layers like those shown on the tree trunk in Figure 12.6a. These formations are **lichens** (LIE-kenz), and are actually symbiotic associations between a fungus and either a green alga (a eukaryote) or a cyanobacterium (a prokaryote).

In a colorful lichen, the fungus forms a dense hyphal mat around the photosynthetic algal or cyanobacterial cells (Fig. 12.6b). The fungal mat provides the photosynthetic cells with some protection and a supply of water, and in turn, the mat filaments are able to obtain some of the photosynthetic cells' newly made carbohydrates. Lichens take almost no nutrients from the substrate on which they live, and they are often the first organisms to inhabit lava flows or other newly exposed rocks. However, they grow extremely slowly. You can estimate the rate of lichen growth by measuring the diameter of the largest lichens found on tombstones of various ages and comparing these to the dates inscribed on the stones. While lichens can survive in bleak environments, they are particularly sensitive to the industrial pollutant sulfur dioxide. Therefore, biologists have found the disappearance of lichens to be an early indicator of air pollution. Since both species in the lichen need to reproduce at the same time, one might expect to see clever evolutionary solutions, and indeed, some exist. Many lichens reproduce by releasing little balls that contain at least one algal cell embedded in a small network of hyphae. Wherever the reproductive structure lands, it has the potential to grow into a new lichen.

lichen
an association between a fungus and an alga, which live together symbiotically

Recent molecular genetic studies suggest that lichenlike associations of fungi and photosynthetic algae existed 1 billion years ago, and colonized Earth several hundred million years earlier than botanists have long accepted based on fossil evidence. These associations may have altered Earth's atmosphere and climate in ways that spurred the explosion of animal evolution in the Precambrian period.

Figure 12.6
Lichens: A Collaboration of Fungi and Algae

(a) Lichen on tree

(b) Hammered shield lichen

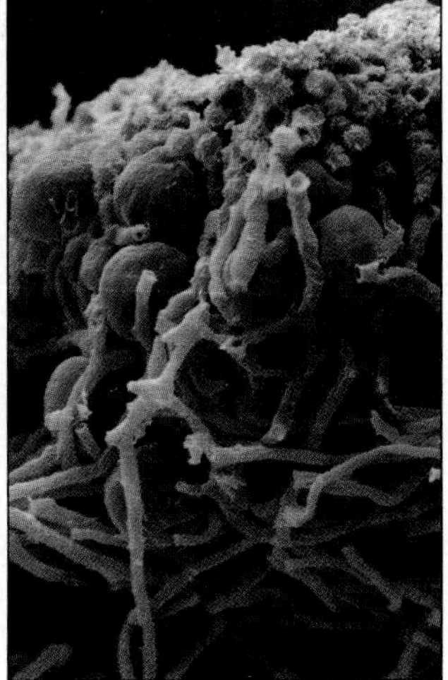

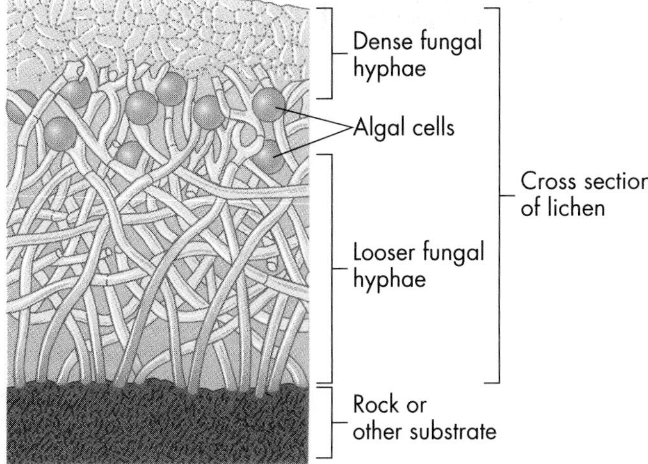

division
a taxonomic group of similar classes belonging to the same phylum, which is often called a *division* in the kingdoms of plants or fungi

Medical Mycology

Fungi are important in medicine, both in sickness and in healing. AIDS patients are susceptible to thrush, a yeast infection of the mucous membranes, and *Pneumocystis carinii*, which is now thought to be a fungus. As we saw earlier, vaginal yeast infections, athlete's foot, and many other skin diseases are common even in people with healthy immune systems. On the plus side, many bacteria-killing antibiotics, such as penicillin, were originally isolated from fungi, as was cyclosporin, which suppresses the rejection of tissue and organ transplants. Medical researchers are continuing to focus on the many kinds of fungi in the hopes of turning up more such medically useful substances.

The Diversity of Fungi

Biologists usually classify fungi by the shapes of their sexual spore-producing structures. Spores are the only means a fungus has of dispersing to new areas, and thus they are of prime importance. Table 12.1 describes several major divisions of fungi, plus the imperfect fungi and lichens. The table focuses on how they form spores and includes other significant characteristics as well. (In the fungal and plant kingdoms, a division is the equivalent of a phylum in other kingdoms.) By reading a field guide to common fungi, you can learn more details about the various groups of fungi, including the bread molds; the yeasts, morels, and truffles; the true mushrooms and other club fungi; and the so-called fungi imperfecti.

Table 12.1
Fungi: The Great Decomposers

Division	Examples	Characteristics and Significance
Lichens	Red, gray, and yellow species found on rocks	A combination of a fungus (ascomycota or basidiomycota) and an alga that live intimately together; degrades rock and helps make new soil; sensitive indicators of pollution
Deuteromycota ("imperfect fungi"; 25,000 species)	Species that produce penicillin	Have lost the ability to reproduce sexually, thus their relationship to other fungi is unclear; reproduce by asexual spores; used in making drugs, cheeses, and soy sauce
Basidiomycota (club fungi; 25,000 species)	Common field mushrooms, giant puffballs, bracket fungi, toadstools, smuts	Produce spores in club-shaped basidia; the fruiting body is the familiar mushroom or toadstool, which can be extremely poisonous
Ascomycota (sac fungi; 30,000 species)	Pink bread mold, brewer's yeast, morels, truffles	Produce spores in an ascus, or sac, borne in a cup-shaped body; because they include the yeasts, they are the most economically useful fungal group; powdery mildews harm fruit trees and grain crops
Zygomycota (bread molds; 600 species)	Common black bread mold	Produce diploid spores and a cottony mat of hyphae on breads, grains, or other foods and organic materials
Chytridiomycota	Species that cause potato wart disease	Sister group to other fungi; mostly aquatic; gametes have a flagellum, allowing them to swim (other fungi have lost flagellae)

LO2 Plants: The Green Kingdom

Plants—chiles, ferns, mosses, and thousands of other types—occupy the branching crown of the eukaryotic tree of life along with fungi, animals, and the Chromista (see Chapter 11). Plants are multicellular eukaryotes that capture energy by photosynthesis and develop from embryos. As we'll soon see, plants also have an alternation of generations, with distinct multicellular haploid and diploid body forms. All plants reside on a single branch of the tree of life. Preceding that branch were the green algae—organisms that are not true plants but that have some similarities to them. We consider both plants and green algae in this chapter because both groups originated from a single common ancestor that did not give rise to other forms of life (see Figs. 12.1 and 12.7). Let's start with the green algae and then move to the plants, which are more complicated, including the chiles.

Green Algae: Ancestors of Land Plants

plant
a multicellular eukaryote that captures energy by photosynthesis and develops from an embryo

green algae
protists with green pigments that are closely related to plants; also called *Chlorophyta*

If you've walked by a lake or stream, or poked around in a tidepool, you've probably seen several species of green algae, although you may not have recognized them. Unlike plants, which are multicellular, green algae, or Chlorophyta, often occur as single cells. Some, however, do occur as multicellular, threadlike filaments; hollow balls; or wide, flat sheets. Most species of green algae live in shallow freshwater environments or on moist rocks, trees, and soil. A few inhabit shallow ocean waters, undulating gracefully in tidepools alongside other kinds of algae. Green algae have the same types of chlorophyll (called *a* and *b*) that occur in plants, together with orange carotenoids; this combination allows both plants and green algae to absorb the sunlight

Figure 12.7
Trends and Milestones in the Evolution of Green Algae and Plants

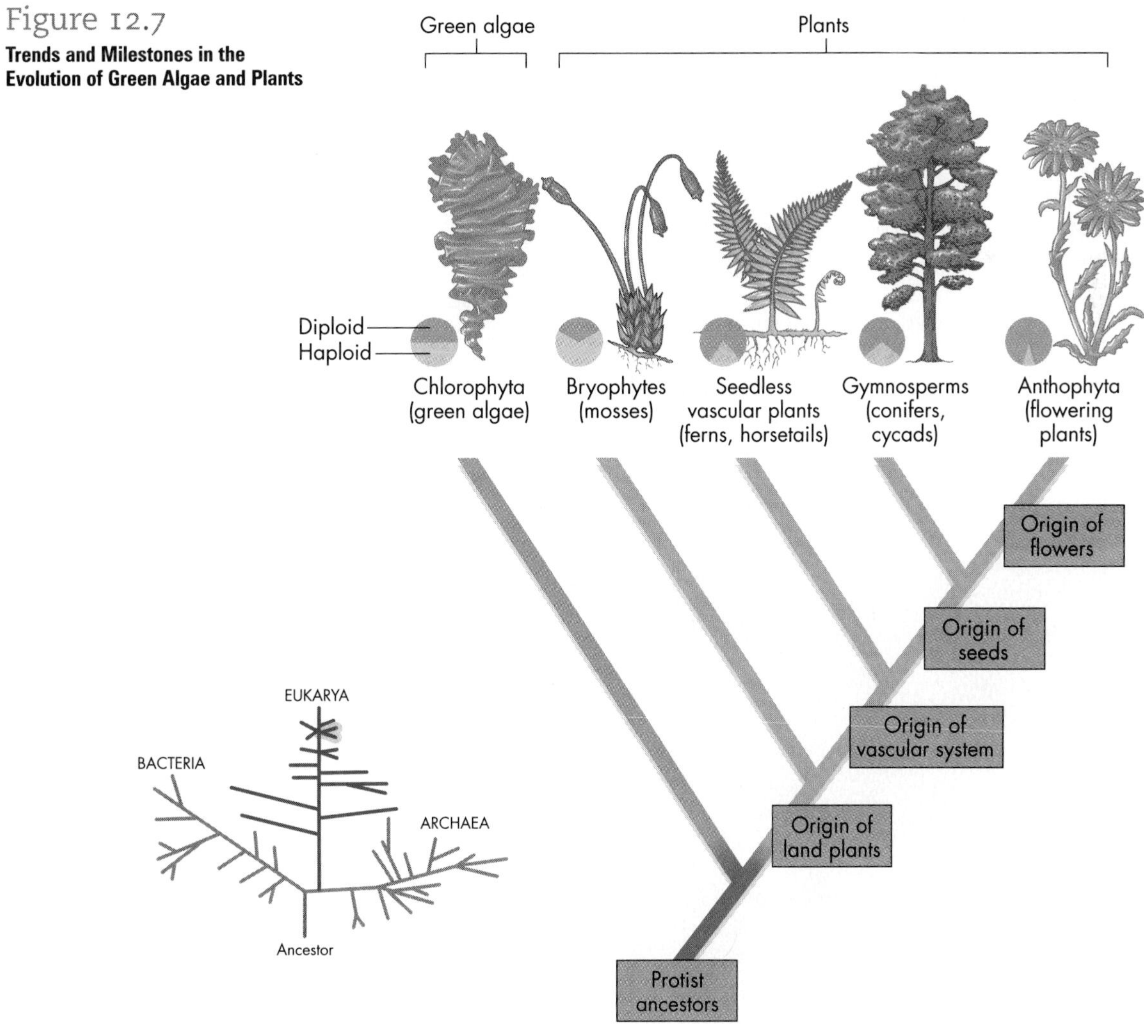

alternation of generations the two different generations of multicellular individuals, one haploid and one diploid, that occur in the life cycle of a plant

sporophyte (Gr. *spora,* seed + *phyton,* plant) the diploid, spore-producing stage in the life cycle of many plants

gametophyte the haploid, gamete-producing phase in the life cycle of plants

embryo an organism in the earliest stages of development; in humans this phase lasts from conception to about two months and ends when all the structures of a human have been formed; then the embryo becomes a fetus

penetrating air or shallow water with maximum efficiency. Green algae, however, do not develop from embryos, as do plants. And the diploid and haploid generations of green algae tend to look the same, while in plants, the haploid and diploid generations have different body forms, as we'll see shortly.

Some green algae, like *Chlamydomonas,* are single-celled. *Chlamydomonas* is a favorite laboratory organism for investigating how genes control mating, how the lashing movements of flagella can propel a cell, and what are the biochemical details of photosynthesis and aerobic respiration. *Chlamydomonas* is an oval cell propelled through freshwater pools and moist soil by two flagellae. Other green algae, like the delicate *Volvox,* grow in colonies. And still others, like *Ulva,* the sea lettuce, inhabit tidepools and grow a delicate, leaflike body, just two cells thick, that resembles sheets of green cellophane (Fig. 12.8). Reproductive spores of *Ulva* bear a striking resemblance to a single individual of *Chlamydomonas,* suggesting close evolutionary ties.

The life cycle of the green alga *Ulva* shows an **alternation of generations**, a regular alternation of multicellular haploid and diploid body forms. *Ulva's* life cycle shows the basic features of this alternation (Fig. 12.8). The diploid phase of the life cycle consists of a leafy sheet of many cells (Step ①). Special cells within the sheet undergo meiosis and produce haploid spores (Step ②). Because the diploid phase produces spores, it is called the **sporophyte** generation (*sporo* = spore + *phyte* = plant, or in this case, plantlike). The haploid spores divide by mitosis into multicellular sheets that have *the same body form* as the diploid generation (Step ③). The haploid organism is called the **gametophyte**, because special cells within the sheets produce haploid gametes (Step ④). In *Ulva*, the two kinds of gametes look the same. This contrasts with the sperm and eggs of plants and animals, which have different forms. When *Ulva's* haploid gametes fuse (Step ⑤), they produce a diploid zygote (Step ⑥), which then grows by mitosis into another leafy sporophyte (Step ①).

The alternation of generations shown by some green algae occurs in a more elaborate form in plants. Botanists think that ancient aquatic green algae probably gave rise to the simplest land plants, which still rely on water for reproduction. Let's move on, now, to explore the physical trends that emerged over time and allowed life to inhabit the land.

Green Algae

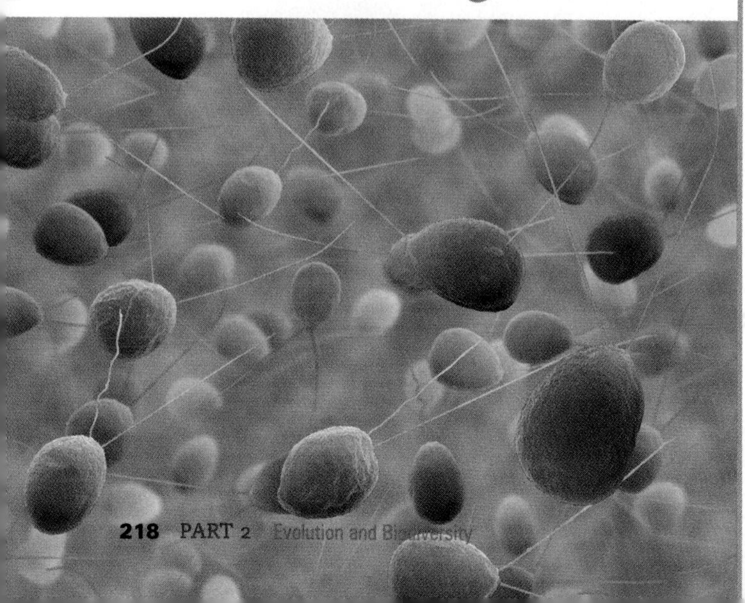

LO3 Plants: Green Embryo Makers

Plants can be small and colorful, like chiles, with their brightly hued pepper fruits. They can be gigantic, like the redwoods of the Pacific coast. They can be minute, like the shimmering green duckweed growing on a pond's surface. Or they can be drably camouflaged, like stone plants. Regardless of differences, however, plants have the common physical characteristics we listed earlier, including this: a plant generally develops from an **embryo**, an immature form of an organism undergoing the early stages of development. The diploid multicellular embryo is supported by a haploid maternal plant. Plant embryos result from sexual fusion, and most plants can reproduce sexually. A brilliantly colored chile pepper, for example, contains within its seeds embryos resulting from sexual reproduction. Plant cells also generally contain chloroplasts (see Chapter 3), green cytoplasmic organelles that transform the energy of sunlight into sugars. Fungi and animals lack chloroplasts or related organelles. The Chromista (the pigmented organisms we discussed in Chapter 11) have chloroplasts whose evolutionary origin may have differed from the chloroplasts in green algae. Plant cells have cell walls made chiefly of cellulose. Recall that although fungal cells have walls, they differ in composition and are usually chitin.

Plants play ecologically crucial roles. They dominate most terrestrial landscapes: just picture a forest or prairie without plants. They sustain the lives of most animals, including humans, through the production of food, fibers, fuel, and shelter. In addition, plants release the oxygen that aerobic organisms, including animals, utilize during aerobic respiration. Finally, plants display an amazing array of substances called **secondary metabolites** that are not directly needed for the organism's energy gathering and reproduction, but defend the plant against animals, fungi, and even other plants. The capsaicins in chiles are secondary metabolites, and so are the caffeine in coffee beans and the nicotine in tobacco leaves. As we saw, capsaicins probably protect chile plants from mammals with their seed-killing digestive tracts. Caffeine and nicotine also protect against certain types of animal predators.

Plant Life Cycles

As in some green algae, the life cycles of plants have a regular alternation of multicellular haploid and diploid generations. This alternation of generations is a key to understanding plant biology. In many plants, the haploid and diploid generations are separate, and each phase is a free-living, multicellular form. In other plants, the gametophyte may consist of just a few cells and be inconspicuous, while the sporophyte is a larger multicellular organism. Figure 12.9 shows the alternating generations in the moss life cycle.

secondary metabolite
substances that are not directly needed for a plant's energy gathering and reproduction, but defend the plant against animals, fungi, and even other plants

The familiar short, fluffy, green moss plant that covers the ground in moist habitats is the haploid gametophyte (gamete-producing) phase of the moss life cycle (Step ①). Female gametophytes produce haploid eggs, and male gametophytes produce haploid sperm (Step ②). Sexual fusion, or fertilization, produces a diploid zygote (Step ③), which grows by mitosis into a diploid embryo protected in the tissues of the maternal gametophyte plant. The embryo grows into a diploid sporophyte, a spore-producing portion of the life cycle (Step ④). The moss sporophyte is often a brown structure that looks like a miniature light post and grows up from the gametophyte. As in *Ulva*, specialized cells in the moss sporophyte produce haploid spores by meiosis (Step ⑤). These spores then germinate and once again grow by mitosis into the gametophyte generation (Step ①). Notice a key feature that distinguishes the life cycle of a moss and other plants from that of your own: In humans and other animals, meiosis produces haploid cells that themselves become gametes, eggs or sperm. In contrast, in plants, the immediate cellular products of meiosis do not become gametes directly, but grow into separate multicellular haploid individuals, which eventually produce the gametes.

Figure 12.8
Green Algae: Plant Ancestors

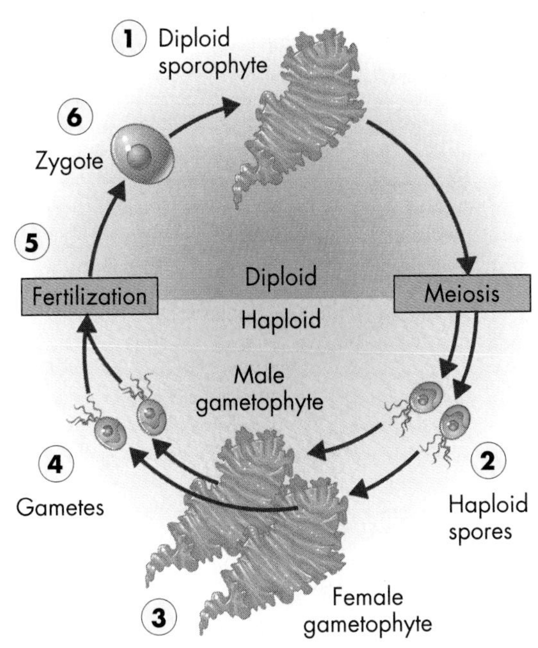

vascular tissue
plant tissue that conducts fluid throughout the plant and helps strengthen roots, stems, and leaves; consists of xylem and phloem cells

bryophyte
the division of the plant kingdom comprising mosses, liverworts, and hornworts

As plants evolved from ancestral green algae, the sporophyte generation gradually took on more and more prominence in plant life cycles (see Fig. 12.7). As we saw in *Ulva*, a living green algae, sporophytes and gametophytes look much the same (see Fig. 12.8). In plants, however—moss, for example—the two phases always look different: The moss gametophyte is green, longer lived, and more complex, and consists of many more cells than the sporophyte. The sporophyte tends to be brown and has a "head" raised on a slender stalk that facilitates the dispersal of haploid spores (Fig. 12.9). In more complex land plants such as chiles, coconut palms, and hibiscus, the diploid sporophyte phase dominates and is the conspicuous plant we recognize.

Figure 12.9
The Plant Life Cycle: Alternation of Generations

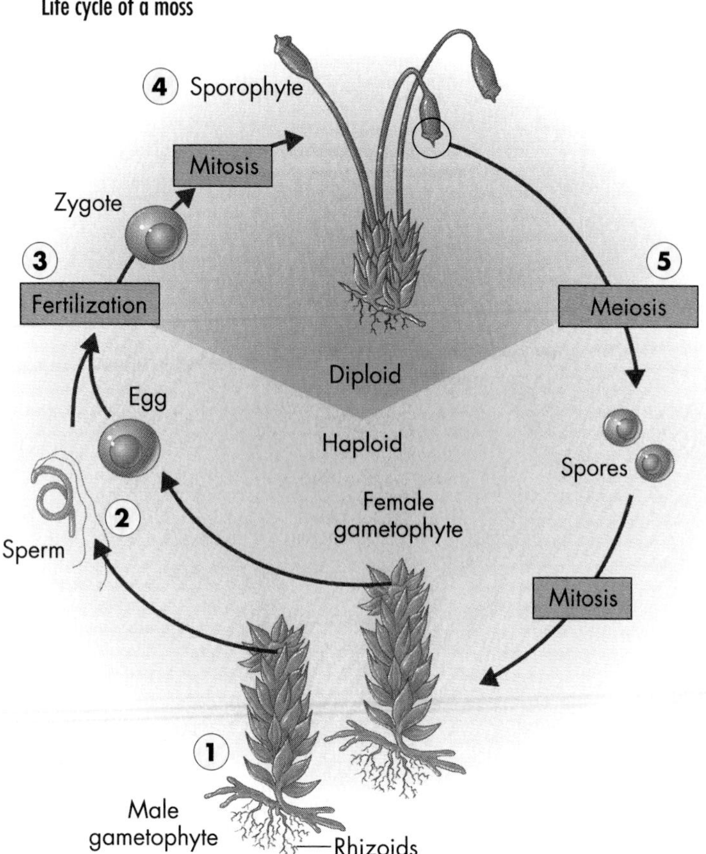

Trends in Plant Evolution

Evidence from fossils as well as from plant genes suggests that plants probably originated from ancient aquatic green algae (Fig. 12.7). The first plants to make the transition from water to land probably survived on the damp fringes of ancient oceans over 450 million years ago (see Table 10.2 on page 222). At that time, the continents were relatively flat. Water levels probably changed dramatically as the seasons changed and when climatic cycles lengthened. Some of the first transitional land plants had lucky combinations of alleles that resulted in adaptations to withstand desiccation (drying out). These drought-resistant individuals would have tended to reproduce more successfully in fluctuating environments than would individuals without the allele combinations. Adaptations to drying include a waxy outer layer perforated by holes that allow gas exchange, and reproductive structures that resist desiccation. Adaptations like these opened up spacious new frontiers in damp areas near oceans. Sunlight was plentiful, the soil held rich mineral nutrients, and initially there were no plant-eating animals on the land.

Biologists think that other evolutionary milestones occurred after plants made that first transition to land. Fluid-conduction tubes or **vascular tissue** evolved and enabled water and nutrients to move efficiently throughout the plant. (We'll see details of this later in the chapter.) Seeds then originated in certain plant types, followed by flowers in other types. Seeds and fruits helped improve the chances of young plants being dispersed to new areas, and flowers helped the chances of successful fertilization of eggs by pollen grains. We already saw that plants underwent a general trend toward the dominance of the diploid generation (the sporophyte). This trend, along with the innovations for resisting drought, for conducting water and nutrients, and for dispersal and fertilization, led to the vast diversity of plants we see today. These form a spectrum from simple, low-growing mosses that lack vascular tissue to showy chile plants with their vascular systems, flowers, seeds, and specialized hot peppers.

Simple Land Plants: Still Tied to Water

Walking through a forest today you can still see plants similar to the earliest survivors on land. All you have to do is look for **bryophytes**: the mosses, liverworts, and hornworts. These plants grow low to the ground and require droplets of water for reproduction. In contrast, the seedless vascular plants, including ferns, horsetails, and club mosses, are stiffened by an internal transport system that allows them to grow taller than the bryophytes. Even with their superior height, though, the seedless vascular plants still require water to reproduce (for example, fern sperm must swim to get to the egg). In time, the seedless vascular plants were surpassed by the seed-forming vascular plants, the gymnosperms (such as pine trees) and angiosperms (*Anthophyta* or flowering plants). Both groups evolved hardy seeds and, for the first time, freedom from environmental water for reproduction. The final major evolutionary innovation was flowers, showy but sterile reproductive structures that attract animals carrying pollen, like bees, moths, and hummingbirds. The inadvertent transport of pollen to individual plant species by specific animal species has helped to rapidly diversify flowering plants into thousands of beautiful species. Genetic and fossil data suggest that an ancient green alga made the transition to land and then later diversified into mosses (bryophytes) and ferns, horsetails, and club mosses (seedless vascular plants). The seedless vascular plants, in turn, gave rise to seed-forming plants, the gymnosperms and angiosperms (Fig. 12.7).

For plants to thrive on dry land, they needed a variety of novel adaptations for the following functions:

1. Avoiding desiccation: Without surrounding water, they had to minimize evaporation while still allowing oxygen and carbon dioxide gases to move in and out of the plant.
2. Support: Without buoyancy from surrounding water, plants needed a strong internal system of support for the plant body.
3. Transport: The visible, above-ground parts of the plant needed structures that absorb water and minerals from the soil, mud, or sand, and a set of tubelike structures that distribute them throughout the plant body.
4. Reproduction: Unable to shed gametes directly into an ocean, lake, or pond, land plants had to develop ways to reproduce in air. As you'll see, the relative success of each group of land plants depends, in part, on how its particular set of adaptations met these challenges.

Bryophytes: Pioneers on Land

The modern bryophytes include mosses, liverworts ("lobed plants"), and hornworts ("horn-shaped plants"). Second only to the flowering plants in number of species, bryophytes are small organisms that generally stand less than about 3 cm (1 1/4 in.) tall (see Fig. 12.9 and Table 12.2). They are often among the first plants to colonize a new area, and in most of the species, the sporophyte (which is diploid) grows like a miniature street lamp from the leafy gametophyte (which is haploid). These small, simple plants have waterproof coatings that help prevent them from drying out. Bryophytes have fairly rigid tissues that help keep the low plants upright. Mosses, liverworts, and hornworts, however, are small enough that most don't have the kind of internal transport system that absorbs or moves water around inside the plant and lends rigidity. Instead of true roots, most bryophytes have hairlike rhizoids that act only as anchors, usually absorbing neither water nor minerals. Water reaches the individual plant cells by slowly diffusing through the entire organism, like a paper towel soaking up a spill. Because of this dependence on diffusion, the plants are limited in size, and are often able to grow only in shady, moist places. In addition, bryophytes can reproduce sexually only where it's at least seasonally wet. This is because their sperm swim about with flagella as do the gametes and certain body cells of so many green algae. Bryophyte sperm can reach and fertilize an egg only when the plant is drenched in water.

The life cycles of most moss species alternate between a conspicuous gametophyte generation—the green, "mossy" plant—and a slightly less obvious sporophyte generation—the red, yellow, or brown "lamp post" in many common mosses. While many bryophyte species have survived in moist habitats, this pioneering group did not itself give rise to other more complex land plants.

seedless vascular plant
a plant with an internal transport system but that requires standing water to reproduce because it does not produce seeds; includes horsetails and ferns

seed-forming vascular plant
a plant with a vascular system that also produces seeds; the gymnosperms and angiosperms

gymnosperm
(Gr. *gymnos,* naked + *sperma,* seed) conifers and their allies; a primitive seed plant whose seeds are not enclosed in an ovary

angiosperm
a flowering plant

flower
a reproductive structure that contains the carpel and at least one stamen

rhizoid
(Gr. *rhiza,* root) in bryophytes, some algae, and fungi, a hairlike structure that anchors the organism to the substrate

Vascular Plants

Bryophytes are generally quite successful in moist shady habitats, but home gardeners know that many kinds of vegetables and herbs "prefer" to grow in raised beds because they "hate wet feet." This is true,

vascular plant a plant that possesses an internal transport system for water and food in the form of xylem and phloem cells

for example, for chile plants. Somehow, chiles have to get water up into their leaves, flowers, and fruits knee-high off the ground, and this goes for any plant with a significant vertical dimension. Upright plants clearly require some method for moving water against gravity, and this method—a system of internal tubes (the vascular system)—evolved a bit more than 400 million years ago, about the time the first insects appeared on land.

The **vascular plants**, which include the ferns, the pine trees (gymnosperms), and chiles (angiosperms) have specialized fluid conduction systems that solve the problem of how to transport water, minerals, and

Table 12.2

Green Algae and Major Plant Divisions

Group	Division	Examples	Characteristics
Angiosperms (Flowering plants)			
Embryo-Producing plants; Vascular plants; Seed plants	Monocots (50,000 species)	Lilies, corn, onions, palms, daffodils	Leaves with parallel veins; seedlings have just one "seed leaf" or cotyledon; flower parts usually occur in multiples of three; seed stores much endosperm
Embryo-Producing plants; Vascular plants; Seed plants	Dicots (225,000 species)	Roses, apples, beans, daisies	Leaves with netlike veins; seedlings have two cotyledons; flower parts in multiples of four or five; seed stores little endosperm
Gymnosperms			
Embryo-Producing plants; Vascular plants; Seed plants	Gnetophyta (70 species)	*Welwitschia*	Low-growing, cone-producing native of southwestern Africa, with long, flat twisting leaves
Embryo-Producing plants; Vascular plants; Seed plants	Coniferophyta (Pinophyta) (600 species)	Pines, sequoias, firs	Naked seeds produced in cones; usually have needlelike leaves or scales; produce pollen; well-developed vascular system; true roots, stems, and leaves; sporophyte is dominant and supports gametophyte; conifers harvested in great numbers for wood products
Embryo-Producing plants; Vascular plants; Seed plants	Ginkgophyta (1 species)	Ginkgos	Round fleshy cones on female trees; leaves turn golden and fall each year
Embryo-Producing plants; Vascular plants; Seed plants	Cycadophyta (100 species)	Cycads	Flourished 200 million years ago; seeds carried on open reproductive surfaces on cones
Embryo-Producing plants; Vascular plants	Seedless vascular plants (Sphenophyta, Pterophyta, and Lycophyta; 13,000 species)	Ferns, horsetails, club mosses	Vascular pipelines; rhizomes, stems, and fronds; gametophyte can be tiny, independent plant or grows from sporophyte; grow on shady forest floors in low-lying damp areas
Embryo-Producing plants	Bryophytes (Bryophyta; 24,000 species)	Mosses, liverworts, hornworts	Waterproof coatings, rigid tissues for upright growth on land, rootlike rhizoids; haploid generation (gametophyte) is dominant; often the first plants to colonize an area
	Green algae (Chlorophyta; 7000 species)	*Ulva*, *Chlamydomonas*	Produce carotene, chlorophyll, like the land plants; many have conspicuous haploid and diploid generations; ancestors to the land plants; found in fresh water

sugars (the products of photosynthesis) throughout the plant (Fig. 12.10a). These transport tubes are strengthened by long cells containing **lignin**, a substance that makes individual plant cell walls rigid enough to collectively support trees as huge as the General Sherman sequoia tree with a trunk wider than a good-sized truck. In addition to the internal transport and support system, early vascular plants had **stomata**, as well: tiny openings in the plant surface that admit carbon dioxide into the plant's interior. These adaptations enabled vascular plants to tolerate drier habitats more efficiently than bryophytes and thus to spread farther inland into unoccupied areas. The sporophytes of vascular plants also evolved the ability to grow from the tips of the plants instead of below the tip, as in bryophytes. This allowed the diploid plant to assume a branching growth habit and with it, the production of many spore-producing organs on a single sporophyte and, in turn, many more spores and greater reproductive potential.

lignin
a plant compound that gives cells and tissues strength and hardness

stoma (pl. stomata)
a tiny hole surrounded by two guard cells in the epidermis of plant leaves and stems, through which gases can pass

The Seedless Vascular Plants: Horsetails and Ferns

Ferns are probably the most familiar of the seedless vascular plants, plants that have a vascular transport system but don't form seeds. Seedless vascular plants include the horsetails, club mosses, and ferns (see Table 12.2). In the tropical climates of the Paleozoic era, a bit less than 400 million years ago, these first vascular plants grew as vast primordial forests, and reached sizes much larger than today's tree ferns. However, with the dawning of the Mesozoic era about 245 million years ago, the climate gradually grew colder and drier, and these giants disappeared, leaving behind smaller representatives of each group, such as

Figure 12.10
The Seedless Vascular Plants

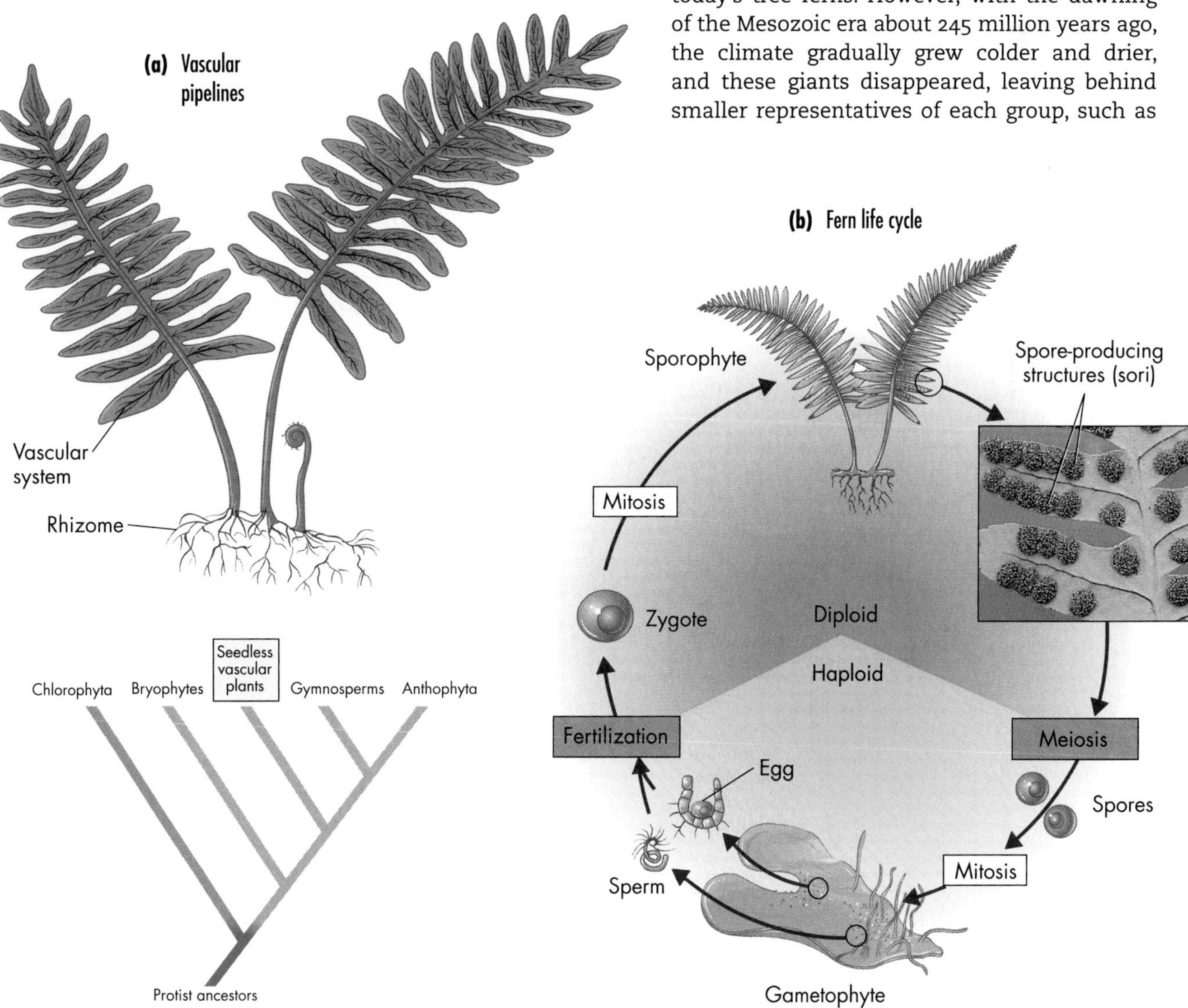

rhizome
(Gr. *rhiza,* root) an elongated underground horizontal stem

true stem
an organ of usually vertical support housing vascular tissue

frond
the leaflike structure of an individual alga that collects sunlight and produces sugars; also refers to the large divided leaf on a fern

the modern horsetails. In a different legacy, the bodies of the fallen giants were only partially decomposed by fungi and bacteria; this left organic compounds that became transformed over millions of years into the enormous coal deposits people mine for fuel.

Some seedless vascular plants have horizontal stems called rhizomes that grow on or just beneath the ground surface and survive from year to year (Fig. 12.10a). Growing up from each rhizome are many erect, leafy stalks. In horsetails and lycopods, the erect parts are true stems; in ferns, they are the central stalks of large leaves called fronds. At the end of each growing season, the stems and leaves die, and are replaced the following season by new aerial parts from new places on the slowly expanding rhizomes. This replacement is the way seedless vascular plants reproduce asexually. Figure 12.10b shows the main stages in a fern's sexual reproduction.

Horsetails, lycopods, and ferns show a continuation of the trend toward dominance by the diploid generation (review Fig. 12.7): In these seedless vascular plants, the sporophyte is the conspicuous adult, and the gametophyte is a small, free-living green plant. Like the mosses, ferns, horsetails, and lycopods retain swimming sperm and require water for sexual reproduction. In a sense, the seedless vascular plants are the botanical equivalents of frogs, salamanders, and other amphibians, which can live on dry land but must have water to lay their eggs. Interestingly, giant amphibians and seedless vascular plants were the dominant land organisms during the warm, swampy Paleozoic era.

Figure 12.11
The Invention of the Seed: A Three-Generation Family

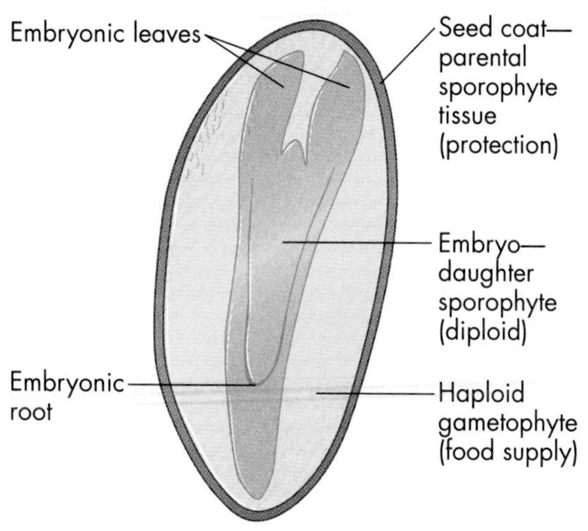

The Invention of Seeds

If you want to grow chiles in your garden or in a clay pot on your patio, you simply visit a gardening store in the spring, pick an envelope off the rack, and at home, shake out the smooth yellowish discs—the chile seeds. We take seeds for granted, but they are amazing little devices. Seeds can survive for several years just sitting on a shelf, yet when they are moistened, they can burst open, allowing the little embryo housed inside—protected by the seed coat and surrounded by stored food—to grow into a new plant (Fig. 12.11). The arrival of seeds spurred a biological revolution that dramatically changed Earth's landscapes.

We've already encountered the two major groups of seed plants, the gymnosperms and angiosperms. In the gymnosperms or naked-seed plants, including yew and pine trees, the seeds are exposed on the surface of the plant's spore-forming organ (as in a pine cone). In the angiosperms or *Anthophyta*, including chiles and other flowering plants, maternal tissues surround the seeds, forming organs called *fruits*. As you might suspect, gymnosperms were the first to evolve.

Gymnosperms: Plants with Naked Seeds

As the Mesozoic era began about 245 million years ago, the great megacontinent Pangaea was breaking apart, the continental seas had retreated, and new masses of high, dry land became open to plants and animals (see Table 10.2). While the seedless vascular plants were tied to swampy lowlands, riverbanks, and coastal lagoons, lines of plants that had already evolved with reproductive innovations—seeds—were able now to dominate the land. These organisms were the gymnosperms ("naked-seed" plants; in ancient Greece, the *gymnasium* was a place where athletic contests were held in the nude). Gymnosperms included the now-extinct seed ferns as well as the cycads, ginkgoes, and conifers, such as yews, pines, firs, sequoias, and redwoods (see Table 12.2). The reproductive innovations of this group include:

1. the pollen grain
2. the seed
3. a further shift toward the dominance of the diploid (sporophyte) generation

The large, familiar pine tree is an example of the sporophyte generation of a gymnosperm (Fig. 12.12). The sporophyte makes small, male cones, which contain hundreds of compartments, each of which contains hundreds of spore-forming cells. In the early spring, these spore-forming cells undergo meiosis and produce small haploid spores. Cell division takes place in each spore and generates a small gametophyte consisting of just four cells; this is the pollen grain. Inside the pollen grain, one of the cells will develop into the sperm. The wind can then airlift the pollen, with the sperm inside, far from the parental plant, where it can meet an egg. Notice how much smaller this gametophyte is—the thickness of a book page—compared to the size of the sporophyte—the huge pine tree. Contrast this with the sporophyte and gametophyte of a moss, wherein the gametophyte is the larger plant (see Fig. 12.9).

A gymnosperm's egg forms in a way roughly similar to the sperm formation in male cones: Cells in the female cone undergo meiosis and produce haploid spores, but these are larger than the ones produced by the male cones. Botanists call the two sizes of spores microspores, which form the male gametophyte, and megaspores, which form the female gametophyte. The production of these different-shaped spores occurs not only in the gymnosperms but carries through in the flowering plants, including the chiles.

The female gametophyte of a pine tree remains microscopic and unable to undergo photosynthesis, but it produces the egg cell. The egg cell—surrounded by cells of the gametophyte mother, which are in turn surrounded by cells of the sporophyte mother—is called the ovule. The ovule becomes the seed when the egg is fertilized (Fig. 12.12). The meeting of the egg and sperm produces the zygote, the start of the next diploid, sporophyte generation. A seed therefore contains cells from three plant generations: the outer protective layers of the parental diploid sporophyte, the nutritive cells of the haploid gametophyte, and the diploid zygote of the next sporophyte generation, the next new pine tree.

It is curious that the dominant land plants and animals in the Mesozoic era, the gymnosperms and reptiles, evolved similar reproductive strategies for life on land. In both groups, water for fertilization comes from moisture in the tissues or reproductive tract of the parent rather than in splashing raindrops or standing water. Furthermore, with their tough seed coats and enclosed nutrients for the embryo, gymnosperm seeds are analogous to reptilian eggs. These too have tough shells and an internal food supply (yolk), and they could be laid and hatched on land. Seeds and shelled eggs freed both plants and animals from the dependence on water that the more ancient bryophytes and amphibians had—and still have today.

Seeds and shelled eggs freed both plants and animals from dependence on water.

pollen grain the male gametophyte of seed plants

microspore a pollen grain that gives rise to the male gametophyte in certain vascular plants

megaspore in certain plants, a haploid spore that gives rise to the female gametophyte

ovule the structure that contains the egg and later matures into the plant seed

Diversity among the Gymnosperms

Gymnosperms share certain important traits, primary among them that their seeds are not enclosed in an ovary. But the members of this group don't occupy a single branch on the tree of life (Table 12.2). Instead, they are a collection of four living divisions

Figure 12.12
Innovations of the Gymnosperms

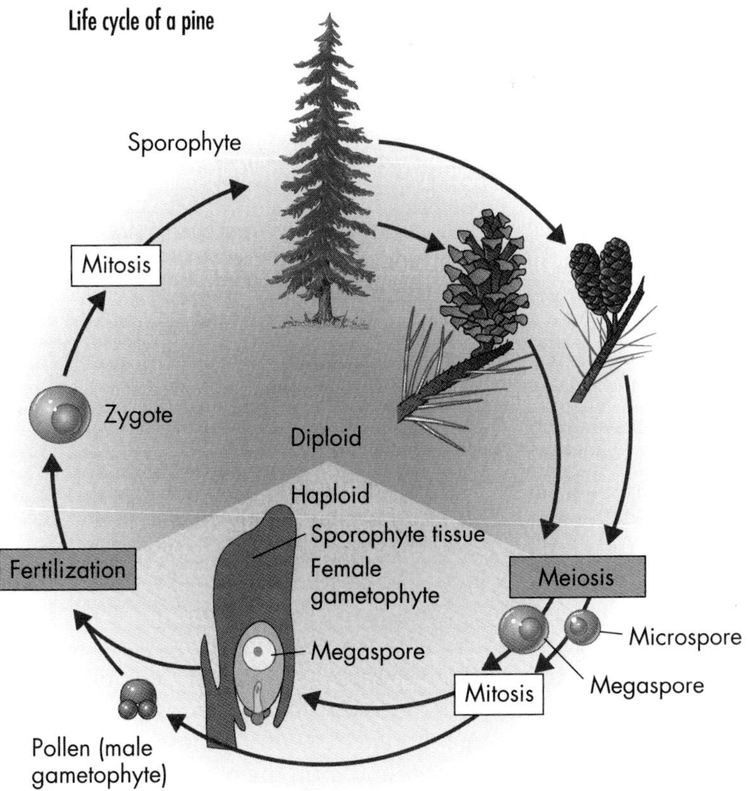

conifer
a cone-bearing tree

cuticle
a waxy or fatty noncellular, waterproof outer layer on epidermal cells in plants and some invertebrates; in parasitic flatworms it prevents the worm from being digested in the gut of the host organism; in insects, the exoskeleton

carpel
a flower's female structure; the carpel is often shaped like a wine bottle and contains the ovary housing the ovules, a necklike style, and a sticky stigma on which pollen grains germinate

ovary
egg-producing organ

stamen
the male or pollen-producing structure within a dicot or monocot flower; each stamen consists of the anther containing pollen sacs and a supporting filament

fruit
in flowering plants, a ripe, mature ovary containing seeds

seed
the fertilized mature ovule of a gymnosperm or angiosperm, containing an embryonic plant

(or phyla): the cycads (SIGH-cads), the ginkgo, the gnetophytes (KNEE-toe-fites), and the conifers. Cycads look roughly palmlike and are found in tropical areas; only one species is native to the United States and was once common but is now rare in Florida's sandy woods. Ginkgo, or maidenhair tree, is an ornamental with fan-shaped leaves that turn a brilliant gold in the fall. Many herbal medicine producers use ginkgo extracts in their preparations for "memory improvement" and other claims. Gnetophytes are a group of naked-seed plants that display some features in common with flowering plants, including some characteristics of their cones and vascular system (see Table 12.2).

Conifers: Familiar Evergreens

The cone-bearing trees called **conifers** are the most familiar and largest remaining group of gymnosperms, and they include many ecologically and economically important species. Millions of acres in mountainous regions are dominated by pine, spruce, fir, and cedar, much of which are harvested for wood, paper, and resins. Smaller conifers, such as yew, hemlock, juniper, and larch, are often used for the graceful landscaping of buildings and parks. Finally, the massive sequoias and redwood trees are conifers as well as living reminders of the Mesozoic era, an age of giants.

Conifers have two distinctive characteristics: their narrow leaves and familiar woody cones. Their needle-shaped leaves are covered by a waterproof **cuticle**, or waxy layer. The needle shape resists drying, having little surface area relative to volume. In spring, at the start of the conifer life cycle, a large pine tree sprouts thousands of soft male cones on lower branches and the more familiar large, hard female cones on upper branches. Cells in these cones produce pollen and eggs and form the pine seeds.

Conifers could colonize higher and drier reaches of the continents during the Mesozoic era because of their numerous attributes:

- Drought-resistant leaves
- Protective seed coats
- Haploid male gametophytes reduced to a few cells that become airborne pollen grains
- Female gametophytes also reduced to a few cells that produce eggs protected in the ovule
- A well-developed vascular system that produces wood and stiffens the trunk, branches, and roots
- The tendency to form the kinds of mycorrhizal associations with fungi we saw earlier
- The ability to survive and produce new generations, sometimes for several centuries

Because of these modifications for life on land, conifers are still a successful group, with more than 600 modern species. Nevertheless, one final set of evolutionary changes gave flowering plants—including our case history plants, the chiles—their still greater success.

LO4 Flowering Plants: A Modern Success Story

When you think of a chile, you usually think not of the plant but of the red, hot pepper. That structure, the chile plant's fruit, is one of the important innovations of the most recent group of plants to evolve, the *Anthophyta,* or flowering plants. Naturally, the flower is the other! These plants (also called angiosperms, or *Magnoliophyta*) appeared as the continents rose and drier, colder conditions became more commonplace on Earth's landmasses about 165 million years ago. During the modern, or Cenozoic, era the flowering plants became—and remain—the dominant land plants (review Fig. 12.7).

As flowering plants came to dominate the landscape, mammals became the dominant land animals. Both groups of organisms evolved with reproductive structures that protect and nourish developing embryos. Mammals evolved the placenta and the womb, while the flowering plants evolved a new reproductive structure, the *flower*, consisting of a female flower part, the **carpel** (containing at least one **ovary**, the maternal tissue that houses and protects the ovules), and at least one **stamen**, the part of the flower that produces the male gametophyte, the pollen. Flowering plants also evolved **fruits**, the mature, ripened ovary containing the **seeds**.

{"We Are the Champions!"}

Today, flowering plants are the most common and conspicuous species in Earth's tropical and temperate regions. It is not surprising, therefore, that when some biologists claimed that the largest living thing is a fungal clone, some botanists, not to be outdone, claimed that a flowering plant is actually the champion. They located a huge stand of quaking aspens in Utah. This 106-acre, 6,000-ton stand arose from a single original seed, and each shoot arises from a single immense root system. This giant clone weighs more than the fungal clones. But the rivalry continues.

Flowers help the process of pollination, and fruits promote seed dispersal. Flowering plants also usually have broad leaves that collect light efficiently. Flowers, fruits, and broad leaves, along with a rapid life cycle, together allowed flowering plants to radiate into more species than the combined numbers of all other plant groups. Figure 12.13 gives a brief overview of the flowering plant life cycle.

From the monumentally successful flowering plants of tropical and temperate regions come virtually all of our crop plants (wheat, rice, corn, soybeans, fruits, and vegetables such as green peppers) and beverages (coffee, tea, colas, and fermented drinks), as well as spices (including paprika and chile powder), cloth, medicines, hardwoods, ornamental plantings, and, of course, flowers—symbols of beauty, affection, and renewal throughout human history. Table 12.2 shows the two major groups of flowering plants—the **monocots**, which have a single embryonic leaf, and the **dicots**, which have two embryonic leaves—and their main characteristics.

monocot
short for monocotyledon, the smaller of two classes of angiosperms (flowering plants); embryos have only one cotyledon, the floral parts are generally in threes, and leaves are typically parallel-veined

dicot
short for dicotyledon, the larger of the two classes of angiosperms (flowering plants); characterized by having two cotyledons (seed leaves), the floral parts are usually in fours or fives, and leaves are typically net-veined

Figure 12.13
Innovations of Flowering Plants

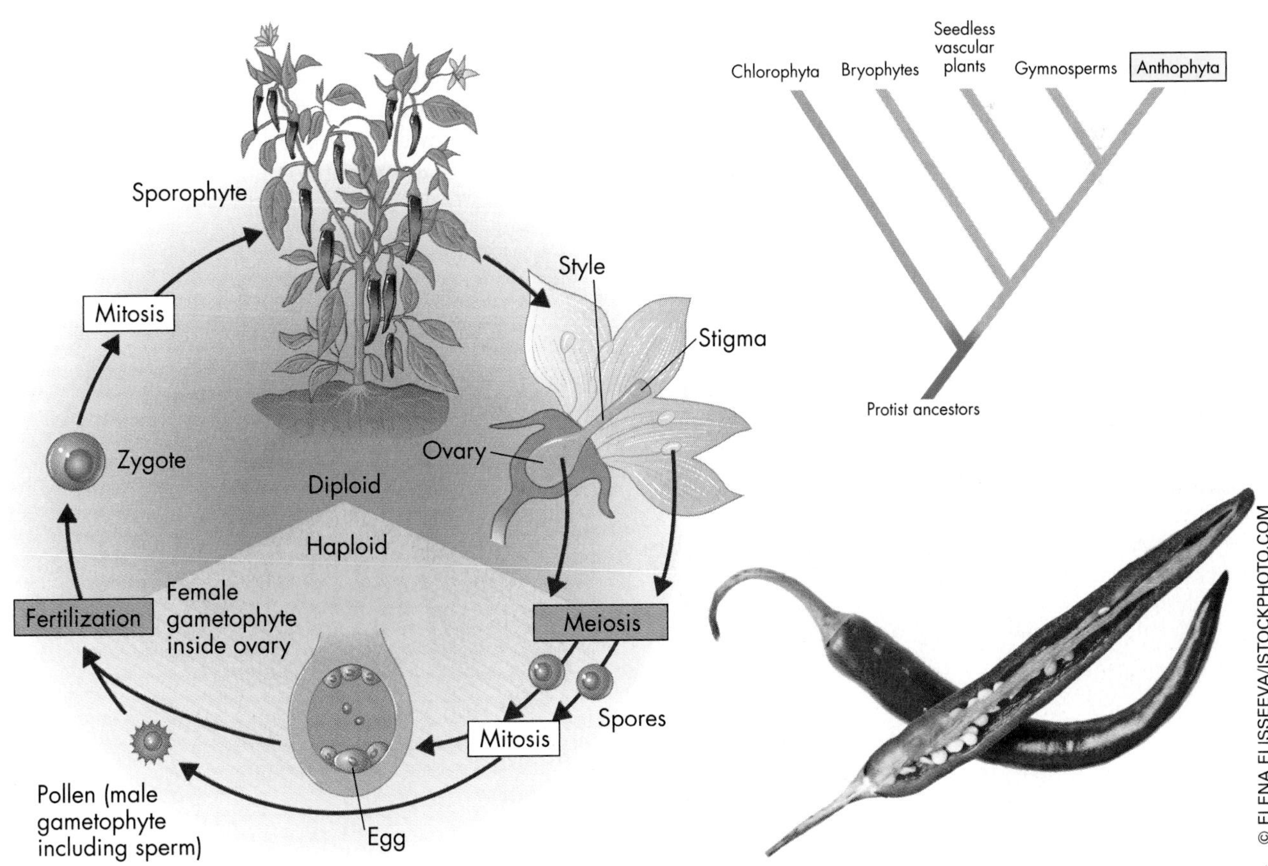

self-pollination
the ability of the pollen from a single flower to fertilize an egg within the same flower; a type of self-fertilization

pollination
the transfer of pollen grain to female flower parts

nectar
a sugary fluid produced by flowers that attracts butterflies, aphids, ants, and other animals

Flowers and Fruits and the Coevolution of Plants and Animals

Chiles are typical flowering plants in many ways, with their flowers of various sizes and hues and their often large, showy fruits in green, gold, orange, red, or brown. Another common trait is **self-pollination**, the ability of the pollen from a single chile flower to fertilize an egg on the same flower. But chile plants have other means of **pollination**, or pollen dispersal for the fertilization of eggs. Chile pollen can also be dispersed by wind from one individual to the next, allowing cross-pollination between different plants. And like so many flowering plants, chile flowers can also attract animals that inadvertently bring about cross-pollination in a way unavailable to naked-seed gymnosperms.

Chile flowers are usually small, but they produce **nectar**, a sugary fluid that attracts butterflies, aphids, ants, and other animals. In general, when you see a plant with bright, showy flowers, you can usually assume that the blossoms attract insect, bird, or mammal pollinators. These "envoys" accidentally carry loads of pollen from one plant to another as they forage for sweet nectar or for the protein-rich pollen grains. A flying or walking animal dusted with pollen and scouting for more of its favorite food will substantially increase the chances of pollination, whether in a chile flower, a wild rose, or an apple tree.

After pollination and fertilization, the seeds begin to develop, and the wall of the ovary eventually matures into a fruit. Fruits not only surround the seeds but also aid in dispersal by attracting hungry animals. Oftentimes, animals attracted to a fruit's colorful skin or pleasing flavor eat the fruit, and then excrete the seeds along with feces in some new location. Chile fruits might have evolved their bright colors as a means of attracting birds. At the same time, though, they probably evolved their fiery capsaicin compounds as repellents for mammals. When a chile seed passes through the digestive tract of a mammal, the seed is killed, but the molten taste of wild and most domesticated chiles tends to keep mammals away (excluding ourselves!). Why do we like the pungency of chiles? The production of pain is often followed by the release of natural brain opiates called *endorphins,* which mask pain and leave a sense of pleasure in its place. Just as people can become addicted to opium, they can become addicted to their own endorphins and will do seemingly irrational things on a regular basis—like eating hot peppers or running long distances—to get it! Birds have a much simpler reason for eating chiles: they can't taste the hot capsaicins due to subtle differences in their pain receptors. And the chile fruits are perfectly good sources of food calories and vitamins. When birds eat pepper fruits and seeds, the little life-units pass through their digestive tracts ready to germinate and are often dropped in new regions suitable for the growth of new chile plants.

© TONY CAMPBELL/ISTOCKPHOTO.COM

Coevolution: Edible Fruit + Hungry Bird = Seed Dispersal

© RUUD DE MAN/ISTOCKPHOTO.COM

Coevolution: Sticky Burr + Wandering Cow = Seed Dispersal

© STEVE BYLAND/ISTOCKPHOTO.COM

Coevolution: Nectar-filled Flower + Thirsty Bird = Cross-Pollination

© DAVID FRANKLIN/ISTOCKPHOTO.COM

© AMIT EREZ/ISTOCKPHOTO.COM

Coevolution:
Nectar-filled flower + Busy Bees =
Cross-Pollination

Who knew?

The primary adaptation of flowering plants is, of course, flowers. Flowers attract pollinators, increasing the success of cross-pollination, which increases genetic diversity. Flowers also develop fruits, which attract dispersers, increasing the likelihood of seeds being deposited farther from the mother plant, reducing competition for light, water, and nutrients, and perhaps increasing the geographic range of the species. If a fruit bearing seeds that have new genetic combinations gets deposited in an area that the new plants can adapt to, then the success of the species has increased. Over time, such incremental increases have added up to the astonishing diversity of the flowering plants.

The mutual dependence of flowering plants on pollinators and seed dispersers and of these animal species on the same plants for nutrition is an example of coevolution. Plants that had genes for flowers with sweet nectar, a strong fragrance, or a bright color might have survived in greater numbers because more animals would have visited them in their constant foraging for food and carried away pollen. The animals that were attracted to the new food sources probably also survived in greater numbers, thus perpetuating the interdependence.

This coevolution has resulted in specialized physical structures in both plants and animals that help ensure the tradeoff of nutrients for pollination and seed dispersal. The mouthparts of many types of bees, butterflies, moths, birds, and even bats are the right shapes for tapping the nectar or pollen of the flowers they visit. Flowers pollinated by hummingbirds, which see best in the red spectrum and have long, narrow beaks and a poor sense of smell, for example, have bright red flowers, such as fuchsia, with a long, slender shape but little fragrance. Conversely, flowers pollinated by bees tend to be yellow or blue and very fragrant, corresponding to the bees' vision and ability to perceive odors.

coevolution
the evolution of one species in concert with another as a result of their interrelationship within a biological community

Plants have evolved other mechanisms of seed dispersal, including "parachutes" that loft the seeds into the wind and hooks for hitchhiking onto animals passing by. The many kinds of seed dispersing mechanisms are almost as diverse as the thousands of color, size, and heat combinations of chile peppers.

© YOKE LIANG TAN/ISTOCKPHOTO.COM

Coevolution:
Hungry Monkey + Sweet Orange =
Seed Dispersal

© FRANK LEUNG/ISTOCKPHOTO.COM

Coevolution:
Seeds for eating or floating
= Seed Dispersal

© IVAR TEUNISSEN/ISTOCKPHOTO.COM

Coevolved?
What do you think?

13

The Evolution *and Diversity of* Animals

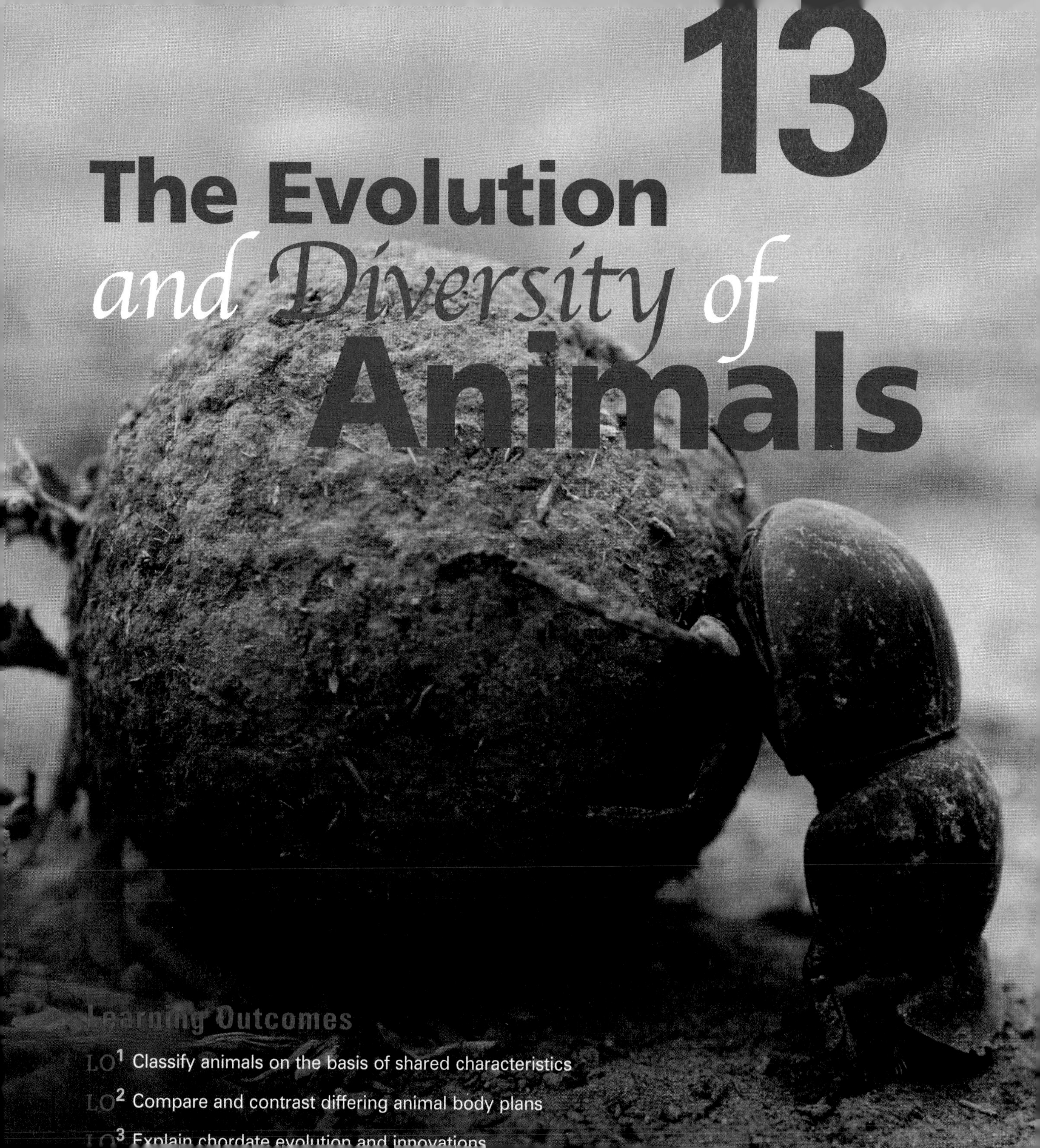

Learning Outcomes

LO[1] Classify animals on the basis of shared characteristics

LO[2] Compare and contrast differing animal body plans

LO[3] Explain chordate evolution and innovations

“No organism alive today is an ancestor of any other living organism.”

Not Your Average Hobby

Bruce Gill has the kind of hobby he can't discuss over dinner: dung beetles. Without dung beetles—nature's own pooper scoopers—the world would be a pretty smelly, foul place. Gill, a Canadian biologist, is best known for discovering evidence of dung beetles among fossilized blocks of feces left by giant plant-eating dinosaurs. Before that evidence was found, scientists had traced the evolution of the beetle lineage back 300 million years, but dated beetles that eat dung to only 40 million years ago. In that relatively recent geological era, mammals were already depositing their droppings as they roamed the continents. Gill's evidence pushed back the age of the oldest dung beetles to 76 million years and established them as the custodians of dinosaur detritus, as well.

What do you know?

Why do many zoologists characterize birds as “flying dinosaurs”?

Oh sure, other decomposers feast on animal droppings, too: earthworms, flies, molds, bacteria. All are crucial to ecological recycling. But none can compare to the dung beetles' speed and flair. One North American rancher who corrals his cattle in one paddock then shifts them to another made an interesting observation: 48 hours after moving out the animals, every speck of manure is gone—the work of 11 native dung beetle species. By contrast, Australian ranchers tried raising cattle on arid grasslands where only kangaroos had grazed. Australian dung beetles were adapted to little marsupial pellets and couldn't handle a great big cow pie. Soon the pies were piling up and plastering the ground into a hard dry pavement that suppressed grass growth. Only by importing and releasing bigger, hungrier dung beetles were the ranchers able to raise cattle in Australia.

Dung beetles are a good case history for this chapter because they remind us that ours is a planet of terrestrial insects: Two out of every three living species so far discovered and named by biologists are “bugs,” and nearly half of all species are beetles. From this group, we can learn a great deal about the animal body, animal evolution, animal diversity, relationships between the major animal groups, and the many threats to Earth's beetles—not to mention other species and ourselves. This chapter will explore animal life, and along the way, you'll find answers to these questions:

- What are the general characteristics that all animals share?
- What are the many different types of animal body plans?
- How did the chordates evolve and what are their innovative features?
- Where do primates, including humans, fit into the spectrum of animal species?

LO[1] The General Characteristics of Animals

What might lead you to classify a dung beetle as an animal and not, say, a plant or a fungus? Biologists look for the following characteristics:

- **Animals are multicellular**. Dung beetles have millions of cells (most prokaryotes and protists are usually single-celled organisms).
- **Animals are heterotrophs**. A dung beetle cannot manufacture its own food (like a plant can) and instead obtains nourishment by consuming other organisms or their byproducts.
- **Animals are self-propelled**. Dung beetles avoid danger or search for mates and food by moving about on their own at some point in their life cycle (plants and fungi have spores that move, but passively, not under their own power).
- **Animal bodies are diploid and develop from embryos**. Dung beetles, like every animal, grow from a zygote, which passes through various distinct stages of embryonic development before becoming an adult.

Innovations in Animal Evolution

One of the fascinating things about studying animals is that as we explore the major groups we can see how various animal features arose: heads, tails, wings, legs, backbones, internal organs, and so on. Fundamental to virtually all evolved animal traits—from simple to complex, and from ancient to recent—were four major anatomical and physiological innovations (Fig. 13.1):

1. Multicellularity. Animals evolved from single-celled protists, and very early on, such cells gained the ability to stick together in colonies.
2. Origin of tissues. Cells within colonies became specialized into inner and outer tissue layers

Figure 13.1
Animal Groups and Evolutionary Innovations

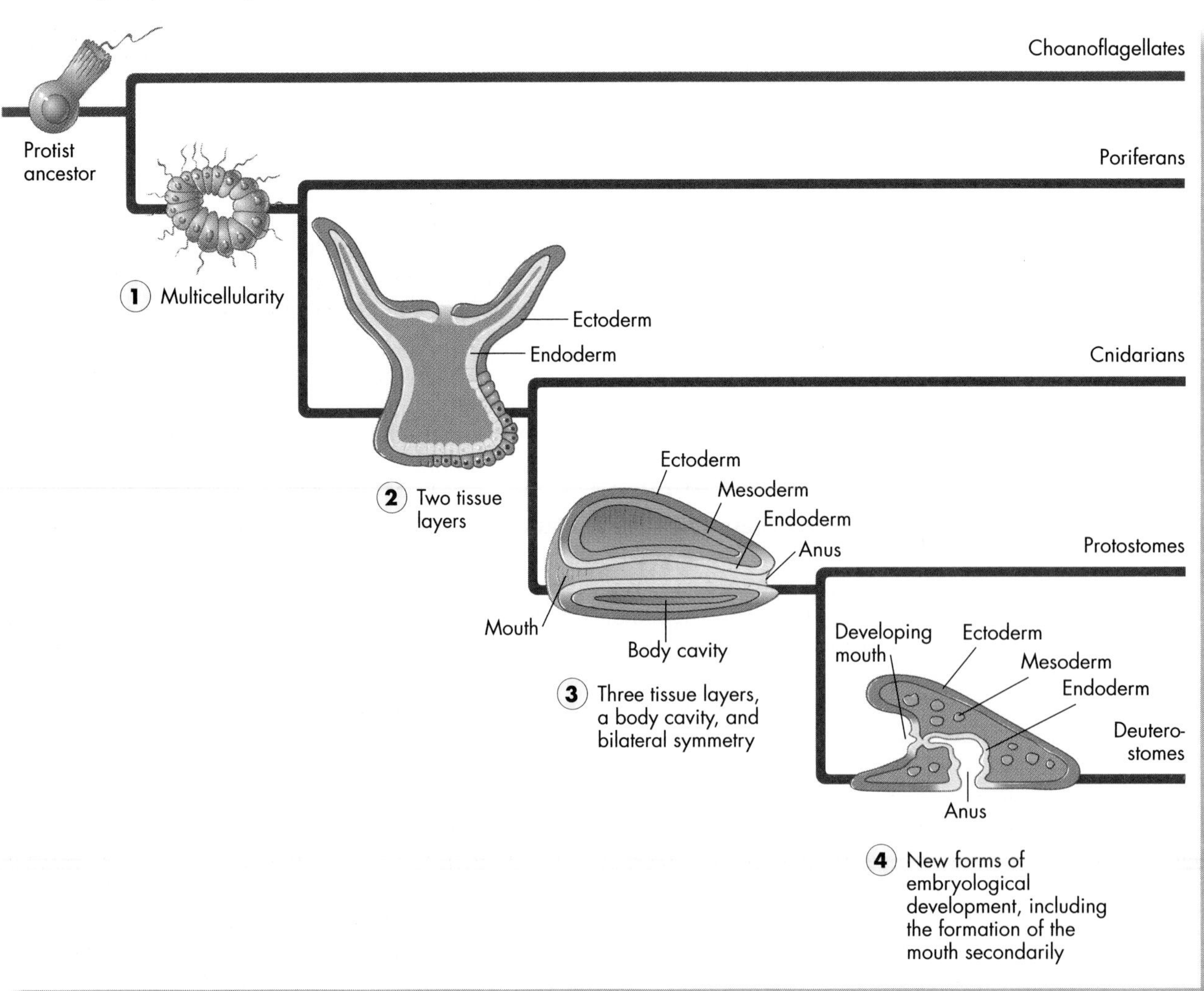

(endoderm and ectoderm, Ch. 8), as well as dedicated nerve cells. This early specialization was associated with an increasingly organized body that was often radially symmetrical (round and often flattened, like a tuna can rather than a ball).

3. Origin of three tissue layers in a bilateral body. An additional middle tissue layer formed between the inner and outer layers (mesoderm, Ch. 8); a body cavity formed in that middle layer; and the body developed **bilateral** (two-sided) **symmetry** with an anterior (front end), a posterior (rear end), and a left and right side, rather than a radial body plan wherein all sides are similar.

4. Origin of specific patterns of embryonic development. New patterns arose including the pattern of cell divisions in the very early embryo (recall cleavage divisions from Chapter 8); the way the middle tissue layer relates to a cavity inside the body; and the way the mouth and anus arise in the embryo (gastrulation, Ch. 8).

Together, these four innovations enabled evolving animals to exploit their environments efficiently and to develop an enormous array of body plans, appendages, and life histories. Today, the innovations serve as a framework that helps us categorize animals, from simple sponges to highly organized dung beetles and millions of others.

Animal Origins

Despite their vast diversity of shapes and complexities, all animals share similarities in cell structure and gene sequences that convince **zoologists** (biologists who study animals) that all animals are derived from a single lineage that originated with an ancient ancestral protist. The protists called *choanoflagellates* (see Chapter 11) are the closest living protistan relatives of animals.

Some time after the first animals evolved, the activities of photosynthetic organisms caused the amount of oxygen in the atmosphere to rise rapidly. This permitted a great evolutionary radiation of the animal kingdom—the rapid origin of large numbers of animal species with a great variety of body forms. All the animals that ever existed were products of this radiation. At least 35 different major body plans evolved, which we categorize into distinct **phyla** (sing., phylum). Most of these still have living members. Together, the animals number more than a million individual species—and that's a conservative estimate. The total could be ten times higher or even more.

Animal Relationships

Biologists investigate the evolutionary relationships between animals by (1) comparing body forms or (2) comparing gene sequences. Comparing body forms is useful but has a complication: animals from separate phyla can be so different that it is often difficult to know what to compare. For example, is the "forehead" of a dung beetle with its antenna more similar to the forehead of a person or the "forehead" of an earthworm? Comparing nucleic acid and protein sequences is a more precise and generally applicable tool, and modern biologists use it frequently to help reconstruct the relationships of various animal groups.

The evolutionary tree shown in Figure 13.2 presents our best current understanding of animal relationships based both on body form *and* molecular data. These figures provide a visual guide for the rest of the chapter.

bilateral symmetry
a body plan in which the right and left sides of the body are mirror images of each other

zoologists
biologists who study animals

phylum (pl. phyla)
a major taxonomic group just below the kingdom level, comprising members of similar classes, all with the same general body plan; equivalent to the division in plants

LO² The Wonderful Variety of Animal Body Plans

This section explores the major kinds of body plans, focusing on evolutionary relationships between the phyla and how new shapes and functions arose. The orderly and linear organization of our story should help you learn the different body plans, but it can leave a misimpression: that one type of living organism gives rise to another. In fact, no organism alive today is an ancestor of any other living organism. All modern organisms, no matter how simple, occupy the tips of branches on the tree of animal life—branches that split off (diverged evolutionarily) lower on the trunk. In fact, most animals on most branches of the animal family tree arose and then became extinct after existing as distinct species for only hundreds of millions of years. Another potential misconception is that evolution is a single directed path toward "bigger and better" organisms. In truth, there are many instances wherein animal lineages lost features present in ancestral predecessors. Today's surviving species may be better adapted to today's environmental conditions, but that doesn't make them "better" overall. Finally, evolutionary innovations have no "plan" or "goal" or "direction." At each point in

Porifera
the phylum containing the sponges

sessile
in animals, the quality of being permanently attached to a fixed surface

time, natural selection works on populations of organisms in ways that allow individuals with certain inherited traits to leave more offspring. We can identify successful innovations in various animal groups, but they usually arose through genetic accidents that happened to benefit survival. With all this in mind, let's begin by looking at the simplest living animals, the sponges (Table 13.1).

The Sponges: Irregular Bodies without Tissues

Sponges are models of simplicity. Their bodies perforated with many tiny openings or pores, sponges belong to the phylum **Porifera** (pore-bearing). They are generally asymmetrical, aquatic organisms lacking distinct tissues. For centuries, people thought sponges were plants because the adults are **sessile** (stationary) and are permanently attached to rocks, pilings, sticks, plants, or other animals. Sponges, however, are multicellular animals with swimming

Figure 13.2
Evolutionary Relationships of the Major Animal Phyla

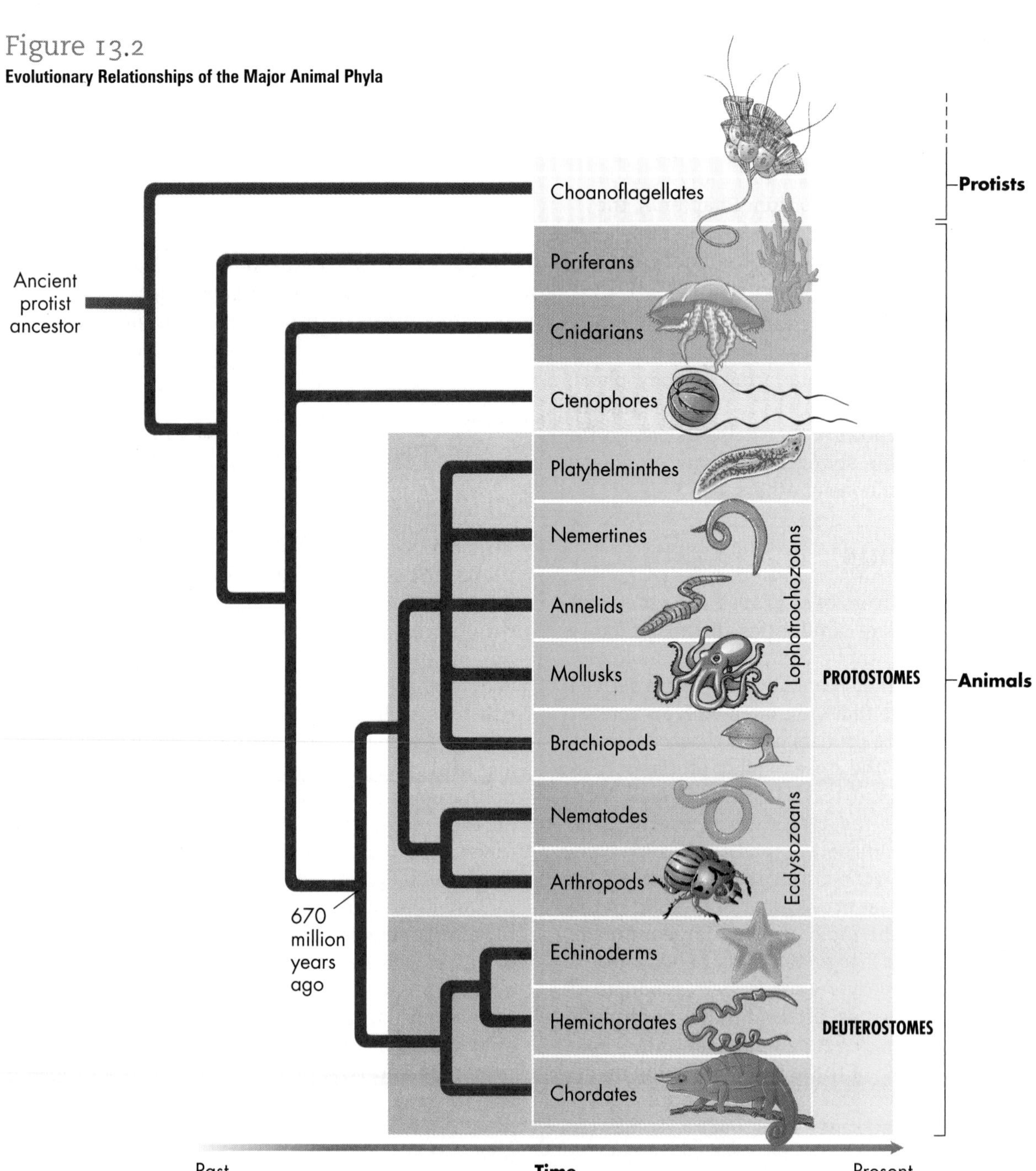

Table 13.1

Some Key Animal Phyla

Phylum	Examples	Number of Species	Notable Features
Porifera	Sponges	10,000	Generally asymmetrical saclike bodies; central opening; body wall perforations; spicules; no evolutionary descendants; no true tissues
Cnidaria	Jellyfish, hydras, corals, sea anemones	10,000	Radial symmetry; grow as polyps or medusae; two tissue layers; gastrovascular cavity; nerve cells, nematocysts, planula larvae; no body cavity
Ctenophores	Comb jellies	100	Radial symmetry; rows of cilia and tentacles; two tissue layers; common plankton species
Platyhelminthes	Flatworms, including tapeworms, flukes	15,000	Bilaterally symmetrical with a head; three tissue layers; true organs and organ systems; life cycles usually complex and include two or more hosts; no body cavity
Nemertea	Ribbon worms	900	Bilateral symmetry; cephalized; organ systems, unique proboscis; regeneration; lives on ocean floor
Annelida	Segmented worms, including earthworms, polychaete worms, leeches	18,000	Bilaterally symmetrical with a head and a gut tube; lined body cavity; segmentation; hydroskeleton; move with bristles pushing against substrate; crop, gizzard; closed circulatory system; earthworms occur widely and help aerate soils
Mollusca	Snails, clams, octopuses, squid, slugs	120,000	Bilaterally symmetrical with a head and a gut tube; three tissue layers with a lined body cavity; gills, mantle; open circulatory system; very common terrestrial, marine, and freshwater organisms
Brachiopoda	Lamp shells	300	Two shells, resemble clams; lophophore feeding apparatus plus coelom; protostomes
Nematoda	Roundworms	20,000	Bilaterally symmetrical with a head and a gut tube; three tissue layers with unlined (false) body cavity; hydroskeleton; extremely common in soils and as parasites on other animals and plants
Arthropoda	Spiders, mites, ticks, scorpions, millipedes, centipedes, insects, lobsters, shrimp	1–30 million	Bilaterally symmetrical with a head and a gut tube; lined body cavity; segmentation; exoskeleton for support and protection; specialized segments and appendages; jointed legs; tracheae or gills; acute senses; most diverse phylum in living world
Echinodermata	Sea stars, sea urchins, sea cucumbers	7000	Gut tube and lined body cavity; a head in some larvae and adults; no segmentation; first endoskeleton; unique water vascular system for locomotion; some have radial symmetry as adults
Hemichordata	Acorn worms	85	Bilateral symmetry; embryos develop as echinoderms do; complete gut; gill slits; dorsal nerve cord
Chordata	Invertebrates like sea squirts, and vertebrates like fish, amphibians, birds, reptiles, and mammals	50,000	Bilaterally symmetrical with head, gut tube, lined body cavity and segmentation; stiff rod of cartilage (notochord); dorsal tubular nerve cord; gill slits

choanocytes
cells that line the pores perforating the body wall of sponges; each cell has a flagellum that draws a current of water through the pore, and a sticky collar, which catches bacteria, protists, and other small organic particles suspended in the water; also called *collar cells*

spicule
a slender, spiky rod of silica or calcium carbonate found in sponges that supports the soft wall and provides some protection from predators

Cnidaria
the phylum containing animals with a radial body plan and nematocysts

radial body plan
a body with a central axis with structures radiating outward like spokes of a wheel

epidermis
the outer layer of cells of an organism

larvae like the one in Figure 13.1 (1), and they filter and consume fine food particles from the water. Figure 13.3 summarizes the important features of sponge biology.

Lacking true tissues, sponges are essentially collections of cooperating cells. Pores lined with special cells (**choanocytes** or collar cells) perforate the body wall of the sacklike animal. These cells have a flagellum, which draws a current of water through the pore, and a sticky collar, which catches the sponge's food: bacteria, protists, and other small organic particles suspended in the water. The water then exits an opening at the top of the sponge. A gelatinous layer inside the body wall contains several different kinds of motile cells, some of which differentiate into gametes. The middle layer also contains skeletal material, including fibers related to hair protein or tiny pointed **spicules** made of silica or calcium carbonate. The irregularly shaped, tan bath sponges sold commercially are the tough, fibrous skeletons left after the animal dies. As we mentioned earlier, collar cells closely resemble choanoflagellates.

Cnidarians and Ctenophores: The Origin of Tissues and Radial Body Plans

Animals "invented" multicellularity. This made a second set of innovations possible: radial symmetry and tissue layers. Two representative phyla, the cnidaria and the ctenophores, show this body plan.

Cnidarians: The Stinging Nettle Animals

Some of the most short-lived and beautiful animals are members of the ancient phylum **Cnidaria** (nih-DARE-ee-ah; Greek *knide* = stinging nettle). This phylum diverged deep in the tree of life and includes the hydras, the jellyfish, the sea anemones, and the corals. Most live in the oceans, but a few, such as the hydras, inhabit freshwater.

Cnidarian Body Plan

Cnidarians have a **radial body plan** consisting of a central axis with structures radiating outward like the spokes of a wheel (Fig. 13.4a on page 238). A cross section of a cnidarian's body wall reveals two tissue layers, the **epidermis** (derived from the embryonic ectoderm) on the outside, and the gastrodermis

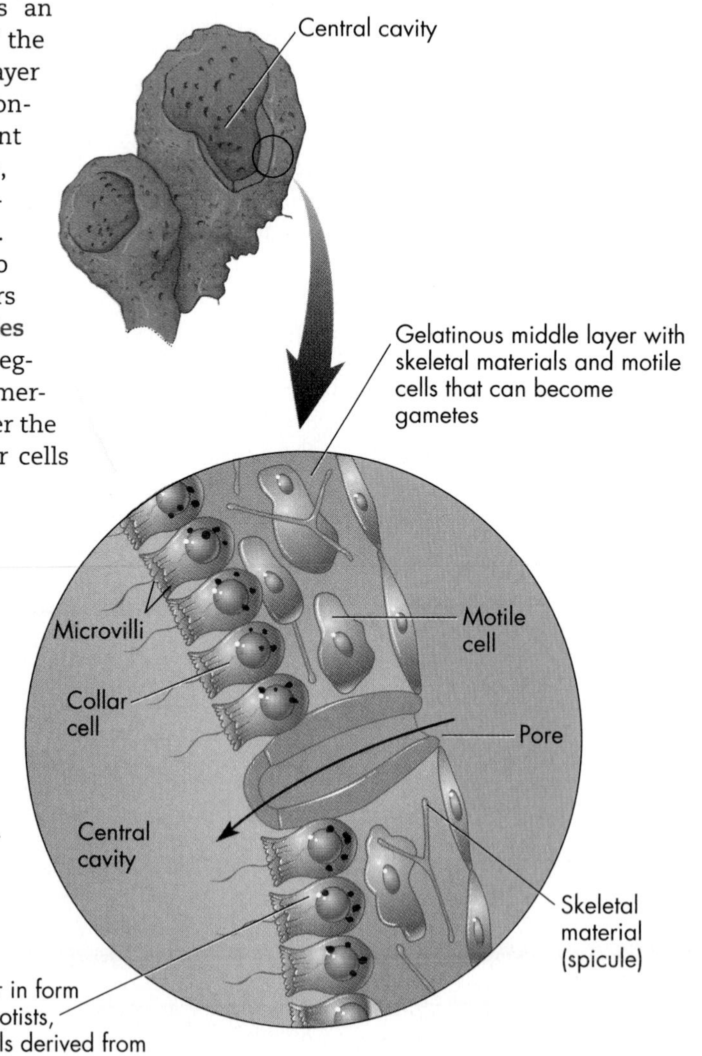

Figure 13.3
Sponges: The Simplest Animals

(derived from the embryonic endoderm) on the inside, lining the central gastrovascular (digestive/vascular) cavity. Sandwiched between the two tissues is a jellylike substance called mesoglea (literally, "middle glue"). Jellyfish resemble jelly because of the extreme thickness of the mesoglea. The gastrovascular cavity has a single opening that serves as both mouth and anus. Tentacles move prey into the cavity, where the food is digested.

Cnidarian bodies have two basic forms: (1) the polyp, a hollow, vaselike body that stands erect on a base and has a whorl of tentacles surrounding a mouth near the top (Fig. 13.4b), and (2), the medusa (pl., *medusae*), an inverted umbrella-shaped version of the polyp, with tentacles and mouth pointing downward (Fig. 13.4b). Sea anemones, corals, and most hydras are polyps as adults, while jellyfish are medusae.

Cnidarian Cell Types

Cnidarians have several specialized cell types not found in sponges. Embedded in the epidermis of cnidarian tentacles are cells that contain remarkable tubelike organelles called nematocysts, or stinging capsules. The slightest contact triggers the tube to evert like a sock turned inside out. The sharply pointed end of this nematocyst can penetrate prey and release a paralytic toxin. Several human deaths have been attributed to nematocysts from *Chironex*, a genus containing many large tropical jellyfish species.

Cnidaria and all animals that diverged more recently from this early animal lineage also have contractile cells equivalent to muscle cells and nerve cells—elongated cells that can conduct electrical signals. These cell types help cnidarians to detect prey, coordinate body movements, and capture victims and move them into the gastrovascular cavity. In cnidaria, nerve cells are arranged in an uncentralized, loose network that conducts information in all directions and coordinates body activities.

Other specialized cells line the gastrovascular cavity and produce enzymes that help break down food extracellularly. Extracellular digestion allows animals to digest larger food pieces, and thus expand the diet compared with the intracellular digestion of very tiny food bits found in most protists and sponges.

Ctenophores: The Comb Jellies

Tow a bottle just below the ocean's surface half a mile from most seashores and you'll probably capture a ctenophore (TEEN-o-for; *cten* = comb + *phore* = bearer), a small, transparent, usually globe-shaped animal with two tentacles and eight rows of cilia that resemble combs. Although there are relatively few species of ctenophores, they can make up a significant portion of the plankton in some regions of the sea.

The Origin of Bilateral Symmetry

Animal evolution saw a third set of basic innovations: the "invention" of three tissue layers as well as of bilateral symmetry (right and left sides). Animals with these characteristics have been amazingly successful and diverse, including the tapeworms, dung beetles, clams, starfish, and, of course, you. Figure 13.2 shows that bilateral animals form two major evolutionary branches, the protostomes and the deuterostomes, distinguished by two different patterns of early embryonic development. Recall from Chapter 8 that early

mesoglea
a jellylike substance lying between the epidermis and gastrodermis in cnidarians such as jellyfish

polyp
(L. *palypus*, many-footed) the sedentary stage in the life cycle of cnidarians; a cylindrical organism with a whorl of tentacles surrounding a mouth at one end; sea anemones and hydras are examples of polyps living alone; corals are examples of colonial polyps

medusa
(Gr. mythology, a female monster with snake-entwined hair) a jellyfish, or the free-swimming stage in the life cycle of cnidarians; an inverted umbrella-shaped version of a polyp, with the mouth and tentacles pointing downward

nematocyst
a stinging capsule found in cnidarians, which, when stimulated, shoots out a tiny barb containing a poisonous substance that immobilizes or kills the prey or predator

ctenophores
the group of animals commonly known as *comb jellies*

Hydras Jellyfish Sea Anemones Coral

protostome
(Gr. *protos,* first + *stoma,* mouth) any bilateral animal whose first opening in the embryo (blastopore) becomes the mouth, while the second opening becomes the anus; also characterized by spiral cleavage during development; includes annelids, mollusks, and arthropods

deuterostome
an animal whose first opening in the embryo (blastopore) becomes the anus, while the second opening to develop becomes the mouth; includes echinoderms and chordates

coelom
(Gr. *koiloma,* a hollow) the main body cavity of many animals, formed between layers of mesoderm, in which internal organs are suspended

animal embryos generally form a hollow ball of cells (the blastula). This ball indents, with the infolding cells becoming the digestive tract (see Fig. 8.6). In the evolutionary branch that contains only invertebrates (animals without backbones) such as dung beetles, the initial indentation of the embryonic ball becomes the mouth (Fig. 13.5a). **Protostomes** (PRO-toe-stomes; "first mouth") are the animals in this evolutionary line. In the other evolutionary line—the one that yielded the vertebrates—the initial indentation (blastopore) becomes the anus, but a second opening becomes the mouth (Fig. 13.5b). These animals are the **deuterostomes** (DUE-ter-oh-stomes; "second mouth").

Besides showing differences in the embryonic origin of the mouth, the major lines of animal descent have two additional differences: First, protostome eggs often have a spiral cleavage pattern, while deuterostome eggs cleave in a radial fashion. Second, the two groups show differences in how the body cavity forms; this cavity is also called the **coelom** (see-LOM) and is an internal space lined at least in part with mesoderm. In protostomes, solid blocks of mesoderm split to form a coelom, but in deuterostomes, the mesoderm forms a pocket from the developmental precursor of the gut, and the enclosed space becomes the coelomic cavity. Radial animals, the cnidaria and ctenophores, lack a coelom. The animals that have one, however, reap several advantages. The space within the coelomic cavity allows the reproductive and digestive organs to evolve more complex shapes and functions than a totally solid body plan. This fluid-filled chamber cushions the gut tube and other organs and thus protects them. And finally, because the gut is suspended in a cavity, digestion can take place undisturbed by the activity or inactivity of the animal's outer body wall.

The rest of this section looks at the two great branches of animal evolution, the protostomes and then the deuterostomes.

Figure 13.4
The Cnidarians: Stinging Nettle Animals

a) Cnidarians have a radial body plan consisting of a central axis with structures radiating outward like the spokes of a wheel.

b) Cnidarian bodies have two basic forms: (1) the polyp, a hollow, vaselike body that has a whorl of tentacles surrounding a mouth near the top, and (2) the medusa, an inverted umbrella-shaped version of the polyp, with tentacles and mouth pointing downward. Cnidarians have two tissue layers and a few specialized cell types.

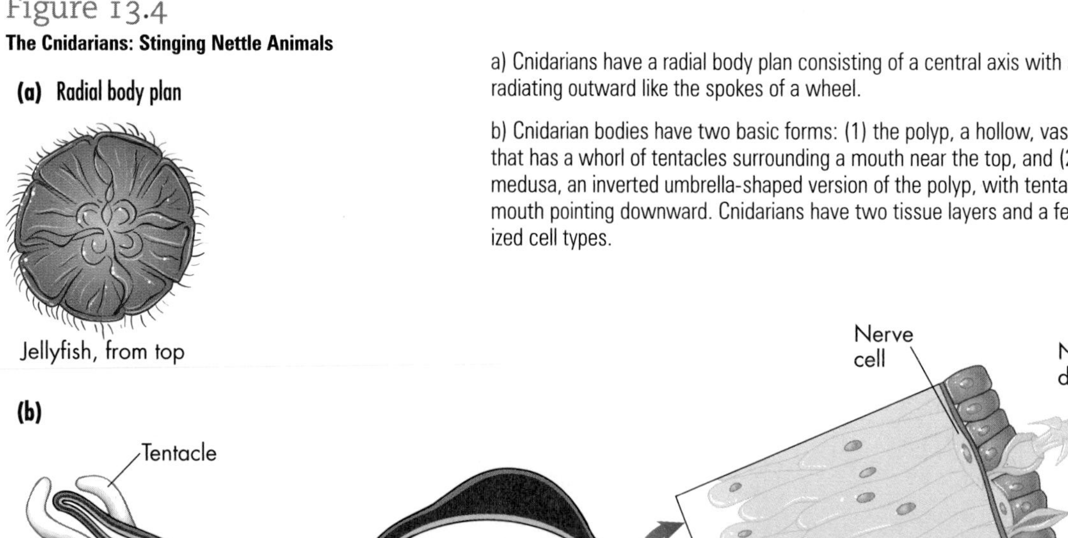

Figure 13.5

Two Lineages of Bilateral Animals: The Protostomes and Deuterostomes

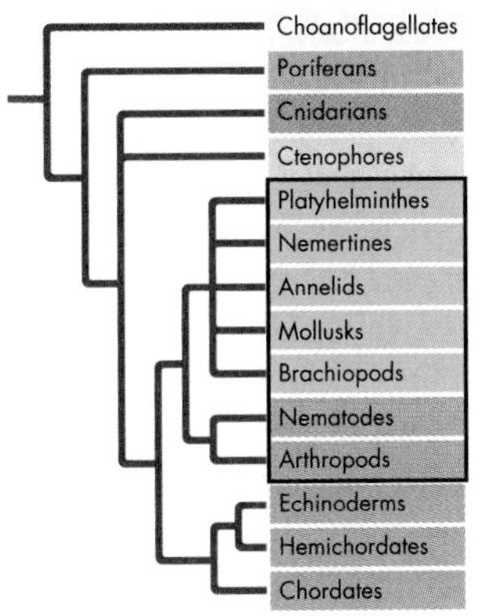

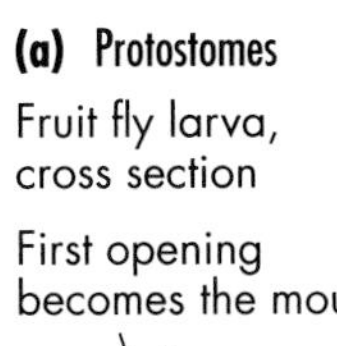

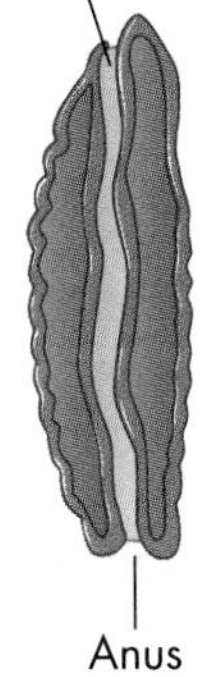

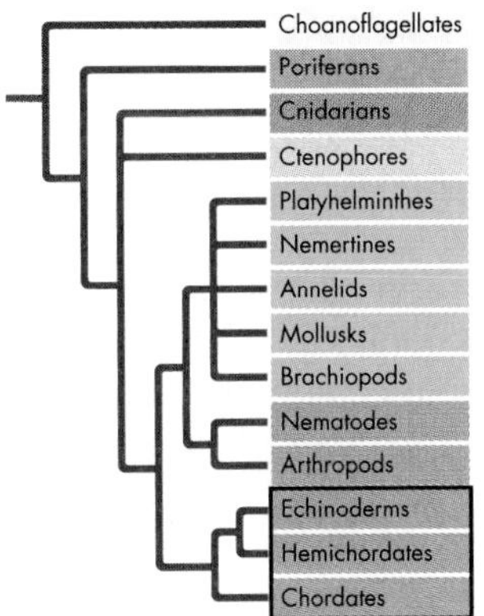

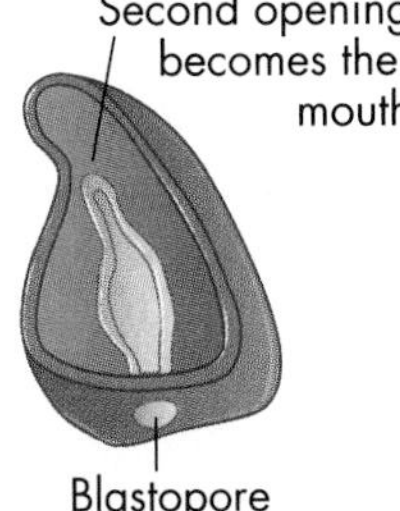

Protostomes: The Diverse Invertebrates

Earth's thousands of dung beetle species, its millions of other insect types, and most of its other immensely varied animal phyla are protostomes. Throughout most of the history of **zoology** (the study of animals), scholars had to infer the relationships between these diverse phyla by comparing gross physical traits. Recent evidence from molecular genetics, however, has allowed zoologists to construct careful geneologies based on genes that code for ribosomal RNA, that control development, or that govern parts of the cytoskeleton. Now, data suggest that one huge set of protostomes includes all animals with specialized feeding tentacles called **lophophores** or a specialized larval form called a **trochophore**. For this reason, they are calling the group the **Lophotrochozoans** (LO-fo-TRO-ca-zoans), and include in it several kinds of worms and all the mollusks (such as snails, clams, and squid). A second set of protostomes, the **ecdysozoa** (eck-DI-so-ZO-ah), includes animals that form a tough external skeleton, which they shed periodically by molting. The Ecdysozoa include the arthropods (such as dung beetles and lobsters), and the nematodes (or roundworms).

zoology
the study of animals

lophophore
specialized feeding tentacles possessed by one group of protostomes

trochophore
a specialized larval form that occurs among many protostome animals, including several kinds of worms and all of the mollusks

Lophotrochozoans
animals with a specialized feeding apparatus called a *lophophore* or a specialized larval form called a *trochophore;* the group includes several kinds of worms and all of the mollusks

Ecdysozoa
animals that form and shed a tough external skeleton; include the arthropods and nematodes

flatworm
a member of the phylum Platyhelminthes, the simplest animal group to display bilateral symmetry and cephalization

planaria
a group of flatworms that inhabit freshwater lakes, rivers, or bodies of salt water

Lophotrochozoans: Specialized Feeding Tentacles or Specialized Larvae

Lophotrochozoans are protostomes in several different, varied phyla that often have either the special feeding tentacles called *lophophore* or the special ciliated larvae called *trochophores*. The feeding tentacles contain an extension of the body cavity (the coelom) as well as cilia that help collect small food particles and bring them to the animal's mouth. The special trochophore larva has bands of cilia in characteristic rows. Zoologists are still trying to understand how the different phyla within the Lophotrochozoan are related to each other. But we can explore what they do know about the groups and their relatedness.

Flatworms: The Platyhelminthes

The phylum Platyhelminthes (PLAT-ee-HEL-min-these; Greek, "flat worms") contains guess what—**flatworms**, with a self-descriptive body plan (Fig. 13.6). Some flatworms are free-living. The **planarians**, for example, inhabit freshwater lakes, rivers, or bodies of salt water. Most flatworms, however, including the flukes and tapeworms, are parasites that live within the bodies of their hosts. Flatworms have a specialized larva that some regard as a modified trochophore. Like many Lophotrochophorans, flatworms have radially cleaving embryos.

cephalization (Gr. *kephale,* head) a type of animal body plan or organization in which one end contains a nerve-rich region and functions as a head

mesoderm the middle cell layer of an embryo; gives rise to muscles, bones, connective tissue, and reproductive and excretory organs

acoelomate in animals, the condition of lacking a coelom

Bilateral Symmetry and Cephalization As in other bilateral phyla, the flatworms have right and left halves that are mirror images (Fig. 13.6a). The anterior and posterior (front and back) ends, however, are different, and so are their dorsal and ventral (top and bottom) surfaces. In addition, flatworms display cephalization (Greek, "head"); one end generally leads during locomotion and contains both a nerve mass that serves as a brain, and specialized regions that can sense light, chemicals, and pressure. When a bilateral animal moves forward, the head, with its sensors and brain, encounters a new environment first. Depending on the data the head collects, an animal can continue forward or back up and try a different direction. This evolutionary adaptation is so successful that heads occur in almost all animals more complex than flatworms.

Figure 13.6

Flatworms: Simple Animals with Primitive Organ Systems

(a) The flatworm body plan

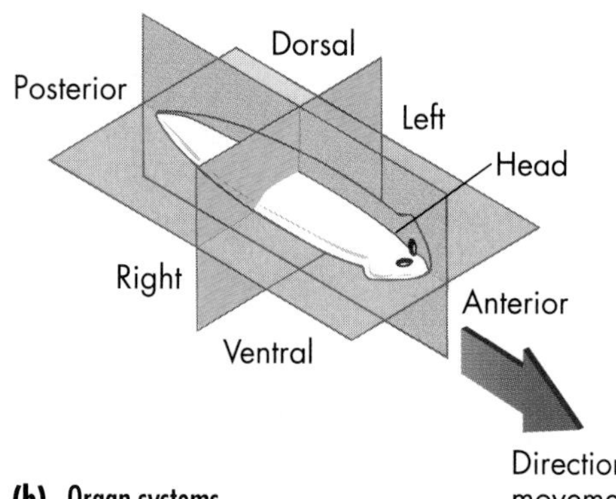

(b) Organ systems

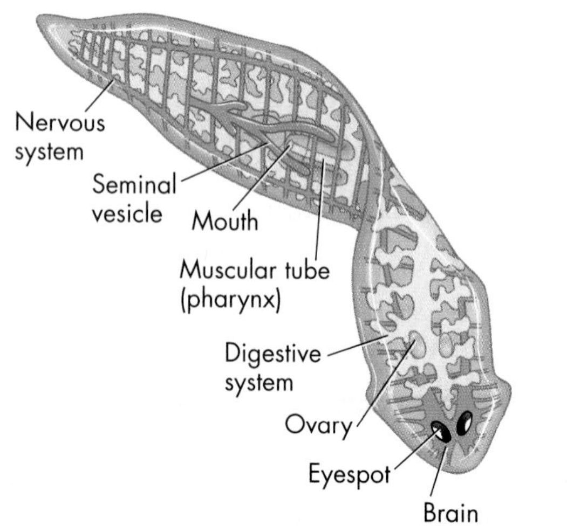

Tissues, Organs, and Organ Systems Flatworms display two additional advances seen in all bilateral animals: three distinct tissue layers and true organs and organ systems. In flatworms and other bilateral animals, the middle tissue layer, or mesoderm, is made up of living cells, not jelly (mesoglea) as in cnidarians, and lies between an outer cell layer, the ectoderm, and an inner cell layer, the endoderm. The mesoderm is important because it gives rise to muscles, blood, and other tissues. Flatworms also have organs and organ systems, described in Figure 13.6b. These multicellular structures help the animal obtain and digest food, dispose of wastes, and reproduce. Flatworms do not have a coelom, a mesoderm-lined body cavity, so instead are termed acoelomate (*a* = not).

Parasitic Flatworms

In contrast to free-living planarians, parasitic flatworms, such as flukes and tapeworms, live a sheltered life, with most of their needs provided for by one or more different hosts. A tapeworm's head is called a *scolex*, and it is little more than a knob with hooks or adhesive suckers around the mouth that attach to host tissues. Their bodies consist mostly of hundreds of individual reproductive units. The tapeworms that infect a cow's intestines shed embryos that bore into its muscles and form protective cysts. If a person consumes undercooked beef bearing these cysts, the cysts can grow into new tapeworms in the person's intestines (the second host).

Nemertea: The Ribbon Worms

Ribbon worms are unfamiliar to most people because they live mainly at the bottom of the oceans. Ribbon worms can be less than one centimeter long or up to 60 m (195 ft) long! Ribbon worms are similar to flatworms in general shape, having bilateral symmetry, cephalization, and organ systems. Nemertines have, in addition, a unique proboscis ("nose") that distinguishes this phylum. A ribbon worm stores its proboscis in a body cavity that represents the animal's coelom, and then shoots it out at prey. Little nail-shaped structures (stylets) can pierce the body of the prey and then the proboscis pulls the prey to the mouth. Nemertines also have the ability to regenerate to impressive proportions: If you cut a ribbon worm into 20 pieces, each fragment will grow a new, complete worm.

Figure 13.7
Mollusks: Variations on One Body Plan

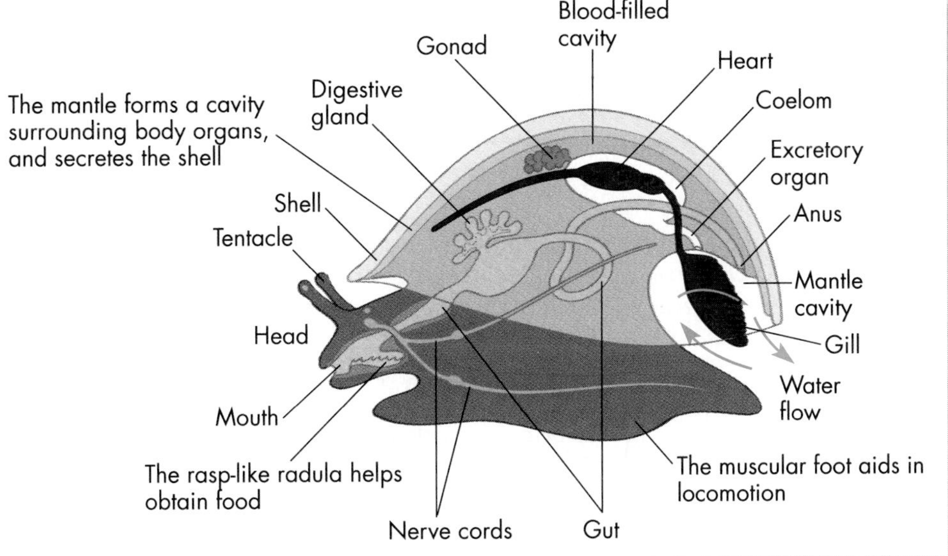

Annelida
the phylum containing the segmented worms, worms with tiny ringlike external segments

segmented worm
a member of the phylum Annelida

mollusk
a member of the phylum Mollusca; includes snails, slugs, clams, oysters, squid, and octopuses

Mollusca
the phylum containing the mollusks

head
the part of an animal that contains a major concentration of sense organs and generally encounters the environment first as the animal moves along

foot
the locomotive organ in mollusks

mantle
a thick fold of tissue found in mollusks that covers the visceral mass and that in some mollusks secretes the shell

visceral mass
most of the body of a mollusk, including the internal organs and excluding the foot and mantle

radula
a rasping organ in mollusks that shreds plant material by rubbing it against the hardened surface of the mouth

gill
[1] a specialized structure that exchanges gases in water-living animals; [2] the plates under the cap of certain fungi

blood-filled cavity
area within the body of a mollusk where internal organs and tissues are bathed by blood

Annelids: Segmented Worms

Marine sandworms, common earthworms, and leeches are all members of the phylum **Annelida**, the **segmented worms**. The term *annelid* means "tiny rings" and refers to the external segments visible on members of this phylum. Annelids belong to one of three classes. The largest class includes colorful marine worms that burrow in the mud or sand and bear such common names as fireworms and feather dusters (class Polychaeta, "many-bristled"). Familiar red earthworms are in the class Oligochaeta ("few bristles"). These ubiquitous inhabitants of moist soils can number 50,000 per acre and literally eat their way through dense, compacted earth, excreting the displaced material in small, dark piles, aerating and mixing the soil.

Leeches, which live mainly in freshwater, are in the class Hirudinea. Leeches parasitize other animals by sucking their blood. A leech can consume three times its weight in blood and go for as long as nine months between meals.

Mollusks: Soft-Bodied Animals

In your mind, picture snails, slugs, scallops, and squid. They seem to be very different types of animals, but because of their structurally similar body plan, they are all **mollusks**, members of the phylum **Mollusca**. The name of this phylum comes from the Latin for "soft," and if you have ever eaten a snail or oyster, you remember that—except for the hard shell—the main sensation they produce is squishy. Mollusks are bilateral protostomes: they produce a trochophore larva, a tiny, fringed stage, as do several other phyla in the Lophotrochophore group of protostome invertebrates (see Fig. 13.2).

Characteristics of Mollusks

The wide variety of mollusks share a distinct body plan. As Figure 13.7 shows, each mollusk has (1) a **head**, housing the mouth, brain, and sense organs; (2) a **foot**, a muscular organ used for gripping or creeping over surfaces or through sand; (3) a **mantle**, a thick fold of tissue that secretes the calcium carbonate that makes up the hard molluscan shell; and (4) a **visceral mass** containing all the internal organs (labeled separately on Fig. 13.7). Most members of the phylum have a special feeding organ called the **radula**. This strap-shaped structure bears rows of tiny teeth and works like a cheese grater, rasping off successive layers of food. Food passes through a complete gut, from mouth through to anus.

Branches of the mollusk's circulatory system flow through the **gills**, the animal's gas exchange surface, and then to the pumping chambers of the heart, and to a **blood-filled cavity**, where internal organs and tissues

open circulatory system a system of fluid transport in spiders, insects, and many other invertebrates in which a clear blood equivalent called *hemolymph* circulates partially in vessels and partially unconfined by tubes or vessels

gastropod a member of the class Gastropoda in the phylum Mollusca; includes snails, garden slugs, and sea slugs

torsion an internal twisting of the body mass during embryonic development

bivalve a mollusk whose body is enclosed within two valves or shells; includes mussels and clams

filter feeders aquatic organisms with feeding structures that can strain out and collect tiny food particles suspended in water

cephalopod a member of the class Cephalopoda, phylum Mollusca, including squids, octopuses, and the chambered nautilus

siphon in certain mollusks, the funnel through which water passes in the mantle cavity

Brachiopoda the animal phylum that contains the lamp shells

roundworm a member of the phylum Nematoda

Nematoda the animal phylum containing roundworms, Earth's most abundant animals

are bathed with blood (see Fig. 13.7). Most mollusks have an **open circulatory system** because the blood is confined within vessels in only certain parts of the body and flows through open spaces, where it bathes body tissues directly, in other regions. Specialized organs and organ systems for respiration and for circulation of blood represent evolutionary advances not found in less complex animals.

Some Classes of Mollusks Let's look at the three main classes of mollusks: gastropods, bivalves, and cephalopods. **Gastropods** (class Gastropoda, "belly foot") include snails, garden slugs, and sea slugs. Most snails are identifiable by their coiled shells, as well as by **torsion**—an internal twisting of the body mass during embryonic development.

Bivalves (class Bivalvia, "two valves") have two half shells that enclose bodies of oysters, clams, mussels, scallops, and relatives. Strong muscles close the valves and stretch an elastic ligament at the hinge. Closed shells protect the animals from most predators. When the muscles relax, the shell snaps open. Rapid closing of the shell in scallops produces a water jet that propels the animal. When a mollusk's shells are open, particles can enter and irritate the soft tissue of the mantle, which responds by secreting layers of nacre (mother-of-pearl) around the offending object. Pearls result from reddish, whitish, or black nacre deposits.

Bivalves are **filter feeders**; they have gills that, in addition to collecting oxygen and releasing carbon dioxide, can strain out and collect tiny food particles suspended in water. Beating cilia draw water across the gills, and a mucous layer traps the particles, which are then passed into the mouth. One biologist has estimated that each week, all the water in San Francisco Bay is drawn through the bivalves that lie half-buried in the silt and sand at the bottom of the bay.

Cephalopods (class Cephalopoda, "head foot") include squids, octopuses, and the chambered nautilus. Cephalopods are the most complex mollusks and evolved as fast-swimming predators of the deep sea. The cephalopod foot bears a circle of 8 or 10 arms, each studded with suckers, and it terminates in a funnel, or **siphon**. Thus, a single organ, the foot, has become specialized for land travel in the gastropods and for hunting, swimming, and feeding in the cephalopods. The cephalopod mantle forms a muscular enclosure, which can expand and draw water into the mantle cavity or contract and force it out of the siphon, jet-propelling the mollusk backward. These explosive bursts can carry the animal to safety or bring its suckered tentacles within reach of prey. Cephalopods have a largely closed circulatory system that pumps blood faster and helps support the active hunter life habit.

Brachiopods: The Lamp Shells

Animals in the phylum **Brachiopoda** ("arm-foot") live as solitary creatures at the bottoms of oceans. Brachiopods have two shells that enclose the body (shells that resemble ancient oil lamps, hence the common name, "lamp shells"). These two shells superficially resemble clam and other bivalve mollusk shells, but the two brachiopod shells are usually shaped differently from each other, and their hinge is very different from a bivalve's. The brachiopods feed with a lophophore, the feeding apparatus consisting of tentacles with cilia on their outside and an extension of the coelom on the inside. Although there are only 300 or so brachiopod species alive today, they have a rich fossil record and were among the most abundant animals in the Paleozoic era.

Ecdysozoans: The Molting Animals

We arrive now at the group that includes dung beetles—the Ecdysozoa—the animals that periodically shed an external skeleton or cuticle. Two very successful phyla molt this way, the roundworms and the arthropods (jointed-leg invertebrates).

Roundworms: The Nematodes

In sheer numbers of individuals, the most abundant animals on Earth are **roundworms**, members of the phylum **Nematoda** (*nema* = thread). Roundworms, as the name implies, are round in cross-section and most are very small. One cubic meter of rich soil can contain 3 billion nematodes. When the environment grows harsh, nematodes can curl up, dry out, and shut down their metabolism for up to 30 years. Then, when conditions improve and water returns, the animals rehydrate and revive: instant nematode!

Characteristics of Roundworms Roundworms are unsegmented, bilateral worms with three tissue layers and a **pseudocoelom** (false body cavity), a body cavity only partially covered with mesoderm. This feature was once thought to be ancestral, but modern molecular genetic data (the first animal genome to be sequenced was a nematode) clearly shows that nematodes diverged higher, not lower, on the tree of life. The roundworm's simplified body plan therefore is likely to be based on a loss of features present in its ancestors.

Roundworms and the Environment Dozens of types of free-living roundworms help consume rotting plant and animal matter. At least 1,000 nematode species parasitize plants, and some ecologists estimate that roundworms consume fully 10 percent of all crops annually. Nearly 50 species parasitize humans, entering in food or contaminated water or through bare skin.

Arthropods: Jointed-Legged Animals

Dung beetles are members of Earth's largest phylum (not just of animals but of *all* organisms): **Arthropoda** (*arthro* = joint + *poda* = leg). Arthropoda is another major phylum in the Ecdysozoa, and its members, the so-called **arthropods** include fossil trilobites, spiders, mites, ticks, scorpions, centipedes, millipedes, lobsters, crabs, beetles, and other insects. The insects, numbering at least 1 million species, make up the vast majority of all animal species. The insects' astounding diversity and success are based on their body plan: they are bilateral, with a head, a one-way gut, and a fully lined body cavity. Arthropods have several other shared characteristics, including an external skeleton (which they periodically shed), specialized body segments, acute sensory systems, and rapid movement and metabolism due to special respiratory structures. All of these contribute to their astonishing success.

The Arthropod Exoskeleton Like the nematodes and the other Ecdysozoa, arthropods have an outer protective cuticle, which they molt. The arthropod **exoskeleton**, or external skeleton, completely surrounds the animal and provides protection, strong support, and rigid surfaces that muscles can pull against. This thick, hard cuticle contains the polysaccharide *chitin* (see Chapter 2), in addition to sugar-protein complexes, waxes, and lipids that make the body covering waterproof. In crustaceans, such as crabs, lobsters, and shrimp, the exoskeleton also contains calcium carbonate crystals, which make it a hard, inflexible armor. Besides providing shieldlike protection from enemies and resistance to general wear and tear, the exoskeleton prevents internal tissues from drying out. This is extremely important for arthropods that live on land.

Exoskeletons do, however, have a major disadvantage: the animal cannot grow larger unless it periodically sheds its constricting armor and produces a larger exoskeleton; this is the *molting* process. During the period between the shedding of one exoskeleton and the hardening of the next, the animal is soft and vulnerable. The arthropod exoskeleton remains thin and flexible at the **joints**, the hingelike areas of the legs and body. The presence of jointed appendages allows arthropods to move quickly and efficiently above the ground or sea floor instead of dragging the body directly along the ground on stubby legs or bristles.

pseudocoelom
a "false" body cavity only partially covered with mesoderm

Arthropoda
Earth's largest phylum; member animals have legs with joints, an exoskeleton made of chitin, and specialized body segments including a head, thorax, and abdomen or a fused cephalothorax and abdomen

arthropod
a joint-legged animal, including insects, spiders, ticks, centipedes, lobsters, and crabs

exoskeleton
the thick cuticle of arthropods, made of chitin

joint
the hinge, or point of contact, between two bones

thorax
the central region of the body of an arthropod or vertebrate between the head and the abdomen

abdomen
the posterior part of an arthropod's body; in vertebrates, the abdomen lies between the thorax and the pelvic girdle

Specialized Arthropod Segments Arthropods have body and leg segments but their segmentation differs from an annelid's. Each annelid segment is basically the same as any other, whereas arthropod segments are different and highly modified, giving the animals a far greater repertoire of activities. The body segments are usually fused into a few major regions; dung beetles and other insects, for example, have three regions: head, **thorax**, and **abdomen**. In spiders and crustaceans, by contrast, the fused body segments are reduced to an abdomen plus one

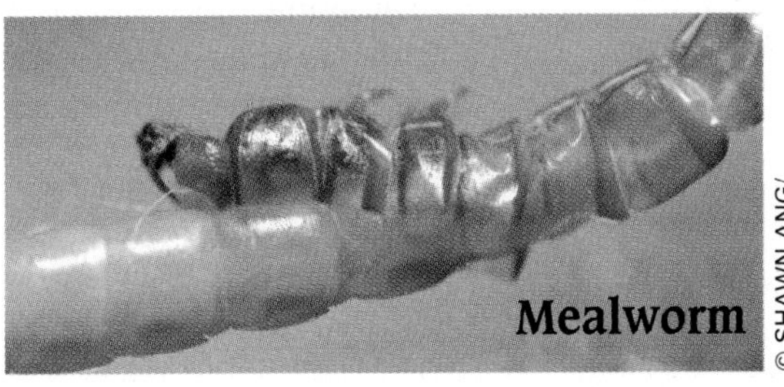
Mealworm

© SHAWN ANG/ ISTOCKPHOTO.COM

cephalothorax
the fused head and thorax of many arthropods

centipedes
arthropods with a series of flattened body segments, each bearing a pair of jointed legs

millipedes
arthropods with a long series of body segments, each bearing two pairs of legs

crustaceans
arthropods with a protective shell and two pairs of antennae, including crabs, shrimp, and lobsters

carapace
the exoskeleton covering the cephalothorax of many arthropods; also refers to the tough outer coverings of the turtle and armadillo

arachnids
arthropods lacking antennae and usually having eight walking legs; includes the spiders, ticks, mites, and scorpions

spinneret
a tubular appendage in spiders and some insects that reels out silk threads

insect
a member of class Insecta; arthropods having three main parts (head, thorax, and abdomen), three pairs of legs, and generally two pairs of wings; the largest class of animals

larva
an immature form of an insect and many other animal types

pupa
in insects with complete metamorphosis, the stage that intervenes between the larva and the adult; in some cases, as in moths, the pupa is encased inside a cocoon

region with fused head and thorax, the **cephalothorax**.

During evolution, arthropods have evolved a veritable Swiss Army knife of tool-like appendages specialized for walking, swimming, and flying, with each species having its own modifications. Some arthropods also developed pincers or palps (feelers), which facilitate hunting and feeding. From the head region grew other appendages: *mouthparts* that allow chewing and sucking and *antennae* that sense odors and vibrations. The dung beetle featured in this chapter's opening photo has specialized cutlerylike mouthparts for carving up dung; long back legs for dung ball rolling; and horns for fighting off competitors. The millions of arthropod modifications like these enable the animals to exploit a wide variety of environments.

Major Arthropod Classes The successful and highly divergent phylum Arthropoda is divided into a number of taxonomic classes, most of which will sound quite familiar.

Centipedes (class Chilopoda) have a series of flattened body segments, each bearing a pair of jointed legs that move the centipede swiftly in search of insects, worms, small mollusks, or other prey. Centipedes kill their prey with venomous claws, which are modified legs on the first body segment. Some huge tropical centipedes are dangerous to humans, but the common varieties that lurk in damp basements are harmless to us.

Millipedes (class Diplo-poda) are slow-moving counterparts to the centipedes. They have, as the term *Diplopoda* suggests, two pairs of legs per segment, a round body, and a preference for decaying vegetable matter instead of live prey.

Crustaceans (class Crustacea) include crabs, shrimp, lobsters, barnacles, crayfish, and sow bugs. These animals are so different from one another that only two generalizations apply:

1. Almost all have an exoskeleton hardened with calcium salts that covers most of the animal as a protective shell, or **carapace**, and
2. all have two pairs of antennae.

Arachnids (class Arachnida) include spiders, ticks, mites, and scorpions. Arachnids lack antennae; have a cephalothorax and segmented abdomen; and usually have six pairs of appendages. The first pair of appendages is modified into *chelicerae,* or venomous fangs, used for killing prey or for self-defense. The second pair holds the prey while the spider injects poison or enzymes. The other four pairs are the spider's eight walking legs. Spiders also have organs called **spinnerets** at the rear of the abdomen that reel out silk threads that are, size-for-size, stronger than steel.

Insects (class Insecta) are the largest—and, by that measure, most successful—class of animals on Earth. Most insects live on land in habitats from the tropics to the poles. Many zoologists attribute insect success to the general arthropod traits we've already considered, but also to the insect's small size and (in many) the ability to fly. Small size enables insects to exploit a vast array of microhabitats—the bark of a tree, the backs of leaves, the dense thickets of an animal's fur, a pile of animal dung, or the universe of midair.

Specific body parts also help account for the insects' success. Many insects have organs for smelling, touch reception, tasting, seeing, and hearing in various parts of the body. An insect's head bears one pair of antennae; its thorax bears three pairs of legs and usually one or two pairs of wings; and its abdomen is usually free of appendages. A series of modified segments, the mouthparts, enables the insect to feed efficiently. We've seen the dung beetles' all-purpose "cutlery" mouthparts. Mosquitoes have pointed stylets that pierce and suck. Others, like locusts and grasshoppers, have chewing mouthparts that can quickly decimate foliage.

Insects have evolved various ways to grow, despite the confining exoskeleton, and various ways to thrive, despite the changing seasons. In some insects, like grasshoppers and cockroaches, the embryo emerges as a miniature version of the adult, but without wings or mature reproductive organs. This organism feeds, grows, and molts five or six times, gradually attaining adult size and characteristics. In most insects, however, including flies, butterflies, and dung beetles, the embryo develops into an immature form, or **larva**, eats voraciously, then forms a transitional stage, or **pupa**, sometimes inside a cocoon. A complete change in form, or metamorphosis, takes place in the body within the pupal exoskeleton—a kind of supermolt. Finally, a nonmolting, reproductively mature adult emerges (Fig. 13.8). In insects that metamorphose, the larvae and adults can be adapted to very different foods and environmental conditions. This successful evolutionary solution allows larvae to specialize in feeding and obtaining energy, and the adult to specialize in reproduction.

Perhaps the most fascinating of the arthropods are the social insects: termites, ants, wasps, and bees. Most species of these insects live in large colonies with labor divided among castes, or subgroups, that differ in appearance and behavior. Such insect colonies are highly successful. Ants, for example, may make up one-third the weight of all the animals on Earth, with 200,000 ants for every person now alive! Insect colonies are highly evolved, functioning, in a sense, as a single well-coordinated organism with the capacity to simultaneously build homes and cities, defend their own, harvest food, and reproduce.

Now that we've investigated the marvelous diversity of protostomes, let's turn to the deuterostomes, the animals whose embryos make a first opening for the anus and a second opening for the mouth.

Deuterostomes: The "Second-Mouth" Animals

Another glance at the animal family tree in Figure 13.2 shows the relationships of the deuterostomes to other groups. After the sponges and radial animals diverged from the trunk of the tree, the bilateral animals evolved and formed the subdivision we just explored, the protostomes, as well as the deuterostome phyla: the echinoderms, the hemichordates, and our own phylum, the chordates.

Figure 13.8
From Grub to Beetle: Hormones Provoke Metamorphosis

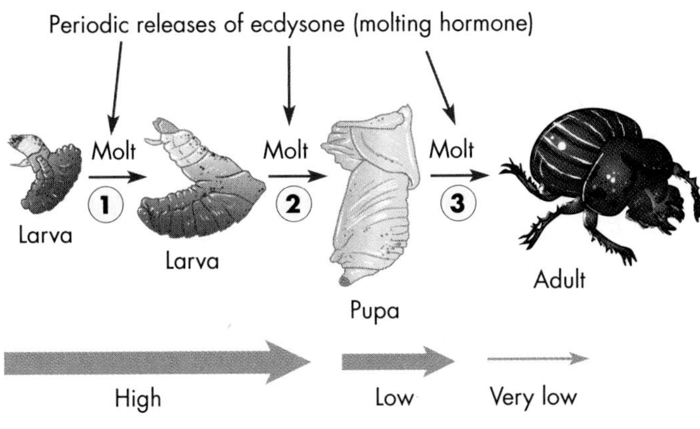

Echinoderms: The First Endoskeletons

The phylum Echinodermata includes sea stars, brittle stars, sea urchins, sea cucumbers, and sea lilies (Fig. 13.9). Echinoderms may constitute 90 percent of the biomass near the ocean floor. The word echinoderm means "spiny-skinned," and virtually all members have spines, bumps, spikes, or unappetizing projections that help to protect these slow-moving marine creatures from predators. These spines are the calcium-based extensions of the animal's endoskeleton, its internal support system. Besides their innovative endoskeleton, echinoderms also have a unique hydraulic pressure system, called the water vascular system, that derives from the coelom and that facilitates locomotion. Echinoderms have an odd mixture of other traits. Some, including the absence of excretory and respiratory systems, may represent a secondary loss from more complex ancestors.

Echinoderm larvae are bilaterally symmetrical, but adults of many species are headless, brainless, and have a fivefold radial symmetry due to the loss of cephalization and bilateral symmetry. Nerve trunks run along each of the adult's arms and unite in a ring around the mouth. This simple system allows coordinated, but slow, movement of the limbs that can generate enough force to open a tightly shut clam shell.

Hemichordates: Wormy Deuterostomes

The small phylum of Hemichordata ("half chordates") consists of a few hundred species of worms that burrow in marine muds, and whose

metamorphosis
(Gr. *meta,* after + *morphe,* form + *osis,* state of) the process in which there is a marked change in morphology during postembryonic development; in insects, the change in body form that takes place as the individual changes from a larva, such as a caterpillar, and emerges as an adult, such as a butterfly; in amphibians, the change from a tadpole to a frog

social insect
a member of any of the insect groups, such as ants, bees, and termites, that lives in colonies of related individuals and exhibits complex social behaviors

castes
subgroups of social insects that differ in appearance and behavior

Echinodermata
a phylum that includes "spiny-skinned" animals like the sea stars, brittle stars, sea urchins, and sea cucumbers

endoskeleton
(Gr. *endos,* within + *skeletos,* hard) an internal supporting structure, such as the bony skeleton of a vertebrate

water vascular system
in echinoderms, a system of fluid-filled canals that includes hundreds of short branches called *tube feet* that can attach to objects and thus aid in locomotion or feeding

Hemichordata
a phylum that includes the acorn worms

Chordata
a phylum that includes animals with a notocord and, often, a vertebral column

vertebral column
the backbone

vertebrate
an animal that possesses a vertebral column, or backbone

notochord
a rod of mesodermal cells in the chordate embryo that marks the location of the backbone in vertebrates

spinal chord
a tube of nerve tissue that runs the length of a vertebrate animal, just above (dorsal to) the notochord

Figure 13.9
Echinoderms: Spiny-Skinned Ocean Dwellers

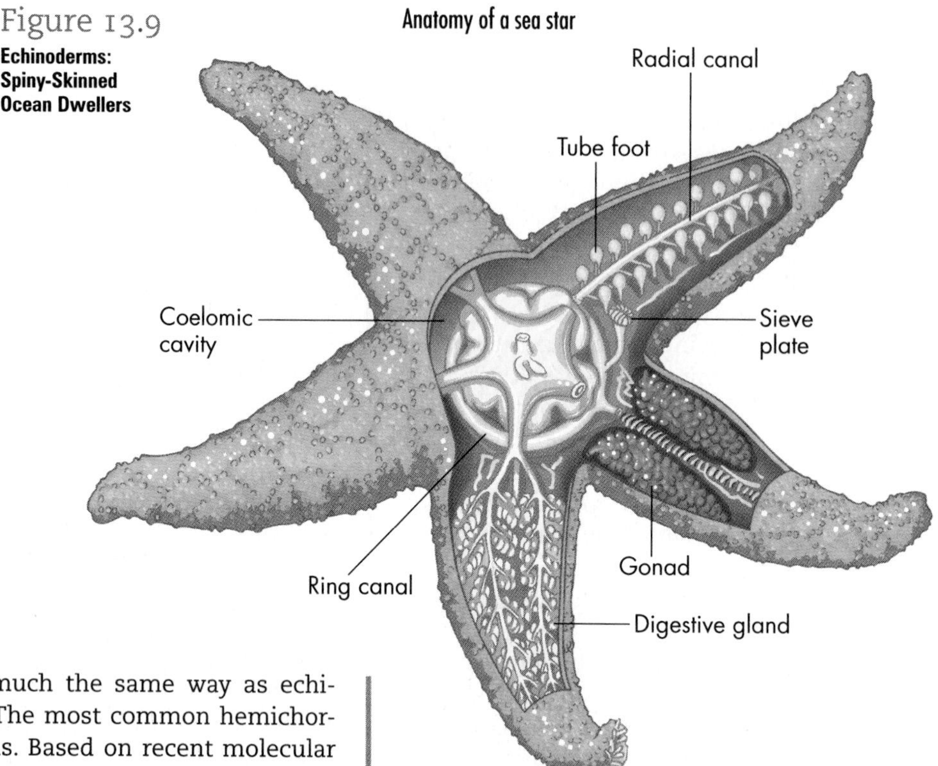

embryos develop in much the same way as echinoderm embryos do. The most common hemichordates are acorn worms. Based on recent molecular genetic evidence, zoologists group echinoderms and hemichordates more closely to each other than to our phylum, the chordates (see Fig. 13.2).

Hemichordates have a complete gut and several other similarities to chordates. Hemichordates have gill slits that open from the pharynx to the water, similar to the gill openings of a fish or the gill arches of a human embryo (Fig. 9.4). Furthermore, they have a dorsal nerve cord (which chordates share), as well as a ventral nerve cord (which chordates lack).

LO3 Evolution of the Chordates

Chordates, or members of the phylum Chordata, are the most familiar and complex animals on Earth (Fig. 13.10). The group includes ourselves, our pets, our livestock, and the subjects of the most popular nature films. Like the sea stars and acorn worms, chordates are deuterostomes. Unlike those simpler animals, however, the chordates have a stiff but flexible cord, the notochord, running down their backs (hence the names *Chordata* and *chordate*). In some chordates this cord is replaced during development by a series of interlocking bones called the *backbone*, or vertebral column, which provides internal support for the body. We call chordates with backbones vertebrates. And while there are only about 50,000 species of vertebrates compared with millions of invertebrate species, the vertebral column has a winning evolutionary design that led to Earth's fastest runners, highest fliers, deepest divers, most agile climbers, and best problem solvers, and to the largest animals that ever lived.

Characteristics of Chordates

All chordates demonstrate the major animal innovations: bilateral symmetry, cephalization, a fully lined body cavity, and a one-way gut tube with an opening at each end. On top of that, chordates have a few novelties of their own that allowed for their radiation into so many active, adroit groups—the notochord, nerve chord, gill slits, myomeres (muscle blocks), and tail.

Notochords and Nerve Cords

Chordates have a new and novel structure, the notochord, a solid, flexible rod of cartilage that provides internal support and generally runs from the brain to the tip of the tail, at least in the embryonic stage (Fig. 13.11). Chordates also have a second innovative structure, the dorsal, hollow nerve cord, or spinal cord, which is a hollow tube of nerve tissue that runs the length of the animal, just above (dorsal to) the notochord. The nerve cord acts like a central "trunk line" carrying impulses that help integrate the body's movements and sensations. In most chordates, the nerve cord is present throughout embryonic and adult life.

Figure 13.10

The Phylum Chordata

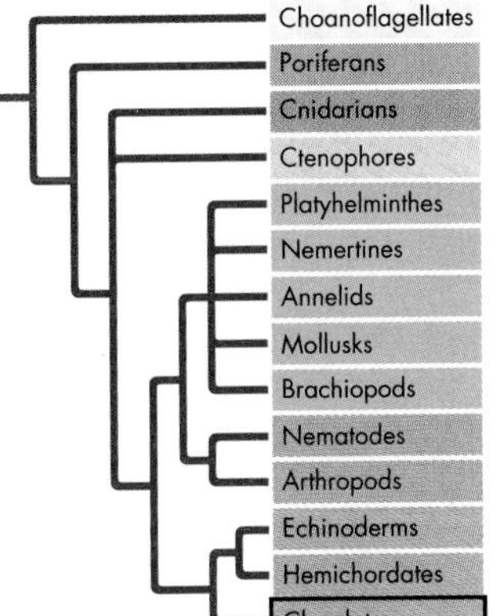

> "The vertebral column has a winning evolutionary design that led to Earth's fastest runners, highest fliers, deepest divers, most agile climbers, and best problem solvers, and to the largest animals that ever lived."

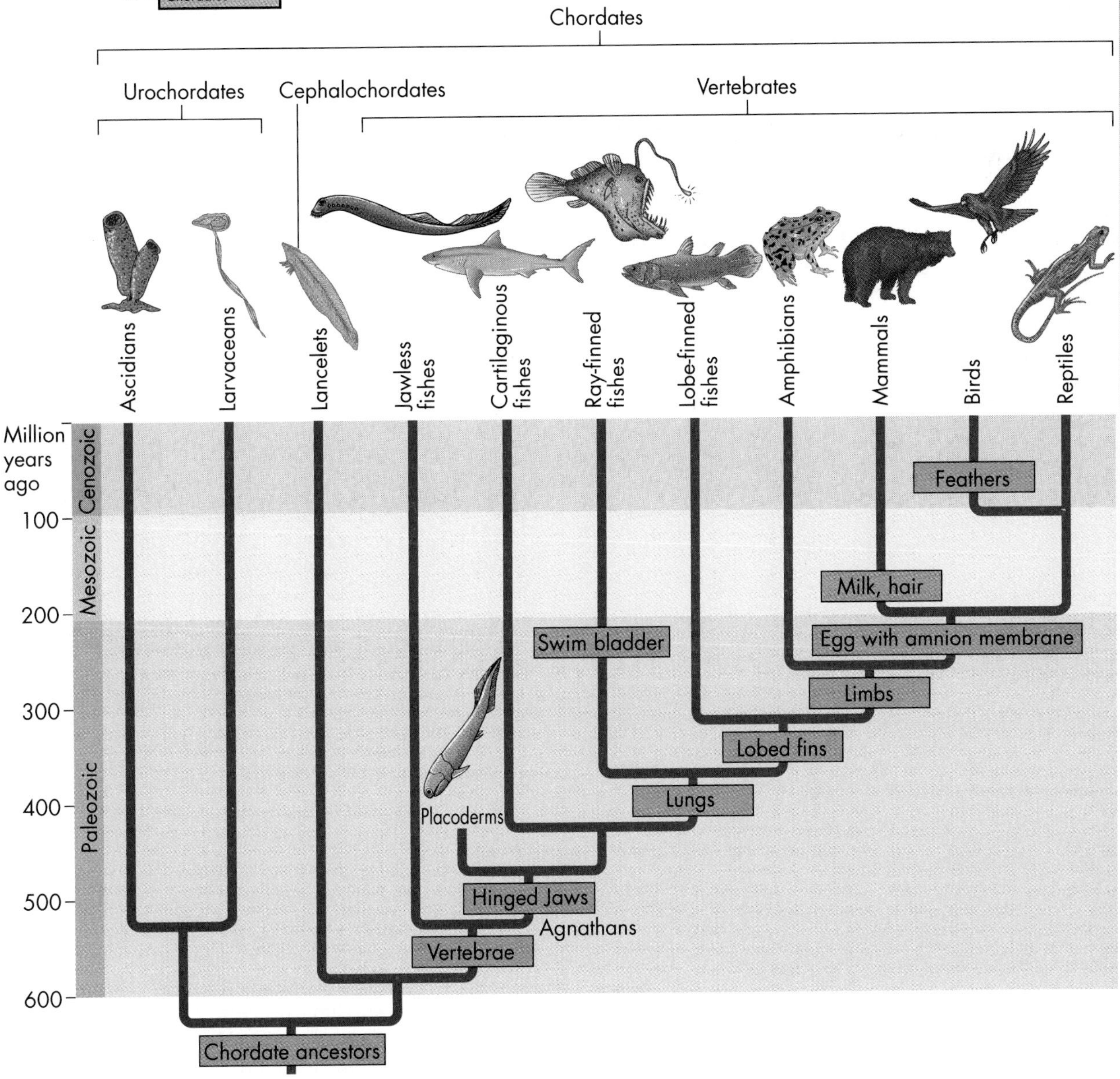

gill slits
in chordates, pairs of openings through the pharynx to the exterior

pharynx
[1] in vertebrates, a tube leading from the nose and mouth to the larynx and esophagus; conducts air during breathing and food during swallowing; the *throat;* [2] in flatworms, a short tube connecting the mouth and intestine

myomeres
blocks of embryonic or larval animal tissue that generate the muscles and bones

Urochordata
the subphlyum containing the sea squirts, which, as larvae, have the five features characteristic of chordates

Cephalochordata
the subphylum containing the lancelets or amphioxus which, as adults, have the five characteristics of chordates

Vertebrata
the subphylum containing the vertebrates or animals with vertebral columns

Gill Slits

Chordates also have a third innovation, gill slits, pairs of openings through the digestive tract's anterior region, the pharynx, or throat. If you still had gill slits, they would be openings from the back of your throat, to the outside of your neck. Simple chordates such as sea squirts use gill slits to filter food particles from the surrounding water. Fish use gill slits for gas exchange. In most reptiles, birds, and mammals, gill slits are either vestigial structures found only in the embryo or they develop into other structures. For example, your eustachian tube is a vestige of one of the gill slits that runs from the back of your throat to your middle ear. This tube allows you to equalize the pressure in your ears with that of the environment. This is particularly important when you drive up a mountain, fly in an airplane, or dive in deep water.

Myomeres

A fourth major chordate trait is usually evident in the larva or embryo: blocks of tissue called myomeres flank the notochord and nerve cord and generate the muscles and bones. Myomeres are the major signs of segmentation in chordates. When you eat fish, you can easily see muscle layers that derive from the myomeres in the stacked units of white or pink meat that flake off under your fork. Or feel your own ribs; each is the product of a single myomere that arose when you were an embryo.

Figure 13.11
Chordate Innovations in the Animal Body Plan

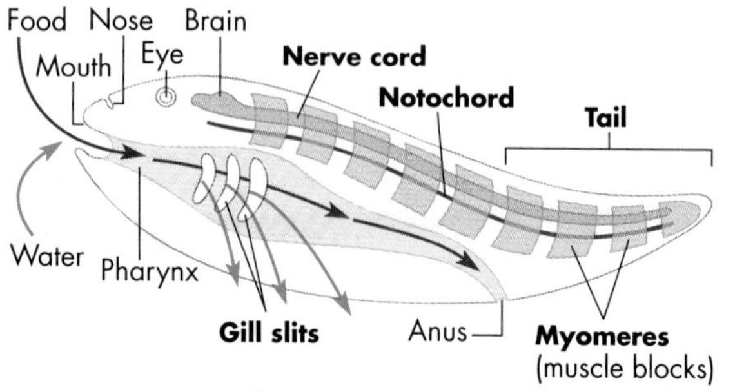

Tails

The notochord, nerve cord, and myomeres extend into the *tail*, the fifth chordate characteristic. In chordates—including humans—the tail protrudes beyond the anus at some point during development. Most chordates have a tail throughout life. But in chordates like ourselves that are tailless as adults, the tail appears only briefly in the embryo.

Importance of Chordate Characteristics

The new chordate characteristics—the notochord, spinal cord, gill slits, tail, and muscle blocks—had dramatic evolutionary implications (Fig. 13.12). The physical support of a notochord, and later of a vertebral column, allowed chordates to grow as large as dinosaurs and whales. The dorsal, hollow nerve cord allowed centralized nerve coordination and the evolution of acute senses and intelligence. From these grew a wide range of behaviors, including agility, and vocal communications. The gill slits could take part in filter feeding or in exchange of respiratory gases. The tail, in its many manifestations, became a major organ of locomotion, balance, and even communication.

Three Major Branches in the Chordate Lineage

Biologists divide all 50,000 or so chordate species into three subphyla: the Urochordata ("tail chordates"), the Cephalochordata ("head chordates"), and the Vertebrata (the chordates with backbones).

Urochordata: The Tunicates

The urochordates, also called *tunicates*, diverged early in the chordate lineage (see Fig. 13.10). Sea squirts (or ascidians) are the squishy tunicates we mentioned earlier; they live as stationary, sacklike adults that attach to submerged rocks or harbor pilings. Sea squirt larvae are free-living, however. Muscle blocks in the trunk and tail power their swimming, and the tail itself is stiffened by the notochord. The nerve cord coordinates their movements, and they filter food from seawater via gill slits (Fig. 13.12a,b). The larvae eventually settle down, attach to a submerged rock, and undergo a dramatic metamorphosis to the stationary adult, resorbing its tail along with noto-

Figure 13.12
Evolution Builds on the Chordate Body Plan

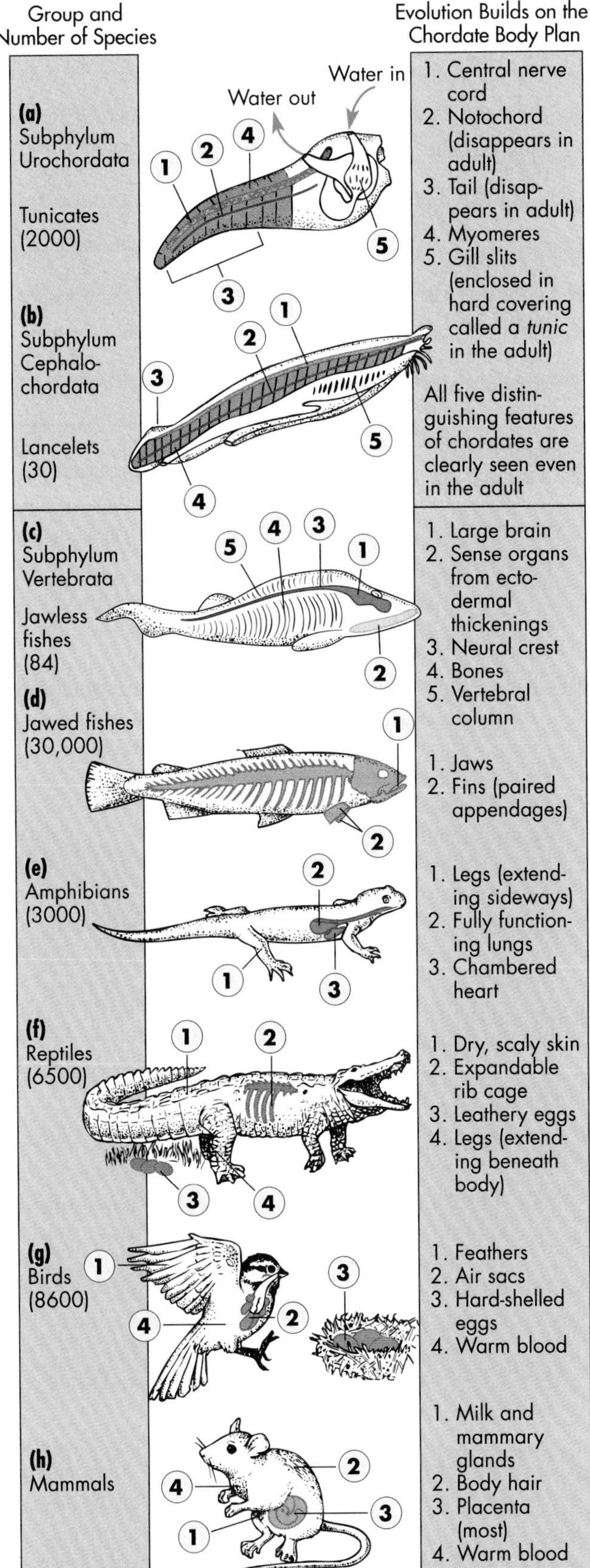

chord and nerve cord. The adult develops an outer *tunic* enclosing a large basket-shaped pharynx perforated by hundreds of gill slits. (This tunic gives the common name tunicates to urochordates.) Sometimes they forcibly squirt water out—to the surprise of many a novice student in zoology lab. Hence the popular name sea squirt.

lancelet
a cephalochordate that lives as a small, streamlined, fishlike marine animal half-buried in sand

In contrast to the sea squirts, another class of urochordates, the larvaceans, retain the five major chordate features even as adults. These urochordates are important to our study of chordates because the ancestor of the fishes probably evolved from a urochordate-like larva. One of the ways biologists study how new traits arise during evolution is to compare the genes of urochordates and fish.

Cephalochordata: Lancelets

In the second group of chordates, the notochord runs all the way to the very front of the head, hence the name *cephalo*chordate. The cephalochordates called **lancelets**, or *amphioxus*, are small, streamlined marine animals about 5 cm (2 in.) long. They live half-buried, tail first, in the sandy bottoms of shallow saltwater bays and inlets. While sea squirts lose their notochord, nerve cord, and tail as adults, lancelets retain the five major chordate traits throughout life. In common with sea squirts, however, lancelets are filter feeders; they have cilialike structures around their mouths and sticky, food-trapping mucous regions in the pharynx. If food becomes scarce, lancelets can pull up anchor and move to more fertile waters. The ability to move as adults gives lancelets access to food supplies that are unavailable to the stationary sea squirts. Note from the tree in Figure 13.10 that cephalochordates are the sister group of the vertebrates.

The Vertebrates: Animals with Backbones

About 550 million years ago, at the dawn of the Paleozoic era, ferocious relatives of the cephalochordates began to swim the ancient seas and leave a fossil record. These new animals became fierce and cunning predators, with enhanced ways to detect food and to snare and devour it. These animals were fishes, the first vertebrates.

The Fishes The earliest fish were streamlined filter feeders that lived in the muddy bottoms

agnathan
a jawless fish

jawless fish
fish that lacked hinged jaws; agnathans

jawed fish
fish with hinged jaws

placoderm
the ancient group of armor-plated fishes that gave rise to the modern fishes

cartilaginous fish
fish whose skull, vertebrae, and other skeletal parts are made of cartilage instead of bone; includes skates and rays

of ancient seas. About 30 cm (1 ft) long, they had fixed, circular, jawless mouths that could filter sediments by muscular pumping rather than solely by ciliary action. These fishes are called **agnathans** (*a* = not, *gnath* = jaw) meaning "jawless." The ancient jawless fishes gave rise to all modern fishes, but they also gave rise to amphibians, reptiles, birds, and mammals. We'll explore those descendant groups later. The four main lines of fish evolution include (1) the jawless fishes, (2) the cartilaginous fishes, (3) the ray-finned fishes, and (4) the lobe-finned fishes.

> "Fish evolved important new adaptations, including the skull, bone, hinged jaws, paired fins, and vertebrae."

Jawless fish had four main features in addition to the usual chordate characteristics: (1) a large, three-part brain; (2) thickenings of the embryonic ectoderm that developed into pairs of highly specialized sense organs (eye lenses, ears, and nose); (3) the neural crest (review Fig. 8.7), a group of migratory cells in the developing embryo that form the gill bars, some hormone-producing cells, and some nerve cells that make the guts work more efficiently; and (4) bone, a meshwork of cells and proteins hardened by calcium, phosphate, and other minerals. The tough bony plates encased and protected the cunning brains of early fish, and served as coats of armor that shielded the ancient fish from dangerous invertebrates such as giant sea scorpions. There are still jawless fish species alive today, like the lamprey, but they have lost the bony plates. Neural crest cells gave rise to bars of bone in the arches between the gill slits, and this support, in concert with the action of strong muscles, helped the fish draw water more forcefully through the gills and into the pharynx than did the cilia of earlier animals. This more powerful current allowed the agnathans to consume greater quantities of food than their earlier cousins did and to grow 6 to 30 times larger. Modern jawless fishes, including the lamprey and hagfish, have very flexible internal skeletons of cartilage. Lampreys are parasites that attach to their prey by suction. Once attached, a lamprey rasps through the victim's body wall with a sharp tongue, digests its tissues, and then consumes the liquified remains. In some lamprey species, however, the adult is not a parasite; once it has metamorphosed into sexual maturity, it does not feed at all—it lives only long enough to reproduce and then dies.

Jawed fishes appeared in the fossil record about 470 million years ago as the **placoderms** (*placo* = flat + *derm* = skin). The placoderms had three basic new characteristics that were so useful they appeared in nearly all the vertebrates that followed. These innovations include hinged jaws, vertebrae, and paired fins. Jaws come from the bones (gill bars) that first appeared in the jawless fishes and support the gills (Fig. 13.13). Jaws allowed placoderms to consume large chunks of food—kelp fronds, clams, and other fishes. This was a big competitive advantage and some became physically huge; a person can stand upright in the fossilized jaws of one placoderm species. The second innovation, the vertebral column, is a series of bones that largely replace the notochord during development and arch over and protect the spinal cord.

The resulting vertebral column provides a site of attachment for muscles, resulting in more powerful propulsion through the water. The third innovation, paired, lateral fins, provides more control over swimming direction, speed, and depth.

Placoderms (Fig. 13.13) gave rise to more modern fishes, and the Devonian period (about 345 to 395 million years ago) is called the Age of Fishes because fishes dominated Earth's lakes and seas. Most of today's fishes fall into two main groups, the cartilaginous fishes and the bony fishes.

Cartilaginous fishes have lost the heavy bony plates of the placoderms, and their skull, vertebrae, and the rest of their skeletons are made entirely of cartilage, a matrix of fibrous proteins in

Figure 13.13

Fishes: The Most Diverse Vertebrates

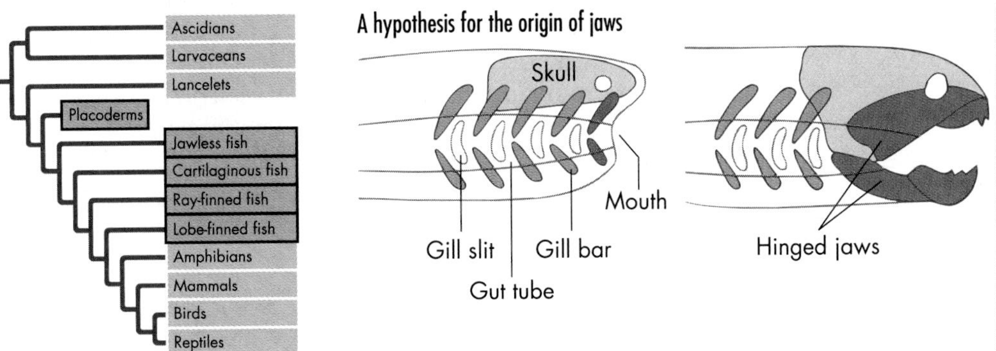

bony fish
fish whose skeleton contains bone

lobe-finned fish
a member of the oldest of the groups of bony fishes; includes lungfishes and coelacanths

lungfish
an air-breathing lobe-finned fish having a lunglike air bladder in addition to gills

coelacanth
a type of lobe-finned fish

ray-finned fish
fish whose fins are made of webs of skin over rays of bone

teleost
a modern bony fish; includes most living fish and characterized by the presence of a swim bladder and spiny fins

amphibian
a cold-blooded vertebrate that starts life as an aquatic larva, breathing through gills, and metamorphoses into an air-breathing adult; includes frogs, toads, newts, and salamanders

a flexible ground substance. Cartilaginous fishes include sharks, skates, and rays. **Bony fishes** include virtually all of today's familiar fresh- and saltwater fishes. Bony fishes have two major groups—the lobe-finned fishes and the ray-finned fishes. The **lobe-finned fishes**, which arose about 400 million years ago, had large, muscular, lobe-shaped fins supported by bones that allowed them to "walk" across the bottoms of shallow bays. They also had *lungs*, or air sacs, that allowed them to breathe air. When water levels fell, these adaptations may well have allowed them to survive by breathing air and migrating overland to other pools or bays. Present-day lobe-finned fishes include the **lungfishes** and the **coelacanths**.

A second group of bony, jawed fishes are the **ray-finned fishes**, or **teleosts**, which lost their ancestors' fleshy fins, but developed much more versatile spiny fins with webs of skin over delicate rays of bone. Their lung sac lost its connection to the exterior and evolved into a swim bladder, an internal balloon below the backbone that can change volume and allow the animal to adjust its swimming depth. Ray-finned fishes radiated into a huge group, and with about 30,000 recorded species, are the largest group of vertebrates. Most fish, however, are relegated by their anatomy to life in water. The great landmasses were to be dominated by other vertebrates—descendants of the early lobe-finned fishes. Nevertheless, fish evolved important new adaptations, including the skull, bone, hinged jaws, paired fins, and vertebrae.

Vertebrate Descendants of the Ancient Fishes

Fossil hunters have found strong evidence that during the Devonian, a new group of animals branched off from the lineage that included air-breathing lobe-finned fishes: this new group included the **amphibians**—vertebrates that can live both on land and in water (*amphi* = both, *bios* = life).

Amphibians About 375 million years ago, the first amphibians overcame the formidable problems of life on land, including more efficient means of walking, breathing, and staying moist. Modern amphibians include approximately 3,000 species of frogs, toads, salamanders, and wormlike apodans.

In early amphibians, the fleshy fins of lobe-finned fish had transformed into front and hind legs containing strong bones and powerful muscles. Extending sideways from the body, the limbs could support the animal's weight far better than lobed fins; this allowed the animal to move about on land for greater distances, even without water's buoyant support (Fig. 13.14). Salamander bones still show this characteristic pattern that emerged in 380-million-year-old lobe-finned fishes.

Laborious walking, however, required a great deal of energy from food. Amphibians needed this energy, along with large quantities of oxygen for aerobic respiration. An amphibian's smooth, moist skin absorbs most of the oxygen and releases most of the carbon dioxide the animals require. Fossil evidence reveals where the rest came from: early amphibians had simple but fully functioning lungs, or air sacs, that provided another site for gas exchange. The first amphibians probably pumped in air by swallowing movements, much as frogs and toads do today. Amphibians also evolved a heart with three chambers plus other circulatory changes that tended to separate oxygenated blood en route to the body tissues from deoxygenated blood bound for the lungs. This

reptile
a cold-blooded, scaly, lung-breathing vertebrate that lays large eggs that usually have a shell; the dominant group of animals in the Mesozoic era; includes crocodiles, lizards, tortoises, and birds

amniote eggs
eggs that encase an embryo in an amnion surrounding a pool of fluid and contain a yolk

separation became complete in crocodilians, mammals, and birds and allowed more and more vigorous activity levels to emerge.

Because an amphibian's skin must remain moist for gas exchange, the animals have always been restricted to life in damp places or near the water's edge. In addition, amphibians lay eggs with a clear, jellylike coating that must also stay moist, or the embryos will die before the fishlike tadpoles emerge and wriggle away. Most amphibians can move away from water at times but must always return to reproduce. In this sense, they are analogous to the mosses, ferns, and other ancient land plants whose sperm must swim to an egg.

Today, frogs, toads, salamanders, and a legless group called the *apodans* are the only remaining members of the class Amphibia.

Reptiles In the vast, steamy Carboniferous swamps, amphibians crawled about in profusion, but they were not alone. Another kind of four-legged land animal—**reptiles**—lumbered about, too (Fig. 13.15). The early reptilians superficially resembled crocodiles and evolved four important innovations for life on land that freed them from dependence on wet environments, much as the early seed plants were free to inhabit areas further from standing water:

1. Their dry, scaly skin provided a barrier to evaporation and sealed in body moisture. This eliminated the skin surface as a major site for gas exchange.
2. Their expandable rib cage could draw in large quantities of air like a bellows. This extra air could be distributed well because modifications in the heart and circulatory system separated oxygenated and deoxygenated blood more fully than in the amphibians.
3. Innovations in reproduction eliminated reliance on open water for reproduction. Females produced so-called **amniote eggs**. These eggs essentially encased the developing embryo in a pool of fluid (the amniotic sac). They also provided a source of food (the yolk) and surrounded both embryo and food with membranes and a leathery shell that prevented

Figure 13.14
The Transition to Land

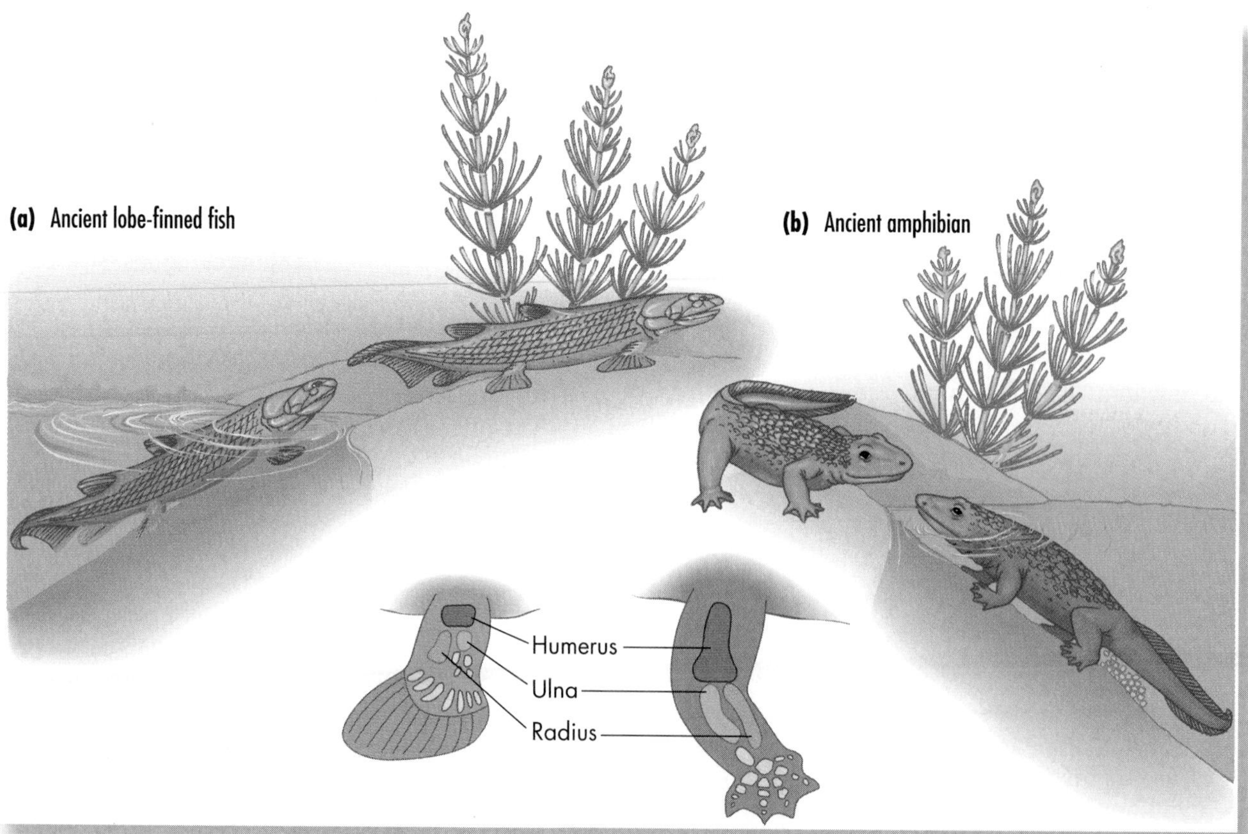

crushing and drying out. Males had copulatory organs that could deliver sperm directly into the female's body (internal fertilization) rather than into the surrounding water (external fertilization).

4. Finally, the legs of the early reptiles extended directly beneath the body rather than out to the side (as amphibians' legs did then and continue to do today). This arrangement provided better support and made walking and running easier.

These new characteristics were crucial to the evolution of all the modern reptiles—the crocodiles, turtles, lizards, tortoises, snakes, and the ancient lizardlike tuatara from islands off New Zealand. They were also, however, pivotal to rise of the birds and mammals, as we'll see shortly.

The earliest crocodile-like reptiles gave rise to two lineages: the **thecodonts**, or small lizards that ran on two legs and led to the dinosaurs, and the **therapsids**, small, heavyset, fiercely toothed animals that led to the mammals.

The biggest, fiercest reptiles that ever lived were **dinosaurs** ("terrible lizards"). The warm Cretaceous period (144 to 65 million years ago) was the heyday of dinosaurs, including the meat-eating *Tyrannosaurus* and the plant-eating *Apatosaurus* (formerly *Brontosaurus*). The grand radiation of reptiles into forms that swam, flew, and lumbered across Earth lasted nearly 150 million years, and inspired the common name for the entire Mesozoic era: the Age of Reptiles. Most of the great and diverse reptiles died out, however, in a massive extinction event at the end of the Cretaceous era.

A few types of small reptiles survived the extinctions and radiated once again during the Cenozoic era (the current geologic era), which began about 65 million years ago. Today, there are only about 6,500 species of reptiles.

thecodonts
small, extinct lizards that ran on two legs and gave rise to the dinosaurs

therapsid
an extinct lineage of fierce, heavyset reptiles that gave rise to the mammals

dinosaur
an extinct giant reptile; the dominant form of land vertebrate during the Jurassic and Cretaceous periods

birds
winged vertebrates with feathers, air sacs, and a four-chambered heart

Birds Except for domesticated animals, the most common chordates we see around us are **birds**. Some biologists have recently suggested that, from a zoological perspective, birds are actually a kind of dinosaur. The first winged vertebrates—the giant, soaring pterosaurs (Fig. 13.15)—might seem like logical ancestors to the birds, but pterosaurs died out long before birds evolved. Instead, small, two-legged, lizardlike thecodonts appear to be the real forerunners of the birds. Paleontologists have unearthed six fossil skeletons of one of the oldest birds, the crow-sized *Archaeopteryx*, at different sites in rocks dated back to the Upper Jurassic period (150 million years ago). The fossil imprints suggest that this animal was a true intermediate: It had scaly skin, curving claws, a long, jointed tail, and sharp teeth like a reptile, but it had birdlike feathers on its forelimbs.

Figure 13.15
Reptiles and Their Descendants, the Birds and Mammals

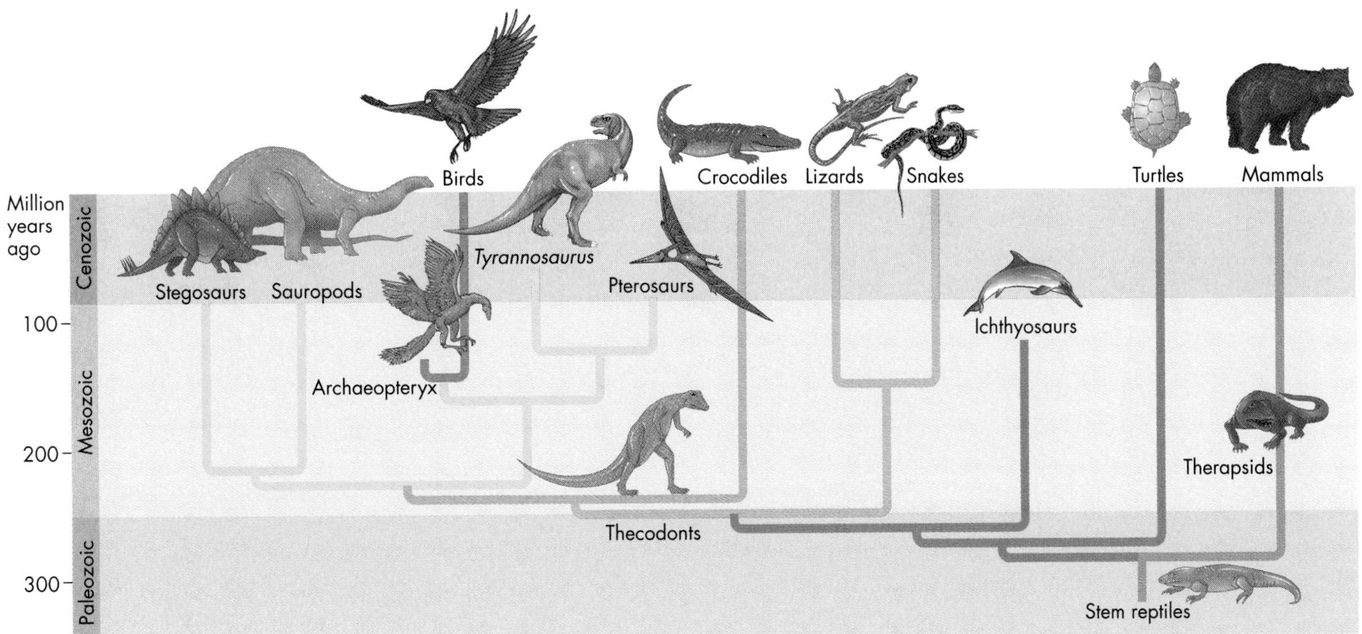

feather
a flat, light, waterproof epidermal structure on a bird; collectively functions as insulation and for flight

homeothermic
"warm-blooded"; able to maintain a constant internal body temperature

air sac
a thin-walled extension of the lungs; in birds, air flows through lungs and air sacs during respiration

four-chambered heart
a heart with four chambers that completely separates oxygenated and deoxygenated blood

Most of the birds' evolutionary adaptations prepared them for efficient flight. Feathers are marvelously lightweight structures made of dead cells containing the protein keratin. Birds also have lightweight, hollow bones and a breastbone, or keel, enlarged into a blade-shaped anchor for the powerful pectoral muscles that raise and lower the wings. The legs are reduced to skin, bone, and tendons and can be folded up like an airplane's landing gear, reducing drag during flight.

Flight is a strenuous activity that requires plenty of oxygen for the aerobic respiration of muscle and other tissues, and birds have three modifications that ensure an adequate supply of oxygen. First, birds are homeothermic (meaning "warm-blooded"). They maintain a constant internal temperature slightly warmer than our own body temperature regardless of changes in the outside air or water temperatures. This constancy helps with the steady production of ATP energy during cellular respiration, which in turn fuels the activities of wing and leg muscles. In fact, the earliest feathers may have been more useful as insulation than as aids to flight. Most fish, amphibians, and reptiles have body temperatures that vary; as a result, they tend to be active in warm environments but sluggish in cold ones. Second, birds have lungs connected to a series of air sacs that exchange oxygen and carbon dioxide in an efficient one-way flow. And third, birds have a four-chambered heart that completely separates oxygenated and deoxygenated blood so only the former reaches body tissues.

Finally, in addition to the evolutionary features of feathers, constant body temperature, and air sacs, birds produce amniote eggs with hard shells rather than leathery coverings. These innovations free birds from dependence on water for reproduction, just like reptiles' leathery eggs, but the shells are more impact resistant and thus more protective for the embryo inside.

Birds' evolutionary innovations were so successful that in the last few million years of the Cenozoic era, birds radiated into a highly diverse class with more than 8,600 species specialized for life in the trees, at the sea shore, in freshwater lakes, in the desert—even on icebergs!

Mammals While the second lineage of reptiles—the fierce, heavyset therapsids—gave rise to the mammals (Fig. 13.16), they developed their own distinct body traits at least 180 million years ago. The very early mammals resembled shrews, and until the end of the Cretaceous, they probably scurried around the great dinosaurs' feet along with the dung beetles. These progenitor mammals survived the mass extinctions of dinosaurs and other animals about 65 million years ago and radiated into 5,000 modern species. Our current geological era, the Cenozoic, is in fact the Age of Mammals.

Figure 13.16
A Proposed Lineage for Mammals

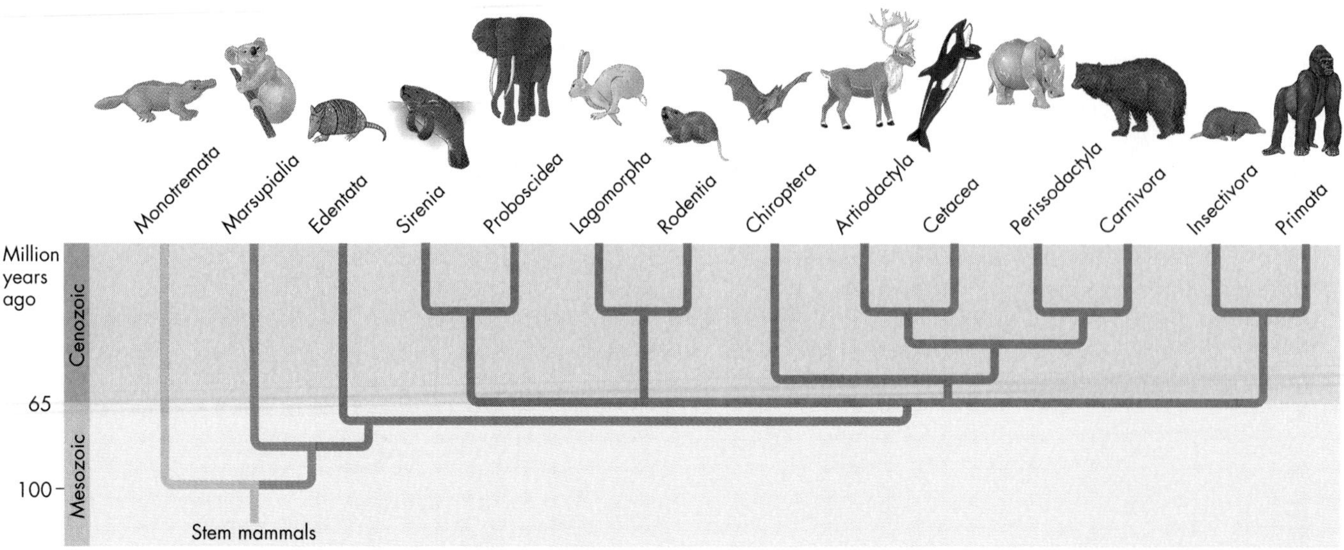

Mammals (Latin, *mamma* = breast) are warm-blooded and have a four-chambered heart, just like birds. These two traits probably evolved independently in the two groups, however. Mammals have two additional traits unique to their class: milk and body hair or fur. A fluid rich in fats and proteins, **milk** is produced in **mammary glands** and is used to nourish newborns. Many mammals have dense carpets of hair called **fur** covering the entire body. Others, such as certain monkeys and humans, have sparse body hair. A few, including the whales and porpoises, have only a few very sparse hairs. These marine mammals rely instead on thick layers of fat called **blubber** for insulation. Significant changes occurred in the mammalian skull, as well.

In addition to milk and to hair or fur, most mammals also have a unique reproductive structure, the **placenta** (Fig. 13.12h). This spongy organ supports the growth of the embryo to a fairly complete stage of development before birth. Placental mammals descended from one branch of the earliest mammals, while two other branches led to nonplacental mammals that nurture embryos in different ways. The **monotremes** lay leathery eggs and warm them until the young hatch, then nurse the young with milk. There are only three species of monotremes: the duck-billed platypus and the two spiny anteaters (or echidnas) of Australia and New Guinea. The **marsupials** give birth to immature live young no bigger than a kidney bean. When the newborn emerges from the birth canal, it crawls upward into an elastic pouch of skin on the mother's ventral surface (the **marsupium**), attaches to a teat, starts to consume milk, and continues to develop inside the pouch for several months. There are dozens of marsupial species, including kangaroos, koalas, and wombats from Australia and oppossums from the Americas.

Biologists sometimes compare the flowering plants to the mammals—especially the placental mammals. With the evolution of flowers as reproductive organs, flowering plants underwent a parallel radiation during the Cenozoic era. Mammals radiated into a large and diverse class that can live on land, in the oceans, and in more specialized niches than any other class of animals except perhaps the birds.

LO4 Human Origins

One branch of the mammalian family tree gave rise to the **Primata**, the order that includes ***Homo sapiens***, the "wise man." We humans are clearly mammals with our body hair, warm blood, mammary glands, and mode of giving birth. But we also have a unique combination of behavioral abilities—including spoken and written language, agriculture, and extensive tool use—that has allowed us to dominate the environment like no other animals before us. Despite our unique abilities, scientists have ample evidence, both fossil and genetic, that humans represent one recent branch of the Primate order within the mammals. Moreover, our branch separated only about 6 million years ago from the lineage leading to chimpanzees. Ethologist Desmond Morris called humans "naked apes," and so, it seems, we are. We are subject to the same kinds of evolutionary forces we discussed in earlier chapters, including natural selection and genetic drift. Because human beings are a branch of the primate evolutionary tree, the shape and size of that tree help us understand ourselves (Fig. 13.17).

The Primate Family Tree

Taxonomists divide the order Primates, with its 150 or so currently living species, into two suborders: the **Strepsirhini** (which means "moist nosed"), and the **Haplorhini** (which means "hairy nosed") (see Fig. 13.17).

Strepsirhini

Strepsirhini are small, mostly arboreal (tree-dwelling) primates, with the claws, long snout, and side-facing eyes common in early mammals. Most species feed on nectars, fruits, leaves, and sometimes insects.

mammal
a vertebrate animal of the class Mammalia, having the body generally covered with hair, nourishing young with milk from mammary glands, and generally giving birth to live young; includes lions, whales, rabbits, and kangaroos

milk
a fluid rich in fats and proteins produced in the mammary glands of mammals that nourishes newborns

mammary gland
gland that produces milk in mammals

fur
a body covering of hair

blubber
a thick layer of fat acting as insulation in such mammals as whales, porpoises, and seals

placenta
in mammals, the spongy organ rich in blood vessels by which the developing embryo receives nourishment from the mother

monotreme
an egg-laying mammal that also has many primitive or reptilian features; the only living forms are the spiny anteater and the duck-billed platypus

marsupial
a mammal having a pouch in which it carries its young, which are born in a small and undeveloped state; found extensively in Australia, with a few representatives in America; includes kangaroos, opossums, and koala bears

marsupium
an elastic pouch of skin that harbors newborn marsupials

Primata
the mammalian order including monkeys, apes, and humans

Homo sapiens
the genus and species of modern humans

Strepsirhini
a suborder of mostly tree-dwelling primates including the lemurs and lorises

Haplorhini
a suborder of primates including the tarsiers and the anthropoids—the New World monkeys, Old World monkeys, apes, and humans

anthropoid
a member of the suborder Anthropoides, order Primata; includes humans, apes, and New and Old World monkeys

New World monkeys
primates with prehensile tails that live in the forests of Southern Mexico and Central and South America

prehensile
grasping (as in the tails of some species of monkeys)

Haplorhini

The Haplorhini include the tarsiers and the anthropoids. The tarsiers have huge, forward pointing eyes. Uniquely, their legs are also elongated due to long tarsal (ankle) bones. The anthropoids (see Fig. 13.17) consist of the New World monkeys, the Old World monkeys, and the hominoids—the apes and humans.

New World monkeys, which inhabit the forests of southern Mexico and Central and South America, have flatish noses with widely separated nostrils oriented somewhat laterally. Some of the larger-bodied New World monkeys, such as spider monkeys and howler monkeys, have grasping, or prehensile, tails. These tails, which have a naked, touch-sensitive pad at the tip, serve as a type of fifth hand, aiding the monkeys in dangerous crossings between branches of adjacent trees.

Old World monkeys, which live in tropical forests and savanna regions of the Eastern Hemisphere from Africa to India and Southeast Asia, have closely set, downward-pointing nostrils and lack prehensile tails (Fig. 13.17).

Hominoids include the gibbons, orangutans, gorillas, chimpanzees, bonobos, and humans.

Trends in Primate Evolution

Fossil evidence suggests that the earliest primates lived high in the canopy of the tropical forest. These early primates evolved several important specializa-

Figure 13.17
Some Living and Extinct Members of the Primate Family Tree

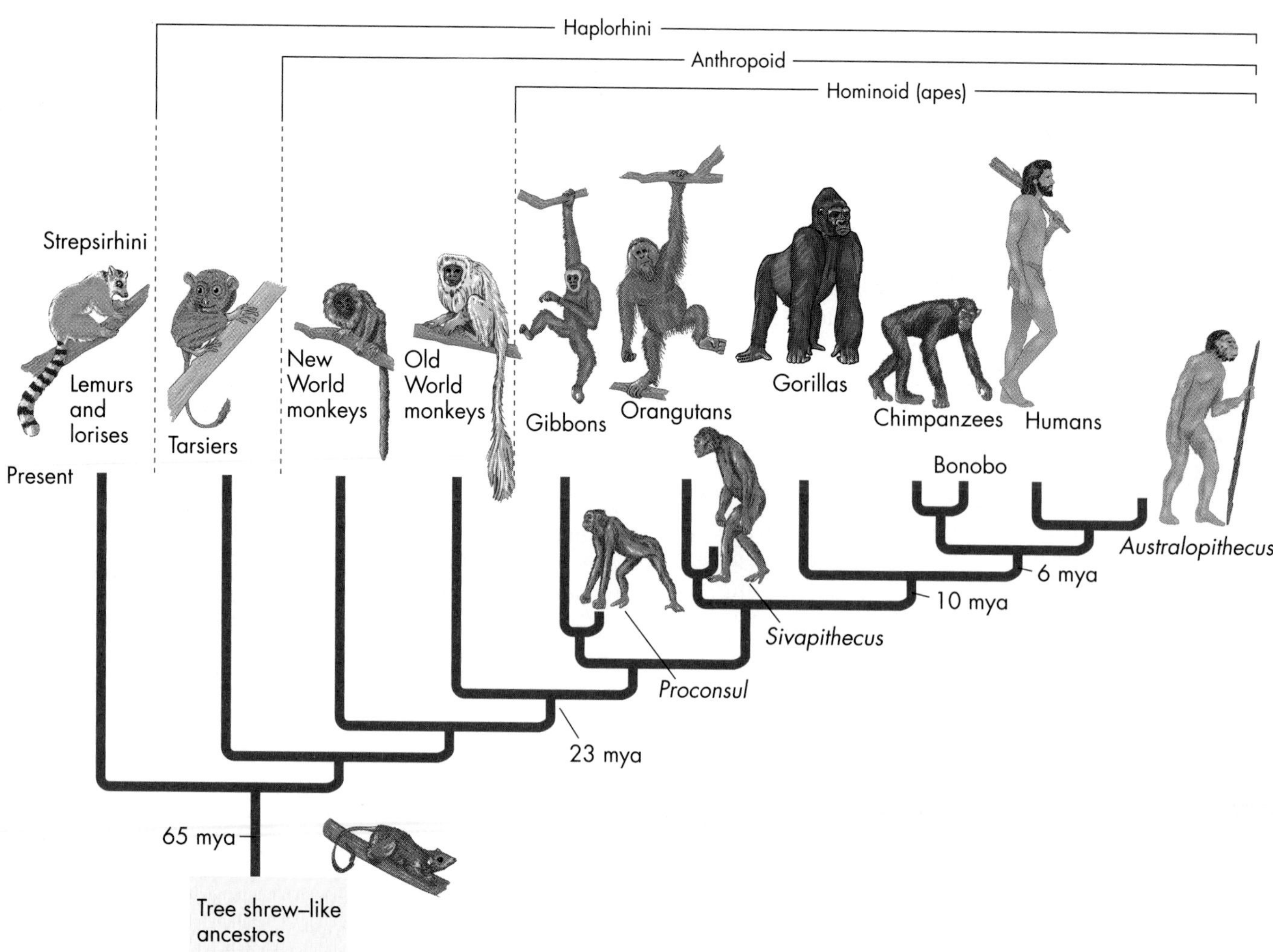

tions that lead to their success as a group and that form the background for human physical features. These traits included manual dexterity, upright posture, acute vision, large brain, extended infant care, and modified teeth.

Dexterity of the Hands

Natural selection tends to eliminate clumsy individuals who bumble awkwardly along high tree branches. Primates evolved an **opposable thumb**, the ability to spread their digits widely and to touch their "thumb" with the tips of each of the other four fingers. This allows them to curl their digits around branches in a power grip, as well as to grasp objects with precision—whether a piece of fruit or a pencil.

Upright Posture

Hands cannot manipulate objects if they are supporting the weight of the body. We humans are the only fully upright primates that walk consistently on two legs (we are **bipedal**). Nevertheless, many monkeys and apes spend large amounts of time in vertical rather than horizontal positions. Even though gorillas and chimpanzees walk on all fours by touching the ground with their feet and the knuckles of their hands, their arms are so long that the angle of the body is still fairly erect. Upright posture improves an animal's ability to see and can leave the hands free for other activities.

Acute Vision

Life in the trees is rich with visual information, such as fluttering leaves, dappled spots of sunlight, and tangled tree limbs. It can take keen vision to focus on fruits and insect foods amidst this backdrop. Furthermore, an error in judging the precise location of a food object on a particular tree limb could result in a fatal fall. Thus, natural selection strongly favored depth perception, or **stereoscopic vision**, in primates. Depth perception occurs because an animal's left and right eyes see the same object from slightly different angles. Try looking at one of your fingers while holding it at arm's length; look first with one eye closed and then the other. Notice how your finger seems to shift position. The brain interprets this difference and creates a realistic three-dimensional picture of the environment. A primate's color vision also helps it recognize members of its species, identify ripe fruits, find edible young leaves, and locate cryptic prey and predators lying in ambush.

Old World monkeys
monkeys lacking prehensile tails that live in Africa, India, and Southeast Asia

hominoids
the primate group including the apes and humans

opposable thumb
a first digit or thumb that can be held opposite the other digits; characteristic feature of some primates

bipedal
walking consistently on two legs

stereoscopic vision
depth perception based on the orientation of the eyes and their brain connections

A Very Large Brain

The evolution of life in the trees was accompanied by the enlargement of specific parts of the brain. The brain of a rhesus monkey, for example, is much larger than the brain of a similar-sized dog. Per gram of body weight, apes and humans have the most complex brains of any mammals—in terms of brain size, number of nerve cells, and numbers of interconnections.

Zoologists have posed several hypotheses to explain the increase in brain size and intelligence during primate evolution. The political hypothesis

great ape
a lineage of animals that includes the orangutans, the gorillas, the humans, and two species of chimpanzees (the common chimp and the bonobo)

holds that primates' complex social interactions were the main driving force behind increased brain size. Larger, more nimble brains might help animals relate better to others in the troop, and perhaps achieve higher social rank and thus leave more offspring. The dietary hypothesis suggests that a larger brain helped primates obtain enough food in the forest canopy, sustain better physical condition, and reproduce more effectively. More agility with grasping and manipulating foods would have helped as well, according to this theory. Both the political and dietary hypotheses make sense and, in fact, could well have played equal roles in increasing primate intelligence.

Infant Care

Compared with most other mammals, primates have smaller litters of young, often just one baby at a time. The infants tend to be helpless and depend on their parents for complete physical care long after birth. In general, the higher a species' intelligence, the more time the parent devotes to the care and nurturing of the young, and the more complex skills the young must master to succeed as an adult. The primates' greater parental investment pays off in their offspring's higher survival rates.

Teeth

Primates have several tooth types, modified for an omnivorous diet of both plant and animal matter. As early primates began to eat more plant foods, individuals with broader, flatter back teeth (molars) could chew these new foods more efficiently. In the line giving rise to humans, the jawline became U-shaped, while in most other primates, it is more V-shaped.

Emergence of Modern Primates

Evidence from genetic studies and from fossilized bones suggests that monkeys, apes, and humans (the anthropoids) arose from early primate ancestors starting about 40 to 35 million years ago (review Fig. 13.17). Monkeys diverged into New and Old World lineages, while the apes and humans (the hominoids) radiated into various niches throughout the Old World.

Some 20 million years ago, a line of primitive Old World apes radiated into many species. One of these apes was *Proconsul africanus*, a tree-living, fruit-eating African ape that walked about on all fours. Adaptive radiation within this group led to a lineage that eventually included all of today's so-called great apes: the orangutans, the gorillas, the humans, and two species of chimpanzees—the common chimpanzee *(Pan troglodytes)* and the smaller pygmy chimp, or bonobo *(Pan paniscus)*.

Biologists have discovered the close relationships between the great apes by comparing their DNAs. The genetic distance between humans and chimpanzees is about the same as, or a little less than, the genetic distance between chimps and gorillas. This means that a chimp is genetically as close or closer to *you* as it is to a gorilla!

The human and chimp lineages split about 6 million years ago—an extremely short span in geologic history. Using a 24-hour analogy, Earth formed at midnight, life appeared at 4:10 a.m., the first vertebrates appeared at 9:35 p.m., and the first humans arose only 38 seconds before the clock struck midnight. Let's explore those last 38 seconds.

The Appearance of *Homo sapiens*

The gene sequences in people and chimps are 98.5 percent the same. The 1.5 percent of genetic differences has a huge impact on our respective capabilities. Human evolution did not follow a ladder from ape to human, instead, the human lineage is the sole surviving branch of an evolutionary bush. To understand our evolution, we need to explore several questions: What changes led to modern humans? How long ago did those changes occur? And what forces brought about those changes?

Innovations in Human Evolution

At least four critical innovations arose as apelike ancestors evolved into humans:

1. the origin of bipedalism, walking on two feet. This is when our lineage diverged from the one leading to modern apes—6 million to 4 million years ago.
2. the invention of toolmaking, the manufacture of stone implements to cut flesh from the bones of game animals and to crush their bones for the nutritious marrow inside. Recent discoveries show that toolmaking began about 2.5 million years ago.
3. the dramatic expansion of the human brain far beyond the relative brain sizes of chimps and gorillas.
4. the use of brain power for abstract thought, including artistic, musical, linguistic, techno-

logical, and other skills—occurring a few tens of thousands of years ago—that have led to our current domination of the world.

The Root of the Bush of Human Evolution

In the mid-1990s, paleontologists discovered 4.4-million-year-old fossils in East Africa belonging to a species they call *Ardipithecus ramidus*. These fossils show a mix of chimpanzeelike and humanlike traits, including large canine teeth but relatively small molars, suggesting a diet of soft fruits and vegetables.

Then paleontologists found a slightly younger fossil (4.2 million years old) with bones that show upright walking. These animals, called *Australopithecus anamensis,* were still not as fully upright as we are today, and their legs were short, so the gait would have been different as well.

Lucy and Her Kin

In the mid-1970s, anthropologists found fossils in East Africa of a species they named *Australopithecus afarensis*. They named one very complete skeleton "Lucy." *A. afarensis* had a very apelike skull and teeth, a brain only slightly larger than a chimpanzee's, and long arms but short legs. Nevertheless, its head sat on top of the backbone like ours does rather than projecting forward like an ape's, and the hands were distinctly humanlike.

Who Were the First Toolmakers?

On an Ethiopian plain overlooking a shallow lake, about 2.5 million years ago, a humanlike individual grasped the hammerstone he had made and carried with him, and struck an antelope's lower leg bone twice, crushing the bone and exposing the fatty, nutritious marrow, which he devoured. He also cut out the animal's tongue, leaving curved parallel tool marks on the jawbone—and perhaps carried the prize off to family members. Paleontologists found the crushed leg bone and scarred jawbone less than a meter from fossils of a new species. The humanlike fossil bones were clearly more advanced than Lucy and presumably belonged to the butcherer of antelope. This new species, *Australopithecus garhi*, was clearly not itself a human. It is a candidate, however, for the ancestor of our own genus, *Homo*. It also shows that tool use had emerged by 2.5 million years ago.

Until *A. garhi* surfaced, *Homo habilis,* the "handy human," was considered the earliest tool user starting about 2 million years ago. These ancient members of genus *Homo* had a large face, big teeth, and trunks and limbs fully adapted for walking upright. At 1.5 m (5 ft) tall, however, *H. habilis* individuals were larger than members of the genus *Australopithecus*, and their brains were 50 percent larger. The "handy human's" brain was still only half the size of a modern human's. Its increased size, however, suggests a real increase in intelligence and probably a heightened

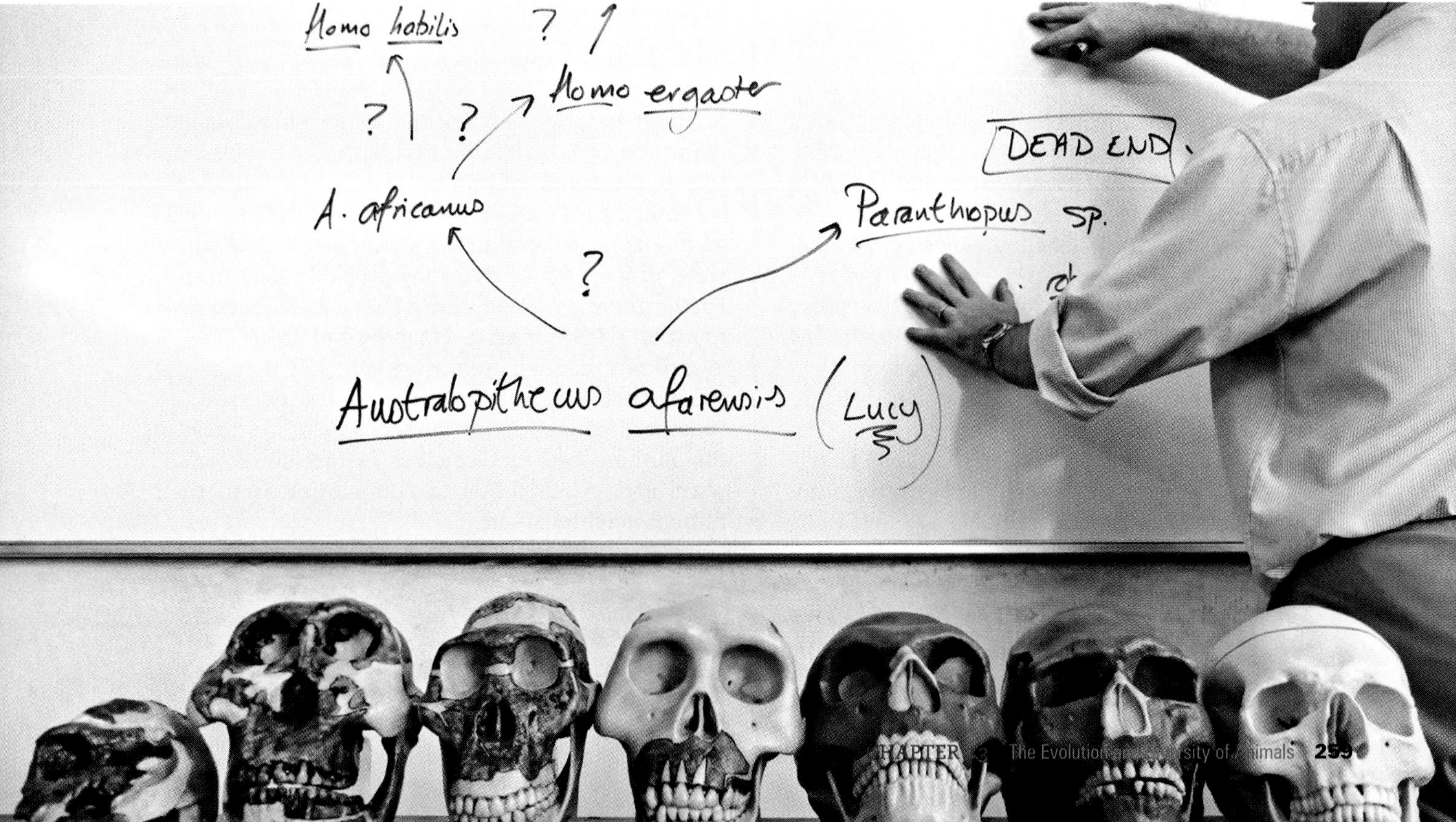

© DON SMITH/ALAMY

Homo erectus
"erect human" species that evolved by 1.8 million years ago, used tools, and spread into northern Africa

degree of manual dexterity and social skills along with it. *H. habilis* individuals appear to have made crude stone tools by cracking rocks and flaking off chips, using the sharp chips to harvest meat from tough animal hides. The ability to make tools—the first emergence of technology—has dominated the rest of human evolution.

Humans Disperse

Tool-using *H. habilis* appears to have remained in African savannas and woodlands, but fossil evidence shows that a new species of human arose in Africa and then migrated away. *Homo erectus* ("erect human") had evolved by 1.8 million years ago and spread into northern Africa and beyond. *H. erectus* fossils are intermingled with a new kind of stone tool, the hand ax. This implement, while still simple, required finer technological skill to manufacture than the hammers, choppers, and scrapers of the earlier *Homo*.

With its expanded geographical range and effective use of new tools, *H. erectus* appears better able than its predecessors to deal with extremes of climate and varied food resources. Fossil evidence suggests that these humans hunted large game in cooperative bands and cooked the meat over campfires. *H. erectus* probably first used fire about 500,000 years ago as the species spread to Europe and Asia.

How and When Did Our Species Supplant *Homo erectus?*

By about 500,000 years ago, *Homo sapiens* had emerged with a smaller face, smaller teeth, and a larger brain than *Homo erectus*. The populations that *H. erectus* established in Africa, Europe, and Asia were eventually supplanted everywhere by *H. sapiens*. Anthropologists are still debating, however, whether *H. sapiens* arose directly from *H. erectus* at the sites where the different groups of humans live today or whether *H. sapiens* arose from *H. erectus* in one place and then migrated throughout the world and developed the regional features of modern Asians, Africans, and Europeans.

Biologists have used mitochondrial DNA to test the two hypotheses. All the DNA in your cells' mitochondria comes from your mother's egg and none comes from your father's sperm. As a result, you can trace your mitochondrial DNA to your mother, your mother's mother, and so on back hundreds of thousands of years. By examining the genetic variation among mitochondrial DNA from people around the world, and estimating rates of mutation, some evolutionary geneticists are convinced that all people alive today are descended through the female line from a small group of people who lived about 200,000 years ago, probably in Africa. (This is how the term "Mitochondrial Eve" was coined.) Most anthropologists currently accept this idea.

Who knew?

Birds are called "flying dinosaurs" partly because of clear similarities between the two groups (and partly because it is satisfying to think that dinosaurs still live among us). Because of taxonomic relations between birds and thecodont reptiles, including the dinosaurs, birds are more similar to the crocodile-type reptiles than those reptiles are to other reptiles! Since taxonomic categories are supposed to reflect degree of evolutionary relatedness, the class Reptilia has fallen out of favor with some biologists. If there is no one group of reptiles, then birds and dinosaurs are more easily classified together.

Neanderthals

The classic movie images of "cave men" come from a separate species, *Homo neanderthalensis*, the Neanderthals, who shared the same general regions of Europe and the Middle East with *Homo sapiens* for tens of thousands of years.

Although *H. sapiens* and *H. neanderthalensis* lived in the same general areas at the same time, there were so few of each that they probably seldom met. Furthermore, modern researchers have been able to isolate DNA from a Neanderthal fossil, and it is different enough from our own that it rules out substantial interbreeding between the two species. Neanderthals appear to have occupied a branch of the human tree that became extinct some 34,000 years ago, perhaps due to competition from their human neighbors—us.

87% of students surveyed feel that LIFE is a better value than other textbooks.

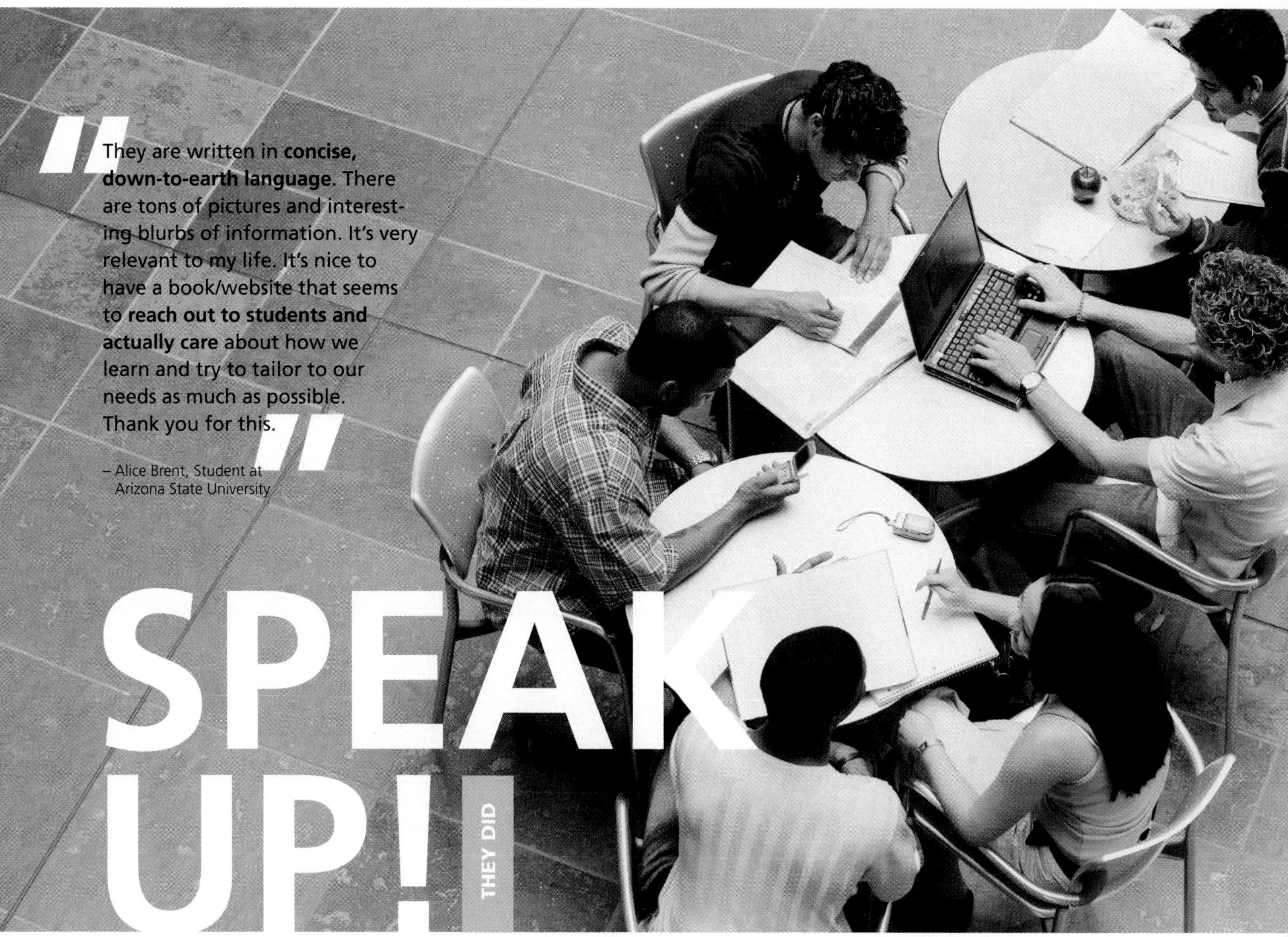

LIFE was built on a simple principle: to create a new teaching and learning solution that reflects the way today's faculty teach and the way you learn.

Through conversations, focus groups, surveys, and interviews, we collected data that drove the creation of the current version of LIFE that you are using today. But it doesn't stop there – in order to make LIFE an even better learning experience, we'd like you to SPEAK UP and tell us how LIFE worked for you.

What did you like about it?
What would you change?
Are there additional ideas you have that would help us build a better product for next semester's non-major biology students?

At **4ltrpress.cengage.com/life** you'll find all of the resources you need to succeed in introductory biology – **animations, visual reviews, flash cards, interactive quizzes,** and more!

Speak Up! Go to **4ltrpress.cengage.com/life**.

14

Ecology *of* Populations *and* Communities

Learning Outcomes

LO[1] Understand the distribution and interaction of organisms in an environment

LO[2] Identify the limiting factors that influence a population's location

LO[3] Identify the limiting factors that influence a population's size

LO[4] Identify long-term human population trends and predict future population growth

LO[5] Identify the relationships among organisms in a community in terms of how they make a living

“The tenets of ecology are crucial to human survival.”

Crashes and Clashes

Until the mid-20th century, no one worried about sustaining or endangering fish populations. In the wide continental shelf system called the Georges Bank offshore from Cape Cod and the coast of Maine, upwelling nutrients have always supported an enormous population of fish. By “trawling,” or dragging nets across the sea floor, fishers were able to catch thousands of pounds of fish at a time. Most of the fish species produce hundreds of thousands of eggs and larvae each year, so the fish populations could recover fairly quickly.

What do you know?

Explain the difference between a yellowtail flounder's habitat and its niche.

In the 1960s, huge foreign fishing vessels started trawling on Georges Bank alongside smaller American boats and harvesting millions of metric tons of fish annually. After years of such enormous catches, some of the fish populations started dwindling—an indication that overfishing was damaging the area's natural ecological balance.

In 1976, the United States extended its territorial limit to 200 miles off shore and the “factory ships” left. In their wake, the American fishing fleet grew, by then with technological improvements such as electronic navigation and satellite weather data to help crew members find large aggregations of target fish such as Atlantic cod or yellowtail flounder. This fleet growth was also spurred by the skyrocketing demand for fish among health-conscious American consumers.

In the early 1990s, once again overfishing took a drastic toll and fish stocks plummeted to record low levels. For a number of fish stocks, at least half or three quarters of the fish in a population was removed every year.

Now management programs limit the access of commercial fishers. They have closed areas in certain seasons, regulated net and mesh sizes, restricted numbers of fishing permits, and limited each day's catch per boat. As a result, stocks of fish are recovering—some quickly, some slowly. The challenge now is in allowing an increased access to those fish without repeating the mistakes of the past.

The subjects of fishing, overfishing, and wildlife conservation in New England waters make an ideal case history for this chapter. Our topic here is the **ecology of populations and communities**, that is, the study of how groups of organisms are distributed in a particular area at a particular time and how they interact with other species coexisting in the same locale. As we explore population and community ecology, you'll discover the answers to these questions:

- At what level do organisms interact in ecology?
- What limits where a species lives?
- What factors limit population size?
- How do species interact in communities?

ecology of populations and communities the study of how groups of organisms are distributed in a particular area at a particular time and how they interact with other species coexisting in the same locale

LO1 Ecology: Levels of Interaction

A fisher's livelihood depends on knowing about the habits of fish: Where do yellowtail flounder or other crop fish live in the oceans? What times of year are best to harvest them? At what depths do they feed? What do they eat? In short, he or she needs as much information as possible to locate and catch the fish efficiently. As you read in Chapter 1, the science of ecology studies the distribution and abundance of organisms and how organisms interact with each other and the nonliving environment. The tenets of ecology are crucial to human survival. That's because our food, our shelter, the quality of our water, and even fundamental factors such as the air we breathe, the temperature of Earth's surface, and, to some degree, its climate depend on where organisms live and how they interact.

A key word in the definition of ecology is *interact*. Organisms interact with other living things and with the nonliving physical surroundings. For a yellowtail flounder, the living environment includes the worms and other small invertebrates it eats, the blackback and witch flounder that competes with it for food, the mackerel that prey on the larval yellowtail, and the roundworms that parasitize and weaken its body. The physical environment includes the temperature, salinity, and depth of the water; whether the bottom is sandy, rocky, or muddy; and the annual cycle of storms and ocean upwellings.

So numerous are each organism's interactions with other living things and the physical environment that biologists organize their study of ecology into a hierarchy of four levels: populations, communities, ecosystems, and the biosphere.

- A *population* is a group of interacting individuals of the same species that inhabit a defined geographical area.
- A *community* consists of two or more populations of different species occupying the same geographical area.
- An *ecosystem* consists of a community of living things interacting with the physical factors of their environment.
- The *biosphere* consists of all our planet's ecosystems and thus is the portion of Earth that contains living species. The biosphere includes the atmosphere, the oceans, and the soil in which living things are found, as well as global phenomena, such as climate patterns, wind currents, and nutrient cycles that affect living things.

This chapter and the next examine ecological interactions at increasingly higher levels of organization. We begin by discussing the factors that influence a population's location and size in time and space. Then we turn to community ecology and discuss how living organisms interact with each other.

LO2 What Limits Where a Species Lives?

Few species are scattered evenly throughout the world; they exist, instead, only in certain spots and are absent entirely from other places. What limits where an organism lives on a global scale and in individual locales?

Limits to Global Distribution

Understanding why an organism's range extends to one part of the world but not to another is important to people when those organisms serve as food or materials, or when they cause disease in people, domestic animals, or crop plants. In general, three conditions limit the places where a specific organism might be found: physical factors, interactions with other species, and geographical barriers.

Physical Factors

Organisms may be absent from an area because the region lacks the proper sunlight, water, temperature, mineral nutrients, or any one of a host of physical or chemical requirements.

Interactions with Other Species

Other species may block survival and limit a population's distribution. If certain species are already firmly established in an area, they may prevent the incursion of new species by monopolizing food supplies or acting as predators or parasites.

Geographical Barriers

A species may be absent from an area because a geographical barrier blocks access. Seas, deserts, and mountain ranges can be so wide or high that an organism cannot crawl, swim, fly, or float across the barrier. For example, bird fanciers from Europe artificially bridged a geographic gap in 1890, when about 80 starlings were introduced into New York City's Central Park. Now millions of the speckled birds chatter throughout America's cities and countrysides. Only their inability to cross the Atlantic had previously stopped their dispersal to North America (Fig. 14.1).

LO3 What Factors Limit Population Size?

One goal of researchers working with the Fishing Management Council in New England is to maintain populations of yellowtail floun-

Figure 14.1

Crossing the Atlantic Barrier Allowed Starlings to Spread throughout North America

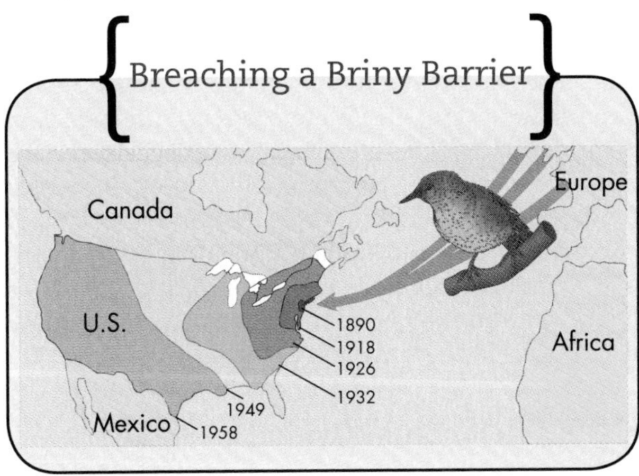

der and other crop fish at optimal levels so the Georges Bank area can sustain a healthy population of the animals. One measure of the population of yellowtail flounder is the weight of fish caught per year. Take a quick look at Figure 14.2. What happened to the yellowtail flounder population between 1962 and 1996? What happened to the populations of skates and spiny dogfish, a type of small shark? Researchers wanted to learn what factors caused this dramatic crash of the yellowtail population, how it might relate to the increase in skates and spiny dogfish, and how to restore populations to their original state to maintain a stable yield of fish.

Figure 14.2

The Population Crash of Yellowtail Flounder

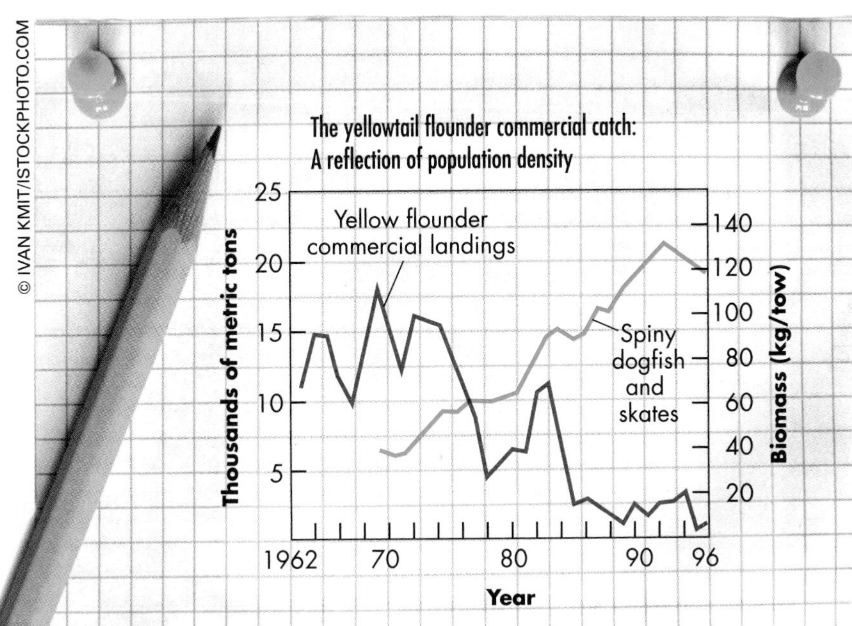

This practical side of ecology is concerned with a population's *density*, the number of individuals in a certain amount of space, for example, the number of metric tons of fish caught per year on Georges Bank. When reporting density, it is often helpful to take into account the species' habitat—that yellowtail flounder on Georges Bank may occupy only the sandy bottoms or that in a certain field a particular species of butterfly might cluster mainly around yellow sulfur flowers. Human populations illustrate quite well the effect of clustering. The density of Earth's population averaged over the whole planet is less than one person per square kilometer (two per square mile). Most people live near ocean coasts, riverbanks, and lake shores, making local densities in those areas much higher than the average.

zero population growth
in a population, the number of individuals gained is exactly equal to the number lost

Factors Affecting Population Size

Curious biologists want to understand both how population density changes over time and the factors that cause it to change. Population size may change when individuals enter or leave the population. Clearly, if more members enter than leave, the population will grow, but if more individuals leave than enter, the population will shrink. Individuals can enter a population by either birth or immigration, and members can leave a population by either death or emigration. If the number of individuals gained is exactly equal to the number lost, then the population shows zero population growth. For people, an average of 2.1 children per couple results in zero population growth.

Let's look now at how populations change as individuals leave a population by death or join a population by birth. We will assume that the effects of immigration and emigration remain constant and equal in the populations we are studying.

How Survival Varies with Age

Death and birth clearly affect population size. For many species, the likelihood that an individual will die depends on its age. For a yellowtail flounder, the chances that it will die

survivorship curve
a plot of the data representing the proportion of a population that survives to a certain age

late-loss survivorship curve
a plot of survivorship data indicating that an organism's life expectancy decreases with each passing year

early-loss survivorship curve
a plot of survivorship data indicating that most individuals in a population die young

life expectancy
the maximum probable age an individual will reach

as a newly hatched larval fish is very high, but after an individual becomes an adult, the likelihood of dying is much smaller. This is reflected in a **survivorship curve**, a plot of data representing the proportion of a population that survives to a certain age (Fig. 14.3). Look first at the human survivorship curve. The chances that a 90-year-old person will live for 1 more year are much slimmer than the chances that a 20-year-old will live for 1 additional year. In contrast, yellowtail flounder die mostly in the first month or two after hatching as they drift in the open ocean and become prey to mackerel, herring, and shrimplike krill. Once they settle on the bottom and become adults, it is less likely that a yellowtail will die in each passing month. The human follows a **late-loss survivorship curve**, whereas the flounder follows an **early-loss survivorship curve**.

A table of numbers used to generate a survivorship curve is called a life table, and it shows the **life expectancy** (average time left to live) and probability of death for individuals of each different age. Insurance companies use life tables to determine policy costs for customers of different ages. From the survivorship curve in Figure 14.3, you can understand why insurance companies charge an 80-year-old man more for insurance than an 80-year-old woman: He is more likely to die in the next year than she is.

How Fertility Varies with Age

Survivorship curves help predict how many individuals of each age class will leave a population through death. The major force that counteracts death is the birth of new individuals. Birth rates, like death rates,

Figure 14.3
Survivorship Curves

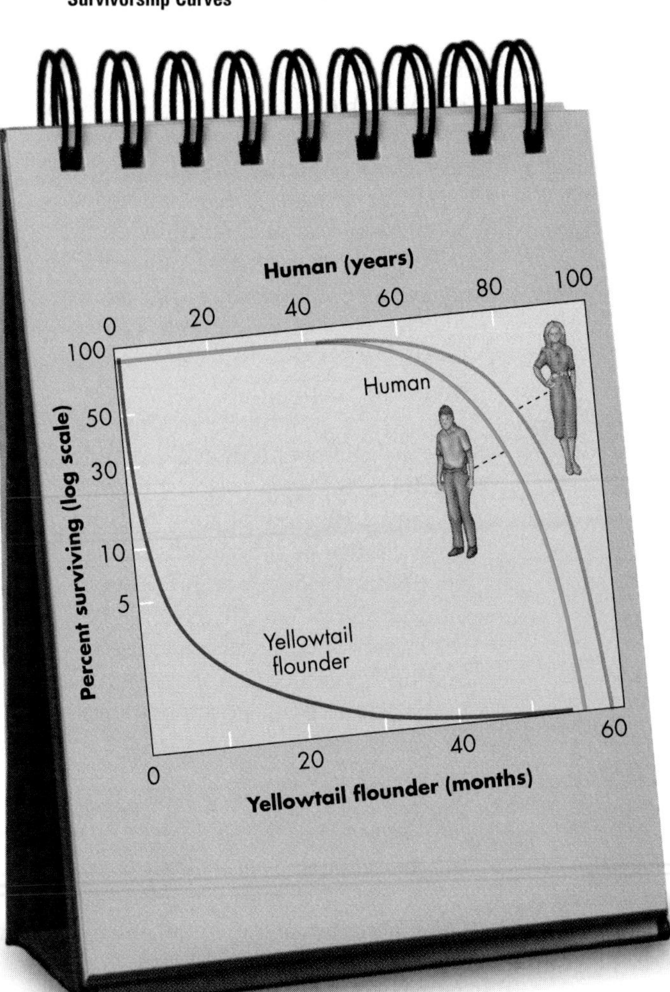

Figure 14.4
Human Fertility Curve

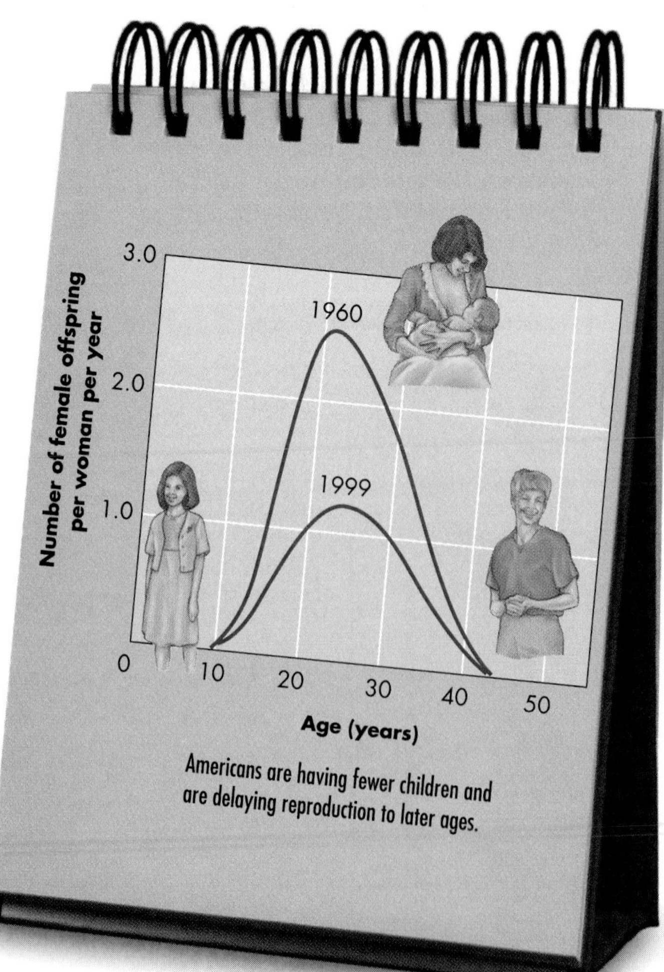

depend on age, and this fact is revealed by **fertility curves**, graphs of reproduction rate versus the age of female population members. (Since only females bear young, ecologists often view populations as females giving rise to more females.)

The human fertility graph in Figure 14.4 shows that American women younger than 20 or older than 30 are less likely to reproduce. Note the difference in the curves for 1960 and 1999. Fertility curves are significant because with them and with a knowledge of a population's age structure, one can predict future population growth. For example, a population with a high proportion of 20- to 30-year-old women is going to grow much more rapidly than a population with few women in these prime childbearing years. Data show that about 12 out of every 100 American women between the ages of 20 and 25 will have a baby girl in any given year.

How Populations Grow

Birth rates and death rates are frequently the major factors influencing changes in population size. By combining the two (but discounting immigration and emigration), population ecologists can make a model for how populations grow. Given plenty of nutrients, space, shelter, water, benign weather, and the absence of predators or disease, every population will expand infinitely, because all organisms have a high innate reproductive capacity under ideal conditions. The capacity for reproduction under idealized conditions, or **biotic potential**, is amazing.

Rapid population growth is easier to visualize when plotted on graphs such as those in Figure 14.5. Let's say that individuals of a long-lived mouse species grow to reproductive age in one year and that, ignoring males, a population initially consists of ten female mice that each produce on average one female offspring per year. At the end of the first year, the population will include 20 female mice (the 10 mothers and each of their daughters). If each of those has 1 more female per year, there will be 40 female mice at the end of 2 years, 80 after 3 years, and so on. The explosive increase results in a **J-shaped curve** representing **exponential growth**.

What factors do you think might affect the shape of the J-shaped curve of Figure 14.5? It turns out that there are two factors: (1) the reproductive rate per individual and (2) the initial population size. If the reproductive rate per individual increases to two female offspring per female mouse rather than one each year, the population would grow much faster. Ecologists represent the reproductive rate per individual by the symbol *r*. For human populations, the reproductive rate per individual can vary widely. Women are capable of having about 30 children each, but they rarely reach that grand potential.

While the reproductive rate per individual is an important factor in the contour of the J-shaped curve, the size of the initial

fertility curve generally, a graph that plots reproduction rate versus the age of female population members

biotic potential an organism's capacity for reproduction under ideal conditions of growth and survival

J-shaped curve a plot of population growth with an upsweeping curve that represents exponential growth

exponential growth growth of a population without any constraints; hence, the population will grow at an ever-increasing rate

Figure 14.5
The J-Shaped Curve of Population Growth

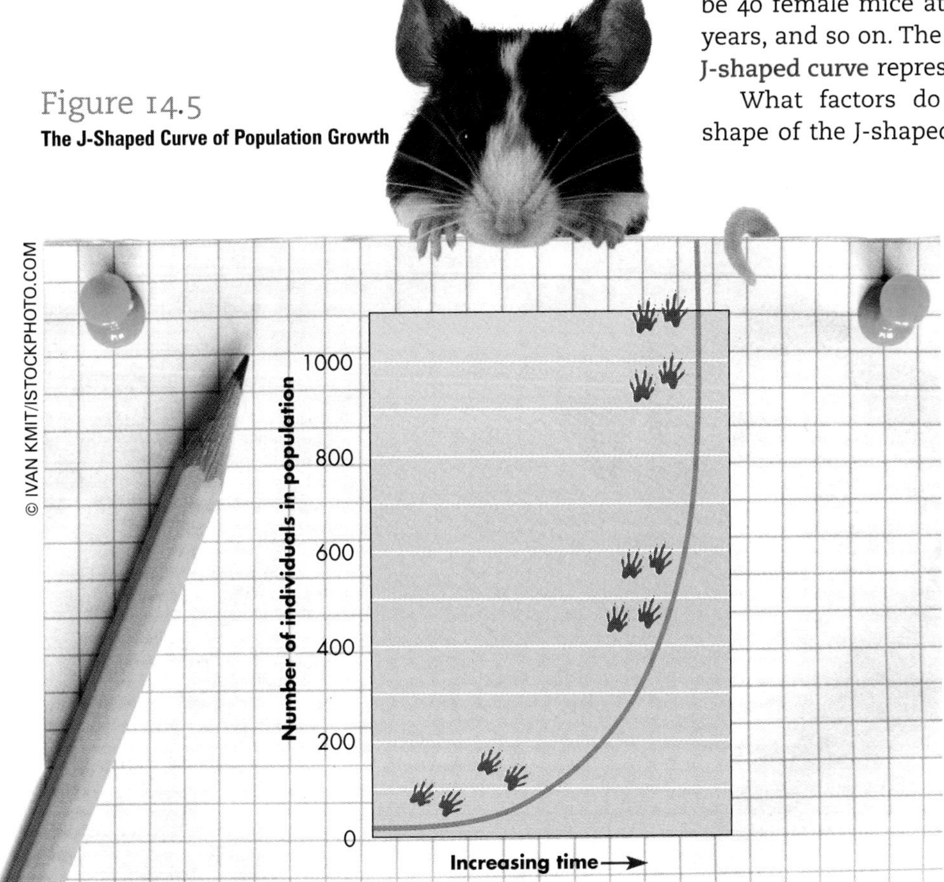

S-shaped curve a plot of population growth with a flat section, a steeply rising section, and then a leveled off section that represents logistic growth

logistic growth growth of a population under environmental constraints that set a maximum population size

carrying capacity the density at which growth of a population ceases due to the limitation imposed by resources

population is also important. If there are only five female mice in the initial population instead of ten, as in the first example, and each produces one female offspring per year, then only five more will be added by the end of the first year rather than ten. This is why small populations add fewer individuals per year than larger ones, all other factors being the same.

Under the artificially ideal conditions we've discussed so far, any population would follow the J-shaped curve of exponential growth if the birth rate exceeded the death rate by even a small amount. Fortunately, real organisms in natural situations do not follow the J-shaped exponential growth pattern—at least not for long. Organisms in nature cannot sustain limitless growth at the full force of their reproductive potential because food supplies and living space are finite. Hence, our planet is not covered with elephants neck-deep in flounders. The realities of supply and demand explain this curb on population growth despite the organism's reproductive capabilities.

Limited Resources Limit Growth

The J-shaped curve gives ecologists an idealized standard against which to measure growth in real populations. A classic case of growth in a real population was that of sheep on the island of Tasmania, south of Australia, in the early 19th century (Fig. 14.6). When English immigrants first introduced sheep to the new environment, resources were abundant and the sheep population expanded nearly exponentially for a couple of decades, following a J-shaped curve during this time. As the density of sheep on the island rose, competition for limited resources increased, and by 1830, each sheep had a smaller share of food and living space. As a result, each individual was less likely to survive and more likely to die, and each had a smaller chance of reproducing. After 1850, the total growth rate decreased, and the population size fluctuated around a mean of about 1.6 million sheep.

As Figure 14.6 shows, the graph representing population growth began like a J-shaped curve, but then flattened into an **S-shaped curve** representing **logistic growth**, a situation in which a large population grows more slowly than a small population would in the same area. The density at which a growing population levels off—1.6 million sheep on Tasmania in the preceding example—is called the **carrying capacity** (ecologists abbreviate this with the letter K). Carrying capacity is the number of individuals an environment can support for a prolonged period of time. At the carrying capacity, individuals are using all of the resources available to them.

Carrying capacity is not a constant number for each species and environment. You can tell that by looking at the fluctuations around the carrying capacity for the sheep shown in Figure 14.6. Because environments change with the season, with alterations in weather

Figure 14.6

Sheep on Tasmania: The Growth of Real Populations Often Follows an S-Shaped Curve

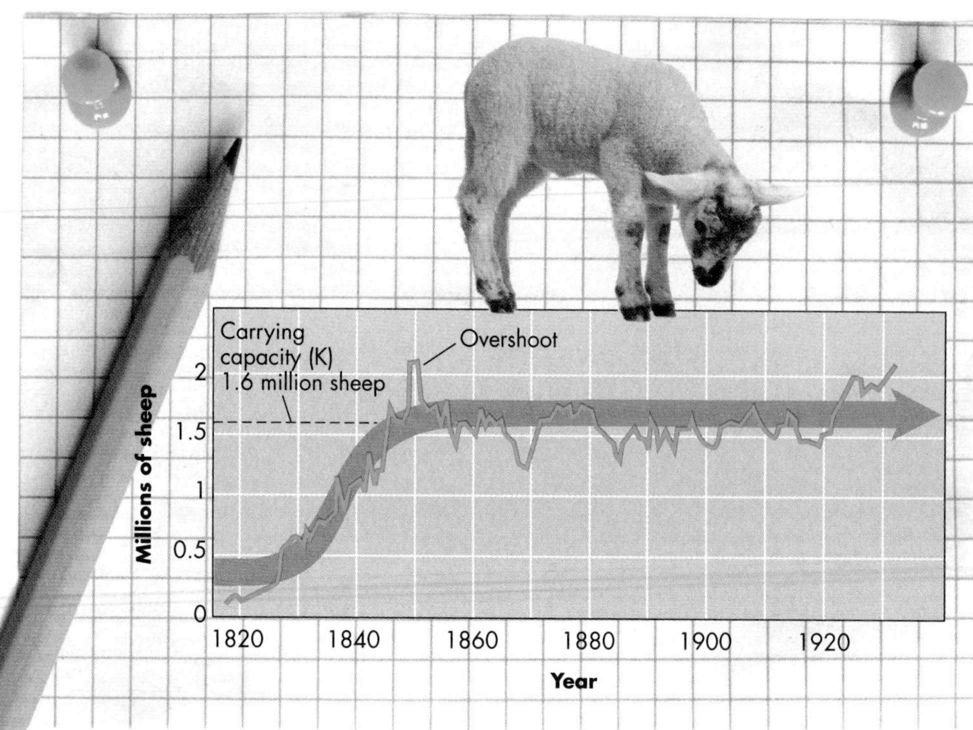

patterns, and with changes of species composition within the community, we should think of carrying capacity as a range of densities toward which populations tend to move from initial densities that can be higher or lower.

Population Crashes

While population growth often slows and reaches a plateau, as it did with sheep in Tasmania, in some species there can be a bust following the boom: a rapid decline following a period of intense population growth. The growth of reindeer introduced onto an island off the southwest coast of Alaska represents a frequently observed pattern (Fig. 14.7). From an initial population of 25 animals in 1891, the herd grew to about 2,000 reindeer in 1938 and then crashed to 8 animals by 1950. The crash can be readily explained on the basis of carrying capacity. When the reindeer were first introduced, lichens and other food sources were plentiful, having accumulated for centuries without predation. Thus, the island's carrying capacity was high. After the reindeer ate the accumulated food, however, new food would appear only as the remaining lichens regrew slowly during each short summer growing season. This change in carrying capacity of that environment is what caused the reindeer population to crash.

How the Environment Limits Growth

The limited growth of real populations is due to such factors as limited food supplies, limited living space, and interactions with other organisms.

The Role of Population Density

Some limits to population growth depend on how dense a population is, and others do not. For instance, in a sparsely populated colony of prairie dogs, the incidence of flea-transmitted bubonic plague is also low. But when the prairie dogs are densely packed, outbreaks of plague often wipe out entire populations. Competition among members of the same species for limited resources is another way density can regulate population size. For example, all the squirrels in a given forest might compete for the same nut crop.

Some factors limit population expansion regardless of population density. A harsh winter, for example, might kill 25 percent of a deer population regardless of population density. In practice, it can be difficult to separate the effects of population density from other factors, and this interaction leads many ecologists to look at whether the mechanism that limits population growth originates outside or inside the population.

Figure 14.7
Overexploiting Limited Resources Can Lead to a Population Crash

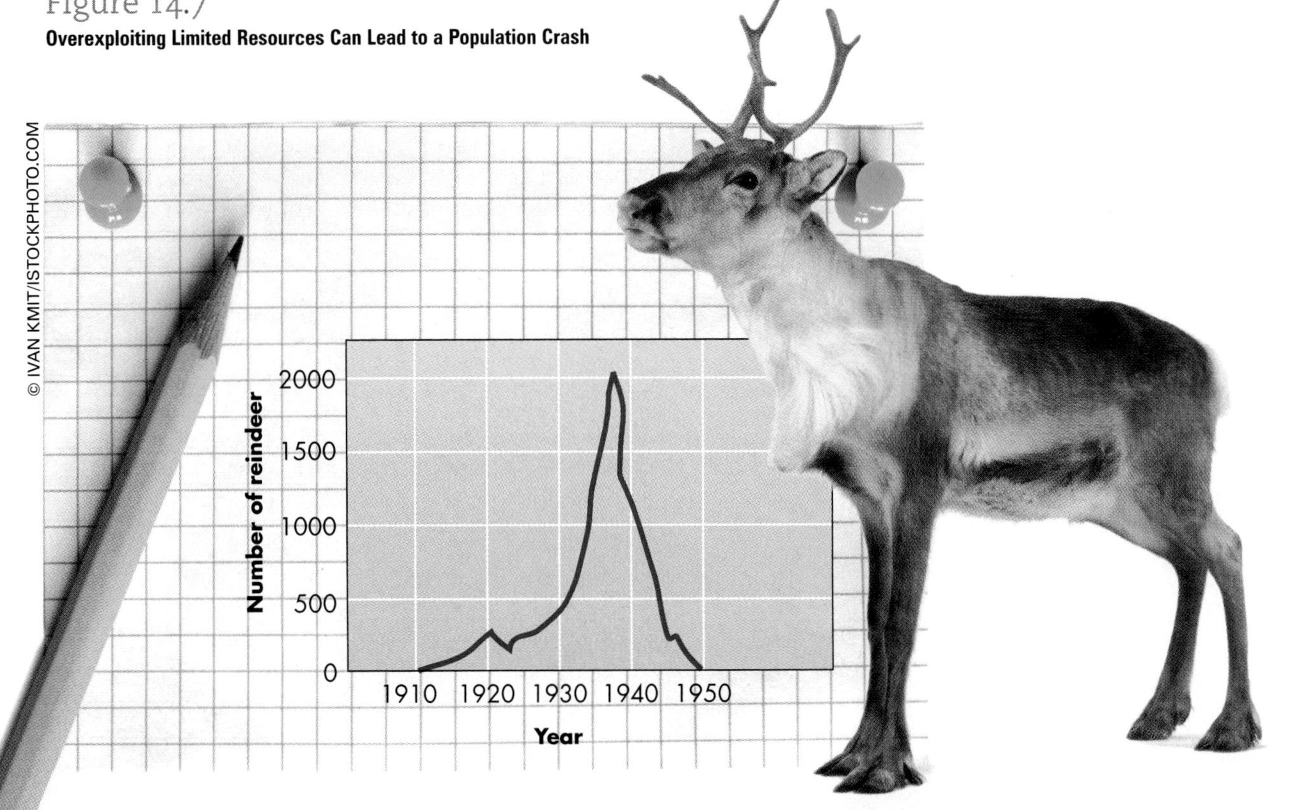

life history strategy the way an organism allocates energy to growth, survival, or reproduction

Extrinsic or Intrinsic Population-Regulating Mechanisms

Extrinsic population-regulating mechanisms originate outside the population and include living factors, such as food supplies, natural enemies, and disease-causing organisms, as well as physical factors, such as weather, shelter, pollution, and habitat loss. In contrast to extrinsic factors that limit population size, intrinsic factors originate within an organism's anatomy, physiology, or behavior. For example, crowded conditions and depletion of resources can cause many marsupials, such as kangaroos and koalas, to absorb their own developing embryos; this intrinsic response lowers the rate of population growth.

Competition

The most important intrinsic population-regulating mechanism is competition among members of the same species for resources that they require but are available in limited supply. The effects of competition among members of a species depend on population density. As the population grows and resources diminish, competition for food and space becomes intense. In a coastal tidepool, for example, where a population of barnacles grows on a rock, other barnacles cannot occupy that same spot, even if food is present in abundance.

While extrinsic and intrinsic factors in the environment limit the growth of all species, they result in populations following either a roughly S-shaped curve or a boom-and-crash curve. Let's look at the factors that dictate which shape growth curve a population follows in nature.

Population Growth and Strategies for Survival

To survive, reproduce, and thus make a genetic contribution to the future, individuals must allocate their limited energy supplies. A fast-growing organism that expends most of its energy enlarging its body may have little energy left over for reproducing. Conversely, an individual that expends a huge amount of energy attracting a mate or producing thousands of eggs may have little energy remaining for day-to-day survival activities. The way an organism allocates its energy is its life history strategy.

To see how different life history strategies work, let's contrast the life history strategy of a dandelion and a rhinoceros. A dandelion reproduces rapidly, based on fast embryonic development and the production of large numbers of small seeds containing few stored nutrients. Dandelions experience an early-loss type of survivorship: most of the light, windborne seeds die shortly after germination. Because of these traits, they can quickly fill a newly plowed field before winter comes or the corn grows and shades them out. Our expression "to grow like a weed" reflects this type of life history strategy.

In contrast, a rhinoceros reproduces only after about 5 years of age, and the development of its embryo (usually single) is very slow (gestation takes about 15 months). Although rhinoceroses have only one calf at a time, the newborns are huge—the weight of an average male college student. Once born, the new individual survives for about 40 years, thus rhinoceroses experience a late-loss type of survivorship. Rhinoceroses are quickly approaching extinction because people slaughter them for their nasal horns. Some people have a superstition that the powdered horn can act as an aphrodisiac, and they are willing to pay more for the material than for gold, ounce for

ounce. Knowing the life strategy of a rhinoceros, you can appreciate how hard it is for their populations to become reestablished once decimated.

Many species have some elements of each strategy and can't be easily pigeon-holed. Which strategy—if either—do you think people follow? Let's look in more detail at the human population.

LO4 The Human Population

Are humans governed by ecological rules? Or have we somehow moved beyond booms, crashes, and growth curves? The answers involve a look at human history, long-term population trends, and some predictions for our future population growth.

Trends in Human Population Growth

You can see in Figure 14.8 the three historical phases of human population growth, and our current staggering rate of population growth.

Hunting and Gathering Phase

Until about 10,000 years ago, the human population grew slowly as people existed by hunting animals, catching fish, and gathering roots and fruits from nature. The worldwide population was probably about 10 million by 8000 B.C. During this early phase, the human life history strategy emphasized slow embryonic development, long lives, large bodies, few offspring, extended parental care, and highly specialized brains that help us compete for resources with cunning efficiency.

Agricultural Phase

Population growth accelerated during a second phase of human history, beginning about 10,000 years ago, when people started planting and tending crops and domesticating animals in the Agricultural Revolution. The shift to agriculture was rapid and worldwide, perhaps because people can transmit their culture, or ways of living, to others. As agricultural techniques spread and improved between about 8000 B.C. and 1750 A.D., world population increased from 10 million to about 800 million. Because agriculture uses some resources more efficiently, its practice increases the environment's carrying capacity for humans.

Agricultural Revolution the transition of a group of people from an often nomadic hunter-gatherer way of life to a usually more settled life dependent on raising crops, such as wheat or corn, and on livestock; it was under way in the Middle East by 8000 years ago

Industrial Revolution the replacement of hand tools with power-driven machines (like the steam engine) and the concentration of industry in factories beginning in England in the late 18th century

Industrial Phase

A third phase of growth began in 18th-century England with the Industrial Revolution. Inventions such as the steam engine triggered vast changes that transformed a populace living mainly as farmers and craftspeople into a population working mainly in factories and living in crowded cities. In the next 250 years, much of the world would follow this pattern of industrialization and social upheaval. A farmer with a steam engine attached to a tractor could accomplish the work of dozens of people in a single day and thus increase food production. A steam-driven train or ship could rapidly distribute food and other necessities of life, and thus blunt the impact of local famine.

In recent times, the rise in human population has been staggering. While it took from the beginning of life until 1950 for the first 2.5 billion people to accumulate on Earth, it took just 40 years—a blink of evolutionary time—for a second 2.5 billion to be added. At current growth rates, by 2025 an additional 5 billion *more* people will join the planet's current population of over 6 billion. How old will you be when the 11 billion figure is reached? What might life be like when there are twice as many people as there are today?

As you look at the graph of human population growth, the towering ascension should unnerve you: It is the familiar J-shaped pattern of exponential growth, much like that of the island reindeer just

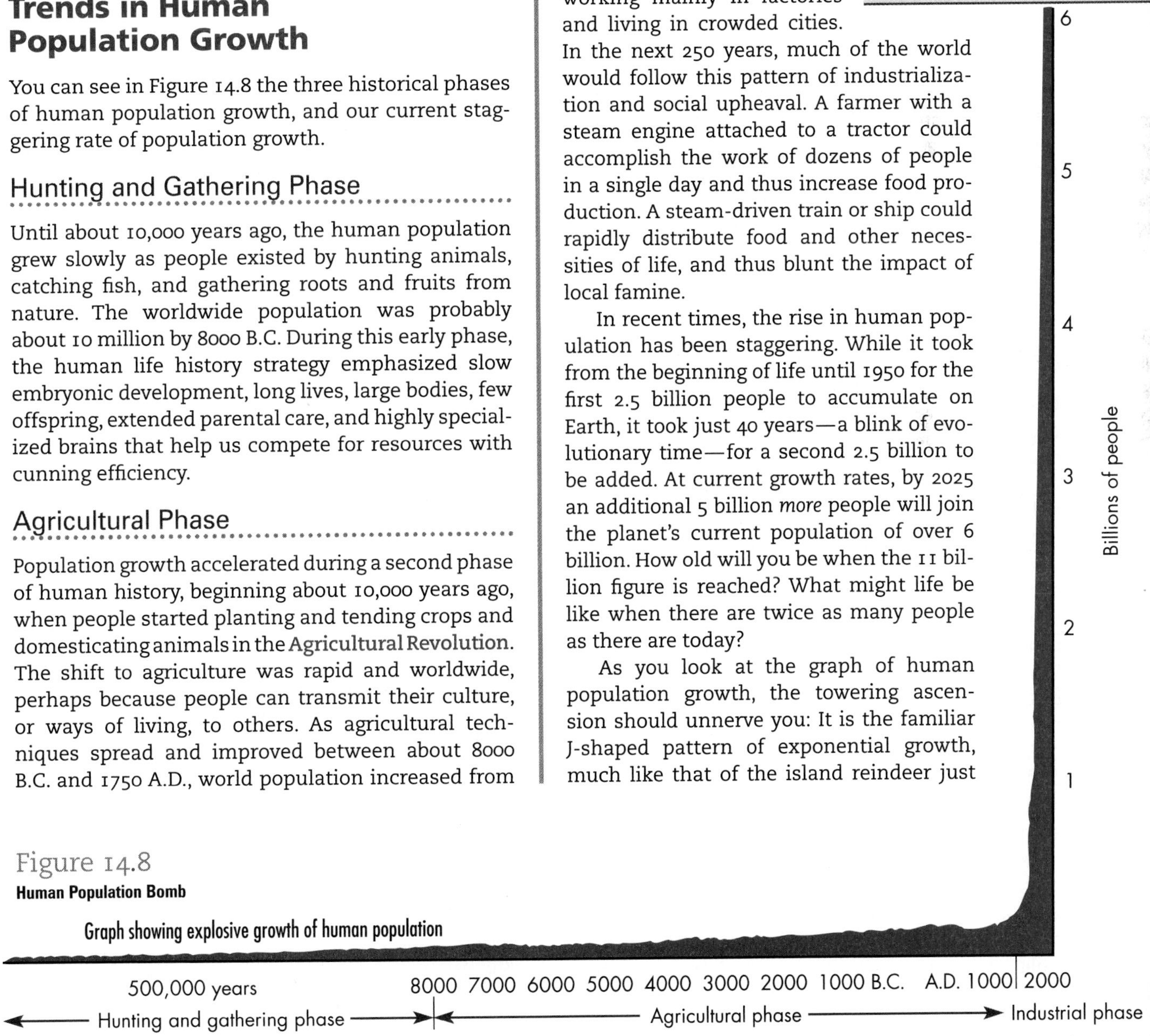

Figure 14.8
Human Population Bomb

demographic transition a changing pattern from a high birth rate and high death rate to a low birth rate and low death rate

age structure the number of individuals in each age group of a given population

before they overexploited their environment and suffered a population crash. By analyzing the causes of our own population boom, ecologists hope to learn how humans can avert a crash in the future.

Change in Human Population Size

How did the agricultural and industrial revolutions quicken the pace of the human population explosion? Did the invention of agriculture *allow* human populations to increase, or were people *forced* to invent agricultural practices to help support population densities that were already exceeding the carrying capacity of the land? Many observers believe the latter and suggest that population growth has been a constant feature of the human experience, continually forcing people to adopt new strategies for increasing the amount of food their land could produce. Carrying capacity, however, cannot be increased forever; the productivity of the land must, at some point, be reached and exceeded.

Birth Rates and Death Rates

To understand the causes of human population increase, particularly the tremendous surge after the Industrial Revolution, we must recall that the population growth rate equals the birth rate minus the death rate. Prior to 1775, the birth rate in developed countries like those of northern Europe was slightly higher than the death rate, and so the population enlarged at a low rate (Fig. 14.9, Step 1). After 1775, as industry expanded, people enjoyed improved nutrition, better personal and public hygiene, protection of water supplies, and the reduction of communicable diseases such as smallpox. These innovations caused a gradual decline in the death rate (Step 2). While the death rate began to decline in 1775, the birth rate did not start to drop in developed countries until about a hundred years later. Consequently, each year many more people were born than died, and this caused an increase in the rate of population growth. By the last decade of the 20th century, both the birth rates and death rates in industrialized nations had dropped to all-time lows, and the gap between them had once again narrowed (Fig. 14.9, Step 3). For example, Japan in 1999 had a birth rate of 9.5 per thousand, and a death rate of 7.3 per thousand, giving a low growth rate. A changing pattern from high birth rate and high death rate to low birth rate and low death rate is called the demographic transition (Fig. 14.9, Step 4).

Growth Rates and Age Structure

A sure sign of a population's growth rate is its age structure: the number of individuals in each age group (Fig. 14.10).

The age structure of a growing Swedish population in 1900—a time when the death rate had already declined substantially but the birth rate had yet to fall—shows a high percentage of people in the younger age classes (Fig. 14.10a). This results in a pyramid-shaped age distribution.

By 1977, the Swedish birth rate had dropped close to the death rate, and each age class was only slightly smaller than the younger one below it (Fig. 14.10b). This results in a bullet-shaped age profile with almost

Figure 14.9

The Difference between Birth Rate and Death Rate Detonates the Population Bomb

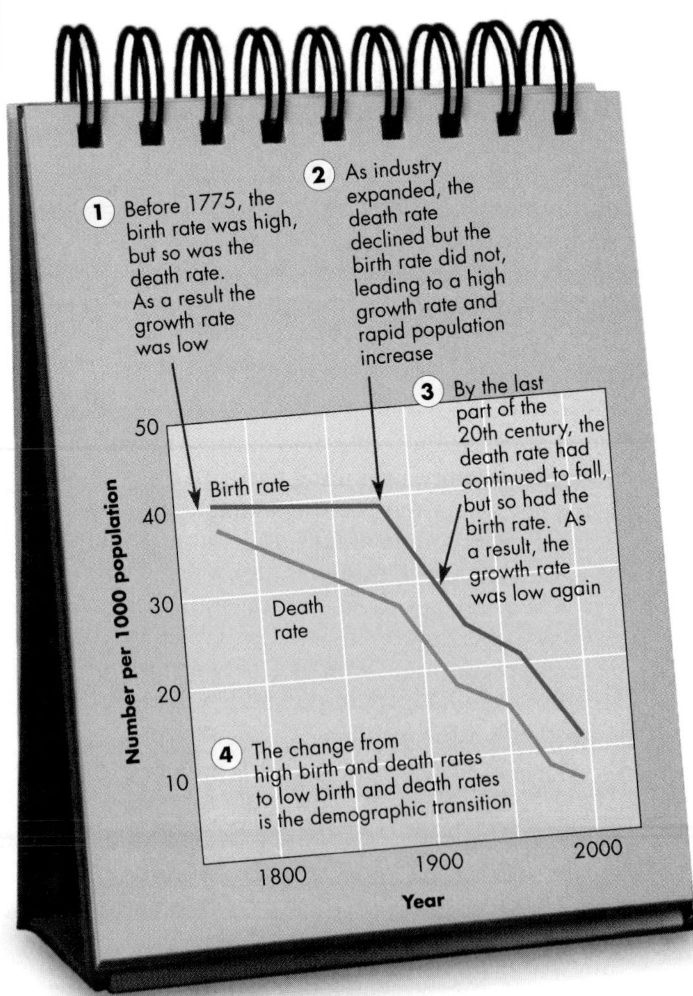

no difference in the size of age classes until the end of the human life span (the point of the bullet).

Age structure is significant because it helps predict a population's growth potential. A pyramid-shaped population with many young people, like the African nation of Uganda in 2002, will grow rapidly (Fig. 14.10d–e). In contrast, a bullet-shaped population, like Sweden in 1977, will be stable or decline. With a look at data graphed this way, you can infer the kinds of social services needed by different populations: schools for pyramid-shaped populations, and health-care facilities for the elderly and pension plans for bullet-shaped ones.

LO5 Where Do Organisms Live and How Do They Make a Living?

The rapidly increasing human population makes it imperative to successfully manage resources such as the fish populations in Georges Bank. Ecologists need to understand not only how fish populations have changed over time, including the disastrous drops that have occurred, but also where the organisms live and how they react with other organisms—their prey, predators, and competitors. These assemblies of different species in a particular area at a particular time are called *communities*.

habitat
the physical place within a species' range where an organism actually lives

niche
the role, function, or position of an organism in a biological community

Habitat and Niche

To understand the intricate web of relationships among organisms in a community, we must discover where the organisms live and how they make a living. The physical place in the environment where an organism resides is its **habitat**. A habitat is like an organism's address or home.

Whereas a species' physical home is its habitat, its functional role in the community is its **niche**. The niche is analogous to the organism's job—how it gets its supply of energy and materials. A niche is what an organism does in and for a biological community.

By 2050, Sweden's population will likely be top-heavy with older people because Swedes are having smaller families.

Figure 14.10

Age Structure Diagrams Reveal Potential for Population Growth

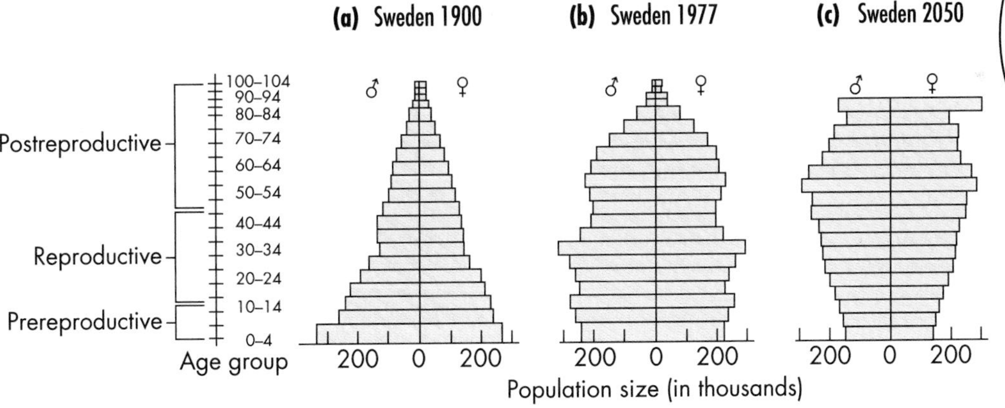

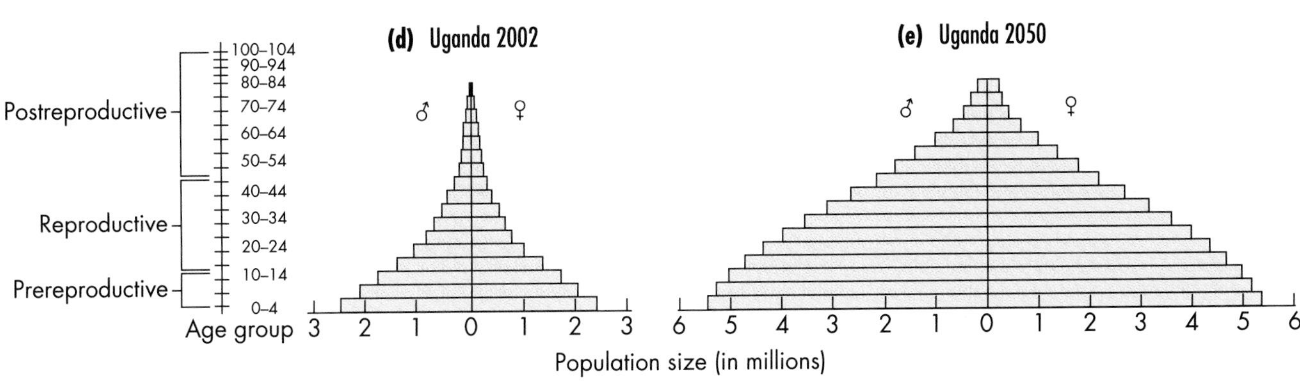

fundamental niche
the potential range of all environmental conditions under which an organism can thrive

realized niche
the part of the fundamental niche that a species actually occupies in nature

interspecific competition
competition for resources—e.g., food or space—between individuals of different species

Limits to Niches

The niche an organism actually occupies is often restricted by other organisms. For example, consider a warbler's niche in the forests of Cape Cod (Fig. 14.11a). The bird has the potential to eat insects wherever they occur in trees, at any height and at any distance from the trunk. The bird might also nest any time in June or July. The potential range of all biotic and abiotic conditions under which an organism can thrive is called its **fundamental niche**. If a warbler could catch insects any place in the tree, it would be operating in its fundamental niche for prey location.

A warbler in eastern forests cannot obtain insects just anywhere, however, because several species compete for food, and each species performs a slightly different role in the community. The different species obtain insects at different heights in the trees, at different distances from the trunk, and their heavy eating comes at slightly different times during the year, depending on when they nest. The myrtle warbler, for instance, eats insects at the base of trees, the bay-breasted warbler specializes in insects in middle branches, and the Cape May warbler seeks insects at the outer edges of the top branches (Fig. 14.11b). Thus, other community residents may force a warbler species from its broader fundamental niche into its narrower **realized niche**, the part of the fundamental niche that a species actually occupies in nature. This kind of niche restriction is rather common. Species interactions like these can be a major factor in defining a species' distribution and abundance.

Figure 14.11
Niche: An Organism's Role in the Community

(a) A warbler in its fundamental niche might find food at any height in the tree, any distance from the trunk, or nest at any time in June or July

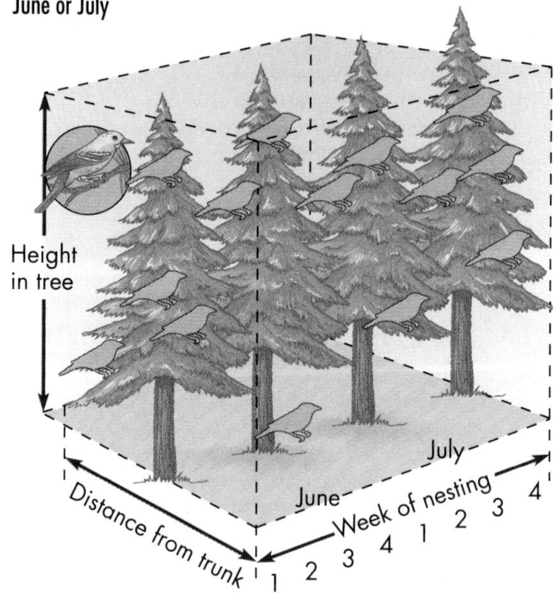

(b) A warbler in its realized niche might be restricted to finding food at certain positions in the tree, and nesting in just a couple of weeks of summer

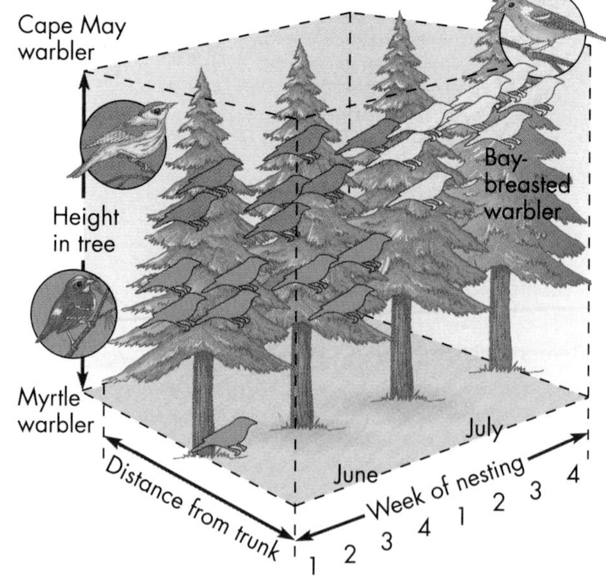

LO[6] How Species Interact

Ecologists have categorized interactions between species into four general types. In (1) competition and (2) predation, one or both of the species suffer. In (3) mutualism and (4) commensalism, neither species is harmed by the interaction. Let's investigate each of these types of species interactions.

Competition between Species

There are never enough good things to go around—sunny spots in which to germinate and put down roots, sheltered places to build nests, or marine worms of a particular size to consume. Because of such restrictions, two different species often compete for the same limited resource, and this interaction restrains the abundance of both species. The key feature of **interspecific competition**, the use of the same resources by two different species, is that one or both competitors have a negative effect on the other's survival or reproduction.

A Model for Species Competition

To see how competition works, imagine populations of yellowtail flounder and witch flounder. If only one species inhabits a region of the Georges Bank, then, in the absence of fishers, that species would increase in numbers until competition between the species' own members limited the population size, as we saw earlier with the Tasmanian sheep. But if both yellowtail and witch flounder inhabit the same area, any individual fish will compete against not only members of its own species, but against individuals from the other species as well. The outcome will depend on the relative strengths of the *inter*specific and *intra*specific competitions.

If a high density of one species (for instance, yellowtail flounder) affects the growth of the other species (witch flounder) more than it affects its own growth, then the first (yellowtail) will eliminate the second (witch flounder). A situation like this, where one species excludes another through competition, is called **competitive exclusion**. In another possible interaction, competition can affect each species' own population growth more than it affects the other species. In this case, the two species could end up coexisting. As a final possibility, each species may slow the growth of the other equally. In this case, the species with the biggest population to start with would take over the community.

Natural Causes of Competitive Exclusion

Laboratory and field experiments with organisms as different as *Paramecium*, beetles, field mice, and aquatic plants typically show that competition does take place and that it can result in either competitive exclusion or in species coexisting.

For example, the Caribbean island of St. Martin has two lizard species that are slightly different in size but eat the same kinds of insects. To find out whether the two species actually compete with each other, researchers fenced off large squares of land that contained individuals of the smaller species, the larger species, or both. They found that where both species coexisted, members of the larger species had less food in their stomachs, grew more slowly, laid fewer eggs, and were forced to perch higher in the bushes than when that species lived alone in an enclosure. Studies like these proved that strong competition does exist in natural populations and that the presence of one species can limit another species to its realized (rather than fundamental) niche.

While competition often limits population size in both interacting species, the second major type of community interaction, predation, is beneficial for one species but harmful to the other.

Predation

Animals that kill and eat other animals are **predators**, their food is **prey**, and the act of procurement and consumption is **predation**. Here we focus on how predation affects the population size of both prey and predator on a short time scale, and how hunter and hunted evolve strategies to outwit each other on a longer evolutionary time scale.

competitive exclusion
a situation in which one species eliminates another through competition

predator
an organism, usually an animal, that obtains its food by eating other living organisms

prey
living organisms that are food for other organisms

predation
the act of procurement and consumption of prey by predators

Do Predator Populations Control Cycles of Prey Populations or Vice Versa?

Many populations of predator and prey rise and fall in cycles like the ripples in a pond. For example, wild populations of the snowshoe hare and lynx periodically rise and fall in phase on about a 10-year cycle. The hare populations occasionally boom when vegetation is abundant and crash when food becomes scarce. When hare populations rise, lynx also increase and devour more of their prey.

In other cases, predators can be the driving force behind the population swings of their prey. To test the effect of predators on populations of pine bark beetles, researchers enclosed some trees with predator-proof cages, while nearby uncaged trees served as controls. They found that populations protected from predation grew better than those exposed to predators. They concluded that predators can indeed help drive the cycles of population growth.

Populations of predator and prey may grow or shrink over the short term, but over the long term, genetic changes can influence the evolutionary balance between hunter and hunted.

The Coevolution of Predator and Prey

Predator-prey interactions lead to a grand coevolutionary race, with predators evolving more efficient ways to catch prey, and prey evolving better ways to escape.

Predators need a way to catch their food, and the two main options are pursuit and ambush. Predators that pursue their prey are selected for speed and often for intelligence as well. Carnivores store information about the prey's escape strategies and make quick choices while in pursuit. In keeping with evolutionary pressures, vertebrate predators generally have larger brains in proportion to their body size than the prey they catch.

camouflage
body shapes, colors, or patterns that enable an organism to blend in with its environment and remain concealed from danger

chemical warfare
a defense strategy of prey species in which these organisms produce distasteful oils or other toxic substances that kill or harm predators

mimicry
the evolution of similar appearance in two or more species, which often gives one or all protection; for example, a nonpoisonous species may evolve protection from predators by its similarity to a poisonous model

parasite
a type of predator that obtains benefits at the expense of another organism, its host; a parasite is usually smaller than its host, lives in close physical association with it, and generally saps its host's strength rather than killing it outright

Some predators ambush their prey. A familiar example is the frog that zaps flying insects by snapping out a sticky tongue. Certain mantis insects ambush their prey by hiding in the open with a camouflaged, plantlike appearance (Fig. 14.12). Those mantises that carry genes for resemblance to the plants they inhabit can be nearly invisible to prey and are thus more effective at ambushing their food and surviving to reproduce than mantises without those genes.

Despite the stealth, athleticism, and cunning of predators, prey species have evolved some remarkably devious tricks—in addition to speed—that help them avoid being eaten. One defense strategy is **camouflage**: the use of shapes, colors, or behaviors that enable organisms to blend in with their backgrounds and decrease the risk of predation. Many insects have evolved shapes that look like twigs, flowers, or leaves, complete with phony leaf veins (Fig. 14.12). Still other insects, and a few amphibians, escape detection by resembling bird excrement on leaves. Behavior plays a role, too; when a flounder nestles to the bottom, it will flap a bit to cover itself with sand. Some flounders can even change their colors to match the color of the bottom on which they rest (Fig. 14.12).

Chemical warfare is another defense strategy. Eucalyptus and creosote bushes, for example, produce distasteful oils or toxic substances that kill or harm herbivores. People sometimes plant oleanders as decorative shrubs because they resist insect pests, but the leaves are so poisonous that chewing a few can kill a child. Animals are not without their own arsenals: Toads, stinkbugs, and bombardier beetles produce highly offensive chemicals that repel attackers. Poisonous prey species often evolve brightly colored patterns, enabling the experienced predator to recognize and avoid them. This is called *warning coloration*. The brilliantly colored but poisonous strawberry frogs of South America have evolved this strategy.

Mimicry

Many nonpoisonous prey species masquerade as poisonous species; this is one form of the process called **mimicry**, wherein one species resembles another. Mimicry could arise in the following way. Individuals of a nonpoisonous butterfly species that by chance contain alleles causing them to resemble the poisonous species even slightly may occasionally escape predation if a hungry animal mistakes them for the poisonous species. A selective pressure like this could, over time, allow the nonpoisonous species to accumulate more and more alleles for resemblance to the poisonous neighbor.

Parasites: The Intimate Predators

Parasites are insidious kinds of predators; they are usually smaller than their hosts, often live in close physical association with individual victims, and generally just sap their strength rather than kill them

Figure 14.12
A Natural Arms Race: The Coevolution of Predator and Prey

outright. Ectoparasites, like fleas, ticks, and leeches, live on the host's exterior, while endoparasites, like tapeworms, liver flukes, and some protozoa, inhabit internal organs or the bloodstream. In a special type of parasitic interaction, certain insects develop inside the body of another insect and inevitably kill it.

Commensalism and Mutualism

The community relationships we have discussed so far—competition and predation—have involved harm to at least one of the species. Sometimes, however, neither species is harmed by their interactions. In commensalism, one species benefits from the alliance, while the other is neither harmed nor helped, whereas in mutualism, both species are helped.

Commensalism

Commensalism is common in tropical rain forests, and the most easily observed examples are the epiphytes, or air plants, that grow on the surfaces of other plants. Using the tree merely as a base of attachment, epiphytes take no nourishment from the host and usually do no harm. Other commensal relationships include birds that nest in trees, algae that grow harmlessly on a turtle's shell, and the small fish that live among the stinging tentacles of sea anemones—unharmed and safe from predators.

Mutualism

In a mutualistic interaction, both species benefit. An example involves the yucca plant that grows in hot, dry regions of the western United States. The yucca plant has a mutualistic relationship with the yucca moth. The moth lays its eggs in the ovary at the base of a yucca flower and then pollinates the flower. Moths depend on the plant for food and reproduction; the plant depends on the moth for pollination. In this relationship, both species benefit: the yucca moth gets a high-energy nectar reward and the plant becomes pollinated. If one species becomes extinct, likely so will the other.

LO7 Organization of Communities

The interactions we have considered—competition, predation, commensalism, and mutualism—affect not just pairs of species, but entire communities consisting of tens to hundreds of species. Communities consist of some species that happened to immigrate into the area and can survive under the available physical conditions, and some species that will grow only if other species are also present. It is a goal of ecologists to learn how the addition or subtraction of a species affects the whole community in the short and long term.

Communities Change over Time

Occasionally, a cataclysm will strip an area of its original vegetation, as can happen after a volcanic explosion or fire, or even after a farmer clears a field. Left to nature, however, a regular progression of communities will regrow at the site in a process called succession.

Soon after a region is denuded, a variety of species begin to colonize the bare ground. These species make up a pioneer community, and they modify environmental conditions, such as soil quality, at the site. Conditions produced by the activities of the pioneer community determine which additional species can establish themselves in the area and form a transition community. Changes are rapid at first, as more and more species join the transition community. Eventually, a particular community of plants and animals becomes relatively stable, and changes take place more slowly over time. Such an assemblage is often called a climax community, but most ecologists today recognize that change is constantly occurring even in old communities, driven by such environmental variables as hurricanes, floods, and fires, as well as global climatic changes.

commensalism a relationship between two species in which one species benefits and the other suffers no apparent harm

mutualism a symbiotic relationship between two species in which both species benefit

succession the process through which a regular progression of communities will regrow at a particular site

pioneer community the species that are first to colonize a habitat after a disturbance such as fire, plowing, or logging

transition community a community of organisms that establish themselves at a particular site based upon conditions produced by the activities of the pioneer community

climax community the most stable community in a habitat and one that tends to persist in the absence of a disturbance

species richness the total number of species in a community

Species Diversity in Communities

The total number of species found in a community is its species richness. In most communities, there are few common species but many rare types of organisms. For example, ecologists captured a group of almost 7,000 moths and identified individuals of 197 different species in one local area. One quarter of the moths belonged to a single species and another quarter belonged to just five other species. The remaining half of the moths fell into 191 species—some represented by just one or two individuals.

What factors allow a region to have large numbers of rare species? Both *latitude* (north-south position) and *isolation* (peninsulas, island chains, or other out-of-the-way locales) influence species richness. Some communities in tropical latitudes, for example, have about 600 types of land birds, while an area of similar size in the arctic tundra may have only 20 to 30 species of land birds (Fig. 14.13).

A single hectare of mainland tropical forest can have 300 different species of trees and tens of thousands of insect species; on a peninsula, however the species richness is diminished. This is because in an isolated area, the many rare species can easily become extinct and their replacement from the mainland is unlikely due to isolation. Species richness on island chains is limited in such specific ways that ecologists are now applying the principles of island ecology to the design of nature preserves, which are, in fact, islands within a sea of human development.

Figure 14.13

Latitude and Isolation Affect Species Richness

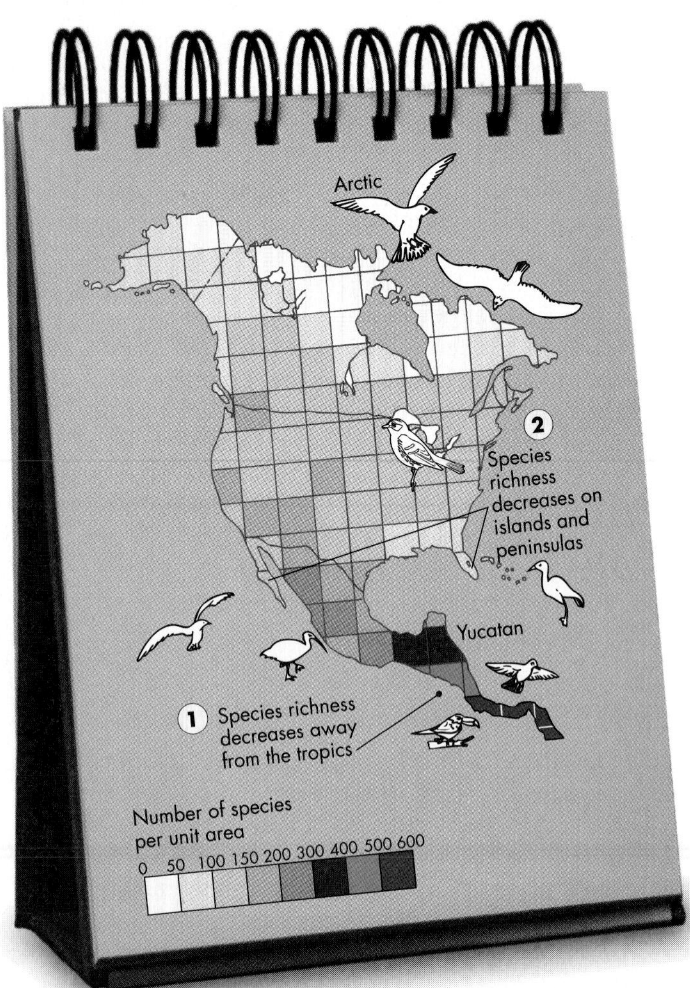

The more resources available in an area—water and solar energy, for instance—the greater the species richness the area can support. Competition and predation also influence species richness. High competition forces smaller niches, and a community can accommodate more species.

Predation can also increase species richness. For example, a sea star preys on barnacles, snails, clams, and mussels. When experimenters removed the sea star, diversity dropped because the mussel population increased and crowded out the other invertebrates. By eating young mussels, the sea star reduces competition for space and so preserves higher species richness. Species like this predatory sea star, whose activities determine community structure, are called *keystone species*.

Species Diversity and Community Stability

Species richness is dwindling all over the globe. The loss of tropical forest ecosystems is one familiar example, and the cutting of old-growth (virgin) forests in the Pacific Northwest is another. Until recently there was little actual data on how species richness affects the health of an ecosystem. The diversity-stability hypothesis suggests that because the species in a community encompass many different traits, a diverse community is more likely to contain at least some species that can survive environmental disturbances such as drought, fire, hailstorms, or overfishing.

In one of the first tests of this hypothesis, researchers measured the amount of living plant material in plots of grassland. These plots contained differing numbers of species, from undisturbed native prairies with more than 20 species per plot to abandoned farmers' fields with less than 10 species per plot. The results showed that the species-rich plots were four times more productive in drought years, and they recovered more quickly than species-poor plots. These results were predicted by the diversity-stability hypothesis; they suggest that a community's ability to resist adverse environmental conditions increases with species richness. The long-term stability of communities may thus depend on the biodiversity of their many interrelating species. These results for grasslands help underscore the urgency of pleas to conserve that biodiversity.

The Future of Species Richness

Perhaps the greatest challenge in the world today is the immense increase in human population and its disastrous disturbance of Earth's biological communities. The collapse of the fishery off the coast of Cape

Cod is an example. But in the species-rich communities of tropical latitudes, human population pressure is threatening the destruction of entire ecosystems. Tropical peoples have traditionally cleared land for agriculture by burning the forest, planting crops for a few years, and then moving on when the soil becomes depleted of nutrients. With enough time, the process of ecological succession can repair these small wounds. Now, however, the J-shaped curve of the human population results in vast areas of the rain forest going up in smoke to provide farmland for crops and pastureland on which to graze cattle for export to developed nations. These fires are so huge and so common that they are clearly visible on satellite photos from space. Many ecologists fear that the plant and animal communities of the tropics may not be able to recover from such extreme disturbances and that early in the 21st century, no tropical rain forests will remain intact. Still worse, many fear that this disruption may cause a distressingly large percentage of all living species to become extinct in our lifetimes.

Who knew?

The yellowtail's habitat is *where* it lives (fingerlings in open water, and adults on the sandy bottoms of the Georges Bank); its niche is *how* it lives (as a bottom predator, eating marine invertebrates).

Ecologists consider it an urgent research priority to learn what makes communities resilient and how they may (or may not) be able to persist in the face of human encroachment. We must solve this problem if our most diverse and interesting communities are to survive through the 21st century.

Nearly two-thirds of deforestation in the Amazon rain forest results from cattle ranches and soybean cultivation; a small proportion results from small-scale agriculture.

Satellite image of Borneo shows smoke from fires started by slash-and-burn agriculture.

Venezuelan children watch as a portion of the tropical rain forest is burned to clear land for cattle grazing.

15

Learning Outcomes

LO[1] Define an ecosystem through its energy flow and material cycling

LO[2] Explain how energy flows through ecosystems

LO[3] Explain how materials cycle through ecosystems

LO[4] Identify global climate and life zones

LO[5] Identify biomes by characteristic plant types

LO[6] Review how the properties of water influence aquatic ecosystems

LO[7] Recognize the impact of human activities on the biosphere

Ecosystems *and the* Biosphere

> "In contrast to the one-way flow of energy, materials cycle through ecosystems."

A Checkered Future

Most meteorologists and climatologists agree that the average global temperature is 0.5°C (or about 1°F) warmer now than in 1900. During that century or so, levels of carbon dioxide (CO_2) and methane (CH_4) in the atmosphere have shot up by 25 percent, along with smaller increases in ozone (O_3) and nitrous oxide (NO). Many scientists are convinced that human activities are causing most of the build-up in atmospheric gases: forest clearing and burning, cattle grazing, operating combustion engines, and releasing industrial pollutants. They also think that those gases, in turn, produce a greenhouse effect: like the glass in a greenhouse, they allow light energy from the sun to pass through toward Earth but trap infrared heat energy that bounces back from Earth's surface toward space. Many scientists predict that by 2100, this greenhouse effect will have raised global temperatures by 2°C (3.5°F) or more and that the consequences will be dire.

What do you know?

How have humans altered the biome you live in?

Camille Parmesan, a trained ecologist and conservation biologist at the University of Texas, Austin, acknowledges the validity of many such predictions and can give a succinct overview of them:

- "The Midwest, our cornbelt, is expected to dry as it gets warmer," she says, and farmers may have to shift away from America's biggest crop, corn, to some other staple that tolerates heat better—or else begin history's biggest irrigation project.
- Warm-adapted tropical diseases like cholera and malaria could spread northward.
- Land set aside for nature reserves could become unsuitable for its inhabitants. As Earth heats and temperature zones shift northward, these small "ecological islands" surrounded by human development could become too hot or dry to support the animals and plants they're supposed to protect.
- As the polar ice caps melt and vast regions of permafrost thaw, newly decomposing organic matter could release huge additional quantities of CO_2 and further accelerate global climate change.
- With that melting, sea level could rise by as much as 1 meter, inundating ports, industrial zones, and coastal housing.

In this chapter, you'll learn how in ecosystems, communities of organisms interact with their immediate physical environments. You'll explore how the water, soil, and air near the planet's surface (the biosphere) support life. And you'll see how human activities (like generating carbon dioxide and methane) threaten to destabilize the delicate balances and interdependencies that have evolved over millions of years. You will also learn about potential solutions to global climate change, to the ozone hole, to pollution, to habitat loss, and to other major environmental issues so that you can act as an informed voter and consumer.

greenhouse effect the result of a buildup of carbon dioxide in the atmosphere (e.g., through the burning of fossil fuels) in which carbon dioxide traps solar heat beneath the atmospheric layers, leading to increased global temperatures and changes in climatic patterns

trophic level
a particular feeding level in a community or ecosystem

primary producer
an organism that produces all the biological molecules required for its growth from nonliving substances taken directly from the environment; autotrophic organisms

primary consumer
in an ecosystem, an organism that eats producers; herbivores (plant eaters like cows and caterpillars) are primary consumers

As you explore this subject, you'll discover answers to these questions:

- Through which pathways does energy flow in ecosystems?
- How do carbon, nitrogen, water, and other materials cycle from the physical environment into organisms, and back again?
- How do global climates affect Earth's life zones?
- What are biomes, Earth's major communities of life?
- What life zones occur in watery environments?
- How are humans changing the biosphere?

LO1 Pathways of Energy and Materials

As you learned in Chapter 1, an *ecosystem* includes the community of organisms in an environment as well as the air, sunlight, water, soils, and other nonliving physical factors that surround them. Ecologists focus on two major interchanges between the living community of organisms and its nonliving environment: the flow of energy, and the cycling of materials.

Energy Flow and Material Cycling

Energy flows through ecosystems in a one-way path: it enters living things from the physical world, passes from one organism to another, and eventually escapes back to the physical environment (yellow arrow, Fig. 15.1).

In contrast to the one-way flow of energy, materials *cycle* through ecosystems. For example, carbon and oxygen atoms move from the nonliving environment into organisms and then back again to the nonliving environment (blue arrow, Fig. 15.1).

Feeding Levels

When organisms take energy directly from sunlight, water, soil, or other aspects of the nonliving environment, it is called *autotrophy* (*auto* = self *trophy* = nourishment). When organisms take in energy first captured by some other organism, the process is *heterotrophy* (*hetero* = other *trophy* = nourishment). Both strategies define the way a species obtains nutrients and, in turn, its place within the community.

Figure 15.1
Pathways of Energy and Materials in Ecosystems

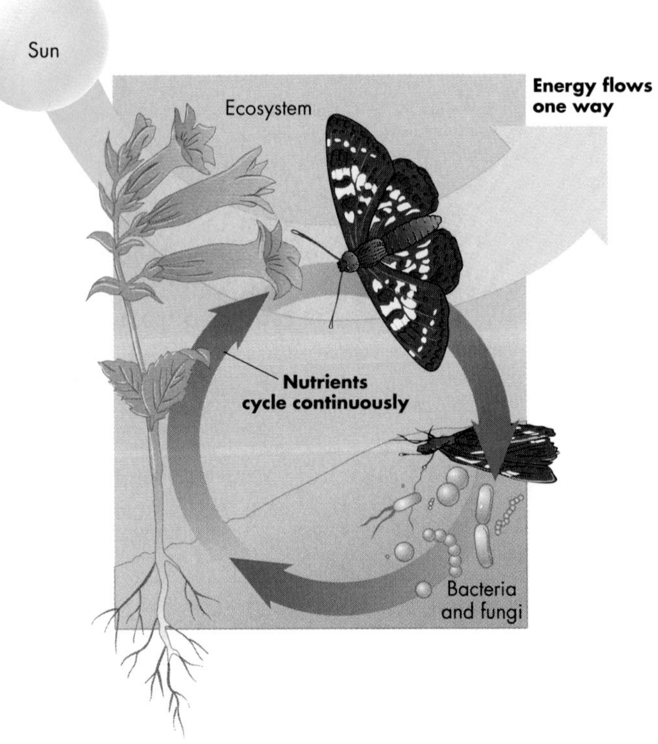

An organism's position in a web of feeding interactions boils down to this: Who eats whom? Ecologists assign every organism in a community to a trophic level, or feeding level, depending on whether it is a producer or a consumer (Fig. 15.2).

Producers

At the lowest trophic levels lie the primary producers—the bacteria and plants that support all other organisms directly or indirectly. In most ecosystems on land, green plants are the producers. By collecting solar energy and carbon dioxide, primary producers build energy-rich biological molecules (review Fig. 3.10). Producers also absorb nitrogen, phosphorus, sulfur, and other necessary atoms and fix them into biological molecules. Ecologists say that producers provide both the energy-fixation base and the nutrient-concentration base for the entire ecosystem.

In most ecosystems, plants are the major producers. In the deep seas, however, as well as deep in the Earth, or otherwise inhospitable environments, prokaryotes are usually the primary producers (review Chapter 11).

Primary Consumers

At the second trophic level are the primary consumers, the organisms that eat the producers. Herbivores (plant eaters) efficiently digest plant matter for energy and serve as ecological links between producers and other levels.

Secondary Consumers

At the next highest level are **secondary consumers**, carnivores (meat eaters) that consume the herbivores.

Tertiary Consumers

In the next trophic level are the **tertiary consumers**, carnivores that eat other carnivores. Finally, a few ecosystems have one more trophic level containing carnivores that eat tertiary consumers, such as a cougar eating a weasel.

Detritivores and Decomposers

A special class of consumers, the **detritivores**, obtains energy and materials from **detritus**, organic wastes and dead organisms that accumulate from higher levels. Earthworms, dung beetles, and carrion feeders are specialized detritus consumers. **Decomposers** break organic molecules into inorganic subunits, which can in turn be recycled by primary producers. Bacteria, fungi, and slime molds are all decomposers critical to nutrient cycling.

The simplified diagram in Figure 15.2 suggests that each trophic level leads directly to the next in a simple chain, and indeed, **food chains** do exist in nature, with groups of organisms involved in linear transfers of energy from producer to primary, secondary, and tertiary consumers. More commonly, however, feeding relationships resemble not chains, but complex interwoven webs.

Feeding Patterns in Nature

Organisms usually consume more than one other species. Some animals feed at several trophic levels. As an omnivore ("all eater"), you yourself might eat vegetables (primary producers), ice cream (from cows, which are primary consumers), tuna fish (secondary consumers), and mushrooms (decomposers). Ecologists call complicated interconnected feeding relationships **food webs** (Fig. 15.3).

Whether an organism is a producer or a consumer, it needs energy for movement,

secondary consumer
in an ecosystem, an organism that consumes herbivores; carnivores (meat eaters) are secondary consumers

tertiary consumer
in an ecosystem, a carnivore that eats other carnivores

detritivore
a consumer organism that obtains energy from dead organisms and/or organic waste matter

detritus
a collective term for dead organisms and organic waste matter

decomposer
a type of consumer, also called a *detritivore,* that obtains energy and materials from organic wastes and dead organisms that accumulate from all trophic levels

food chains
the levels of feeding relationships among organisms in a community

food webs
complex, interconnected feeding relationships between all the species in a community or ecosystem

Figure 15.2
Feeding Levels and Food Chains

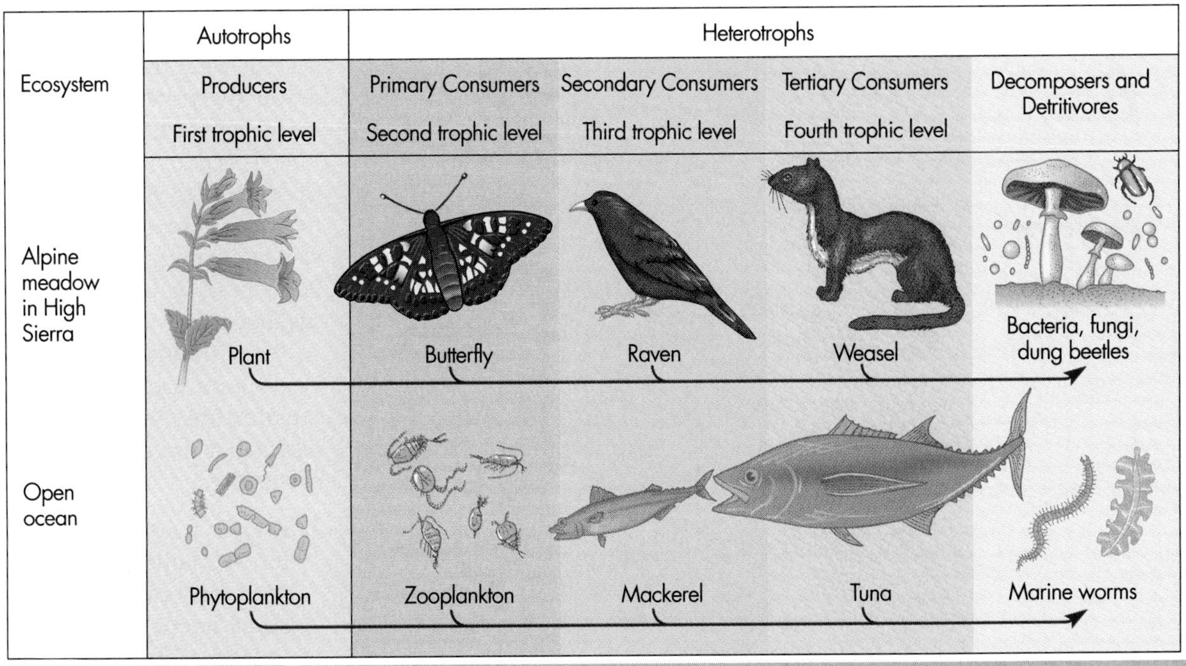

	Autotrophs	Heterotrophs			
Ecosystem	Producers First trophic level	Primary Consumers Second trophic level	Secondary Consumers Third trophic level	Tertiary Consumers Fourth trophic level	Decomposers and Detritivores
Alpine meadow in High Sierra	Plant	Butterfly	Raven	Weasel	Bacteria, fungi, dung beetles
Open ocean	Phytoplankton	Zooplankton	Mackerel	Tuna	Marine worms

gross primary productivity
the percent of sunlight that producers convert to chemical energy in the form of organic compounds

net primary productivity
the amount of chemical energy producers actually store as organic molecules in the form of new leaves, roots, stems, flowers, fruits, and other structures and compounds

active transport of nutrients and ions, and synthesis of proteins and nucleic acids, and it needs other large molecules for growth and repair. Keep in mind that regardless of the trophic relationship, a key factor is the flow of energy from one organism to the next.

LO2 Energy Flow through Ecosystems

Because producers obtain their energy directly from the environment, their activities set a limit for the amount of energy that can be captured and channeled throughout the entire ecosystem.

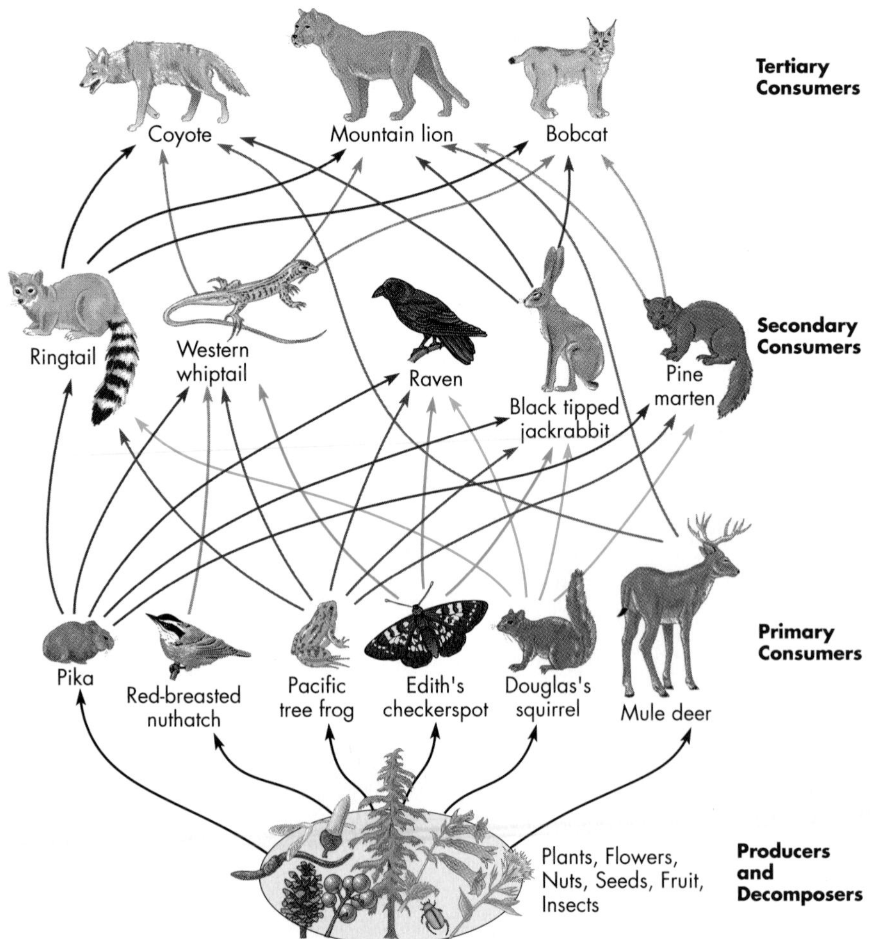

Figure 15.3
A Food Web: Interdependencies in the Living World

Energy Budget for an Ecosystem

By studying the input of energy, water, mineral nutrients, and organic matter into zones within a forest and tracing their incorporation into both living organisms and the physical environment, researchers have learned in detail how energy flows through an entire ecosystem.

The summer sun brings a huge amount of solar energy into the forest. However, most of this energy reflects back into the atmosphere as light or heat, or causes water to evaporate from the soil and plants. In total, only about 2 percent of the sunlight is converted by producers to chemical energy in the form of organic compounds (via photosynthesis; Fig. 15.4). Ecologists call this small fraction an ecosystem's **gross primary productivity**. The gross primary productivity limits an ecosystem's structure, including how fast birch trees will grow, for example, and how many butterflies will thrive there.

Plants use about half of the gross primary productivity to fuel their own cellular respiration, eventually losing most as heat. The amount of energy remaining after respiration is called the **net primary productivity**, the amount of chemical energy that is actually stored in new leaves, roots, stems, flowers, and fruits. Of all the energy impinging on the ecosystem, only the net primary productivity—about 1 percent of the light energy striking the forest—is potentially available to consumers. But most of the new leaves and twigs end up as dead material on the forest floor, and this fuels the detritus food web.

The information in Figure 15.4 allows ecologists to formulate three general principles about the energy budget of the forest:

1. Even in a lush, leafy green forest, plants and other producers convert only a small fraction (2 percent or less) of the solar energy

that enters the ecosystem into stored chemical energy.

2. Animals ingest an even smaller amount (in this case, 0.01 percent) of the energy in the grazing food web.

3. As energy flows through the trophic levels of the ecosystem, metabolic activities (mostly aerobic respiration) release it back into the air, where it ultimately returns to space as heat, a form of energy that does little or no work.

Pyramids of Energy and Biomass

Ecologists have discovered another important fact in their experiments on energy flow in ecosystems: food chains on land rarely have more than four or five links (excluding parasites, decomposers, and detritivores), and ocean food chains based on plankton are generally limited to about seven links. When there are hundreds or even thousands of species in an alpine meadow, a patch of forest, or a coral reef, why do food chains tend to be so short? Consider the California ground squirrel. This rodent eats seeds, leaves, or berries, assimilates some of the energy, then excretes some in liquid and solid wastes. The ground squirrel uses most of the assimilated energy in aerobic respiration and to maintain and repair its body, storing less than 2 percent of ingested food energy in new tissues or offspring. The consequences of a huge loss through respiration and a small net increase in growth is that ground squirrels store very little energy in a form that can be used at the next trophic level, say, by a coyote. That small amount of stored energy is what keeps food chains short.

energy pyramids
the energy relationships between different trophic levels in an ecosystem

Energy Pyramids

Ecologists draw **energy pyramids** with building blocks proportional in size to the amount of energy available at different trophic levels in an ecosystem.

Figure 15.4
Energy Budget of a Hardwood Forest

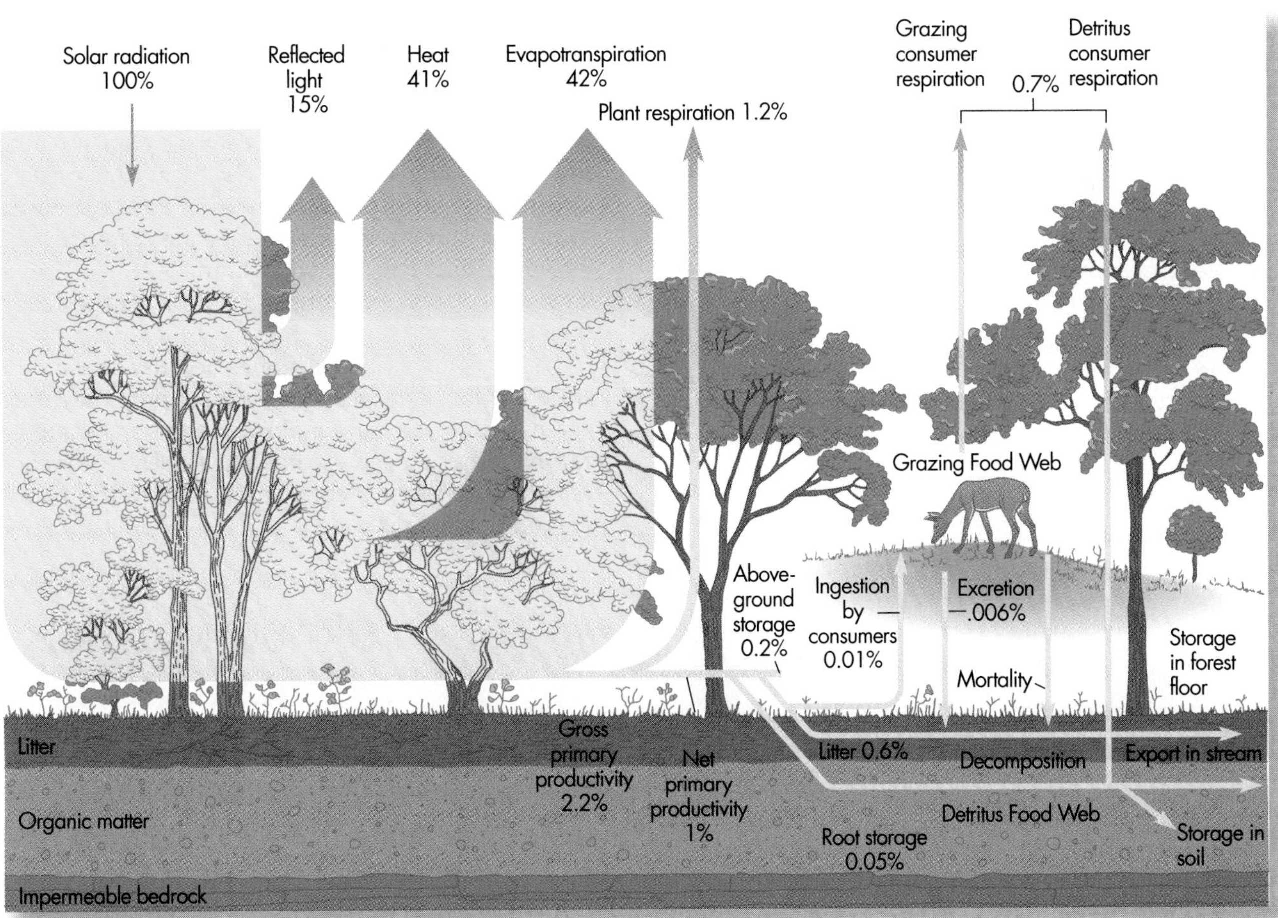

biomass
the total dry weight of organic matter present at a particular trophic level in a biological community

pyramid of biomass
the relationship between the total masses of various groups of organisms in a food chain, in which there is usually less mass, hence less stored energy, at each successive trophic level

biological magnification
the tendency for toxic substances to increase in concentration as they move up the food chain

Figure 15.5
Energy Pyramid in a Florida River Ecosystem

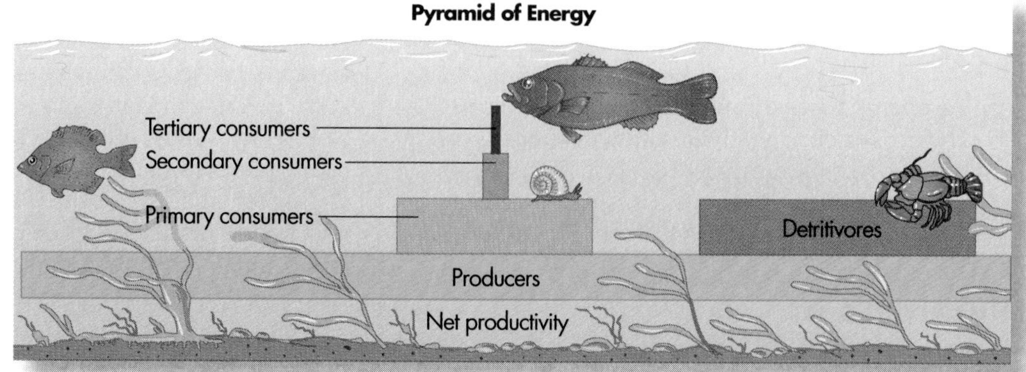

At each trophic level, the energy stored by the organisms is substantially less than that of the level below it (Fig. 15.5).

Biomass Pyramids

Stepwise energy decline explains why food chains usually have only four links: The small amount of energy available at the top is too difficult to collect. Ecologists have invented a handy way to measure the diminishing returns through biomass, the dry weight of organic matter at a particular level. Biologists collected organisms in various trophic levels in a river ecosystem, and then dried them to remove their water content and weighed the remaining material. They then displayed the data as a pyramid of biomass, which showed that each trophic level contained, as a rough approximation, only about 10 percent of the biomass in the level just below it (Fig 15.6).

Ever hear the phrase, "Eat low on the food chain"? This is based on energy and biomass pyramids, and refers to the fact that it takes 10 kg of grain to build 1kg of human tissue if the person eats the grain directly, but it takes 100kg of grain to build 1 kg of human tissue if a cow eats the grain first, and the person eats the beef. Eating lower on the food chain—eating producers, not consumers—saves precious resources on a small planet.

Biological Magnification

Energy pyramids in ecosystems have an important ramification: **biological magnification**, the tendency for toxic substances to build up in progressively higher levels of a food chain. Many chemical insecticides, such as DDT, resist breakdown in the environment and, when eaten, tend to be stored in body fats. If farmers spray DDT on their cabbage plants to control caterpillars, some of the chemical runs off into streams and lakes. There, instead of breaking down, some of it may enter water plants later eaten by herbivorous fish. Fish and other animals cannot break down or excrete the toxin, and it is instead stored in their body fats.

The magnification continues because consumers store all the DDT present in all the prey they have eaten. The concentration of DDT in the body of a top carnivore, say an osprey that eats the fish, can reach levels 10,000 times greater than in the water plants that originally took it up.

Figure 15.6
Pyramid of Biomass

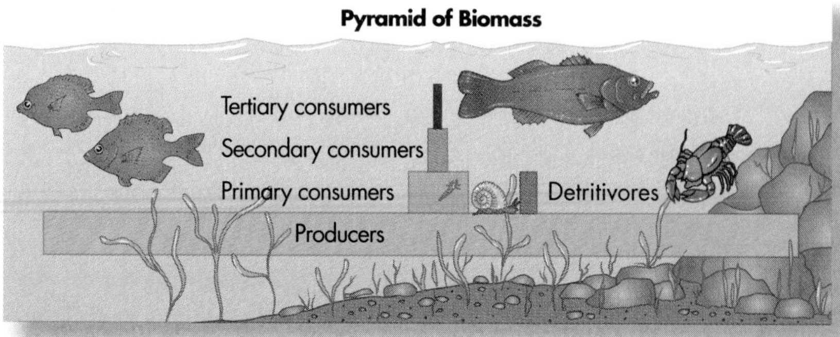

LO3 How Materials Cycle through Global Ecosystems

At the start of this chapter, we saw how human disruptions of the carbon cycle are leading to increasing levels of CO_2 and

Figure 15.7
The Water Cycle

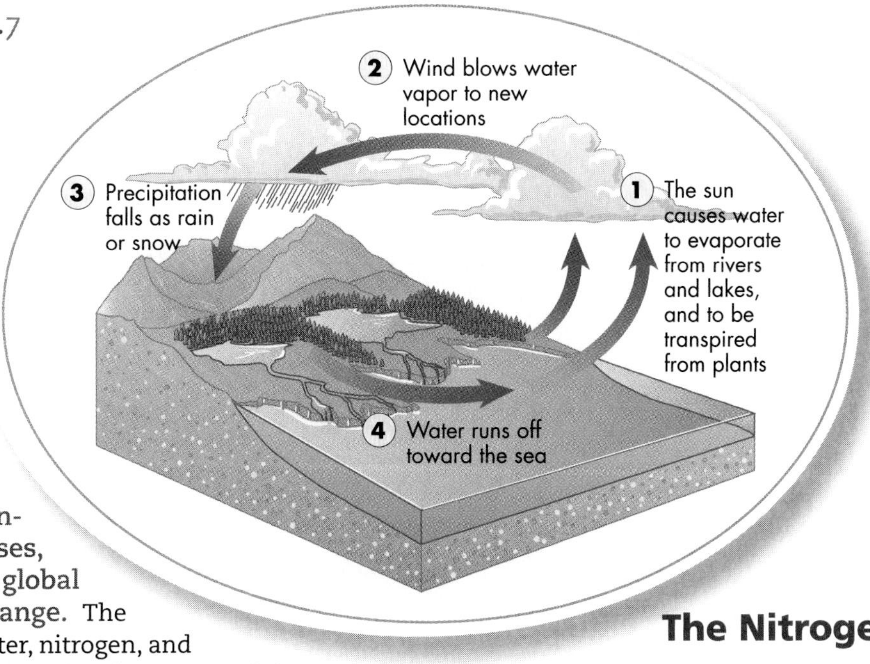

other greenhouse gases, and to global climate change. The cycles of water, nitrogen, and phosphorus are also important globally. To understand how potential changes in these cycles might affect the functioning of ecosystems, let's see how these materials cycle through the living and nonliving world.

Biologists consider materials stored temporarily in the atmosphere, in soil, or in living organisms to be in *pools,* or reservoirs. In general, the pool in organisms is much smaller than the pool in the physical environment. Pools remain constant in size as long as a substance's rate of entry equals its rate of departure. Let's explore the cycling among the major "pools" of water, nitrogen, phosphorus, and carbon.

The Water Cycle

In the global water cycle, water moves from the atmosphere to Earth's surface as rain or snow, and back again to the atmosphere as the sun causes water to evaporate from puddles, ponds, rivers, oceans (Fig. 15.7) and from the leaves of plants in a process called transpiration. In some ecosystems on land, more than 90 percent of the moisture passes into plants and evaporates from their leaves, and only 10 percent evaporates directly from surfaces in the environment. In ecosystems like these, which include large tracts of tropical forests, the plants literally create their own rain: moisture moves from plants to air to clouds and back to Earth in the form of rain wherever the clouds blow. When people cut down a forest's trees for timber or agriculture, water runs off to the sea rather than evaporating; as a consequence, clouds fail to form downwind, rainfall decreases, and climate patterns change. The global climate change we have been experiencing will have a strong effect on the water cycle.

water cycle
a sun-driven global exchange involving evaporation, precipitation, runoff, and transpiration that cycles water from the atmosphere to Earth's surface, through organisms, and back again

transpiration
the loss of water from plants by evaporation, mainly through the stomata on stems and leaves

nitrogen cycle
the cycling of nitrogen from the atmosphere to Earth's surface, through organisms, and back again

The Nitrogen Cycle

Nitrogen gas makes up about 79 percent of our atmosphere, but ironically, most organisms can't use the gaseous form of nitrogen. Instead, they depend on a few species of nitrogen-fixing bacteria to trap nitrogen in biologically useful forms such as ammonium (NH_4^+) and nitrates (NO_3^-). Other bacterial species return nitrogen to the atmosphere as nitrogen gas and complete the nitrogen cycle (Fig. 15.8).

Because available nitrogen often determines how well crop plants will grow, farmers have long fertilized their fields to increase the amount of ammonia and nitrate in the soil. Many farmers add useful nitrogen to their soils by practicing *crop rotation:* they plant *legumes*—crops such as beans, clover, or alfalfa—one year and corn, wheat, or sugar beets the next. This takes advantage of the nitrogen-fixing bacteria in the legumes' root nodules and the nitrogen released naturally into the soil.

Today, most large agribusiness farms around the world depend on nitrogen fertilizers produced through an industrial process. In fact, nitrogen fixed in chemical factories may now represent 30 percent of the input to the global nitrogen cycle—a truly colossal shift in the natural cycle. Industrial nitrogen fixation requires tremendous heat and pressure, which is usually produced by burning huge quantities of fossil fuels. Since reserves of fossil fuels are limited, this enormous reliance on industrially fixed nitrogen fertilizers represents an energy drain that can't go on indefinitely.

The Phosphorus Cycle

Because nitrogen gas, water vapor, and carbon dioxide are airborne, they are blown anywhere by

phosphorous cycle the cycling of phosphorus between organisms and soil, rocks, and water

eutrophic an aquatic environment with high phosphorous and other nutrient levels, characterized by dense blooms of algae and other aquatic plants and a decrease in dissolved oxygen

eutrophication in an aquatic environment, an oversupply of nutrients that support primary production

carbon cycle the global flow of carbon atoms from plants through animals to the atmosphere, water, soil, and back to plants

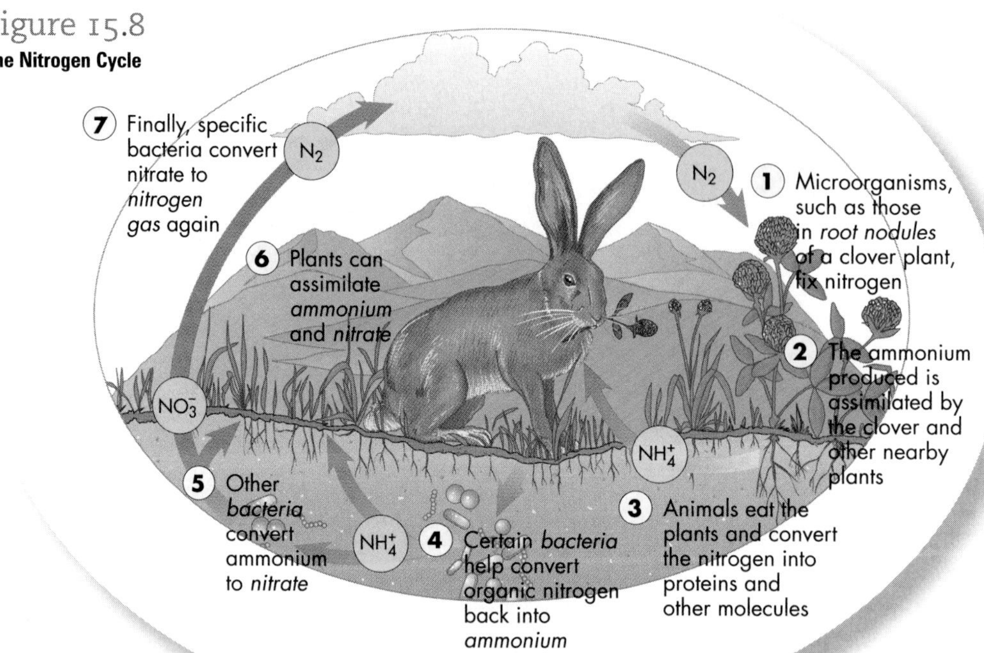

Figure 15.8
The Nitrogen Cycle

the wind. This mobility makes the nitrogen, water, and carbon cycles truly planetary. In contrast, certain other substances, such as calcium and phosphorus, lack a gaseous phase and thus cycle locally. Nevertheless, local cycles of these elements can cause great fluctuations in the populations of local organisms. Phosphorus is essential for life. It is a component of cell membranes, nucleic acids, and ATP, the energy currency of cells. The **phosphorus cycle** consists of two interlocking circuits, one that acts locally during short stretches of time, and another that operates on a more global scale over vastly longer time periods.

Eutrophication

As with other cycles, our human activities have altered the dynamics of the phosphorus cycle, especially in aquatic ecosystems, where phosphorus is often the factor limiting primary productivity. Phosphates are major ingredients of agricultural fertilizers and until recently were also main components of detergents. For years, phosphorus-rich water from fertilized fields ran directly into lakes. The additional phosphates allowed algae and other aquatic plants to grow faster, and the ecosystem became **eutrophic** (*eu* = true or truly *trophic* = fed), or overly supplied with nutrients that support primary production. This process, termed **eutrophication**, continued as algae and other plants "overgrew" and then died in the lake. Decomposing bacteria then fed on the dead algal cells and used up so much dissolved oxygen that fish suffocated in massive fish kills. People can improve the health of lakes and streams by restricting the runoff of phosphorous-containing substances such as detergents and fertilizers.

The Carbon Cycle

As with water and nitrogen, carbon atoms move globally in a vast **carbon cycle** from the physical environment through organisms and back to the nonliving world. The carbon cycle is closely linked to energy flow, because producers—including the photosynthetic plants of meadows, forests, and oceans—trap not only light energy but also carbon in sugar molecules. The trapped carbon comes from carbon dioxide in the surrounding air or water (Fig. 15.9). As the cycle proceeds, consumers ingest organic carbon compounds synthesized by producers. Then, via respiration, both consumers and producers return carbon to the nonliving environment in the form of carbon dioxide.

Carbon accumulated in wood eventually returns to the atmosphere as a result of fires or through consumption and respiration by decomposers, such as fungi and bacteria, and by detritivores. Organic carbon can leave the cycle for even longer periods of time when sediments bury organic litter, which decomposes only partially and gradually transforms into coal or oil. Carbon also leaves the cycle when the cast-off calcium carbonate shells of marine organisms sink to the ocean floor and become covered with sediments that compress them into limestone. Eventually, however, even these carbon deposits recycle into atmospheric carbon dioxide as the limestone

Figure 15.9
The Carbon Cycle

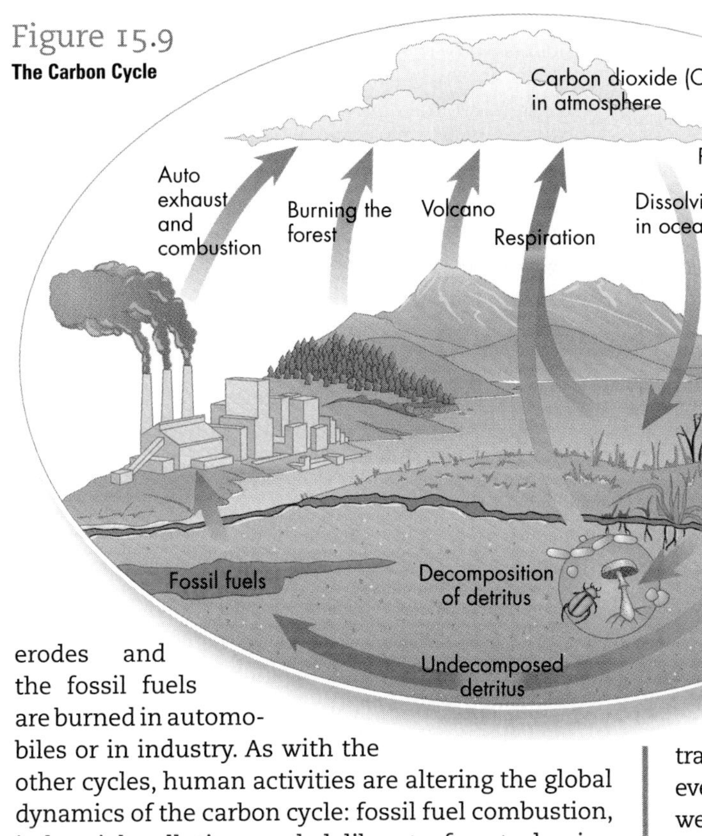

weather
the condition of the atmosphere at any particular place and time, including its temperature, humidity, wind speed, and precipitation

climate
the accumulation of seasonal weather patterns in an area over a long period of time

erodes and the fossil fuels are burned in automobiles or in industry. As with the other cycles, human activities are altering the global dynamics of the carbon cycle: fossil fuel combustion, industrial pollution, and deliberate forest clearing and burning, particularly in the tropics, are releasing more and more carbon dioxide into the atmosphere. Some of this becomes fixed by photosynthesis but some contributes to the greenhouse effect and global climate change.

LO4 Global Climates and Earth's Life Zones

What gives different parts of the world their unique combinations of temperature and rainfall? And how do these physical environments influence the types of organisms that live in Earth's many zones?

This global view of biology involves the *biosphere* which is, you'll recall from Chapter 1, the portion of our planet that supports life. The biosphere includes:

- every body of water;
- the atmosphere to a height of about 10km (6mi);
- Earth's crust to a depth of many meters; and
- all the living things within this collective zone.

This section explores how global physical factors—particularly worldwide currents of air and water—produce regional climates. The following section then discusses how climates determine the abundance and distribution of living organisms.

What Generates Earth's Climatic Regions?

Weather is the condition of the atmosphere at any particular place and time, including its humidity, wind speed, temperature, and precipitation. In contrast, climate is the accumulation of seasonal weather events over a long period of time. Within the biosphere, weather has temporary, local effects, whereas climate is the major physical factor determining the abundance and distribution of living things. What, then, causes climate? The answers involve the uneven way that sunlight heats our planet, the behavior of water and air at different temperatures, and Earth's rotation.

The Earth Heats Unevenly

The tropics are warm and the poles are cold because the sun heats Earth's surface unevenly. Like a flashlight beam shining directly down onto a table from above versus one coming in obliquely from the side, sunlight hitting Earth directly is more intense than sunlight striking at an angle. The precise angle the sun makes at a point on the globe depends on the distance between the point and the equator (the *latitude*) and the season.

Temperatures fluctuate with the seasons because our planet spins on an axis tipped at a constant tilt relative to the sun. Consequently, during the summer the Northern Hemisphere inclines maximally toward the sun and receives more sunlight, while in winter it tips away from the sun and receives less sunlight. The seasons are reversed in the Southern Hemisphere.

Where Does Rain Come From?

Sunlight striking Earth's tilted sphere heats the air. Warm air has different properties than cool air, and these account for the formation of rain and snow.

trade winds predictable surface winds caused by a moving air mass that interacts with the rotating Earth

Cold air weighs more per unit volume than warm air, and so tends to sink through lighter, warmer air.

Dense, cool air holds less moisture than light, warm air. This principle ultimately brings about rain and explains why a region like the tropics is so wet. Powerful sunlight at the equator heats the air, which picks up moisture by evaporation from land surfaces and from plant leaves. The hot, moist air rises, and once aloft, it cools, releasing the water it can no longer hold as rain onto the lush tropical forest.

Currents of Air and Water

Uneven heating of the globe generates circular currents of air and water that create different climates. Hot air ascending near the equator cools, and then travels north and south at high altitude (Fig. 15.10). At about 30° latitude (the approximate latitude of Egypt to the north and Australia to the south), this cold, dry air falls to Earth. The dry, descending air creates the great deserts of Australia, North and South Africa, and North America.

The dry air from the deserts moves toward the equator, replacing the ascending hot air (Fig. 15.10). This moving air mass interacts with the rotating Earth, and causes surface winds. These predictable breezes, the **trade winds**, propelled traders' sailing ships in past centuries. Analogous currents in other circulating coils produce winds from west to east over much of North America.

Figure 15.10

Massive Air Coils Create Earth's Climatic Zones

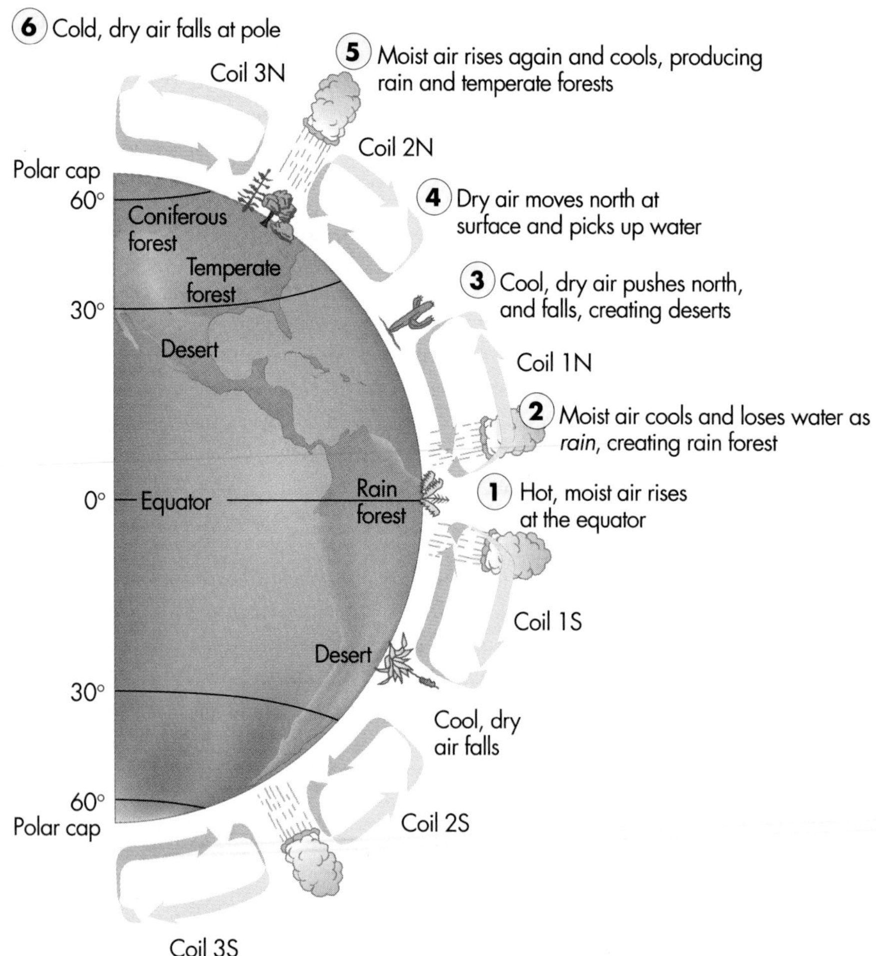

Ocean Currents

The heating and rotation of the Earth not only affect the winds, they also drive ocean currents. Continents redirect the flow of ocean waters in slow, circular patterns in the North and South Atlantic and the North and South Pacific (Fig. 15.11). Ocean currents, like air currents, circulate heat and hence influence climate and the distribution of plants and animals. The Gulf Stream, for example, carries warm water from the tropics up the eastern coast of North America, then across the Atlantic, warming northern Europe. Compare ocean temperatures off the coast of Ireland and southern Alaska, which are about the same latitude (Fig. 15.11).

Short-term irregularities do occur in wind and ocean currents, such as the El Niño event, a temporary reversal of ocean currents in the South Pacific that occurs about every five years or so. This event leads to weather shifts on several continents. Nevertheless, in general, the sun and Earth's tilt and rotation create stable patterns of rainfall and temperatures that determine the general character of Earth's major communities of plants and animals.

LO5 Biomes: Earth's Major Communities of Life

Wherever similar climatic conditions exist, such as in a desert or rainforest, plants have evolved similar adaptations that help them exploit the climate's benefits and minimize its drawbacks. Biomes are large terrestrial geographic regions containing distinctive plant communities (Fig. 15.12). Major plant types characterize biomes because plants best reflect adaptations to rain, temperature, light, and wind. Furthermore, as primary producers, plants influence the consumers and decomposers that coexist in the biome.

Two major climatic features—temperature and moisture—set boundaries within which life will flourish. Tropical rain forests, for example, appear in regions with high amounts of precipitation and high temperatures, while tundra appears in regions with cold and moderately dry climates. The differences in temperature and moisture cause differences in productivity in the different biomes. For example, productivity is high in temperate and tropical rain forests, and low in tundra and desert regions.

In this section, we cover Earth's major biomes roughly in order from equator to poles—tropical rain forests, savannas, deserts, temperate grasslands, chaparral, temperate forests, coniferous forests, and tundra. And though not strictly a biome, we discuss the polar ice caps, as well.

biomes
large terrestrial geographic regions containing distinctive plant communities; major ecological community types, such as a desert, rain forest, or grassland

tropical rain forest
a lush forest biome that occurs near the equator in Central and South America, Africa, and Southeast Asia, where rainfall is 200 to 400 cm (80 to 160 in.) per year and temperatures average about 25 °C (77 °F)

Tropical Rain Forest

Nearly half of all living species reside in the world's warm, wet tropical rain forests occurring near the equator. Heavy rains tend to leach nutrients from tropical soils; thus, the biomass of the living forest itself is the biggest source of nutrients. Rain forests have an upper story, or emergent layer of tall trees that capture direct sunlight; a canopy of shorter forest trees where leaves, flowers, fruits and arboreal animals abound; an understory penetrated only by

Figure 15.11
Ocean Currents Flow in Four Circular Patterns, Redistributing Heat

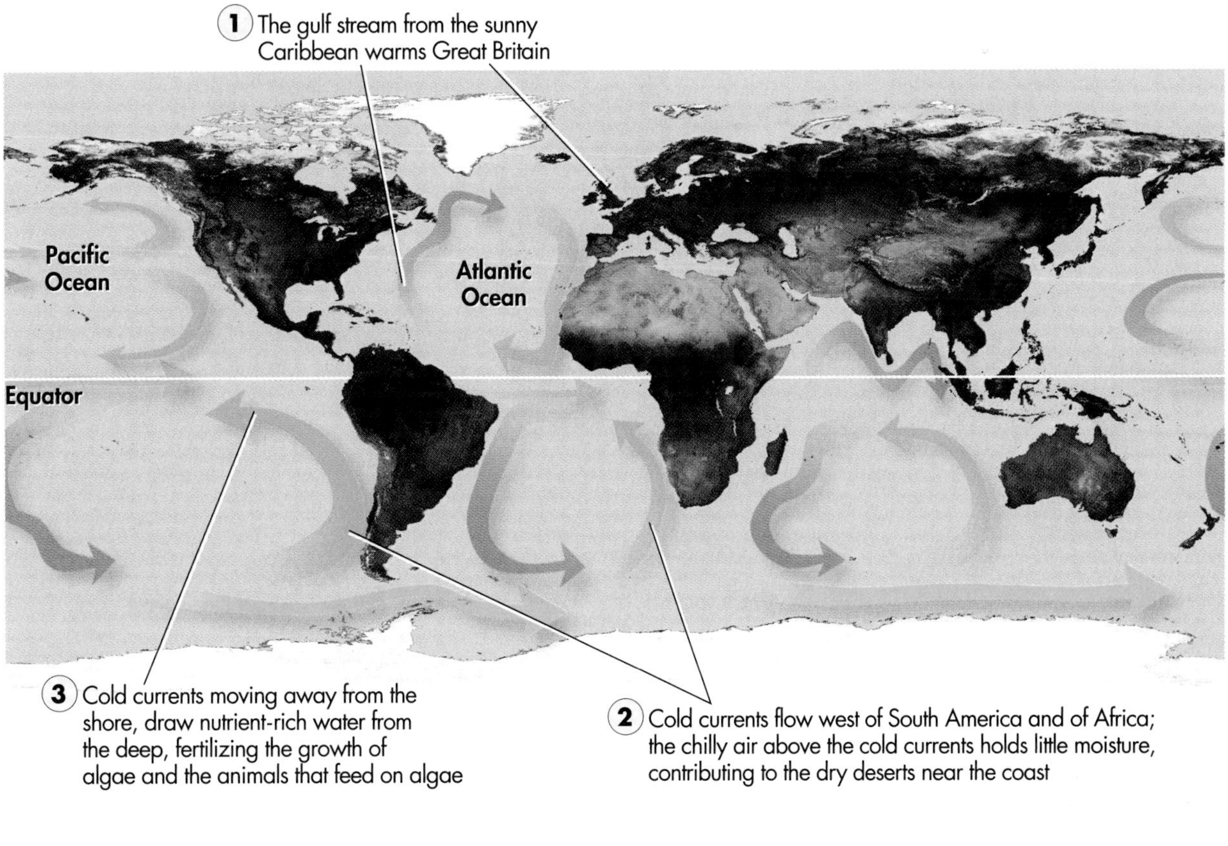

savanna
a tropical grassland biome, containing stunted, widely spaced trees, that is situated between tropical forests and deserts

dim light; and a dark forest floor where plants often have huge, deeply green leaves. Species diversity is not quite as rich in the adjacent biome called *tropical dry forest*, where the organisms have evolved a tolerance for longer dry seasons.

Savanna

Year-long warmth with an extended dry season results in a tropical savanna, or dry, open grasslands with sparse tree

Figure 15.12
Map of the World's Major Biomes

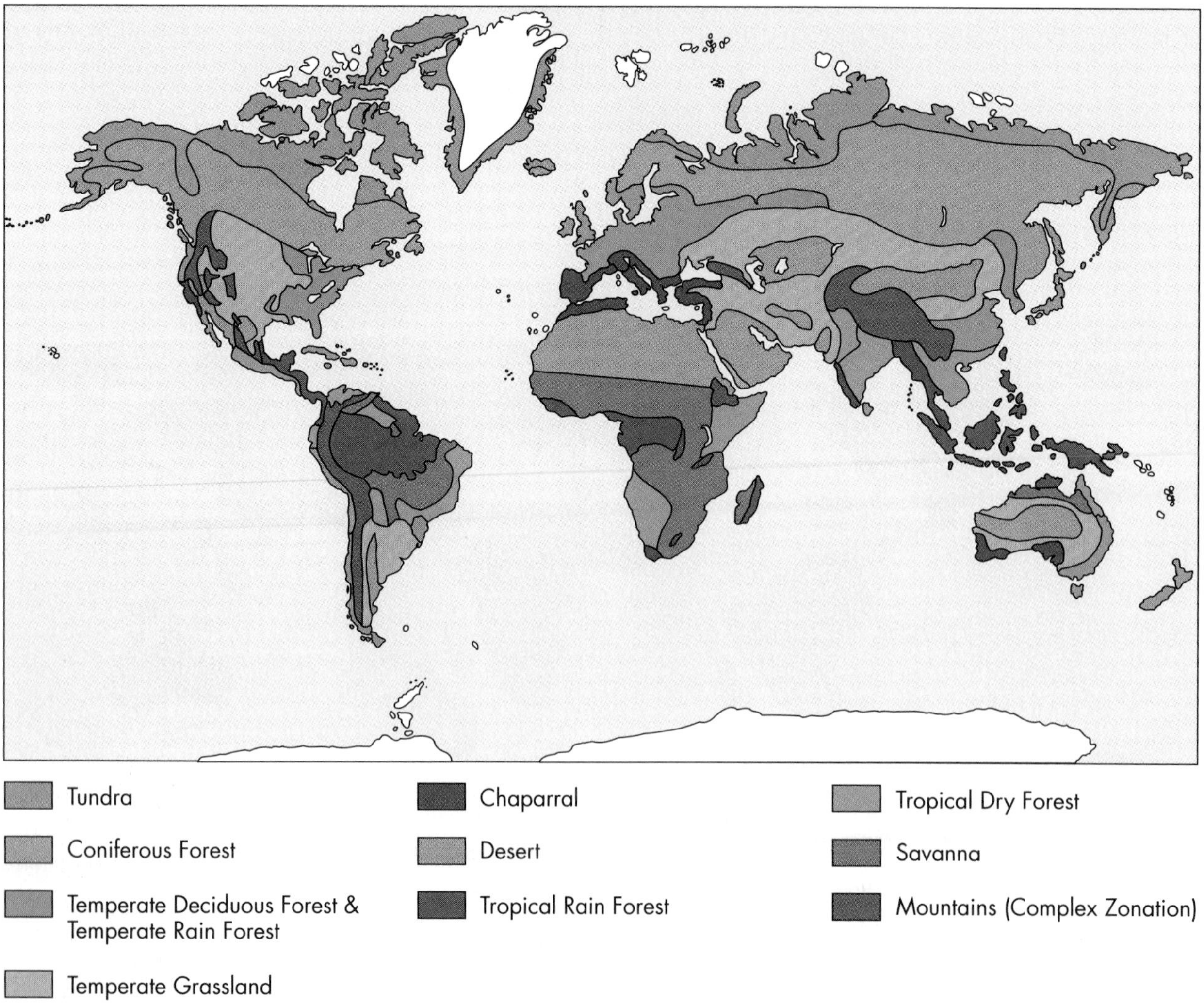

Tundra
Coniferous Forest
Temperate Deciduous Forest & Temperate Rain Forest
Temperate Grassland
Chaparral
Desert
Tropical Rain Forest
Tropical Dry Forest
Savanna
Mountains (Complex Zonation)

coverage. Situated between tropical forests and deserts, savannas contain stunted, widely spaced trees with tall grasses growing in between that support grazing animals. These herbivores, in turn, support numerous carnivores.

Desert

Desert regions receive less than one-tenth of the annual rainfall of tropical rain forests; hence, desert plants are widely spaced and cover less than one-third of the ground surface. Daytime temperatures can soar to 49°C (120°F), while at night, heat tends to radiate and leave a biting chill to the air. Many desert plants have adaptations such as fleshy stems and spines. Desert animals tend to be active only in the evening or early morning hours, avoiding the heat of the day.

desert
a very dry, often barren biome characterized by temperature extremes and by widely spaced plants with thick, waxy leaves and often protective spines

grassland
a treeless temperate region dominated by grass species; known as *prairies* in North America, *pampas* in South America, *steppes* in Asia, and *veldt* in Africa; this biome is wetter than deserts but drier than forests

Temperate Grassland

Bordering many of the world's deserts are grasslands, treeless regions dominated by dozens of grass species. Known as *prairies* in North America, *pampas* in South America, *steppes* in Asia, and *veldt* in Africa, these regions are wetter than deserts, drier than forests, and have seasonal extremes of hot and cold rather than wet and dry. Frequent wildfires in the dry grasses prevent grasslands from turning into forests. Grassland soils are richer in organic matter than the soils of other biomes.

chaparral
a biome that borders deserts and grasslands; characterized by hot, dry summers and cool, wet winters, and low woody shrubs that are often fragrant and have generally thick, waxy, evergreen leaves

temperate forest
a biome that occurs north or south of subtropical latitudes; characterized by generally a mild climate and varied populations of evergreen and deciduous trees

coniferous forest
a biome that occurs across much of Canada, northern Europe, and Asia with vast forests of coniferous trees growing at latitudes with cold, snowy winters and short summers

Chaparral

A biome called the chaparral (shap-uh-RAL; Spanish, "thicket"), or temperate scrublands, borders grasslands and deserts. Chaparral has hot, dry summers and cool, wet winters. Chaparral plants are generally less than 2 m (6.5 ft) tall and have small, leathery, often hairy leaves that stay green all year. Because grasses grow during the wet winters and then die and dry out during the hot summers, this biome, like the grasslands, experiences frequent fires.

© STACEY PUTNAM/ISTOCKPHOTO.COM

Temperate Forest

Temperate forests are dominated by broad-leaved trees and have intermediate amounts of rainfall and fairly moderate temperatures that fluctuate between summer highs and winter lows. The dominant trees drop their leaves and become dormant until spring. The fallen leaves allow for the recycling of nutrients and produce excellent topsoil.

Coniferous Forest

At latitudes with cold, snowy winters and short summers, vast forests of cone-bearing pine, fir, spruce, and hemlock trees grow in the coniferous forest.

© DON WILKIE/ISTOCKPHOTO.COM

© CECELIA HENDERSON/ISTOCKPHOTO.COM

Conifer leaves, which are needle-shaped and have thick, waxy cuticles, combat water loss and are not shed each winter, thus they can begin collecting sunlight as soon as the short growing season begins. Coniferous forests contain the world's greatest lumber reserves. Traditionally, loggers have harvested this timber by *clear-cutting,* the practice of sawing down all standing trees, clearing the land, and then planting seedlings of a single tree species. Biologists have found that the major source of fixed nitrogen for old-growth forests is probably lichens. Lichens don't thrive until the forest canopy is about 100 years old. Young replanted fir forests grow and are recut long before this, and the forest soils often become depleted of nitrogen, jeopardizing future tree crops. Some modern foresters are now harvesting only a few of the mature trees at a time.

Polar ice caps are not considered true biomes because they lack major plants.

Tundra

At the northern boundary of the coniferous forest, fragrant, giant conifers give way to the low vegetation of the tundra, a cold, treeless plain. With annual temperatures of -5°C (23°F) or less in the tundra, soil thaws to only about 1 m (39 in.). The deeper, permanently frozen soil, or *permafrost,* prevents most trees from growing. Plants in the tundra grow low, where they can absorb warmth reradiated from the solar-heated ground and minimize the effects of wind and desiccation. During summer, clouds of mosquitoes and flies fill the air, reproducing on the soggy ground. Many birds migrate to the tundra in the summer, feast on the insects, and breed during the long summer days.

tundra
a biome at the northern boundary of the coniferous forest characterized by low vegetation; this cold, treeless plain has annual temperatures of 5°C or less

polar ice caps
icy, treeless regions at our planet's highest latitudes

marine
characteristic of oceans and seas

Polar Caps

The polar ice caps are icy, treeless regions at our planet's highest latitudes. The arctic ice cap is a vast frozen ocean covering the planet's North Pole. The Antarctic ice cap encompasses the entire Antarctic continent, which is surrounded by frigid seas. Together the polar ice caps take up millions of square kilometers of land and ocean surface. These icy regions are not considered true biomes because they lack major plants; only animals and microbes survive in these regions. In the Arctic, polar bears hunt seals and other aquatic animals; in the Antarctic, penguins and seals inhabit the continent's ice shelf.

LO6 Life in the Water

Most of us are familiar with at least some of the biomes we just explored on dry land. When viewed from space, however, Earth is a blue planet because three-fourths of its surface is covered by water. Most of that area is marine—salty oceans and seas—with less than 1 percent of the surface covered by freshwater lakes, rivers, streams, swamps, and marshes. This freshwater area is tiny compared to the oceans, but it still amounts to tens of thousands of square kilometers. Ecologists do not call marine or freshwater communities biomes. Their studies, nevertheless, show that aquatic ecosystems differ from each other in significant ways, including their productivity, with coral reefs and estuaries being far more productive than lakes or the open ocean.

Properties of Water

As we saw in Chapter 2, water has unique physical characteristics, and these properties strongly influence aquatic organisms. For example, it requires more energy exchange to heat or chill water than it does to change air temperature. Aquatic organisms therefore suffer fewer rapid temperature changes than their land-dwelling counterparts. In addition,

estuary the area where a river meets an ocean; estuaries are some of Earth's most productive ecosystems

intertidal zone a region that lies at the ocean's edge that is underwater at high tide, exposed to air at low tide, and pounded by waves and wind

air grows steadily denser as it cools, but water has its greatest density at 4°C (39°F). Because of this, ice at 0°C (32°F) floats. If it didn't, lakes would freeze from the bottom up, only the upper layers would thaw in summer, and bodies of water outside of the tropics would support little life.

The next few sections describe the different water habitats, beginning with freshwater ecosystems, including running and standing waters, and then various saltwater habitats, including estuaries and oceans.

Mountain Streams

A stream bubbling from a mountain spring or from melting snow is cold and clear but contains few nutrients. As the water tumbles downhill and over rocks, it picks up oxygen and more nutrients. Species that live in these turbulent zones include algae, mosses, and trout, which thrive only where oxygen is plentiful and water temperatures are low. As nutrients accumulate and the stream widens and is less shaded, the photosynthetic productivity of algae and green aquatic plants increases. A river's middle stretch has the greatest species diversity. In the lower, slower-moving section of a large river, the water becomes cloudy and enriched with nutrients. The decreased light lowers the rate of photosynthesis, and primary productivity drops once again. In these areas, microbes and invertebrates live on detritus in the bottom sediments; catfish and bass replace trout.

Freshwater Lakes

Lakes have different life zones that depend on light penetration and depth. In shallow areas, where light can reach the lake bottom, abundant producers like water lilies, cattails, and algae exist and support consumers such as insects, snails, amphibians, fishes, and birds. Farther from shore, the light-penetrated top layer of water supports huge populations of photosynthetic algae and small crustaceans. The darker bottom region supports mainly insect larvae, scavenger fishes, and decomposers. In lakes, dissolved nutrients are as important as light; excess phosphorus, for example, can cause a bloom of plant growth in a lake or pond. As the seasons change, changing water temperatures stir up the nutrients in bottom sediments.

Estuaries

Ecologists use the term **estuaries** for the areas where rivers meet oceans, fresh and salt water mingle, and temperatures and salt concentrations vary widely with the tides and seasons. Estuary organisms like brine shrimp can often tolerate wide ranges of salinity. Constant water movements stir up nutrients, making estuaries some of the Earth's most productive ecosystems and nursery grounds for many fish species.

Tide Pools

At the ocean's edge lies the **intertidal zone**, a region that is underwater at high tide, exposed to air at low tide, and pounded by waves and wind. Most intertidal producers (kelp and other algae) have structures that anchor them to rocks, while most intertidal animals have tough bodies or shells as well as underwater "glues" that help fasten the organisms to the rocks. The ocean's most productive region extends from the intertidal zone to the edge of the *continental shelf,* the submerged part of the continents. Huge populations of phytoplankton and zooplankton serve as a nutrient base for the rest of the ocean's consumers.

Coral Reefs

In the tropics, the shallow ocean zone is home to reef-building corals and the photosynthetic microorganisms that inhabit them and give coral reefs their fantastic colors. The crannies and caves in these stony colonies create sheltered nesting sites and hiding places for the most diverse and productive communities in the seas.

The Open Ocean

Beyond the continental shelf and covering the deep abyssal plane, the water in the oceanic zone or open ocean is nearly deserted, even in the sunniest upper regions near the surface. The open ocean lacks nutrients such as phosphate and usable nitrogen, and overall is less productive than the arctic tundra. The ocean's most productive areas occur mainly where currents cause nutrients to well up from very deep waters. The open oceans help modulate climate and maintain favorable concentrations of oxygen and carbon dioxide in the atmosphere.

LO7 Change in the Biosphere

coral reefs underwater structures, sometimes enormous, produced by reef-building corals in the shallow ocean zones of the tropics

oceanic zone (open ocean) water beyond the continental shelf and covering the deep abyssal plane, which is nearly deserted, even in the sunniest upper regions near the surface

Recall from Chapter 10 how the evolution of photosynthesis about 2.8 billion years ago revolutionized the atmosphere, causing oxygen to accumulate in the air and the oceans to "rust." Today, human activities are causing equally pervasive modifications in the biosphere—but the modifications are coming far faster than organisms usually adapt to change via evolution. These alterations include global climate change, the degradation of the Earth's ozone layer, and the loss of habitat as humans convert more of the planet to their own use.

Global Climate Change

Spring temperatures have come earlier, birds and frogs are now laying their eggs earlier, leaves are appearing on trees earlier—in fact, meteorologists have found that the 20th century was by far the hottest 100 years in the last 1,000. What's more, about half of the century's 0.5°C (1°F) average warming has come in just the past 30 years. After weighing all the possible causes, the United Nations Intergovernmental Panel on Climate Change determined that people are mostly to blame for the greenhouse effect.

Greenhouse gases include carbon dioxide, methane, chlorofluorocarbons, and nitrous oxide. Carbon dioxide alone, however, is responsible for about half the human-induced greenhouse gases. Not only do we release much more CO_2 than the other gases, explains Dr. Parmesan, but the carbon dioxide is stable in the atmosphere for over 100 years. Methane (CH_4) is generated by decomposer bacteria in flooded rice fields and in the guts of cattle and termites, as well as from the burning of coal and natural gas. Far less methane is released than carbon dioxide, however, and CH_4 breaks down 10 times faster in the atmosphere. Chlorofluorocarbons (CFCs) are industrial chemicals that contain atoms of chlorine, fluorine, and carbon, and are used as coolants in refrigerators and air conditioners, as well as in plastic foam and insulation materials. Like carbon dioxide, CFCs remain in the atmosphere for up to 100 years, but they also break down ozone. Laws restricting their use have started to help, but decades of accumulated CFCs still hang above the Earth. Nitrous oxide (N_2O_2), or laughing gas, is produced by microbes in

ozone layer
a zone encircling Earth several miles above its surface that absorbs about 99 percent of the ultraviolet light that would otherwise penetrate and destroy many biological molecules, including DNA

sustainability
balancing the human population and economic growth against the abilities of the environment to provide a constant level of resources

soils, by the burning of forests and fossil fuels, and by the production of chemical fertilizers. It is a less abundant but still significant contributor to global climate change.

The global production of carbon dioxide has increased by roughly 25 percent in the past century as our burning of coal, oil, and gasoline has accelerated. The clearing of temperate forests and tropical rain forests and the burning of felled trees is another major source of the gas. Some of the excess carbon dioxide dissolves in the ocean. Much of it, however, enters the atmosphere—much more, unfortunately, than can be removed by plants during photosynthesis.

Many atmospheric scientists predict that sometime between the years 2025 and 2075—well within many of our lifetimes—the accumulating blanket of greenhouse gases will send global temperatures up an additional 1 to 3.5°C (2 to 5°F) on average. A change this big could have disastrous effects. Ice near the poles would melt, and sea levels would rise like mercury in a thermometer, inundating many of the world's most populous coastal cities. Dust bowl conditions could occur in the world's great grain-producing regions. Irrigation would probably not solve the problem because groundwater reserves would run out quickly. Ominously, even the coldest years of the last decade have been warmer than nearly every year of a century ago, and we are beginning to see widespread droughts, forest fires, and crop losses as an apparent result.

Most ecologists believe that we must decrease our consumption of fossil fuels through an emphasis on energy conservation and a commitment to renewable energy sources such as wind power and solar energy. Farmers could employ agricultural practices that better sustain the levels of organic matter in the soil, and people in tropical regions could reduce clear-cutting and burning of forests and step up efforts to replant denuded areas. Finally, we must all become well-enough informed to take an active role as voters and consumers in helping governments and ourselves make ecologically sound decisions.

The Ozone Hole

Chlorofluorocarbons not only act as greenhouse gases, they also attack *ozone* (O_3) molecules in Earth's protective **ozone layer**, a zone encircling the planet several miles above its surface that absorbs about 99 percent of the ultraviolet light that would otherwise penetrate and destroy many biological molecules, including DNA. Ozone forms naturally when ultraviolet light streaming from the sun strikes atmospheric oxygen gas, with its two atoms of oxygen.

Experts have shown that CFCs started accumulating in the atmosphere in the middle of the 20th century, proving that they are generated by humans. These long-lived compounds act as catalysts that convert ozone (O_3) to oxygen (O_2), with each CFC molecule continuing to destroy ozone molecules over and over again for decades.

{The Ozone Hole}

Earth's ozone layer absorbs most of the sun's ultraviolet rays. This colored satellite image reveals low atmospheric ozone levels over Antarctica. Ozone levels are color-coded, from purple (lowest) through blue, cyan and green to yellow (highest). At the time this image was taken (2006), the hole set records for extent—over 29.5 million square kilometers.

Sustainability

Experts predict a near doubling of world population during this century, from 6 billion in late 1999 to more than 10 billion by 2100.

Our future depends on **sustainability**—balancing the levels of human population and economic growth against the quantities of available resources and the quality of the physical environment. In the past, economic development has come at the expense of environmental quality. High prosperity in certain countries has rested, in part, on environmental changes in other countries. To ensure a stable global environment over the long term, say experts like William Ruckelshaus, former administrator of the Environmental Protection Agency (EPA), all people must have a reasonable level of prosperity and security, and this will require a change in human attitudes and actions as great as those that brought about the agricultural and industrial revolutions of the past.

The sustainability revolution will require industries to use fewer materials and less energy, to rely on wind and solar energy instead of fossil fuels wherever possible, and to recycle their wastes into other products and processes. Sustainable agriculture will involve

- Rotating crops to increase yields
- Building up rich soil by preventing erosion and by using natural fertilizers, such as animal manure, nitrogen-fixing crops, and blue-green algae
- Controlling weeds, insects, and plant diseases through the use of integrated pest management, through genetically altered crop plants, and through the planting of traditional and nontraditional crops.

Creative economic solutions will also play a part in global sustainability. Among the many ideas now under consideration are debt-for-nature swaps and conservation easements. In the former, wealthy nations would cancel the debts of developing nations in exchange for the promise to preserve natural habitats. In the latter, rich nations would pay the equivalent of mineral rights to secure forested areas and nature preserves against destruction.

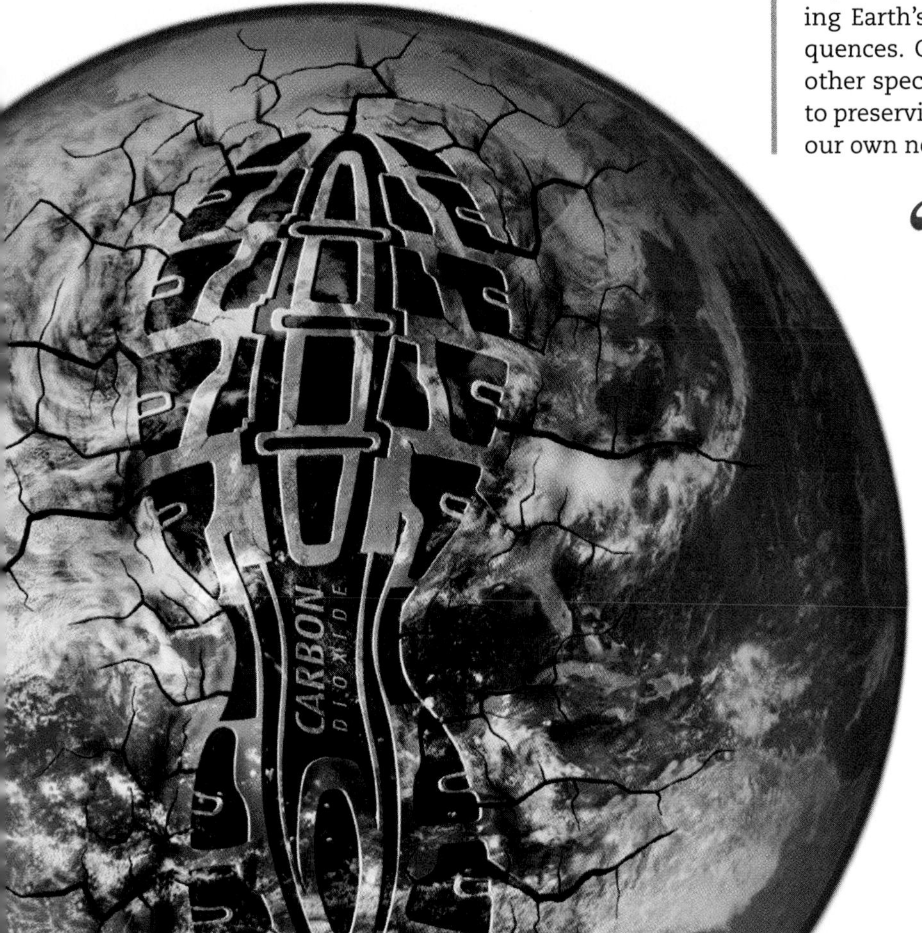

Who knew?

In temperate North America, many native grasslands have been converted to agriculture, and most of the original forests have been harvested for timber; human habitations encroach on more forest lands every year. Deliberate and accidental introduction of nonnative plants and animals causes an increasing, and irreversible, displacement of native species.

In addition to such approaches, say various experts on sustainability, all nations, rich and poor, will have to provide their citizens with information on sustainability techniques and birth control. They will have to fund international agencies to manage global environments. And they will have to be fully cooperative and committed to the sustainability revolution. The alternative—the crash of the human population following its present boom—is just too dismal a prospect.

Our species has changed the global environment in a stunningly short time period. But we are also capable of understanding that we're mismanaging Earth's resources and of foreseeing the consequences. Our own survival and that of millions of other species depend on working out new solutions to preserving our environment while also supporting our own needs.

> "To ensure a stable global environment over the long term, all people must have a reasonable level of prosperity and security . . . this will require a change in human attitudes and actions as great as those that brought about the agricultural and industrial revolutions of the past."

81% **of students surveyed find that LIFE helps them prepare better for their exams**

LIFE puts a multitude of study aids at your fingertips. After reading the chapters, check out these resources for further help:

- **Chapter in Review cards**, found in the back of your book, include all learning outcomes, definitions, and visual summaries for each chapter.
- **Online printable flash cards** give you three additional ways to check your comprehension of key biology concepts.

Other great ways to help you study include **interactive biology games, podcasts, audio downloads, and online tutorial quizzes with feedback**.

You can find it all at **4ltrpress.cengage.com/life**.

Index

C

D

E

F

G

P

92% of students surveyed believed the Interactive Quizzes are a useful study tool.

Discover your **LIFE** online experience at **4ltrpress.cengage.com/life**.

You'll find everything you need to succeed in your class.

- Interactive Quizzes
- Printable and Online Flash Cards
- Animations
- Visual Reviews
- And more

4ltrpress.cengage.com/life